Reproductive Biology and Phylogeny of Anura

Volume edited by
BARRIE G.M. JAMIESON
Department of Zoology and Entomology
University of Queensland
St. Lucia, Queensland
Australia

Volume 2 of Series:
Reproductive Biology and Phylogeny

Series edited by
BARRIE G.M. JAMIESON
Department of Zoology and Entomology
University of Queensland
St. Lucia, Queensland
Australia

Science Publishers, Inc.
Enfield (NH), USA   Plymouth, UK

Cover photographs reproduced by courtesy of Richard M. Lehtinen, Ronald A. Nussbaum and William Duellman.

SCIENCE PUBLISHERS, INC.
Post Office Box 699
Enfield, New Hampshire 03748
United States of America

Internet site: *http://www.scipub.net*

sales@scipub.net (marketing department)
editor@scipub.net (editorial department)
info@scipub.net (for all other enquiries)

Library of Congress Cataloging-in-Publication Data

Reproductive biology and phylogeny of Anura/volume edied by Barrie G.M. Jamieson.
    p. cm.--(Reproductive biology and phylogeny; v. 2)
    Includes bibliographical references.
    ISBN 1-57808-288-9
    1. Anura--Reproduction. 2. Anura--Phylogeny. I. Jamieson, Barrie G.M. (Barrie Gillean Molyneux) II. Series.

QL668.E2R494 2003
597.8--dc21

                                                       2003045478

ISBN (Set)   1-57808-271-4

ISBN (Vol. 2) 1-57808-288-9

Published by Science Publishers, Inc., Enfield, NH, USA
Printed in India

Preface to the Series

This series was founded by the present series editor in consultation with Science Publishers, Inc., in 1991. Whereas its precursor, 'Reproductive Biology of Invertebrates' published by Oxford & IBH in association with John Wiley & Sons, dealt with a given reproductive topic in a volume for all, or most, invertebrate groups, it was decided here to attempt to cover, as far as is feasible, the full panoply of reproduction in each volume for a specific taxonomic group and to extend the coverage to vertebrates. The new series bears the title 'Reproductive Biology and Phylogeny' and this title is followed in each volume with the name of the taxonomic group which is the subject of the volume. B.G.M. Jamieson is the founding series editor and each publication will have one or more invited volume editors and a large number of authors of international repute. The level of the taxonomic group which is the subject of each volume will vary according, largely, to the amount of information available on the group, the advice of proposed volume editors, and the interest expressed by the zoological community in the proposed work. The order of publication will reflect these concerns, and the availability of authors for the various chapters, and it is not proposed to proceed serially through the animal kingdom in a presumed phylogenetic sequence. Nevertheless, a second aspect of the series will be coverage of the phylogeny and classification of the group, as a necessary framework for an understanding of reproductive biology. Evidence for relationships from molecular studies will be an important aspect of the phylogenetic sections.

It is not claimed that a single volume can, in fact, cover the entire gamut of reproductive topics for a given group but it is hoped that the series will be considered to have given a good and up to date coverage of reproduction and will provide a general text rather than being a mere collection of research papers on the subject. Nor is it likely that coverage in different volumes will be uniform in terms of topics, though it is clear from the first volumes that the standard of the contributions by the authors will be uniformly high. The stress will vary from group to group; for instance, modes of external fertilization, or vocalization, important in one group, might be inapplicable in another. A further, perhaps inevitable reason, for some unevenness in coverage may be failure, in the time-frame which is reasonable for the production of a volume, of a given author to complete a contribution owing to factors beyond his or her control. It is hoped that the reader will bear with such deficiencies as may be evident while recognizing that each volume is a major contribution in the field of reproductive biology and phylogeny.

The first volumes reflect the above criteria and the interests of certain research teams. Thus, in view of the recent great increase in our knowledge of the ultrastructure of spermatozoa of the Amphibia, it was decided to commence with volumes on the reproductive biology and phylogeny of Amphibia. We were fortunate that Professor David Sever accepted the invitation to be volume editor for the first volume in the series, that on Urodela, in which role he has set a standard for all volumes to come. The second volume, on Anura, is edited by myself, reflecting the present interests of my laboratory. Other volumes in preparation are on Gymnophiona (J.-M. Exbrayat), Annelida (G. Rouse and F. Pleijel), Mollusca (J. Healy) and Crustacea (C. Tudge).

While volume editing is by invitation, reproductive biologists who consider that a given taxonomic group should be included in the series and may wish to undertake the task of editing a volume should not hesitate to make their views known to the series editor.

The editor is grateful to the publishers for their support in producing this series. Sincere thanks must be given the volume editors and the authors, who have freely contributed their chapters, in very full schedules. The editors and publishers are confident that the enthusiasm and expertise of these contributors will be reflected by the reception of the series by our readers.

16 December 2002

Barrie Jamieson
Department of Zoology and Entomology
University of Queensland
Brisbane

Preface to this Volume

This is the second of what it expected to be a long series of volumes which record major aspects of our knowledge of reproductive biology and phylogeny in the Animal Kingdom.

The volume could not have a more distinguished beginning than its first chapter, an overview of anuran phylogeny, classification and reproductive modes by William E. Duellman whose text, with Linda Trueb, *Biology of Amphibians*, (McGraw-Hill) is a *sine qua non* for students of the Amphibia. It is followed by a succinct account of reproductive anatomy by Michael J.Tyler, whose name is also a byword in the study of frogs.

Sara S. Sánchez and Evelina I. Villecco in a valuable and original chapter on oogenesis deal not only with the stages of oogenesis, commencing with embryonic germ plasm and primordial germ cells and ending with the mature oocyte, but also the importance of the progressive cellular and molecular interactions within follicles, gene expression and accumulation of transcripts during primary oocyte growth, the significance of different RNAs in early development and the process of vitellogenesis.

Silvia N. Fernández and Inés Ramos present a broad and deep treatment of the endocrinology of reproduction. They show that control of reproduction and ultimately the release of a gamete capable of fertilization and successful embryonic development resides in the hypothalamus, which controls pituitary and therefore gonadal functions. Seasonal profiles of the circulating steroids show that different hormones, acting concurrently or sequentially, synchronize gonadal processes such as folliculogenesis, oocyte maturation, and oviductal processes involving growth and specific secretions.

A large portion of the volume is devoted to an account by David M. Scheltinga and Barrie G.M. Jamieson of spermatogenesis, the ultrastructure of the mature spermatozoon, their form, function and phylogenetic implications. This chapter attempts to review all anuran spermatozoal literature and reveals the quantum leap in our knowledge of anuran sperm ultrastructure in the unpublished Ph.D. thesis of the first author.

Breeding glands, so characteristic an aspect of reproduction in the Anura, receive an original and profound treatment by Rossana Brizzi, Giovanni Delfino and Silke Jantra. They conclude that selective pressures due to environmental and social constraints produced the rich variety of secretory structures observed in the skin of the various body regions; that their study supports the view that the structural and histochemical similarities of anuran breeding glands and urodele courtship glands reflect a common evolutionary process; and that morphological and physiological characters within the integumentary apparatus are essential for understanding the evolutionary etiology and the functional significance of the glands.

Whereas internal fertilization is obligatory in gymnophionans and common in urodeles it is exceptional, though not, perhaps, as rare as is supposed, in the Anura. In an authoritative account of internal fertilization with special reference to mating and female sperm storage in *Ascaphus*, David M. Sever, William C. Hamlett, Rachel Slabach , Barry Stephenson and Paul A. Verrell examine the structure and function of the unique penial structure of *Ascaphus*. Among other topics, they discuss sperm storage in the context of the numerous structural and physiological differences which exist between the spermathecae of salamanders and the sperm storage tubules of *Ascaphus*.

A masterly account of parental care, in a phylogenetic perspective is given by Richard M. Lehtinen and Ronald A. Nussbaum. This account appeals greatly to the naturalist in one, and the various forms of parental care illustrated add to what the most objective of us must see as the endearing qualities of frogs. However, it includes rigorous cladistic analyses and cogent discussions of the evolution of parental care, and despite necessary caveats, shows that a historical approach is a powerful tool in identifying patterns and testing evolutionary hypotheses.

Ronald Altig, whose name is inseparable from the study of tadpoles, succeeds in giving an account of anuran development which is not only highly informative at the level of the unfolding of structure but is thought provoking in directions which go far beyond development. We receive, near the end of the volume, a cautionary recommendation of tadpoles as an antidote to "the intoxication of adult frogs". It is timely, too, that a call is made for descriptive work, so sadly neglected in all areas of biology, to the detriment of science, at the present time while acknowledging that large strides in understanding developmental mechanisms probably will be made only with the inclusion of molecular techniques.

The present editor, though an unashamed devotee of descriptive biology, also recognized the necessity to include molecular studies by inviting Brian Key to contribute a chapter on molecular development. Professor Key presents a dynamic picture of the cooperative activity of biochemical networks involving both activators and repressors of gene expression as well as of protein activity and shows how the interacting components are intricately intertwined, acting in separate and overlapping pathways to simultaneously sculpture germ layers and to set the stage for the subsequent differentiation of specific tissue types.

The generosity of authors and publishers who have allowed illustrations to be reproduced is most gratefully acknowledged. Earlier stages of research work on anuran spermatozoal ultrastructure, which forms such a large component of this volume, were supported by the Australian Research Council and the support of the Department of Zoology and Entomology, University of Queensland has facilitated its preparation. The kind collaboration of all the authors, who have borne uncomplainingly the requests of an editor overseeing the gestation of this work, has been greatly appreciated. Finally the efficient and courteous participation of the publishers was indispensable to production of this volume.

16th December 2002   **Barrie Jamieson**
                                                                Brisbane

Contents

CHAPTER 1

An Overview of Anuran Phylogeny, Classification and Reproductive Modes

William E. Duellman

1.1 INTRODUCTION

The discipline of systematics is the study of relationships of organisms; the classification of organisms should reflect our knowledge of their relationships given that each category in the hierarchical classification is an evolutionary lineage. This is a dynamic field of investigation. Consequently, as new knowledge is gained about the relationships of a group of organisms, we can anticipate that the classification is likely to change. Non-systematists often are befuddled by changing classifications, which commonly result in recognition of new genera and families, and even more confusing, the rearrangement of taxa at different hierarchical levels. Classification is not haphazard; it is based on philosophical concepts and governed by the *Code of Zoological Nomenclature* (Anonymous 1999). To be meaningful in an evolutionary framework, the interpretation of biological data, whether morphological, physiological, ecological, or behavioral, must be viewed in a phylogenetic context.

The hypothesized phylogenetic relationships and resulting classification of frogs and toads (Order Anura of the Class Amphibia) have been changing for nearly 250 years; the seemingly frenetic level of present research will result in many more changes which will aid biologists to interpret diverse kinds of data with respect to histories of evolutionary lineages. In order to place our present classification in perspective, I provide a brief review of anuran classification prior to discussing the relationships of anurans with other amphibians and the phylogenetic relationships within the Anura. I conclude by presenting a classification of the Anura and a synopsis of the family groups and their

Natural History Museum and Biodiversity Research Center, The University of Kansas, Lawrence, Kansas 66045, USA reproductive modes. I do this with the realization that by the time this book is published new data and interpretations will result in many changes. For details beyond the scope of this chapter, an online checklist of amphibians is available (Frost 2002); the most recent summary of families of anurans is in Hutchins *et al.* (2003).

1.2 DEVELOPMENT OF ANURAN CLASSIFICATION

A natural classification should reflect the phylogenetic relationships of organisms in a hierarchical manner; each taxonomical level in the hierarchy should contain all of the descendents in an evolutionary lineage. Classification of anurans has changed dramatically since Linnaeus' (1758) recognition of the class Amphibia to contain, not only amphibians as we know them today, but also reptiles and a group of cartilaginous fishes; Linnaeus recognized 17 species of anurans, all of which were placed in the genus *Rana*. The recognition of Reptilia stems from Laurenti (1768). In their thorough review of the history of amphibian classification, Duellman and Adler (2003) noted that Oppel (1811) was the first worker to place caecilians with amphibians and that Blainville (1822) was the first author to recognize unequivocally Amphibia and Reptilia as two separate classes; the former contained the groups now recognized as the orders Anura (frogs and toads), Caudata (salamanders), and Gymnophiona (caecilians).

Most early classifications were based solely on external morphology, but even in the 18th and early part of the 19th centuries some knowledge of development and internal anatomy corroborated the distinction of amphibians from reptiles and distinguished groups of anurans. For example, Wagler (1830) divided Anura (his Salientia) into Aglossa (tongueless frogs, now the family Pipidae) and Phaneroglossa, which contained all other anurans. Cope (1864, 1865) discovered that anurans possessed two kinds of pectoral girdles; he placed those in which the two halves of the girdle overlap midventrally in Arcifera and those in which the two halves of the girdle are fused along the midline in Firmisterna. Subsequently, Nicholls (1916) noted differences in vertebral structure, such as the nature of the centra, sacrococcygeal articulation, and presence of free ribs; he divided the Anura into four groups (Opisthocoela, Anomocoela, Procoela, and Diplasiocoela). Noble (1922) introduced characters of thigh musculature into anuran classification and noted that *Ascaphus* and *Leiopelma* have nine, instead of the usual eight, presacral vertebrae, but that the vertebrae were amphicoelous and thus created a fifth group, Amphicoela.

The first work on life cycles of amphibians was that by Roesel von Rosenhof (1758) on common species of European anurans. Subsequently, life histories, including mode of amplexus, type of fertilization, and morphology and development of larvae, have played an important role in the classification of amphibians. Life-history data were important ingredients in Noble's (1931) classification. Orton (1953, 1957) used the characteristics of the larval mouth and spiracle to define four larval types. However, earlier attempts to classify anurans solely on larval characters (e.g., Starrett 1973) have failed, as demonstrated by Sokol (1975).

Subsequent to Noble's (1931) classification of anurans in which he recognized 11 living families and based his phylogeny primarily on vertebral characters, the classification of anurans has changed dramatically. Reig (1958) proposed the subordinal names Archaeobatrachia and Neobatrachia for the so-called archaic and advanced families, respectively. Osteological trends were reviewed by Trueb (1973), as was a suite of 38 primarily morphological characters by Lynch (1973). Based primarily on their data, Duellman (1975) proposed a classification of Anura, in which he recognized six living and one extinct family in three superfamilies (Discoglossoidea, Pipoidea, and Pelobatoidea) in the suborder Archaeobatrachia and 14 living families in three superfamilies (Bufonoidea, Microhyloidea, and Ranoidea) in the suborder Neobatrachia. Pelobatodiea and Pipoidea were considered by Lynch (1973) and Duellman (1975) to be transitional between archeobatrachians and neobatrachians, and Laurent (1979) proposed the suborder Mesobatrachia to include those subfamilies. Duellman and Trueb (1986) recognized one extinct and 21 living families but did not use subordinal or superfamilial categories. In the Pelobatoidea, spermatozoa of the families Megophryidae and Pelobatidae are similar to those of Hyperoliidae, whereas the sperm of the Pelodytidae are similar to those of the Bufonidae, Centrolenidae, Hylidae, Leptodactylidae, and Rhinodermatidae. The spermatozoa of the Pipidae are similar to those of the Petropedetidae, Ranidae, and basal Rhacophoridae (see Scheltinga and Jamieson, Chapter 5 of this volume).

Ford and Cannatella (1993) reviewed existing morphological data on the monophyly of family-groups and proposed a phylogenetic classification of anurans, in which they recognized 22 families and different rankings of taxa without referring to subordinal or suprafamilial levels in the hierarchy; they considered the Archaeobatrachia to be paraphyletic and used the informal term "archaeobatrachian" for those groups that are not included in the Neobatrachia. However, molecular evidence supports the monophyly of Archaeobatrachia (Hay *et al.* 1995). The spermatozoa of Leiopelmatidae, Ascaphidae, Bombinatoridae, Discoglossidae seem to be the most basal in Anura; but the ultrastructure does not support monophyly of the Archaeobatrachia (see Chapter 5).

Thus, we have witnessed a fantastic increase in the numbers of taxa at all levels during the past 250 years. Linnaeus (1758) recognized 17 species of anurans in a single genus; Duméril and Bibron (1841) placed 167 species in 47 genera in four families, whereas those numbers increased to 859, 116, and 14, respectively, in Boulenger (1882). Extensive work on the systematics of anurans beginning in the latter half of the 20th century brought the number of species to 3483 in 301 genera and 23 families in Frost (1985) and to 3967 species in 334 genera and 25 families in Duellman (1993). The most recent compilation (Frost 2002) lists 4837 species in 352 genera and 29 families.

1.3 RELATIONSHIPS OF ANURANS WITH OTHER AMPHIBIANS

The hypothesized relationships among the three living groups of amphibians (anurans, salamanders, and caecilians) have been controversial for more than a century (Trueb and Cloutier 1991a). A thorough cladistic analysis of morphological characters of fossil and Recent amphibians by Trueb and Cloutier (1991b) resulted in the recognition of a monophyletic group, the Lissamphibia, which includes all living amphibians, as well as some extinct relatives. In their analysis, caecilians are the sister group to anurans and salamanders. Without a cladistic analysis, Carroll (2000) argued that caecilians represent an independent lineage related to gonorhynchid microsaurs, a group of Permian lepospondyls.

Independent phylogenetic analyses of DNA sequences of the 12S ribosomal RNA gene (Hedges and Maxson 1993) and the 16S ribosomal RNA gene (Hay *et al.* 1995) support the monophyly of the Lissamphibia and each of the three orders. In these analyses, the anurans are placed as the sister group to salamanders and caecilians in contrast to the arrangement hypothesized by Trueb and Cloutier (1991b), but a phylogenetic analysis of the complete mitochondrial genomes of one species in each order by Zardoya and Meyer (2001) placed caecilians as the sister group to anurans and salamanders, thereby supporting the morphological analysis by Trueb and Cloutier (1991b). Thus, the living amphibians (and lissamphibian fossils) can be classified as follows (Fig. 1.1):

Class Amphibia
   Subclass Lissamphibia
      Infraclass Gymnophiona (Caecilians and Jurassic *Eocaecilia*)
         Order Apoda (Caecilians)
      Infraclass Batrachia
         Superorder Urodela (Salamanders and Jurassic Karauridae)
            Order Caudata (Salamanders)
         Superorder Salientia (Frogs and Triassic *Triadobatrachus*)
            Order Anura (Frogs)

Although the monophyly of living amphibians and the relationships of the three living groups seems to be well supported, controversy remains regarding the relationships of the Lissamphibia with extinct groups of amphibians. Trueb and Cloutier (1991a) concluded that lissamphibians are part of the Dissorophoidea, a group of temnospondyl amphibians from the Upper Carboniferous and Permian. One of their most persuasive characters was the nature of the teeth, which are pedicellate in living amphibians and possibly also in at least some dissorophoids, especially *Doloserpeton*, but Holmes (2000) argued that because of the small size of *Doloserpeton* it is not possible to determine if the teeth are truly pedicellate. A cladistic analysis by Laurin and Reisz (1997) resulted in their suggesting that lissamphibians are lepospondyls, but they did not include in their analysis some of the dissorophoids analyzed by Trueb and Cloutier (1991b). Thus, the extralimital relationships of lissamphibians remains clouded in controversy.

1.4 PHYLOGENETIC RELATIONSHIPS WITHIN ANURA

The last decade has witnessed significant strides in attempts to unravel the

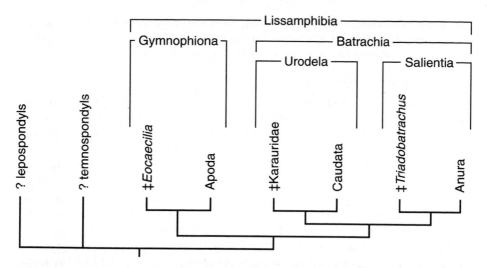

Fig. 1.1 Hypohesized phylogenetic relationships of the Lissamphibia, based primarily on Trueb and Cloutier (1991b). ‡ denotes fossil taxa: ‡*Eocaecilia,* a caecilian-like creature with small limbs from the Lower Jurassic of Arizona, U.S.A.; ‡ Karauridae contains primitive stem salamanders including ‡*Karaurus* from the Upper Jurassic of Kazakhstan, ‡*Kokartus* from the Middle Jurassic of Kirghizstan, and ‡*Marmorerpeton* from the Middle Jurassic of England (Milner 2000); ‡*Triadobatrachus,* a frog-like creature with a short tail from the Lower Triassic of Madagascar. These fossils are hypothesized sister taxa of modern caecilians, salamanders, and anurans, respectively.

phylogenetic relationships among anurans, a uniquely derived group of tetrapods. Most studies have included cladistic analyses in an effort to determine monophyletic groups. Major morphological studies of adults are those by Ford and Cannatella (1993) and Maglia (1998), whereas other studies (e.g., Haas 2001, Pugener *et al.* 2003) focused on larval features. Spermatological evidence for phylogenetic affinities within the Anura is reviewed in Chapter 5. Major contributions to higher level relationships involving molecular data are those by Hedges and Maxson (1993), Hillis *et al.* (1993), and Hay *et al.* (1995); a few studies (e.g., Cannatella and Hillis 1993) combined molecular and morphological data. A better understanding of some morphological characters has been gained from ontogenetic studies on selected taxa by de Sá and Trueb (1991), Trueb and Hanken (1992), Larson and de Sá (1998), Maglia and Pugener (1998), Trueb *et al.* (2000) and Maglia *et al.* (2001). The increasingly expanding fossil record of basal anurans has provided important material for cladistic analyses of fossils and associated Recent anurans (e.g., Báez and Basso 1996; Báez and Trueb 1997).

Whereas some phylogenetic analyses of molecular data have clarified relationships among genera (e.g., de Sá and Hillis 1990; Austin *et al.* 2002), most analyses of higher level relationships have suffered from insufficient taxon sampling (e.g., Richards and Moore 1996; Emerson *et al.* 2000, Bossuyt and Milinkovitch 2001). Furthermore, deep-rooted relationships apparently can be resolved most effectively by analysing slow-evolving genes (Alrubaian *et al.* 2002). Despite these shortcomings, phylogenetic analyses of molecular data are providing new insights into the evolutionary relationships of anurans.

1.4.1 Archaeobatrachia

Basal anurans or archeobatrachians are comprised of 168 living species placed in 27 genera relegated to nine families (Fig. 1.2). Among these are the two living genera (*Ascaphus* with two species and *Leiopelma* with three species placed in their own families) having nine presacral vertebrae and other primitive morphological characters that separate them from all other living anurans. However, the Jurassic ‡*Vieraella* has 10 presacral vertebrae, and the Jurassic ‡*Notobatrachus* is like *Leiopelma* and *Ascaphus* in having nine presacral vertebrae but differs in several other features. According to the analysis by Báez and Basso (1996), ‡*Vieraella* is the sister group to ‡*Notobatrachus* and the rest of the Anura. Spermatological studies reveal that *Leiopelma* apparently is the most plesiomorphic extant anuran, whereas *Ascaphus* is the next most basal (see Scheltinga and Jamieson, Chapter **5** of this volume).

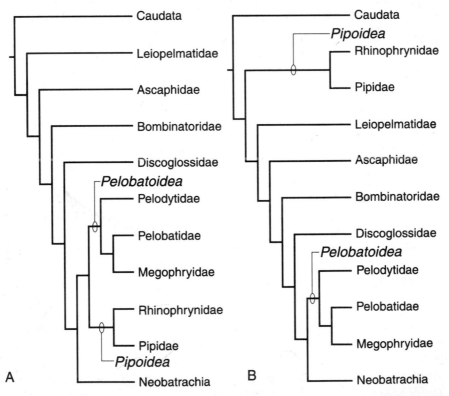

Fig. 1.2 Alternative phylogenetic hypotheses of the relationships among the families of "Archeobatrachia." A. Arrangement proposed by Ford and Cannatella (1993) with pipoids as the sister group to pelobatoids. B. Arrangement proposed by Pugener *et al.* (2003) with pipoids as the sister group to all other anurans.

Morphological and molecular evidence (Lathrop 1997; Maglia 1998) support the monophyly of the superfamily Pelobatoidea with Pelodytidae as the sister group to Pelobatidae + Megophryidae. However, the spermatozoa of Pelobatidae and Megophryidae are similar to those of Ranoidea, whereas those of Pelodytidae are similar to those of Bufonoidea (see Chapter 5). Thorough analyses of morphological data on fossil and Recent frogs of the families Pipidae and Rhinophrynidae support the recognition of the superfamily Pipoidea (Báez and Trueb 1997; Maglia *et al.* 2001). Most phylogenetic analyses of archaeobatrachians (e.g., Ford and Cannatella 1993) place the pipoid and pelobatoid families as "advanced" with respect to other families of basal anurans and commonly Pelobatoidea and Pipoidea are arranged as sister groups (Fig. 1.2A). However, based on an extensive suite of larval characters, Pugener *et al.* (2003) placed Pipoidea as a sister group to all other Anura (Fig. 1.2B). The spermatozoa of Pipidae share significant apomorphies with those of Mantellidae, Petropedetidae, Ranidae, and Rhacophoridae but differ by possessing a plesiomorphic short mitochondrial collar (see Chapter 5).

Within Archaeobatrachia, ‡Palaeobatrachidae from the Tertiary of Europe contains many species that are were aquatic and had larvae resembling those of the pipid *Xenopus* (Spinar 1972). ‡Palaeobatrachidae usually has been placed as a sister group of Pipidae and thus included in Pipoidea (Báez and Trueb 1997; Ford and Cannatella, 1993); however, those Tertiary frogs may represent an independent lineage of aquatic anurans not related to pipids.

1.4.2. Neobatrachia

Most anurans (more than 4600 species) are placed in the so-called advanced frogs or Neobatrachia. Both morphological data (Ford and Cannatella 1993; Pugener *et al.* 2003) and molecular evidence (Hillis and Davis 1987, Hay *et al.* 1995, Ruvinsky and Maxson 1996) support the monophyly of neobatrachians. Nonetheless, considerable controversy exists regarding relationships within the neobatrachians. The monophyly of the Neobatrachia *sensu stricto* does not seem to be supported by sperm ultrastructure; Mesobatrachia + Neobatrachia apparently is monophyletic, but some mesobatrachians (Pelobatidae and Megophryidae) are like Ranoidea, whereas Pelodytidae shares features with Bufonoidea (see Chapter 5). One superfamily, Ranoidea, is defined as having a firmisternal girdle with the epicoracoid cartilages completely fused (Ford and Cannatella 1993). This is in contrast to the arciferal condition of the pectoral girdle in all other neobatrachians that are allocated to the superfamily Bufonoidea (= Hyloidea of some authors), in which a cluster of three families having arciferal pectoral girdles and intercalary elements between the terminal and penultimate phalanges has been noted as hyloids (Duellman 2001).

Although morphological evidence has been the primary criterion used for definition of families, it has not had a significant impact in phylogenetic hypotheses, except for defining Ranoidea. For example, in the phylogenetic tree proposed by Ford and Cannatella (1993) most families and Ranoidea form a basal polytomy, and the families in Ranoidea form another polytomy. In contrast, an analysis of 12S and 16S rRNA gene sequences by Ruvinsky and

Maxson (1996) not only supported Ranoidea as the sister group to Bufonoidea but implied relationships among many lineages within Bufonoidea and suggested that Hylidae and Leptodactylidae were polyphyletic; however, this analysis was based on only 20 taxa of bufonoids, and four of ranoids, with four archaeobatrachians as outgroups. Spermatological evidence for affinities is discussed in Chapter 5.

Within Bufonoidea, many aspects of the phylogenetic relationships are unresolved. Neither somatic (Ford and Cannatella 1993), spermatological (see Chapter 5), nor molecular (Ruvinsky and Maxson, 1993) evidence supports the monophyly of Leptodactylidae. Likewise, molecular evidence suggests that Hylidae is polyphyletic with Hemiphractinae not associated with other hylids (T. W. Reeder and J. J. Wiens, unpublished). Pseudidae has been recognized as a distinct family; but on morphological evidence, Duellman (2001) placed pseudids as a subfamily of Hylidae, an arrangement also supported by molecular data (T. W. Reeder and J. J. Wiens, unpublished). However, the spermatozoa of pseudids are more like those of microhylids than other hylids (see Chapter 5), but this similarity might be homoplastic. Both morphological (Duellman 2001) and molecular data (Austin *et al.* 2002) place the monotypic Allophrynidae as the sister group to Centrolenidae + Hylidae. Bufonidae, Brachycephalidae, and

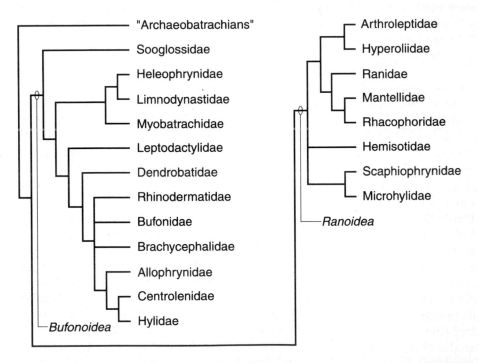

Fig. 1.3 Suggested phylogenetic relationships among the families of Neobatrachia. This arrangement follows no particular published cladogram but is an amalgamation of relationships hypothesized by Emerson *et al.* (2000), Ford and Cannatella (1993), Hay *et al.* (1995), Ruvinsky and Maxson (1996), and Vences *et al.* (2000).

Rhinodermatidae are well defined by morphological characters, but their placement within Bufonoidea is equivocal; the molecular analysis by Ruvinsky and Maxson (1996) revealed a polytomy including Rhinodermatidae, Bufonidae, and Hylidae + Centrolenidae. The spermatozoa of the Bufonidae are very similar to those of the centrolenids, hylids, some leptodactylids, and rhinodermatids (see Chapter 5).

Australian frogs of the families Limnodynastidae and Myobatrachidae, as well as the African *Heleophryne* (Heleophrynidae) were considered to be primitive leptodactylids by Lynch (1971). Molecular data (Ruvinsky and Maxson 1996) placed these three families near the base of Bufonoidea with Mybobatrachidae as the sister group to Heleophrynidae + Limnodynastidae. In Ruvinsky and Maxson's (1996) analysis, Sooglossidae, which is endemic to Seychelles, is the most basal bufonoid, but a tree based on molecular data by Emerson *et al.* (2000) shows Heleophrynidae and Myobatrachidae more basal than Sooglossidae. Leptodactylidae are heterogeneous spermatologically, and the family seems to be paraphyletic, whereas Myobatrachidae and Limnodynastidae are monophyletic but spermatozoal differentiation is not strong (see Chapter 5).

Controversy has existed regarding the placement of Dendrobatidae. Various interpretations of morphological evidence have associated dendrobatids with cycloramphine (= hylodine) leptodactylids in Bufonoidea (Lynch 1971, Noble 1931) or petropedetine ranids and arthroleptids in Ranoidea (Griffiths 1959; Ford 1993; Ford and Cannatella 1993). Ruvinsky and Maxson's (1996) analysis resulted in Dendrobatidae being imbedded in Bufonoidea. A more thorough analysis using the same genes from many species of dendrobatids, several bufonoids and ranoids, including four arthroleptids and one petropedetine ranid strongly supported the placement of Dendrobatidae in Bufonoidea (Vences *et al.* 2000).

Within Ranoidea, morphological data (Ford and Cannatella 1993) strongly support the recognition of Microhylidae as a sister group to other ranoids, but scant molecular evidence (Hay *et al.* 1995; Richards and Moore 1998) is equivocal in that in the analysis by Hay *et al.* (1995), Hyperoliidae is the sister group to Microhylidae, which in turn is the sister group to Ranidae + Mantellidae. Microhylid sperm are like those of Mantellidae, Ranidae, Rhacophoridae, Pipidae, Pseudidae and some Leptodactylidae in the apomorphic absence of fibres associated with the axoneme, but at least some homoplasy in this character is suspected (see Chapter 5). The only group of microhylids of questionable affinity is Scaphiophryninae, which has tadpoles with a unique combination of microhylid and ranoid characters (Blommers-Schlösser 1975; Wassersug 1984). Morphological data suggest that Hemisotidae is related to Hyperoliidae or Ranidae (Ford and Cannatella 1993), but in an analysis of the 12S tRNA gene, *Hemisus marmoratum* is nested within six species of microhylids (Emerson *et al.* 2000). The relationships of Hyperoliidae are questionable; Emerson *et al.* (2000) concluded that the family is paraphyletic with respect to Arthroleptidae, because in their analysis arthroleptids are the sister group hyperoliids, exclusive of leptopeline hyperoliids. Spermatologically, hyperoliids are similar to megophryids and pelobatids, but arthroleptids have not been examined for sperm ultrastructure (see Chapter 5).

Old World treefrogs of the family Rhacophoridae are members of Ranoidea, and the monophyly of most of the group is supported by morphological (Wilkinson and Drewes 2000) and molecular (Emerson et al. 2000) evidence. However, the relationships of one group of genera in Madagascar are equivocal. These genera have been placed in Mantellinae, a subfamily of ranids by Dubois (1981) and Blommers-Schösser (1993), regarded as a sister group to rhacophorids (molecular analysis by Emerson et al. 2000), or imbedded in Rhacophoridae (morphological analysis by Wilkinson and Drewes, 2000). Morphological data were used by Glaw et al. (1997) to transfer the mantelline *Aglyptodactylus* to Ranidae, but this is contrary to molecular evidence presented by Emerson et al. (2000).

Molecular analyses by Emerson et al. (2000) place Ranidae as the sister group to Rhacophoridae, but so few taxa of the highly diverse Ranidae have been included in molecular analyses that these results are equivocal. The two families share some derived morphological features (Ford and Cannatella 1993), and differ primarily in the presence of an intercalary element between the distal and penultimate phalanges in Rhacophoridae (if mantellids are not considered to be ranids). Rhacophorid sperm possess the apomorphic character of auxiliary microtubules in the tail; this feature is known elsewhere only in Mantellidae and the ranid *Limnonectes*. Spermatologically, the Ranidae is clearly paraphyletic (see Chapter 5)

1.6 CLASSIFICATION OF ANURA

In the following list, living families are arranged alphabetically within higher categories; in those families in which subfamilies are recognized, the subfamilies are arranged alphabetically within the family. Names in parentheses are probably paraphyletic. Numbers in parentheses following the family group names are number of genera/number of species. Numbers in brackets refer to numbered reproductive modes in Table 1.1. An abbreviated statement of distribution is given for each family and subfamily; for more explicit distributions and analyses thereof, see Frost (2002) and Duellman (1999), respectively. The families recognized below essentially are the same as in Frost (2002), except that Petropedetidae is considered to be a subfamily of Ranidae, which according to Dubois (2003) is made up of 11 subfamilies, and two additional subfamilies of Leptodactylidae are recognized.

Suborder "Archaeobatrachia"
    Superfamily "Discoglossoidea"
        Ascaphidae. (1/2). North America [2].
        Bombinatoridae. (2/10). Eurasia [1, 2].
        Discoglossidae. (2/10). Western Eurasia [2, 26].
        Leiopelmatidae. (1/4). New Zealand [17].
    Superfamily Pelobatoidea

Megophryidae. (11/107). Southeast Asia and associated islands [1, 2].
Pelobatidae. (3/11). North America and western Eurasia [1].
Pelodytidae. (1/3). Western Eurasia [1].
Superfamily Pipoidea
    Pipidae. (5/30). Two subfamilies:
        Dactylethrinae. (4/23). Sub-Saharan Africa [1].
        Pipinae. (1/7). South America [11, 12].
    Rhinophrynidae. (1/1). Mesoamerica [1].
Suborder Neobatrachia
Superfamily Bufonoidea
    Allophrynidae. (1/1). South America [1].
    Brachycephalidae. (2/6). South America [19].
    Bufonidae. (33/344). Cosmopolitan except Madagascar and Australo-Papuan Region [1, 2, 19, 30, 31].
    Centrolenidae. (3/136). South America and Mesoamerica [20].
    Dendrobatidae. (9/207). South America and Mesoamerica [15, 16, 17].
    Heleophrynidae. (1/6). Southern Africa [2].
    Hylidae. (42/854). Five subfamilies:
        Hemiphractinae. (5/75). South America and Mesoamerica [27,28,29].
        Hylinae. (26/561). New World and Eurasia [1, 2, 3, 4, 5, 10, 20].
        Pelodryadinae. (3/159). Australo-Papuan Region [1, 2, 20].
        Phyllomedusinae. (6/50). South America and Mesoamerica [20,21].
        Pseudinae. (2/9). South America [1].
    Leptodactylidae. (51/1106). Five subfamilies:
        Ceratophryinae. (3/12). South America [1].
        Cycloramphinae. (8/44). South America [1, 2].
        Eleutherodactylinae. (12/745). New World tropics and subtropics [19, 22, 30].
        Hylodinae. (3/34). South America [14].
        Leptodactylinae. (9/152). New World tropics and subtropics [1, 8, 23, 24].
        Odontophryinae . (3/27) South America [1].
        Telmatobiinae. (11/92). South America [1, 2].
    Limnodynastidae. (9/48). Australia and New Guinea [2, 8, 9, 14, 23, 24].
    Myobatrachidae. (12/73). Australia and New Guinea [1, 7, 13, 19, 20].
    Rhinodermatidae. (1/2). Southern South America [18].
    Sooglossidae. (2/3). Seychelles [18, 19].
Superfamily Ranoidea
    Arthroleptidae. (8/11). Two subfamilies:
        Arthroleptinae. (3/50). Sub-Saharan Africa [1, 19].
        Astylosteninae. (5/27). Sub-Saharan Africa [2].
    Hemisotidae. (1/8). Sub-Saharan Africa [13].
    Hyperoliidae. (19/240). Four subfamilies:
        Hyperoliinae. (12/171). Sub-Saharan Africa and Madagascar [1, 20, 25].
        Kassininae. (5/19). Sub-Saharan Africa [1].

Leptopelinae. (1/49). Sub-Saharan Africa [1, 14].
Tachycneminae. (1/1). Seychelles [20].
Mantellidae. (5/134). Three subfamilies.
Boophinae. (1/46). Madagascar [1, 2].
Laliostominae. (2/4). Madagascar [1].
Mantellinae. (2/84). Madagascar [14, 19, 20].
Microhylidae. (67/364). Eight subfamilies:
Asterophryinae. (8/60). Indonesia to New Guinea [19].
Brevicipitinae. (5/19). Sub-Saharan Africa [19].
Cophylinae. (7/36). Madagascar [21].
Dyscophinae. (2/9). Madagascar and southeast Asia and associated
islands [1].
Genyophryninae.(11/118).Indo-AustralianArchipelago to northern
Australia [19,22].
Melanobatrachinae. (3/4). Central Africa and southern India [2, 21].
Microhylinae. (30/116). North and South America, Indo-Australian
Archipelago, eastern Asia, and Madagascar [1, 17, 19].
Phrynomerinae. (1/5). Sub-Saharan Africa [1].
Ranidae. (51/686). 11 subfamilies:
Cacosterninae. (10/37). Sub-Saharan Africa[1, 19].
Dicroglossinae. (15/162). Sub-Saharan Africa, southern Asia, and
Indo-Australian Archipelago [1, 2, 14, 19].
Lankanectinae. (1/1). Sri Lanka [1].
Micrixalinae. (1/11). Southern India [1].
Nyctibatrachinae. (1/11). Southern India [1].
Occidozyginae. (1/12). Southeast Asia and Indo-Australian
Archipelago [1].
Petropedetinae. (5/77). Sub-Saharan Africa [1, 2, 6, 22].
Ptychadeninae. (3/51). Sub-Saharan Africa [1, 2].
Pyxicephalinae. (2/5). Sub-Saharan Africa [1].
Raninae. (11/309). Cosmopolitan except for southern South
America, Madagascar, and most of Australia [1, 2, 3, 19].
Ranixalinae. (1/10). Southern India [2, 14].
Rhacophoridae. (8/207). Two subfamilies:
Buergerinae. (1/5). Taiwan to Japan [2].
Rhacophorinae. (7/202). Sub-Saharan Africa, southern Asia, and
Indo-Australian Archipelago [19, 21, 25].
Scaphiophrynidae. (2/9). Madagascar [1].

1.7 CONCLUSIONS

Frogs and toads comprise the monophyletic order Anura within the subclass
Lissamphibia in the class Amphibia. There are two suborders of Anura:
"Archaeobatrachia" contains 168 species in nine families, of which four families
are in the superfamily "Discoglossoidea," three in Pelobatoidea, and two in
Pipoidea; Neobatrachia contains more than 4600 species in 20 families, of

which 12 families are in the superfamily Bufonoidea and eight Ranoidea. In the past, most phylogenetic analyses (and resulting classification) were based on morphological characters. More recently, many phylogenetic analyses are based on molecular data and there is an increasing use of spermatological characters. Significant changes in hypothesized phylogenetic relationships and in classification can be expected with greater taxon sampling (morphological and molecular), sequences of more nuclear genes, and analyses of combined morphological and molecular data with life-history traits.

Table 1.1 Outline of reproductive modes in anurans. Modes always involving parental care are designated by an asterisk (*)

I.	Eggs aquatic

A. Eggs deposited in water
1. Eggs an tadpoles in lentic water (e.g., most *Rana*)
2. Eggs and tadpoles in lotic water (e.g., *Ansonia* and *Atelopus*)
3. Eggs and early larval stages in natural or constructed basins; subsequent to flooding, feeding tadpoles in ponds or slow-moving streams (e.g., *Hyla rosenbergi*)*
4. Eggs and herbivorous or insectivorous tadpoles in phytotelmata (e.g., *Hyla bromeliacia*)
5. Eggs and oophagous tadpoles in phytotelmata (e.g., *Anotheca spinosa*)*
6. Eggs and nonfeeding tadpoles in water-filled tree holes (e.g., *Phrynobatrachus guineensis*)
7. Eggs deposited in stream and swallowed by female; eggs and tadpoles complete their development in the stomach (e.g., *Rheobatrachus*)*

B. Eggs in foam or bubble nest
8. Foam nest on surface of a pond; feeding tadpoles in pond (e.g., most *Leptodactylus*)
9. Foam nest in pool and tadpoles in stream (e.g., *Limnodynastes interioris*)
10. Bubble nest on surface of a pond; feeding tadpoles in pond (e.g., *Chiasmocleis leucosticta*)

C. Eggs imbedded in dorsum of aquatic female
11. Eggs hatch into feeding tadpoles in ponds (e.g., *Pipa carvalhoi*)*
12. Eggs hatch into froglets (e.g., *Pipa pipa*)*

II. Eggs terrestrial or arboreal

D. Eggs on ground or in burrows
13. Eggs and early tadpoles in excavated nest; subsequent to flooding, feeding tadpoles in ponds or streams (e.g., *Pseudophryne*)
14. Eggs on ground or rock above water or in depression or excavated nest; upon hatching tadpoles move to water (e.g., *Mixophyes*)
15. Eggs hatch into feeding tadpoles that are carried to water by adult, where they feed on detritus (e.g., *Colostethus*)*
16. Eggs hatch into feeding tadpoles that are carried to phytotelmata by adult, where they are oophagous (e.g., some *Dendrobates*)*
17. Eggs hatch into nonfeeding tadpoles that complete their development in nest (e.g., *Leiopelma*)
18. Eggs hatch into nonfeeding tadpoles that complete their development on dorsum of adult (e.g.,) or in pouches in adult (e.g., *Rhinoderma darwini*)*
19. Eggs hatch into froglets (e.g., *Eleutherodactylus*)

E. Eggs arboreal
20. Eggs hatch into tadpoles that drop into ponds (e.g., *Phyllomedusa*) or streams (e.g., *Centrolene*)

Contd.

Table 1.1 Contd.

    21. Eggs hatch into tadpoles that drop into water-filled cavities in trees (e.g., *Nyctixalus*)

    22. Eggs hatch into froglets (e.g.,*Phrynodon sandersoni*)

  F. Eggs in foam nest

    23. Nest in burrow; subsequent to flooding, feeding tadpoles in ponds or streams (e.g., some *Leptodactylus*)

    24. Nest in burrow; nonfeeding tadpoles complete development in nest (e.g., *Adenomera*)

    25. Nest arboreal; hatchling tadpoles drop into ponds or streams (e.g., *Rhacophorus*)

  G. Eggs carried by adult

    26. Eggs carried on legs of male; feeding tadpoles in ponds (e.g., *Alytes*)*

    27. Eggs carried in dorsal pouch of female; feeding tadpoles in ponds (e.g., some *Gastrotheca*)*

    28. Eggs carried on dorsum or in dorsal pouch in female; nonfeeding tadpoles in phytotelmata (e.g., *Flectonotus*)*

    29. Eggs undergo direct development into froglets on dorsum or in dorsal pouch in female (e.g., *Hemiphractus*)*

III. Eggs retained in oviducts

  H. Ovoviviparous

    30. E.g., *Nectophrynoides tornieri*

  I. Viviparous

    31. E.g., *Nectophrynoides occidentalis*

1.8 ACKNOWLEDGEMENTS

I thank J. J. Wiens and T. W. Reeder for some molecular data, K. Adler for details of some publications, and L. Trueb for critically reviewing the manuscript.

1.9 LITERATURE CITED

Alrubaian, J., Danielson, P., Walker, D., and Dores, R. M. 2002. Cladistic analysis of anuran POMC sequences. Peptides 23: 443-452.

Anonymous. 1999. *International Code of Zoological Nomenclature.* International Trust for Zoological Nomenclature, London. 306 pp.

Austin, J. D., Lougheed, S. C., Tanner, K., Chek, A. A., Bogart, J. P. and P. T. Boag. 2002. A molecular perspective on the evolutionary affinities of an enigmatic neotropical frog, *Allophryne ruthveni*. Zoological Journal of the Linnean Society 134: 335-346.

Báez, A. M. and Basso, N. G. 1996. The earliest known frogs of the Jurassic of South America: review and cladistic appraisal of the relationships. Pp. 131-158. In G. Arratia (ed.), *Contributions of Southern South America to Vertebrate Paleontology.* Münchner, München, Germany. 342 pp.

Báez, A. M. and Trueb, L. 1997. Redescription of the Paleogene Shelania pascuali from Patagonia and its bearing on the relationships of fossil and Recent pipoid frogs. Scientific Papers Natural History Museum University of Kansas 4: 1-41.

Blainville, H.-M. D. de. 1822. *De l'Organisation des Animaux*. Volume 1. Levrault, Paris. 574 pp.

Blommers-Schlösser, R. M. A. 1975. Observations on the larval development of some Malagasy frogs, with notes on their ecology and biology (Anura: Dyscophinae, Scaphiophryninae, and Cophylinae). Beaufortia 24: 7-26.

Blommers-Schlösser, R. M. A. 1993. Systematic relationships of the Mantellinae Laurent

1946 (Anura Ranoidea). Ethology, Ecology and Evolution 3: 199-218.

Bossuyt, F. and Milinkovitch, M. C. 2001. Amphibians as indicators of Early Tertiary "out-of-India" dispersal of vertebrates. Science 292: 93-95.

Boulenger, G. A. 1882. Catalogue of the Batrachia Salientia s. Ecaudata in the Collection of the British Museum. Second edition. British Museum, London. 495 pp.

Cannatella, D. C. and Hillis, D. M. 1993. Amphibian relationships: phylogenetic analysis of morphology and molecules. Herpetological Monographs 7: 1-7.

Carroll, R. L. 2000. *Eocaecila* and the origin of caecilians. Pp. 1402–1411. In H. Heatwole and R. L. Carroll (eds), *Amphibian Biology*, Volume 4, Paleontology. Surrey Beatty and Sons, Chipping Norton, NSW, Australia, 523 pp.

Cope, E. D. 1864. On the limits and relations of the Raniformes. Proceedings of the Academy of Natural Sciences Philadelphia 16: 181-183.

Cope, E. D. 1865. Sketch of the primary groups of Batrachia Salientia. Natural History Review, new series 5: 97-120.

de Sá, R. O. and Hillis, D. M. 1990. Phylogenetic relationships of the pipid frog *Xenopus* and *Silurana*: an integration of ribosomal DNA and morphology. Molecular Biology and Evolution 7: 365-376.

de Sá, R. O. and Trueb, L. 1991. Osteology, skeletal development, and chondrocranial structure of *Hamptophryne boliviana* (Anura: Microhylidae). Journal of Morphology 209: 311-330.

Dubois, A. 1981. Liste des genres et sous-genres nominaux de Ranoidea (Amphibiens Anoures) du monde, avec identification de leurs especes-types: consequences nomenclaturales. Monitore Zoologico Italiano (N.S.). Supplemento 15: 225-284.

Dubois, A. 2003. True frogs. Ranidae. In Hutchins, M., Duellman, W. E. and Schlager, N. (eds), Grzimek's Animal Life Encyclopedia, 2nd edition. Volume 6: Amphibians. Gale Group, Farmington Hills, Michigan, U.S.A. (in press).

Duellman, W. E. 1975. On the classification of frogs. Occasional Papers of the Museum of Natural History, University of Kansas 42: 1-14.

Duellman, W. E. 1985. Reproductive modes in anuran amphibians: phylogenetic significance of adaptive strategies. South African Journal of Science 81: 174-178.

Duellman, W. E. 1989. Alternative life-history styles in anuran amphibians: evolutionary and ecological implications. Pp 101-126. In M. N. Bruton (ed.), *Alternative Life-history Styles of Animals*. Kluwer Academic Publishers, Dordrecht, Netherlands. 616 pp.

Duellman, W. E. 1993. Amphibian species of the world: additions and corrections. Special Publication, Museum of Natural History, University of Kansas 21: 1-372.

Duellman, W. E. (ed.), 1999. *Patterns of Distribution of Amphibians*. Johns Hokins University Press, Baltimore, Maryland, U.S.A. 633 pp.

Duellman, W. E. 2001. *Hylid Frogs of Middle America*. Society for the Study of Amphibians and Reptiles, Ithaca, New York, U.S.A. 1158 pp.

Duellman, W. E. and Adler, K. 2003. The evolution of amphibian systematics: an historical perspective. In Heatwole, H. and Tyler, M. J. (eds), *Amphibian Biology*, Volume 7. Taxonomy and Systematics. Surrey Beatty and Sons, Chipping Norton, NSW, Australia (in press).

Duellman, W. E. and Trueb, L. 1986. *Biology of Amphibians*. McGraw-Hill, New York. 670 pp (reprint, Johns Hopkins University Press, 1994).

Duméril, A.-M.-C. and Bibron, G. 1841. *Erpétologie Générale ou Histoire Naturelle Complète des Reptiles*. Volume 8. Roret, Paris. 792 pp.

Emerson, S. B., Richards, C., Drewes, R. C. and Kjer, K. M. 2000. On the relationships among ranoid frogs: a review of the evidence. Herpetologica 56: 209-230.

Ford, L. 1993. The phylogenetic position of the dart-poison frogs (Dendrobatidae) among anurans: an examination of the competing hypotheses and their characters. Ethology, Ecology and Evolution 5: 219-231.

Ford, L. and Cannatella, D. C. 1993. The major clades of frogs. Herpetological Monographs 7: 94-117.

Frost, D. R. (ed.) 1985. *Amphibian Species of the World*. Association of Systematic Collections and Allen Press, Lawrence, Kansas, U.S.A. 732 pp.

Frost, D. R. 2002. Amphibian species of the world: an online reference. [http://research.amnh.org/herpetology/amphibia/index.html] V2.21 (15 July 2002)

Glaw, F., Vences, M. and Böhme, W. 1997. Systematic revision of the genus *Aglyptodactylus* Boulenger, 1919 (Amphibia: Ranidae), and analysis of its phylogenetic relationships to other Madagascan genera (*Tomopterna, Boophis, Mantidactylus*, and *Mantella*). Journal of Zoological Systematics and Evolution Research 35: 1-21.

Griffiths. I. 1959. The phylogeny of *Smithillus limbatus* and the status of the Brachycephalidae (Amphibia Salientia). Proceedings of the Zoological Society of London 132: 457-487.

Haas, A. 2001. Mandibulr arch musculature of anuran tadpoles, with comments on homologies of amphibian jaw muscles. Journal of Morphology 247: 1-33.

Hay, J. M., Ruvinsky, I., Hedges, S. B. and Maxson, L. R. 1995. Phylogenetic relationships of amphibian families inferred from DNA squences of mitochondrial 12S and 16S ribosomal RNA genes. Molecular Biology and Evolution 12: 928-937.

Hedges, S. B. and Maxson, L. R. 1993. A molecular perspective on lissamphibian phylogeny. Herpetological Monographs, 7: 27-42.

Hillis, D. M., Ammerman, L. K., Dixon, M. T., and de Sá, R. O. 1993. Ribosomal DNA and the phylogeny of frogs. Herpetological Monographs 7: 118–131.

Hillis, D. M. and Davis, S. K. 1987. Evolution of the 28 Sribosomal RNA gene in anurans: regions of variability and their phylogenetic implications. Molecular Biology and Evolution 4: 117-125.

Holmes, R. 2000. Paleozoic temnospondyls. Pp. 1081-1120. In H. Heatwole and R. L. Carroll (eds), *Amphibian Biology*, Volume 4, Paleontology. Surrey Beatty and Sons, Chipping Norton, NSW, Australia, 523 pp.

Hutchins, M., Duellman, W. E. and Schlager, N.(eds) 2003. *Grzimek's Animal Life Encyclopedia, 2nd edition. Volume 6: Amphibians.* Gale Group, Farmington Hills, Michigan, U.S.A. (in press).

Larson, P. M. and de Sá, R. O. 1998. Chondrocranial morphology of *Leptodactylus* larvae (Leptodactylidae: Leptodactylinae): its utility in phylogenetic reconstruction. Journal of Morphology 238: 287-305.

Lathrop, A. 1997. Taxonomic review of the megophryid frogs (Anura: Pelobatoidea). Asiatic Herpetological Research 7: 68-79.

Laurent, R. F. 1979. Esquisse d'une phylogenèse des anoures. Bulletin de la Societe de Zoologique de France 104: 397-422.

Laurenti, J. N. 1768. *Specimen Medicum, Exhibens Synopsin Reptilium.* J. Thomae Trattnern, Vienna (reprint, A. Asher, 1966). 214 pp.

Laurin, M..amd Reisz, R. R. 1997. A new perspective on tetrapod phylogeny. Pp. 9–59, In S. Sumida and K. Martin (eds), *Amniote Origins — Completing the Transition to Land*. Academic Press, London. 510 pp.

Linnaeus, C. 1758. *Systema Naturae.* Editio decima. L. Salvii, Stockholm (reprint, British Museum [Natural History], 1956). 823 pp.

Lynch, J. D. 1971. Evolutionary relationships, osteology, and zoogeography of leptodactyloid frogs. Miscellaneous Publications, Museum of Natural History University of Kansas 53: 1-238.

Lynch, J. D. 1973. The transition from archaic to advanced frogs. Pp. 133-182. In J. L. Vial (ed.), *Evolutionary Biology of the Anurans*. University of Missouri Press, Colombia, Missouri, U.S.A. 470 pp.

Maglia, A. M. 1998. Phylogenetic relationships of extant pelobatoid frogs (Anura: Pelobatoidea): evidence from adult morphology. Scientific Papers Natural History Museum University of Kansas 10: 1-19.

Maglia, A. M. and L. A. Pugener. 1998. Skeletal development and adult osteology of *Bombina orientalis* (Anura: Bombinatoridae). Herpetologica 54: 344-363.

Maglia, A. M., L. A. Pugener, and Trueb, L. 2001. Comparative development of frogs: using phylogeny to understand ontogeny. American Zoologist 41: 538-551.

Milner, A. R. 2000. Mesozoic and Tertiary Caudata and Albanerpetontidae, Pp. 1412-1444. In H. Heatwole and R. L. Carroll (eds), *Amphibian Biology*, Volume 4, Paleontology. Surrey Beatty and Sons, Chipping Norton, NSW, Australia, 523 pp.

Nicholls, G. C. 1916. The structure of the vertebral column in Anura Phaneroglossa and its importance as a basis of classification. Proceedings of the Linnean Society of London, Zoology 128: 80-92.

Noble, G. K. 1922. The phylogeny of the Salientia. I—The osteology and the thigh musculature; their bearing on classification and phylogeny. Bulletin of the American Museum of Natural History 46: 1-87.

Noble, G. K. 1927. The value of life history data in the study of the evolution of the Amphibia. Annals of the New York Academy of Sciences 30: 31-128.

Noble, G. K. 1931. *The Biology of the Amphibia*. McGraw-Hill, New York. 577 pp.

Oppel, M. (1811) "1810." Mémoire sur la classification des reptiles. Annales Museu d'Histoire Natural, Paris 16: 254-295, 376-393.

Orton, G. L. 1953. The systematics of vertebrate larvae. Systematic Zoology 2: 63-75.

Orton, G. L. 1957. The bearing of larval evolution on some problems in frog classification. Systematic Zoology 6: 79-86.

Pugener, L. A., Maglia, A. M., and Trueb. 2003. Larval characters and phylognetic relationships of basal anurans: a value added contribution. Zoological Journal of the Linnean Society (in press).

Reig, O. A. 1958. Proposiciones para una nueva macrosistemática de los anuros. Nota preliminar. Physis 21: 109-118.

Richards, C. and Moore, W. 1996. A phylogeny for the African treefrog family Hyperoliidae based on mitcochondrial rDNA. Molecular Phylogenetics and Evolution 5: 522-532.

Richards, C. and Moore, W. 1998. A molecular phylogenetic study of the old world treefrog family Rhacophoridae. Herpetological Journal 8: 41-46.

Roesel von Rosenhof, A. J. 1758. *Historia Naturalis Ranarum Nostratium/Die Natürliche Historie der Frösche Hiesigen Landes*. J. J. Fleischmann, Nuremberg. 115 pp.

Ruvinsky. I. And Maxson, L. R. 1996. Phylogenetic relationships among bufonoid frogs (Anura: Neobatrachia) inferred from Mitochondrial DNA sequences. Molecular Phylogenetics and Evolution 5: 533-547.

Sokol, O. M. 1975. The phylogeny of anuran larvae: a new look. Copeia 1975: 1-24.

Spinar, Z. V. 1972. *Tertiary Frogs from Central Europe*. Dr. W. Junk, The Hague, Netherlands, 286 pp.

Starrett, P. H. 1973. Evolutionary patterns in larval morphology. Pp. 251–271. In J. L. Vial (ed.), *Evolutionary Biology of the Anurans*. University of Missouri Press, Colombia, Missouri, U.S.A. 470 pp.

Trueb, L. 1973. Bones frogs, and evolution. Pp. 66-132. In J. L. Vial (ed.), *Evolutionary Biology of the Anurans*. University of Missouri Press, Colombia, Missouri, U.S.A. 470 pp.

Trueb, L. and Cloutier, R. 1991a. Toward an understanding of the amphibians: two centuries of systematic history. Pp. 175-193. In H.-P. Schultze and L. Trueb (eds), *Origins of the Higher Groups of Tetrapods*. Cornell University Press, Ithaca, New York, U.S.A. 724 pp.

Trueb, L. and Cloutier, R. 1991b. A phylogenetic investigation of the inter- and intrarelationships of the Lissamphibia (Amphibia: Temnospondyli). Pp. 223-313. In H.-P. Schultze and L. Trueb (eds), *Origins of the Higher Groups of Tetrapods*. Cornell University Press, Ithaca, New York, U.S.A. 724 pp.

Trueb, L. and Hanken, J. 1992. Skeletal development in *Xenopus laevis* (Anura: Pipidae). Journal of Morphology 214: 1-41.

Trueb, L., Pugener, L. A., and Maglia, A. M. 2000. Ontogeny of the bizarre: an osteological description of *Pipa pipa* (Anura: Pipidae), with an account of skeletal development in the species. Journal of Morphology 243: 75-104.

Tyler, M. J. 1976. *Frogs*. Collins, Sydney. 256 pp.

Vences, M., Kosuch, J., Lötters, S., Widmer, A., Jungfer, K.-H., Köhler, J. and Vieth, M. 2000. Phylogeny and classification of poison frogs (Amphibia: Dendrobatidae), based on mitochondrial 16S and 12S ribosomal RNA gene sequences. Molecular Phylogenetics and Evolution 15: 34-40.

Wagler, J. G. 1830. *Natürliches System der Amphibien, mit Vorangehender Classification der Säugthiere und Vögel*. J. G. Cotta, München. 354 pp.

Wassersug, R. J. 1984. The *Pseudohemisus* tadpole: a morphological link between microhylid (Orton type 2) and ranoid (Orton type 4) larvae. Herpetologica 40: 138-149.

Wilbur, H. M. 1977. Propagule size, number, and dispersion pattern in *Ambystoma* and *Asclepis*. American Naturalist 111: 43-68.

Wilkinson, J. A. and Drewes, R. C. 2000. Character assessment, genus level boundaries, and phylogenetic analysis of the family Rhacophoridae: a review and present day status. Contemporary Herpetology 2000 (1); 1-27 [only online: http://www.calacademy.org/research/herpetology/ch/]

Zardoya, R. and Meyer, A. 2001 On the origin of and phylogenetic relationships among living amphibians. Proceedings of the National Academy of Science 98: 7380-7383.

CHAPTER 2

The Gross Anatomy of the Reproductive System

Michael J. Tyler

2.1 ANATOMICAL CONSTRAINTS AND EVOLUTIONARY OPPORTUNITIES

In several countries the frog has been associated with human fertility symbols. The simplest interpretation is that laying as many as thousands of eggs and fertilizing them all is good reason for this association. To what extent the reproductive system, rather than anuran behavior, contributes to successful fertilization is uncertain, but it can be concluded that the emerging ova and spermatozoa have been well equipped for fertilization to be such a reliable process.

The number of eggs released at any one time is related to the reproductive mode (Salthe and Duellman 1973), and ranges from five or six in several microhylid species, to 30,000 in *Bufo marinus*. Nevertheless, the gross structural changes in the nature of the reproductive system are not substantial. Clearly the system should be viewed as conservative in an evolutionary context.

The nature and form of the reproductive system of anuran amphibians is a consequence (perhaps benefit) of the overall structure of these animals. The significant anatomical feature of anurans is the absence of a diaphragm and, as a consequence, organs normally confined to the thorax can extend into the abdomen. For example, hyperinflation of the lungs permits them to extend for the entire length of the body cavity, as occurs in behavioral postures designed to deter predators (Williams *et al.* 2000).

An additional advantage of the lack of a diaphragm is the capacity of other organs to expand far beyond what, in mammalian terms, would be considered normal. Being unconstrained the abdominal organs can move anteriorly into the thoracic region, ultimately occupying almost the entire abdomino-thoracic cavity. There is no mammalian equivalent. Rather, it is necessary to focus upon the fact that anurans have unique anatomical opportunities, which have

Department of Environmental Biology, The University of Adelaide, South Australia 5005 influenced the evolution of the reproductive system. As a result, mature eggs may occupy more than 50% of the volume of the total body cavity of a gravid female.

The male reproductive system is less intrusive, the only significant variation being in the relative size of the testes and the presence of a penis in two species of *Ascaphus* that have internal fertilization (Sever *et al.* this volume, Chapter 7), and the presence of Bidder's Organ in bufonids.

2.2 THE OVARIES

As noted by Spengel (1876), Badhuri (1953) and Horton (1984) the ovaries vary considerably in the number of lobes. Horton attributes this phenomenon as directly related to exposing individual oocytes to the epithelial lining for physiological efficiency; a fact that she demonstrated graphically (Fig. 2.1). It follows that egg diameter (which is linked to reproductive mode) should be

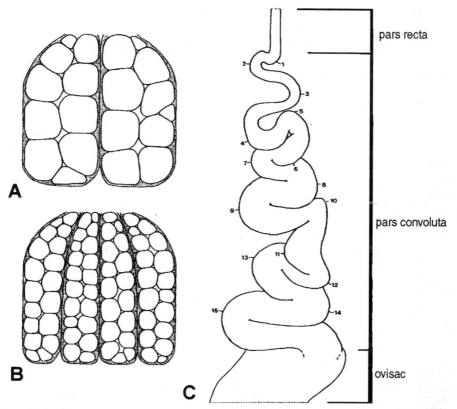

Fig. 2.1 A,B. Diagrammatic representation of the manner in which paired ovaries of similar volume accommodate ova, with the requirement that each one is in contact with the epithelium enveloping them. **A**. Paired lobes of macrolecithal eggs. **B**. Smaller eggs accommodated in four lobes. **C**. An anuran oviduct with 15 convolutions numbered at the apex of each. The ostium is not shown. After Horton P. 1984. Ph.D. Thesis, Figs 2 and 13.

reflected in the number of lobes and that the diameter will decrease as ovidiameters increase. Species with direct development or which retain developing embryos in the oviduct (see Severs *et al.*, this volume, Chapter **7**) customarily have only a single lobe on each side. The condition may be conservative: Wake (1978) and Townsend *et al.* (1981) report it in two species of *Eleutherodactylus* Dumeril and Bibron, which have internal fertilization, whilst Bhaduri (1953) reports that three congeners which lack this habit also have single lobes.

Normally the ovaries are bilaterally symmetrical in lobe number. The maximum number of lobes reported in the literature are "8 or 10" for *Pleurodema cinerea*, "9 or 11" for *Hyla gratiosa* and *Pachymedusa* (as *Phyllomedusa) dacnicolor* (Bhaduri 1953), and 21 in *Litoria caerulea* (Horton 1984). Van den Broek (1933) suggested that as age progresses there might be a fusion of lobes, so reducing the number. However, there is no evidence in support of this assertion; in fact there is evidence that the number of lobes in juveniles is identical to that of adults (Horton 1984). The same author noted the only reported instance of significant intraspecific variation: an individual of *L. caerulea* with 21 lobes on one side and 12 on the other; another individual had 13 lobes on each side. Horton attributed the disparity to a reproductive abnormality, which she considered exceptional.

2.3 THE OVIDUCTS

In almost all vertebrate animals the oviducts, which are derived from Müllerian or pronephric ducts, are lost ontogenetically in males. The exception is in the Anura where the Müllerian ducts often persist to some degree in males. Nevertheless, intraspecific variation can be considerable. Johnston and Gillies (1919) report that in males of the hylid species *Litoria caerulea*, the range extends from a total lack of ducts to a degree of development comparable to that in females.

The oviducts are paired, tubular organs, which are divisible into four separate components (Fig.2.2). On a level with the liver the anterior component of the oviduct is the dilated and funnel-shaped opening through which ova are released from the ovaries into the coelomic cavity and, driven by ciliated epithelia, enter the tube discharging them. The opening is variously named an ostium (used here) or infundibulum. The second region is a short, thin-walled and straight tube: the *pars recta*, which leads to the longest and convoluted region, termed appropriately the *pars convoluta*. The *pars convoluta* terminates, and discharges into, a broadly dilated region termed the ovisac (Fig.2.2), which is termed the "uterus" by some authors.

Each paired ovisac may discharge ova separately into the cloaca prior to their emergence, or else the ovisacs coalesce above the cloaca to form a single chamber. The latter condition may well have had an influence upon the adoption of the term uterus.

One of the variables in oviduct structure that can be quantified is the number of convolutions of the *pars convoluta*. Horton (1984) undertook a

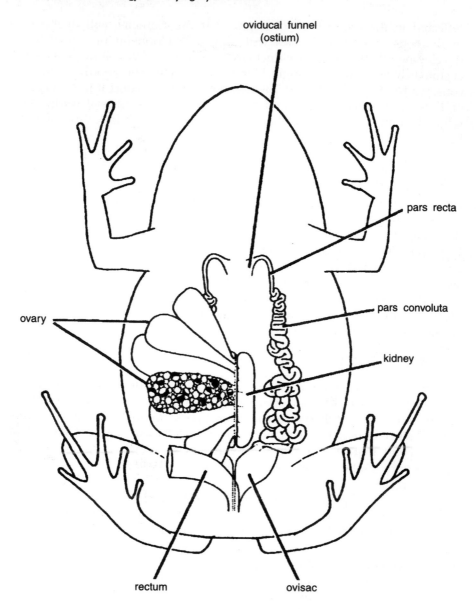

oviducal funnel
(ostium)

pars recta

pars convoluta

kidney

ovary

rectum

ovisac

Fig. 2.2 Generalized female anuran reproductive system. The ovary has been removed on the left side. After Horton P. 1984. Ph.D. Thesis, Fig.1.

detailed study of the female reproductive system and found considerable variation in 108 species. There was a distinct trend (amongst the predominantly Australian species in the sample) for the highest number of convolutions to be found in the largest species. Thus *Litoria caerulea* (179.8 ± 44.5) and *Cyclorana*

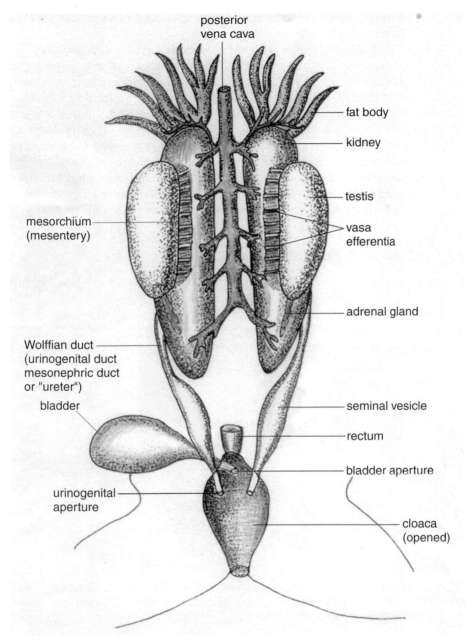

Fig. 2.3 Semidiagrammatic representation of the anuran male reproductive system. From Jamieson, B. G. M., unpublished, after various sources.

australis (179.0 ± 14.4) topped the list, whereas *L. iris* (32.3 ± 2.6) and *L. bicolor* (42.3 ± 3.5) were small species near the bottom of the list. Nevertheless, species of intermediate size had even lower numbers of convolutions: *L. pratti* (14.7 ± 5.5) and *L. rheocola* (19.0). What is unquestionable is the fact that high numbers of convolutions are to be found only in species of large size i.e. 80 – 100 mm snout to vent length.

Histologically the oviduct wall comprises both circular and longitudinal muscle fibers to drive the ova peristaltically down the duct. It is imperfectly known, biochemically, which portions of the oviduct are responsible for investing the ovum with the different constituents detected in the surrounding jelly. One factor detected in significant amounts in homogenates of diverse species, is Prostaglandin E2, which may prove to be the capacitation factor (Tyler *et al.* 1983). The role of the various regions of the oviduct in forming the definitive egg is discussed in Fernández and Ramos (chapter 4 of this volume).

The presence of convolutions increases the transit time of ova but this does not explain satisfactorily the variability in number or increase in length. Further data on the diversity of the structure of oviducts are provided by Wake (1985) and Wake and Dickie (1998).

2.4 THE MALE REPRODUCTIVE SYSTEM

The male reproductive system is illustrated in Fig. 2.3 and the relationship between the testes, kidneys and Wolffian ducts in Fig. 2.4.

The testes (see below) are supported by the mesorchium which is a mesentery

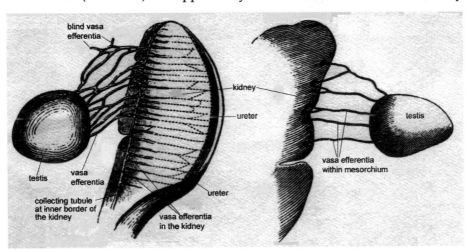

Fig. 2.4 Testes and their connections with the kidneys and "ureters" (Wolffian ducts). After Ecker, A. 1882. *Die Anatomie des Frosches*. Friedrich Vieweg und Sohn, Braunschweig, Figs 21 and 22.

attaching them to the kidneys. The ducts from each testis (vasa efferentia) pass along the mesorchium, enter the kidney and lead to the Wolffian or mesonephric duct (Fig. 2.4), often termed the ureter, which is both a urinary and seminiferous tract.

The ureter has not been examined by many authors, but Ecker (1882) notes that the mucous membrane lining has longitudinal folds. Weidersheim (1886) stated that he found mucous glands in the wall of the ureter of *Rana temporaria*, but this remains to be confirmed.

By far the most exhaustive study of anuran testes is that of Blair (1972) who was concerned with the phylogenetic significance of testis color and size in the genus *Bufo*. He found that he could group species into five cohorts on color characteristics, and further subdivided groupings according to whether the testis width/length relationship was less than or greater than 30%. Because they are attached to the ventral surface of the kidneys and extend along their length, they are always longer than they are broad. Blair reports instances of hypertrophy in Africa and South American species, stating "the body cavity is literally crammed with the hypertrophied testes", as in an individual of *B. haematiticus* which had a total head and body length of 52mm and testes 28 mm long.

G. F. Watson (pers. comm.) examined the testes of Australian hylid and myobatrachid species in the course of hybridization experiments, and stated that the testes of *Crinia georgiana* were "huge". He also described the testes of *Paracrinia haswelli* as "very large". Hoffman (1931) described the testes of *C. georgiana* as "very small in comparison with those of *Bufo* and are unique in being black in color". The identity of Hoffman's material must be questioned.

Bhaduri (1953) measured the length and width of each kidney and testis in numerous species and concluded that those of *Eleutherodactylus rubicola* and *E. alticola* are "enormous" and "extremely large" respectively. The significance of these and other examples of hypertrophy remains unknown. There are no published reports of gonadal hypotrophy but it is possible that, in the absence of an established data set, it might not be recognized.

2.5 BIDDER'S ORGAN

Bidder's Organ is a small, round aggregation of oocytes upon the anterior end of the testis of *Bufo* species and constitutes a feature unique to that genus. The organ is perceived as an incipient ovary in males and it has attracted considerable interest about its function. Duellman and Trueb (1986) describe it as cortical remnants of the germinal ridge and point out that the oocytes remain in an immature state. The circumstances under which sex reversal could take place remain uncertain.

2.6 LITERATURE CITED

Bhaduri, J.L. 1953. A study of the urogenital system of Salientia. Proceedings of the Zoological Society of Bengal 6(1): 1-75

Blair, W.F. 1972. Characteristics of the testis. Pp. 324-328. In W. F. Blair (ed.), *Evolution in the genus Bufo*. University of Texas Press, Austin and London.

Duellman, W.E. and Trueb, L. 1986. *Biology of Amphibians*. Mc-Graw Hill Book Company, New York. 670 pp.

Ecker, A. 1882. *Die Anatomie des Frosches*. Friedrich Vieweg und Sohn, Braunschweig. 95 pp.

Hoffman, A.C. 1931. On the general anatomy of the genus *Heleophryne*. South African Journal of Science 28:399–407.

Horton, P. 1984. The female anuran reproductive system in relation to reproductive mode. Ph. D. Thesis, The University of Adelaide, Adelaide, South Australia.

Johnston, T.H. and Gillies, C.D. 1919. A note on the occurrence of Mullerian ducts in the male of *Hyla caerulea* White. Journal and Proceedings of the Royal Society of New South Wales 52: 461-462.

Salthe, S.N. and Duellman, W.E. 1973. Quantitative constraints associated with the reproductive mode in anurans. Pp.229–249. In J. L. Vial (ed.), *Evolutionary biology of the Anurans. Contemporary Research on Major Problems.* University of Missouri Press, Columbia.

Spengel, S.W. 1876. Das urogenitalsystem der Amphibien. Arbeit   aus dem Zoologische Zootomische Institut, Wurzburg 3: 77 –114.

Tyler, M.J., Shearman, D.J.C., Franko, R., O'Brien P., Seramark, R.F. and Kelly, R. 1983. Inhibition of gastric acid secretions in the gastric brooding frog *Rheobatrachus silus.* Science 220: 609-610

Wake, M.H. 1985. Oviduct structure and function in non-mammalian vertebrates. Pp. 427-435. In H.-R. Duncker and G. Fleischer (eds), *Vertebrate Morphology.* Gustav Fischer, Stuttgart.

Wake, M.H. and Dickie, R. 1998. Oviduct structure and function and reproductive modes in amphibians. Journal of Experimental Zoology 282: 477-506.

Wiedersheim, R. 1886.   *Elements of Comparative Anatomy of Vertebrates.* Macmillan, London, 345 pp.

Williams, C.R., Brodie, E. Jr, Tyler, M.J. and Walker. S.J. 2000 Antipredator mechanisms of Australian frogs. Journal of Herpetology 34 (3): 431-443.

CHAPTER 3

Oogenesis

Sara S. Sánchez and Evelina I. Villecco

3.1 GERM PLASM AND PRIMORDIAL GERM CELLS

All sexually reproducing animals originate from the fusion of a male and a female gamete, the sperm and the egg. The result of this union, the zygote, is a cell with an immense potential to construct a new individual that maintains and propagates the characteristics of the species. The zygote inherits not only genetic material but also its cytoplasm from the egg. This maternal cytoplasm supports to varying degrees the development of the early embryo.

The female gamete is the cellular link between one generation and the next. This gamete is the result of a complex process known as oogenesis, which is characterized by a series of cellular and molecular events that take place in the nucleus and in the cytoplasm of the developing oocytes. The successive stages of oogenesis are: diploid primordial germ cells; several generations of diploid oogonia; diploid primary oocytes which undergo the first, reductional meiotic division; and haploid secondary oocytes which undergo the second, non-reductional (equational) meiotic division and are capable of fertilization.

The gamete is thus the ultimate product of a precursor stem cell population, the primordial germ cells (PGCs). These cells originate extragonadally early in embryogenesis and migrate through somatic tissues to reach the developing gonads, where they are surrounded by somatic gonadal cells, divide and differentiate into definitive gametes. The origin of PGCs, how these cells become specific and develop separately from the somatic lineage was, and still is, a fundamental question in developmental biology that has let to the discovery of a defined cytoplasmic area of the egg, the germ plasm (Ikenishi 1998).

In many animals, germ plasm material, present in the egg, can be found in PGCs and in germ cells throughout the life of the organism and is thought to act as a determinant of germ cell fate.

Departamento de Biología del Desarrollo, Instituto Superior de Investigaciones Biológicas (INSIBIO), Consejo Nacional de Investigaciones Científicas y Técnicas (CONICET) y Universidad Nacional de Tucumán (UNT), Chacabuco 461, 4000-San Miguel de Tucumán, Argentina

Germ plasm is morphologically similar in all organisms and can be identified by the presence of a fibrillar germinal cytoplasm and large aggregations of mitochondria intermingled with electron-dense germinal granules and ribosomes (Czolowska 1969, 1972; Williams and Smith 1971).

The germ plasm of anuran amphibians is formed during oogenesis. In *Xenopus laevis* eggs, for instance, it is present as numerous islands at the vegetal pole.

The early movements of the germ plasm in amphibians have been analyzed in detail by Savage and Danilchik (1993), who labeled the germ plasm with a fluorescent dye. They found that the germ plasm of unfertilized eggs consists of tiny islands that appear to be tethered to the yolk mass near the vegetal cortex. These germ plasm islands move with the vegetal yolk mass during the cortical rotation of fertilization.

After fertilization, the islands are released from the yolk mass and begin fusing together and migrating to the vegetal pole. Their aggregation depends on microtubules and the movement of these clusters to the vegetal pole requires a kinesin–like protein, Xklp–1 (Roob *et al.* 1996).

3.1.1 Germ Plasm Segregation and Primordial Germ Cells Migration

During the first embryonic divisions the germ plasm is segregated asymmetrically and the cells that inherit this localized cytoplasm give rise to PGCs (Eddy 1975; Whitington and Dixon 1975; Dixon 1994).

The experimental removal of the germ plasm or its components from the oocyte causes abnormal germ cell migration, a decrease in germ cell number, or complete sterility. When ultraviolet light is applied to the vegetal surface of frog embryos, the resulting frogs are normal but lack germ cells in their gonads (Smith 1966; Buehr and Blacker 1970). Very few primordial germ cells reach the gonads. Those that do have about one tenth of the volume of normal PGCs and have aberrantly shaped nuclei (Züst and Dixon 1977). Savage and Danilchik (1993) found that UV light prevents vegetal surface contractions and inhibits the migration of germ plasm to the vegetal pole.

The introduction of a vegetal pole cytoplasm into irradiated eggs can restore fertility. This observation led Smith (1966) to conclude that germ plasm contains germ cell determinants.

Germ cells rarely differentiate into gametes at the place where they first appear. Instead, in most metazoans, PGCs search out special somatic cells with which to form the gonad.

The germ plasm of anuran amphibians collected around the vegetal pole in the zygote is brought upward during cleavage by periodic contractions of the vegetal cell surface that push this germ plasm along the cleavage furrows of the newly formed blastomeres, enabling it to enter the embryo. At the blastula stage, the germ plasm becomes associated with the endodermal cells that line the floor of the blastocoel (Resson and Dixon 1988; Kloc *et al.* 1993) (Fig. 3.1).

PGCs are first discernible at the margin of the embryo proper. As gastrulation proceeds, germ cells are carried rather passively inside the

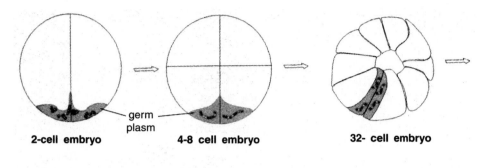

germ plasm

2-cell embryo   **4-8 cell embryo**   **32- cell embryo**

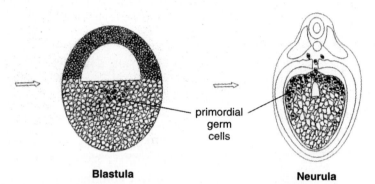

primordial germ cells

Blastula   **Neurula**

Fig. 3.1 Segregation of *Xenopus* germ plasm and primordial germ cell migration during early embryogenesis. Originally collected around the vegetal pole of the uncleaved egg, the germ plasm is segregated asymmetrically and advances along the cleavage furrows during the first embryonic divisions. At the blastula stage germ plasm is associated with endodermal cells that are positioned near the floor of the blastocoel. At the neurula stage germ plasm is concentrated in the posterior endoderm and migrate along the dorsal mesentery and into the genital ridge to reach the gonadal primordium. In all drawings the animal pole is at the top, and the vegetal pole at the bottom.

Shaded: Germ plasm. Spheres: Primordial germ cells. Modified after Ikenishi K. 1998. Development Growth and Differentiation 40: 1-10, Fig. 2.

embryo, where they are found in close association with the developing gut and are thought to undergo two or three divisions. At the late tail bud stage (stage 32/33) PGCs concentrate in the posterior region of the larval gut and, while the abdominal cavity is being formed, they migrate along the dorsal aspect of the gut.

Around the early tadpole (stage 40) PGCs accumulate in the dorsal crest of the posterior endoderm and are subsequently incorporated into the lateral plate mesoderm that forms the dorsal mesentery. In later stages, PGCs migration continues to the dorsal body wall and then laterally to the forming genital ridges (Wylie and Heasman 1976). The PGCs migrate up the dorsal mesentery until they reach the somatic gonadal primordium (Fig. 3.1). Contact guidance in this migration seems likely, as both PGCs and the extracellular matrix over which they migrate are orientated in the direction of migration (Wylie *et al.* 1979). Furthermore, PGCs adhesion and migration can be inhibited if the mesentery is treated with antibodies against *Xenopus* fibronectin (Heasman *et al.* 1981). Thus, the pathway for germ cell migration in these frogs appears to be composed of an orientated fibronectin-containing extracellular matrix. The fibrils over which PGCs travel lose their polarity soon after migration ends.

The mitotic activity of PGCs in the course of migration varies among species. In *Xenopus*, PGCs undergo about three division cycles, so that 25 to 30 PGCs colonize the gonads (Whitington and Dixon 1975; Wylie and Heasman 1993). In the frog *Rana pipiens*, PGCs follow a similar route, but may be passive travelers along the mesentery rather than actively motile cells.

3.1.2 Oogonial Proliferation

In the gonadal primordium, PGCs divide to form germ cells which then begin an active proliferation and become gonial cells. Female gonial cells, termed oogonia, undergo four incomplete mitotic divisions, forming a nest or cyst of sixteen secondary oogonia (Pepling *et al.* 1999). These groups of cells are cysts since individual cells are interconnected by intercellular bridges located in the middle of the nest. Ovarian nest formation is a conserved and widespread event that probably gives specific advantages to female germ cells. The presence of intercellular bridges could permit a direct communication and transference of signals and/or nutrients between the cells of the nest. These intercellular connections support the formation of oocytes at the same maturation stage.

In *Xenopus*, where the nest stage takes place in the ovaries of tadpoles, the pear-shaped cells of the nest develop synchronously (Coggins 1973; Gard *et al.* 1995). The nest is surrounded by the prefollicular cells, which are characterized by their electron-dense nuclei. These cells extend processes between the contiguous meiotic oocytes (late pachytene/early diplotene) and each developing primary oocyte is also surrounded by prefollicular cells.

In the nest stage, components that will later originate the germinal plasm can be observed. Each cell of the nest (secondary oogonium) contains a primordium of a mitochondrial cloud called a mitochondrial aggregate. After the secondary oogonia of the nest separate into individual early prestage I oocytes, the mitochondrial aggregate acquires additional mitochondria and becomes a premitochondrial cloud. The mitochondrial aggregate and the premitochondrial cloud are composed of a centrally located centriole surrounded by mitochondria and electron-dense mitochondrial cement (nuage material) (Fig. 3.2) that is probably a precursor of the granulo-fibrillar material (GFM). The coalescing of the GFM forms the germinal granules (Heasman *et al.* 1984; Kloc *et al.* 2001a).

From late prestage I to stage I, primary oocytes contain a large mitochondrial cloud with mitochondria, GFM, and fully developed germinal granules. The mitochondrial cloud has two distinct regions. The one closer to the vegetal pole of the oocyte, called Messenger Transport Organizer (METRO), contains a

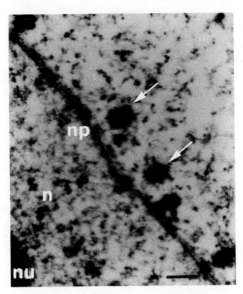

Fig 3.2. Electron micrograph of part of the nuclear membrane from *Bufo arenarum* primordial germ cells showing aggregations of electron-dense granular material or nuage (arrow), associated with the pores of the nuclear envelope. n, nucleus; np, nuclear pores; nu, nucleolus. Scale bar 0.2 μm. Original.

concentration of germinal granules and is a vehicle for the localization of various RNAs. The other, closer to the nucleus, is devoid of germinal granules (Heasman *et al.* 1984; Kloc *et al.* 1998, 2000). Between stages II and V of oocyte growth, the mitochondrial cloud breaks apart and its fragments migrate to the vegetal cortex of the oocyte (Fig. 3.3). In stage VI oocytes, the germinal granules are located in the apical region of the vegetal cortex. During first cleavages, the germinal granules and mitochondria form the islands of the germ plasm at the vegetal tips of vegetal blastomeres.

During cleavage, the germinal granules formed during oogenesis change their morphology and small germinal granules coalesce into more complex aggregates (Ikenishi and Kotani 1975). The germ plasm at the eight-cell stage embryo contains a few large irregularly shaped germinal granules (about 2000 nm in diameter) with a complex and variable morphology. Between cleavage and the neurula stage, the large germinal granules disappear and only a few small granules (approximately 300 nm in diameter) remain in the PGCs from the neurula (Fig. 3.1).

3.1.3 Molecular Composition of Germ Plasm

The analysis of the molecular composition of germ plasm and germinal granules has shown that they contain several species of RNAs (King *et al.* 1999; Kloc *et al.* 2002). The study of the distribution of RNAs within the germ plasm of cleaving embryos has demonstrated that, out of 11 different RNAs, only two RNAs remain associated with the germinal granules, Xcat2 RNA and Xpat

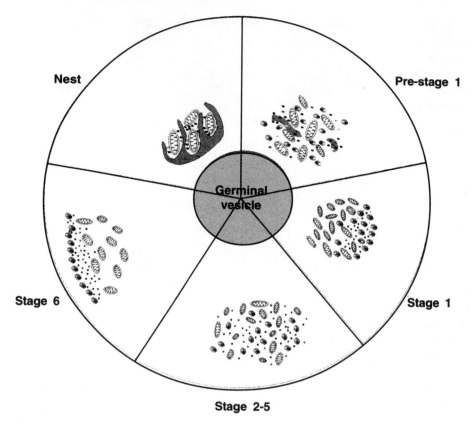

Stage 2-5

Fig. 3.3 Formation of the germ plasm in *Xenopus* oogenesis. In the nest stage, the oogonium has a juxtanuclear mitochondrial aggregate formed by mitochondria and electron-dense mitochondrial cement (nuage). The mitochondrial aggregate increases in early prestage I oocyte becoming a premitochondrial cloud. In late prestage I to stage I, oocytes present a large mitochondrial cloud that contains granulo-fibrillar material (GFM), fully developed germinal granules, (probably both derived from nuage), mitochondria, and germ plasm matrix. In stage I oocyte the germ plasm is located at the vegetal tip of the mitochondrial cloud in the METRO region. The analysis of mRNA distribution shows that Xcat2, Xpat, Xdazl and DSouth are present in GFM, Xcat2 and Xpat in the germinal granules and Xpat, Xdazl, Xlsirts, Xwn11, DSouth, fatVg, fingers and XFACS in the matrix. Between stage II and V of oocyte the mitochondrial cloud breaks down and its fragments migrate to the vegetal cortex of the stage VI oocyte. Irregular structure, mitochondrial cement; elongated structure, GFM; middle-sized spheres, germinal granules; small spheres, mRNAs. Modified after Kloc M. *et al.* 2002. Developmental Biology 217: 221-229, Fig. 1.

RNA. The former is sequestered inside the germinal granules while the latter is present at a lower level in the granule periphery. These RNAs are associated both with the small and the large aggregated granules. All other RNAs are present at different levels in the germ plasm matrix, between the granules and the mitochondria (Kloc *et al.* 2001b).

In *Xenopus*, nest stage oogonia and early prestage 1 oocytes, Xcat2 mRNA and Xlsirts RNA, are located in the mitochondrial aggregate between the mitochondria and the cement.

The RNAs present in the germ plasm may play a structural role and/or may be involved in the germ plasm assembly or in germ cells determination, differentiation and migration, while the germ plasm matrix could be an anchor for the molecules that are functional in oogenesis or in embryogenesis. Kloc (2002) proposed germinal granules as sequestration organelles that protect certain molecules to be used in early embryogenesis.

At present, the data on the possible functions of germ plasm localized RNAs in *Xenopus* are limited to Xlsirts and Xdazl. The former plays a structural role in anchoring Vg1RNA in the vegetal cortex of the oocyte while the latter is critically involved in PGCs development in the *Xenopus* embryo. In the absence of Xdazl, PGCs do not successfully migrate (Houston and King 2000) (Fig. 3.3).

3.2 OOGENESIS AND FOLLICLE FORMATION

Once the gonad is assembled, the gonial cells reach a well-defined number in the embryonic ovaries of several vertebrates. Then all oogonia either degenerate or enter meiosis as primary oocytes and are released from the ovary by the time the adult female becomes reproductively active (Peters 1978; Tokarz 1978). However, in amphibians, oogonia persist in the adult ovary either individually or arranged in the nest and continuously enter meiosis.

In *Xenopus*, as stated above, primary oocytes in the nest stage develop synchronously and are found from the premeiotic interphase stage to the late pachytene stage. All sixteen cells develop into individual oocytes, each developing oocyte being covered by prefollicular cells. Oocytes separate from one another and are no longer found in nests. They develop asynchronously, probably because of the disappearance of the intercellular bridges of the nest with the beginning of folliculogenesis, as suggested by Coggins (1973).

In most anuran amphibians, when the oocyte is in the diplotene stage, the prefollicular cells become organized into the follicular epithelium which, originating from the coelomic epithelium, envelops the oocyte, creating an ensemble called a follicle. Newly formed follicles, surrounded by a monolayer of squamous follicle cells closely apposed to the oolemma are also enveloped by a thin connective tissue sheath (thecal layer) derived from the mesenchymal tissue.

Amphibian oogenesis is an amazing process that varies in different species, resulting in different size and number of eggs formed. These variations are due to the need of the species for adaptation to the various environments in which they must lay their eggs and thus secure their development in order to perpetuate themselves (Del Pino 1989a). These differences agree with the different patterns of reproduction observed among amphibian species, some of which reveal a fast early development (e.g. *Xenopus laevis, Rana pipiens, Bufo arenarum, Ceratophrys cranwelli*) and others a slow one (e.g. *Gastrotheca riobambae, Flectonotus pygmaeus, Stefania evansi, Hyla cinerea*).

In amphibians, the ovary contains oogonia that can generate a new cohort of oocytes each year. When the oogonia enter the first meiotic division, they become primary oocytes. In this stage, oocytes progress through the first meiotic prophase and remain arrested at the diplotene stage (Jorgensen 1973); the primary oocytes undergo a complex process of cytodifferentiation and division that results in the production of the female gamete.

This process can be divided into three main stages: previtellogenesis, vitellogenesis and postvitellogenesis or maturity.

Oogenesis in *Xenopus* is a continuous, asynchronous process, and oocytes in all stages of development are scattered in the ovary (Dumont 1972). In spite of this oocyte distribution, in the ovary of the leptodactylid *Ceratophrys cranwelli* oocytes are distributed in two zones: a proximal zone which contains previtellogenic and early vitellogenic oocytes and a distal zone with vitellogenic and fully grown oocytes (Villecco *et al.* 2002) (Fig. 3.4A, B).

Oogenesis in *Xenopus laevis* can be divided into six stages on the basis of the anatomical appearance of unfixed intact oocytes. Each stage of oocyte development is correlated with histological, ultrastructural, physiological and biochemical characteristics (Dumont 1972). This classification serves as a framework for most studies in anuran oogenesis and is useful for comparing this process in different species.

3.3   PREVITELLOGENESIS

During primary growth, the diameter of the primary oocyte grows from approximately 10-20 μm at the leptotene stage to 100-200 μm at the diplotene stage. The formation of cytoplasmic organelles (including energy-producing organelles or mitochondria) (Callen *et al.* 1980 a, b), the accumulation of precursors for DNA, RNA and proteins synthesis and the storage of messenger RNA and of structural proteins are events responsible for the increase in oocyte size as well as for the decrease in the nucleo-cytoplasmic ratio. Growing oocytes are active in the transcription of genes, the products of which are necessary for oocyte metabolism, meiotic maturation and early development.

During the diplotene stage, amphibian oocytes are very active in the synthesis of RNAs. In some amphibian species, this synthesis is facilitated by the formation of multiple nuclei or germinal vesicles (4 to 3,000) that disappear gradually until only one remains in the oocyte before the last step of oogenesis. Del Pino (1989a) called this particular situation multinucleate oogenesis, and mononucleate oogenesis that in which only one germinal vesicle is present throughout the whole oogenetic process (Del Pino and Humphries 1978).

Amphibian oocytes usually have only one nucleus, which can grow to a considerable size as in the case of *Xenopus*, whose full grown primary oocyte nucleus has a diameter of about 0.5 mm.

During previtellogenesis, the main components of the amphibian oocytes nucleus are lampbrush chromosomes, amplified nucleoli, the nuclear envelope and the nucleoplasm or nuclear sap, which contains the majority of the nuclear proteins (Scheer and Dabauvalle 1985).

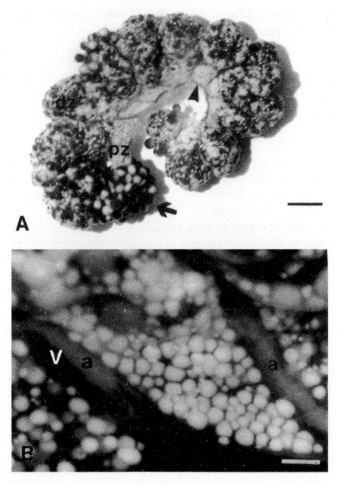

Fig. 3.4 Anatomical characteristics of the *Ceratophrys cranwelli* ovary. **A**. Stereoscope micrograph of the whole ovary showing 13 lobes. pz, proximal zone; dz, distal zone. The lobes are separated by branches of the ovarian artery and vein (arrow). Scale bar 4000 μm. **B**. Proximal zone of the lobe limited by the interlobal vein (V) and artery (a), containing previtellogenic oocytes of different sizes. Scale bar 600 μm. From Villecco E.I. *et al.* 2002. Zygote 10: 163-173, Fig. 1.

3.3.1 Lampbrush Chromosomes

The chromosomes of the diplotene stage of the meiotic prophase are the most remarkable constituents of amphibian primary oocyte nuclei. They gradually unfold lateral loops of DNA, giving rise, to lampbrush chromosomes.

The precise stage at which lampbrush chromosomes are formed is difficult to establish because the unfolding of the lateral loops is a gradual process. However, it is clear that the lampbrush form develops very early in the diplotene stage, shortly after the pachytene stage, and lasts until the oocyte approaches maturity. Hybridization *in situ* reveals these chromosomes as the site of RNA synthesis (Old *et al.* 1977). The lateral loops are the manifestation of ongoing transcription. Numerous lateral fibrils attached to the loop axes have been identified as nascent ribonucleoprotein (RNP) transcripts, each containing a growing RNA chain associated with proteins.

Lampbrush chromosomes, which become engaged in an intense transcription of heterogeneous RNA (hnRNA), are generally assumed to generate a complex set of maternal as well as polysomal mRNA, and are the synthesis sites for functional mRNAs (Miller and Hamkalo 1972; Sommerville 1977; MacGregor 1980; Davidson 1986).

In frog oocytes the rate of synthesis and accumulation of stable cytoplasmic hnRNA has been quantified and found to be in agreement with the structural evidence that numerous lampbrush chromosomes are closely packed with transcribing RNA polymerases.

Synthesis of hnRNA occurs very rapidly in vitellogenic oocytes. The high rate of synthesis of mRNA in *Xenopus* is probably necessary to accumulate and maintain the large amount of mRNA present in full grown oocytes and the eggs (Anderson and Smith 1977).

hnRNA and mRNA synthesis continues at a constant rate throughout oocyte growth. However, in *Xenopus*, polyadenylated RNA accumulates during the previtellogenic stage and the total amount found in the egg is already present in early oocyte growth (Anderson and Smith 1978; Anderson *et al.* 1982).

3.3.2 Oocyte Localization of Maternal mRNAs

hnRNA, which is processed into poly (A)-containing RNA ("maternal message" mRNA), moves from the germinal vesicle into the cytoplasm and occupies different spatial positions during oogenesis (Capco and Jeffery 1982; Melton 1987). Cytoplasmic localization has long been recognized as an important mechanism by which cells derived from the totipotent egg become determined to specific fates. In *Xenopus* the vegetal cortical cytoplasm of the oocyte is the site of localization of several important maternal mRNAs (Mowry and Cote 1999).

There are two main pathways for getting mRNA into the vegetal cortex during oogenesis:

The METRO or early pathway contains transcripts localized in the mitochondrial cloud in the early oocytes before redistribution to the germ plasm at the vegetal cortex. These messages are transported in a manner that appears to be independent of the cytoskeleton.

The late pathway transcripts are not localized in early oocytes, do not associate with the mitochondrial cloud and become distributed to a broad domain of the vegetal cortex in full grown oocytes.

The early pathway includes the nanos family member Xcat2 (Mosquera *et al.* 1993), which may be involved in axial patterning, the germ cell associated protein Xpat (Hudson and Woodland 1998), the *Xenopus* ortholog of the *Drosophila* orthodenticle gene Xotx1 (Pannese *et al.* 2000) and the signaling factor Wnt11 (Ku and Melton 1993).

The late pathway includes the T-box transcription factor VegT (Stennard *et al.* 1999), the TGFb-related factor Vg1 (Melton 1987), involved in mesoderm induction, and the RNA-binding protein Bicaudal- C (Wessely and De Robertis 2000).

In the Vg1 pathway, messages are first seen through the oocyte but they are translocated through a microtubule-driven system to the vegetal hemisphere and, in a second phase, the microfilaments are responsible for anchoring the Vg1 message to the vegetal cortex (Yisraeli *et al.* 1990).

Recently, RNAs such as fat Vg have been shown to follow a pathway using both the mitochondrial cloud and elements of the late pathway in order to reach their localization in the vegetal cortex (Chan *et al.* 2001).

Although much is known about the processes by which these mRNAs are translocated to the vegetal cortical cytoplasm, the mechanism by which mRNAs remain anchored to this site is not yet clear.

3.3.3 Amplified Nucleoli

In addition to mRNA synthesis, the patterns of ribosomal RNA (rRNA) and transfer RNA (tRNA) transcription are also regulated during oogenesis. In *Xenopus* primary oocytes, transcription appears to begin during the diplotene stage in early stage I. At this time, all the rRNAs and tRNAs needed for protein synthesis until the mid-blastula stage are formed and all the maternal mRNAs needed for early development are transcribed.

The rate of rRNA production is enormous. Although the amplification process begins in premeiotic oogonia, the main period for the selective replication of rRNA is the pachytene of the meiotic prophase, followed by completion in early diplotene. The synthesis of rRNA reaches its maximal values only after the onset of yolk deposition.

rRNA genes are amplified in amphibian oocytes (Brown and Dawid 1968). As a consequence, a great number of extrachromosomal rRNA genes occur (in addition to the hundreds of rRNA genes clustered at the chromosomal nucleolus organizer regions) in numerous nucleoli that are not associated with lampbrush chromosomes.

In *Xenopus*, the amount of extrachromosomal rDNA per oocyte nucleus is about 30 pg, corresponding to approximately two million copies of rRNA genes that are distributed in about 1,000 amplified nucleoli (MacGregor 1972). The amplification of rRNA enables amphibian oocytes to support unusually high rates of rRNA synthesis. Thus, in maximal growth stages a single *Xenopus* oocyte synthesizes about 300,000 ribosomes per second (Scheer 1973).

Xenopus oocytes accumulate large numbers of ribosomes stored in the cytoplasm as monosomal particles for their subsequent use in early embryogenesis. To satisfy the need for rRNAs, the genes encoding the large rRNAs (28s, 18s, and 5,8s rRNA) are amplified in the oocyte to yield about 2×10^6 extrachromosomal copies per nucleus. The synthesis of 18s and 28s rRNA is low in young oocytes and their accumulation begins in early vitellogenesis.

In previtellogenic oocytes 5s RNA and tRNA each represent about 40 percent of the total cellular RNA. The multiple 5s and tRNA sequences are rapidly transcribed and accumulated in large amounts. The great stock of ribosomes present in the egg is enough to support early development. In order to produce this amount of rRNA within a reasonable period of time, the rDNA in the oocytes is amplified about 1,000 fold, and the extrachromosomal nucleoli are transcribed during vitellogenesis.

In growing oocytes, the amplified nucleoli are distributed toward the periphery of the nucleus and are often firmly attached to the inner nuclear membrane by a network of fibrillar strands (Franke and Scheer 1970).

Amplified nucleoli appear either as compact spheroidal bodies 2-10 μm in diameter (Fig. 3.5) or as ringlike structures. The transition between both nucleolar forms has been observed in various stages of oogenesis in several amphibian species.

3.3.4 The Nuclear Envelope

The nuclear envelope of amphibian oocytes is another main component during the first phase of growth. This membrane is perforated by numerous pore complexes which in freeze fractured or negatively stained preparations appear as ringlike structures with diameters of 70-80 nm. In transverse section, the

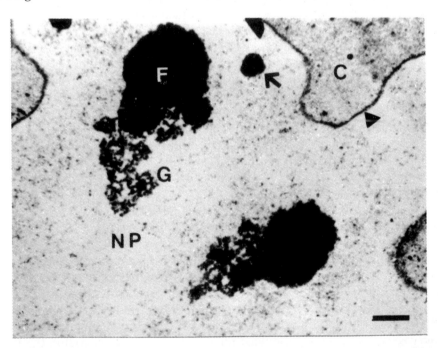

Fig. 3.5 Ultrastructural features of the nucleolus of the diplotene *Ceratophrys cranwelli* oocytes. The nucleoli display the classical fibrillogranular constitution. F, fibrillar mass; G, granular components; C, cytoplasm; NP, nucleoplasm; arrow, micronucleolus. Scale bar 1.3 μm. Original.

pores are the site of local fusion of the inner and outer membrane and have a very characteristic and highly regular architecture, the most prominent nonmembranous components being eight symmetrically arranged annulus subunits on each pore margin.

In amphibian oocytes the number of pore complexes per unit area is high (50-60 pores/μm^2). This high density, together with the huge size of the germinal vesicle, results in a total number of about 38 million pore complexes per nucleus of mature *Xenopus* oocytes. They are formed during oogenesis at the rate of eight pores/sec (Fig. 3.6).

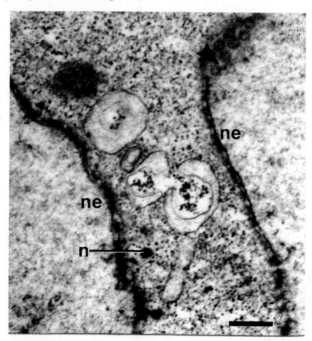

Fig. 3.6   Electron micrograph of a portion of a nuclear envelope from *Ceratophrys cranwelli* oocytes. Note the nuclear envelope perforated by numerous pore complexes (arrowhead). ne, nuclear envelope; n, nuage. Scale bar 0.34 μm. Original.

In *Xenopus* growing oocytes, two to three rRNA molecules are transferred every minute through a pore complex to the cytoplasm (Scheer 1973). However, it is unlikely that the main function of the millions of pore complexes on the surface of oocyte nuclei should be simply to provide sufficient export site for ribosomal ribonucleoprotein material.

3.4   VITELLOGENESIS

During the vitellogenesis stage, the major event is the accumulation within the oocyte of nutrients in the form of yolk proteins. In most anuran amphibians, stage III oogenesis represents the beginning of the vitellogenic process, characterized by the incorporation and accumulation of yolk (Wallace 1985), which comprises up to 85 percent of the total egg proteins and is responsible for the remarkable increase in oocyte growth. Yolk accumulation provides the nutritive materials needed for embryogenesis.

Yolk is not a single substance and vitellogenin (VTG), the macromolecular precursor of yolk proteins, is normally a female-specific protein selectively sequestered by growing oocytes. Much of the work on VTG in amphibian oocytes has demonstrated that VTG is incorporated by the oocyte and proteolytically cleaved into yolk proteins: lipovitellin 1 α–γ and 2 α–γ, phosvitin and phosvette 1 and 2 (Wiley and Wallace 1981; Opresko and Karpf 1987).

Vitellogenins are phosphoglycoproteins composed of two subunits of molecular weight that range from 180,000 to 240,000 according to species. They seem to have been conserved throughout evolution because the same sequence appears in birds (chickens), amphibians (*Xenopus laevis*) and fish (salmon). Consequently, it is evident that vitellogenins from different species are closely related both functionally and structurally.

3.4.1 Hormonal Regulation of Vitellogenesis

Gonadotropins are known to promote vitellogenesis through the mediation of estradiol, synthesized by follicles (Redshaw 1972). In general, estrogens, especially estradiol-17β, enhance the synthesis of the yolk precursors in the liver. The liver VTG synthesis and its hormonal requirements have been well elucidated in *Xenopus* and constitute a model for understanding steroid regulation of protein synthesis (Wallace and Dumont 1968; Ho 1987; Varriale *et al.* 1988).

VTG is transported via the blood stream to the ovary, where it successively passes through the endothelium, the basal lamina, the follicular epithelium and the vitelline envelope until it reaches the oocyte, where it is recognized by receptor-mediated endocytosis (Opresko and Wiley 1987 a, b). This process is promoted by gonadotropins but does not seem to be mediated by estradiol-17β (Wallace and Bergink 1974).

The internalized products are then stored in a stable form, the yolk, until embryogenesis occurs.

3.4.2 Vitellogenin Pathway

The specific transport pathway of VTG in amphibian oocytes has been investigated in several anuran species such as *Xenopus laevis* (Brummett and Dumont 1977; Wallace 1985; Opresko and Karpf 1987), *Bufo marinus* (Ritcher 1987), *Ceratophrys cranwelli*, *Bufo arenarum* (Villecco *et al.* 1999) and *Bufo paracnemis* (Villecco *et al.* unpublished data). The endocytic process traversed by VTG is a unidirectional pathway that transports receptor-mediated VTG from the oocyte cell surface to oocyte storage compartments and to specialized organelles, the yolk platelets. The maternal protein transport during vitellogenesis has been confirmed by many authors using different macromolecular tracers, e.g. horseradish peroxidase (HRP) or ferrolabeling and ferromagnetic sorting, coupled with transmission electron microscopy (TEM) visualization (Ritcher and Bauer 1990).

The injection of labeled VTG or macromolecular tracers into *Xenopus*, *Ceratophrys* and *Bufo* females after estradiol-17β and gonadotropin stimulation to induce vitellogenesis demonstrated that the reaction products are found within the perifollicular capillaries and within the intercellular space between

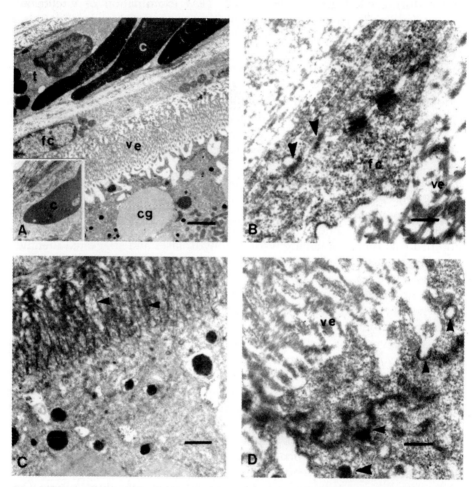

Fig. 3.7 Electron micrographs of the outer region of *Ceratophrys cranwelli* vitellogenic follicles from HRP-injected animals. **A**. After 1 h, HRP injection products were located within the vascular components of the theca. Insert: control follicle showing no reaction products. c, capillaries; ve, vitelline envelope; fc, follicle cell; cg, cortical granules; t, theca. Scale bar 2 μm. **B**. HRP reaction products after 2 h of injection are located within the intercellular space between follicles cells (arrowheads). Scale bar 0.29 μm. **C**. HRP reaction products (arrowheads) visible in vitelline envelope tunnels 3 h after injection. Scale bar 0.8 μm. **D**. HRP reaction products are located within coated pits, endocytic vesicles and smooth-surfaced tubules of the oocyte cortex 4 h after injection. Scale bar 0.29 μm. From Villecco E.I., *et al.* 1999. Zygote 7: 11-19, Fig. 4.

adjacent endothelial cells (Fig. 3.7A). Tracers seem to be also distributed among the fibroblasts and collagen fibers of the theca. Macromolecular materials, which apparently pass through the basal lamina and the overlying follicular epithelium, are localized within the intercellular spaces between the follicular cells (Fig. 3.7B) and also within the pores of the vitelline envelope (Fig. 3.7C). Thus, the maternal blood circulating proteins have ready access to the growing oocyte during vitellogenesis (Fig. 3.7D). TEM examination of vitellogenic oocytes (especially stages IV and V) in *Xenopus* and *Ceratophrys* reveals a folded surface with numerous coated pits (Fig. 3.8A). The fusion of the coated vesicles present in the cortical ooplasm both with each other and with cortical endosomes can be seen in the cytoplasm. After fusion, the small clathrin-coated vesicles become completely incorporated and inserted into the growing endosomes, transforming them into multivesicular endosomes (Fig. 3.8B). This feature confirms that the receptor-mediated endocytic mechanism is involved in VTG incorporation.

In the multivesicular endosomes of *Ceratophrys* and *Xenopus* oocytes, VTG condensation and crystallization occur, forming electron-dense masses which gradually grow and originate multivesicular bodies that represent the initial yolk platelet precursor (Fig. 3.8B). These bodies have been observed in the early stage of vitellogenesis. Based on various experimental data from *Xenopus* oocytes, Wall and Patel (1987), demonstrated that these organelles are modified lysosomes containing lysosomal hydrolases and suggested that they might be necessary for VTG cleavage into yolk proteins to form mature platelets. Opresko and Karpf (1987) suggested that multivesicular bodies could function in the uncoupling of VTG from its receptor and demonstrated that when the proteolysis of VTG is inhibited, VTG-containing yolk platelets precursors neither fuse nor develop into mature yolk platelets.

The multivesicular body develops into two nascent yolk organelles (primordial yolk platelets Types I and II) (Fig. 3.8C). Type I yolk primordial platelets are small and develop in the cortical cytoplasm of vitellogenic amphibian oocytes (stages III to V). They are located in regions of high pinocytic activity near the membrane crypts of the oocyte surface. In these platelets, lipoproteins irregularly condense and stratify in whorls which are packed in a random para-crystalline lattice at different angles to each other. Crystals are perforated by yolk-free tubular cavities. The organelle membrane is in close contact with endoplasmic reticulum vesicles, suggesting that vitellogenin-transporting vesicles and endoplasmic reticulum tubules are involved in the construction of Type I yolk platelets.

Primordial yolk platelets Type II, which originate from the assembly of small precursors fused together, usually develop in the subcortical cytoplasm and contain only one crystalline core and the lipoprotein sheets are uniformly stratified. Precursors Type II are frequently attached to another yolk platelet precursor (Fig. 3.8C).

Once Type I and II yolk precursor platelets are formed, they migrate toward the subcortical cytoplasm, where they fuse together to form the totally grown

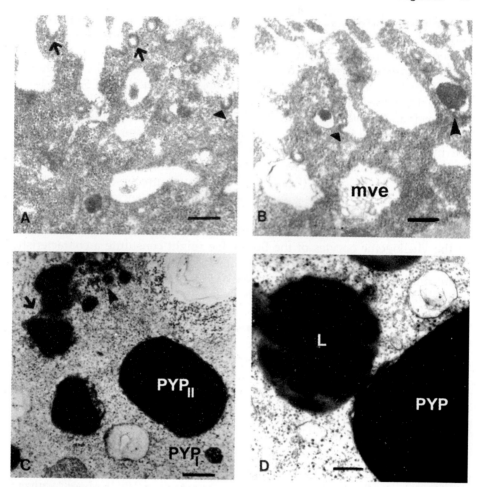

Fig. 3.8 Electron micrographs of *Ceratophrys cranwelli* follicles. **A**. Oocyte cortex showing a folded surface. Coated pits and coated vesicles (arrows); fusion of coated vesicles with cortical endosomes (arrowheads). ce, cortical endosome. Scale bar 0.29 µm. **B**. Oocyte cortex showing a multivesicular endosome and a multivesicular body. Fusion of coated vesicles with cortical endosome (small arrowhead); multivesicular body (large arrowhead); mve, multivesicular endosome. Scale bar 0.3 µm. **C**. Oocyte cortex showing yolk platelet precursors. Multivesicular body (large arrowhead); fusion of Type II primordial yolk platelets (arrow); PYP_I, Type I primordial yolk platelet; PYP_{II}, Type II primordial yolk platelet. Scale bar 0.39 µm. **D**. Oocyte cortex showing fusion of a lipid droplet with a primordial yolk platelet. PYP, Type II primordial yolk platelet; L, lipid droplet. Scale bar 0.29 µm. From Villecco E.I. *et al.* 1999. Zygote 7: 11-19, Fig. 1.

platelet. During fusion, lipid droplets become attached to the platelet membranes and will probably be completely incorporated into the main body of existing platelets Fig. 3.8D).

The precursor of yolk organelles can be considered as a "fluid crystal", probably due to the high content of initially unbound lipids and the 15-30 percent water in the crystalline core (Wallace 1963).

The ultrastructural images described above can be seen in the vitellogenic oocytes of *Xenopus laevis, Rana pipiens, Ceratophrys cranwelli* and *Scinax nasica*. However, in the vitellogenic oocytes of the Bufonidae (*Bufo arenarum, B. paracnemis* and *B. marinus*), lipid droplets that occupy a considerable volume of the ooplasm are found in intimate contact with endosomes (Fig. 3.9A) and are included in the multivesicular bodies (Fig. 3.9B). In these structures, lipoproteins are condensed, forming a para-crystalline core either near the periphery or in its center (Fig. 3.9C). These yolk precursors fuse together forming membrane limited yolk platelets that always enclose lipid droplets (Fig. 3.9D).

The presence of both lipids and proteins inside the multivesicular bodies in the vitellogenic oocytes of the Bufonidae might constitute a characteristic feature of this family.

In spite of the differences observed by TEM in the vitellogenic oocytes concerning the incorporation of lipids in the yolk precursors, all amphibian oocytes finally have the same characteristic fully grown yolk platelet (Fig. 3.10). Mature yolk platelets contain one single main body where lipoproteins are packed in a crystalline lattice (Fig. 3.11A, B), in sheets orientated at the same angle and surrounded by amorphous material and a membrane.

Fully grown yolk platelets are the destination of most of the proteinaceous nutritive material to be internalized by vitellogenic oocytes.

3.4.3 Vitronectin-like Protein in Amphibian Yolk Platelets

Another estrogen-regulated component synthesized by the liver and found in yolk platelets during *Bufo arenarum* oogenesis is a vitronectin-like protein (VN). VN is a multifunctional cell-adhesive glycoprotein present in vitellogenic oocytes (stages III to V). By immunofluorescent staining, VN has been localized in precursors of yolk platelets and in fully grown yolk platelets in the oocyte cortical cytoplasm (Fig. 3.12A). Immunoelectron microscopy analysis has shown that the transport pathway of VN in vitellogenic oocytes is similar to vitellogenin (Fig. 3.12B). Although the role of VN is not known, it is possible that it could act as an adhesive material during morphogenesis (Aybar *et al.* 1996).

It is believed that yolk platelets formed during the vitellogenic stage provide the organic components and/or precursors that are utilized during events of morphogenesis and differentiation. They could also supply essential nutritive and informational substances that could be used after oocyte activation to support the whole embryogenesis.

3.5 POSTVITELLOGENESIS OR MATURITY

3.5.1 General Morphology of Full Grown Oocyte

In most amphibian species, the general anatomical characteristics of the fully

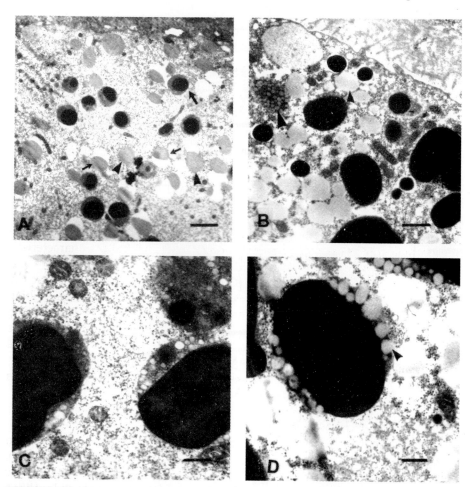

Fig. 3.9 Electron micrographs of *Bufo arenarum* vitellogenic follicle. **A**. Oocyte cortex showing abundant lipid droplets in close proximity to endosomes and multivesicular bodies. Lipid droplets (arrowheads); lipid droplets in endosomes (small arrow); multivesicular body (large arrow). Scale bar 1.4 μm. **B**. Oocyte cortex of an unstimulated female showing numerous lipid droplets and yolk platelet precursors. Multivesicular bodies (large arrowhead), lipid droplets (small arrowhead). Scale bar 0.5 μm. **C**. Multivesicular bodies in different degrees of yolk condensation. Scale bar 1.4 μm. **D**. Primordial yolk platelet with lipid droplets. Lipid droplets (arrowhead). Scale bar 0.29 μm. Villecco E.I. *et al*. 1999. Zygote 7: 11-19, Fig. 2.

grown primary oocytes are similar. The salient feature is that oocytes and eggs are polarized along an animal-vegetal axis made visible by the differential localization of pigment granules in the animal cortex. In most species studied, this characteristic makes possible the distribution between a pigmented black animal hemisphere and a white to yellowish vegetal one (Fig. 3.13). However, in the marsupial frog *Gastrotheca riobambae*, oocytes are uniformly yellow. Consequently, in this species, the animal pole cannot be identified by its

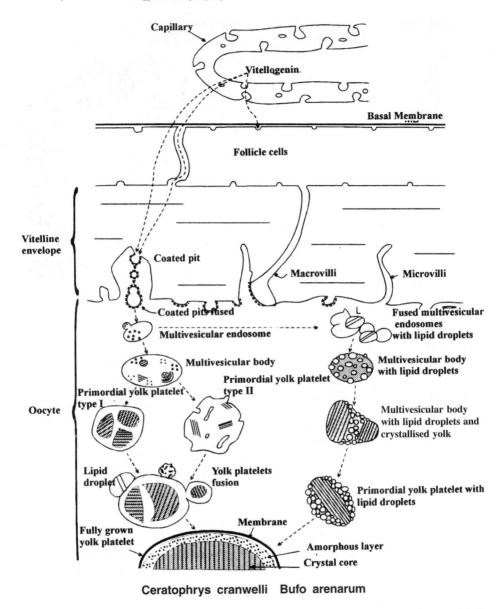

Ceratophrys cranwelli   Bufo arenarum

Fig. 3.10 Schematic illustration of the transport pathway of vitellogenin from blood capillaries to the full grown yolk platelets, in *Ceratophrys cranwelli* and *Bufo arenarum* oocytes. Modified after Richter P.H. and Bauer A. 1990. European Journal of Cell Biology 51: 53-63, Fig. 6.

pigmentation but by the external visualization of a translucent spot that corresponds to the germinal vesicle (Del Pino 1989a).

The oocyte polarity extends to many other characteristics of the oocyte, including the eccentric position of the germinal vesicle in the animal pole, the

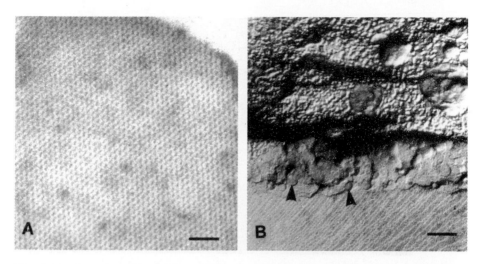

Fig. 3.11 Structure of fully grown yolk platelet and primordial yolk platelet. **A**. Thin section of the crystalline lattice of *Bufo arenarum* fully grown yolk platelet (that of *Ceratophrys cranwelli* is identical). Scale bar 0.042 μm. **B**. Freeze-fracture replica of a *Bufo arenarum* primordial yolk platelet, showing the absorption of lipid droplets (arrowheads) into yolk precursors. Scale bar 1.0 μm. Villecco E.I. *et al.* 1999. Zygote 7: 11-19, Fig. 3.

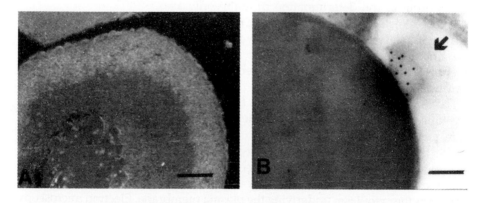

Fig. 3.12 Immunolocalization of vitronectin-like protein in *Bufo arenarum* vitellogenic oocytes. **A**. immunofluorescent staining of vitronectin-like protein in vitellogenic oocytes (stage IV). Fluorescence is observed in the cortical cytoplasm localized in yolk platelets. The protein is detected by an indirect immunofluorescence method using anti-human VN antibodies (mouse monoclonal anti-human vitronectin IgM, Clone VIT-2, (Sigma) and Clone VN5-3, Panvera Corp.). Scale bar 100 μm. **B**. Immunoelectron microscopy of vitellogenic oocytes. Ultrathin section of stage IV oocytes treated with anti-human vitronectin mouse monoclonal antibody coupled with gold particles. Gold particles are seen on endosomes closely related to primordial yolk platelets undergoing yolk crystallization (arrow). Scale bar 0.1 μm. Aybar M.J. *et al.*, 1996. International Journal of Developmental Biology 40: 997-1008, Figs. 4C and 7B.

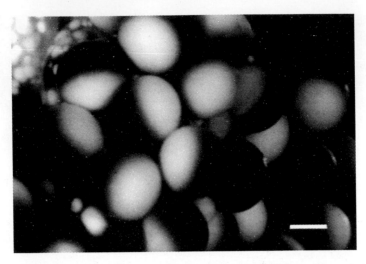

Fig. 3.13 External view of *Ceratophrys cranwelli* oocytes. The dark animal hemisphere and the white vegetal one are seen. Scale bar 1000 μm. Original.

distribution of the yolk with the heavier platelets concentrated at the vegetal pole, the localization of specific mRNAs stored in certain regions of the oocyte cytoplasm and the cytoskeletal organization. Many of these animal-vegetal asymmetries persist during oocyte maturation and after fertilization, contributing to the determination of embryonic axes.

Yolk platelets are the predominant organelles in the fully grown oocyte. Small yolk platelets are present in the animal pole and gradually increase in size toward the vegetal hemisphere. Mitochondria are distributed in the cytoplasm of both hemispheres, frequently forming aggregates present in the perinuclear region.

The nucleus is a large oval structure surrounded by a highly folded envelope on the vegetal side. The nuclear envelope is regularly perforated by large nuclear pores (Fig. 3.6).

The nucleoplasm is homogenous, with several nucleoli of different sizes occupying a central location. Chromosomes are condensed and genes are no longer transcribed. The oocyte cortex is devoid of large yolk platelets and represents the cytoplasm underlying the plasma membrane. Electron microscopy studies reveal, in this actin-rich cortex, several aligned cortical granules, clusters of mitochondria, pigment granules, coated vesicles and rough endoplasmic reticulum. Numerous Golgi complexes are found near the cortex and in the subcortical cytoplasm.

The plasma membrane is invaginated with crypts and microvilli which penetrate the vitelline layer (Fig. 3.21A).

Elaborate membranous structures are also present in amphibian oocytes, distributed in the perinuclear as well as in the cortical and subcortical cytoplasm. They are stacks of lamellae with pores either in a parallel array or with a concentric arrangement (Fig. 3.14). Each lamella consists of two roughly parallel membranes that enclose a space and are continuous at the margins of pores. Annulate lamellae have a morphology similar to that of the nuclear envelope and they are often found in juxtaposition. This characteristic led Kessel (1968) to suggest that the lamellae originate in the nuclear envelope and could represent a storage form of nuclear membranes (see also spermiogenesis, Scheltinga and Jamieson, chapter 5 of this volume).

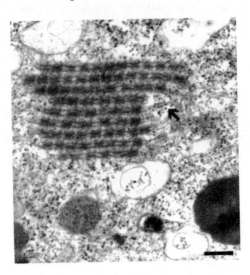

Fig. 3.14 Electron micrograph of section of *Ceratophrys cranwelli* oocyte illustrating a stack of annulate lamellae. Note rough endoplasmic reticulum associated with lamellae (arrow). Scale bar 0.34 μm. Original.

Recent studies have provided data concerning the organization and regulation of the cytoskeleton during *Xenopus laevis* oogenesis. It has been established that the cytoskeleton contains a complex interconnected network consisting of three major cytoplasmic filaments: microtubules, composed of tubulin and its associated proteins; microfilaments, composed of actin and its associated proteins, and intermediate filaments, composed of simple epithelial keratins and cross-linking proteins that interconnect them such as plectin (Gard and Klymkowsky 1998).

It has been demonstrated that microtubules and microfilaments play important roles in the formation and maintenance of the animal-vegetal axis of the *Xenopus* stage VI oocytes. Both elements are required for the transport of maternal mRNAs to the vegetal cortex (Yisraeli *et al.* 1990).

3.5.2 Oocyte Metabolism

During the postvitellogenic period, the fully grown oocytes have already completed vitellogenesis and are arrested at the prophase of the first meiotic division. Although these oocytes have stored numerous structural and regulatory components, they remain in a state of metabolic inhibition in terms of their rate of protein synthesis (Maller and Krebs 1980). In relation to their intermediary metabolism, *Bufo arenarum* ovarian oocytes are affected by seasonal changes in response to modifications of the pituitary activity. During the winter season, *Bufo* oocytes exhibit a metabolic behavior similar to adult tissues of the same species, defined as "adult metabolism" (Legname *et al.* 1972). This type of metabolism is characterized by an active carbohydrate breakdown via the glycolytic pathway followed by the classic Krebs cycle. However, during the breeding season, ovarian oocytes exhibit an "embryonic metabolism", utilizing sugars through the pentose-phosphate cycle, while the tricarboxylic acid cycle operates as its variant known as the glutamic-aspartic cycle (Salomón de Legname 1969; Salomón de Legname *et al.* 1971).

These metabolic changes probably play an important role in cytoplasmic maturation. According to Iwamatsu and Chang (1972), this process implies the acquisition of the capacity to segment and duplicate DNA. The embryonic metabolism in *Bufo* segmenting eggs is mainly directed toward nucleotide synthesis (Salomón de Legname *et al.* 1975). Consequently, it is possible to assume that the biochemical changes transforming the adult oocyte metabolism into an embryonic one would constitute an important factor in cytoplasmic maturation.

3.5.3   Oocyte Maturation

Fully grown (primary) oocytes can remain for months in the diplotene stage of the meiotic prophase, not being able to either be fertilized or support embryonic development unless they undergo oocyte maturation. The quiescent metabolic state is broken when the process of oocyte maturation is initiated by gonadotropin hormones (see Fernández and Ramos, chapter 4 of this volume). It has been well established that the in vivo epithelial layer of follicle cells surrounding each oocyte is stimulated by a pituitary-derived luteinizing hormone (LH-like) to synthesize and secrete progesterone, which acts directly on the oocyte to induce maturation.

Oocyte maturation involves a complex differentiation program that transforms the fully grown primary oocyte into a secondary oocyte that is fertilization competent and capable of supporting early embryonic development (Yamashita *et al.* 2000). The amphibian oocyte represents one of the best characterized models for oocyte maturation at the biochemical as well as the morphological level (Bement and Capco 1990).

In response to progesterone stimulation, primary oocytes reenter the meiotic cell cycle, complete the first meiotic division and, as secondary oocytes, are arrested in the second metaphase. The secondary oocyte is released from the ovary by a process called ovulation.

During maturation, oocytes undergo germinal vesicle breakdown (GVBD), the nuclear envelope dissolves, nucleoli disintegrate and chromosomes contract and migrate to the animal pole. Meiosis I is completed with the extrusion of the first polar body and the oocytes are arrested at the metaphase of the second meiotic division (Masui and Clarke 1979). The second polar body is extruded after telophase following penetration of the spermatozoon.

3.5.4 Regulation of the Meiotic Cycle

Oocyte maturation is initiated in response to progesterone, which induces a signal transduction cascade culminating in the activation of the maturation-promoting factor (MPF), a complex composed of two subunits of a serine/threonine protein kinase (p34^{cdc2}) and a regulatory cyclin B subunit (Nebreda and Ferby 2000).

The activity of MPF is subject to a complex and tight regulation that includes both specific phosphorylation events and protein-protein interactions. During the meiotic cell cycle of *Xenopus* oocytes, MPF is first activated at meiosis I, then transiently inactivated between meiosis I and II and finally reactivated at meiosis II.

MPF activation is necessary for GVBD and for meiosis resumption (Smith 1989). Considering the fact that the subunits of MPF are present in amphibian oocytes, it has been suggested that progesterone transforms a pre-MPF complex into an active MPF.

The pre-MPF is a complex of cdc2 kinase and cyclin B and is kept in an inactive form by the phosphorylation of cdc2 kinase on threonine-14 (Thr14) and tyrosine-15 (Tyr15). These inhibitory phosphorylations are probably catalyzed by a protein kinase, whereas dephosphorylation of these residues requires a cdc25c. Thus, the activation of MPF may be brought about by the activation of cdc25c phosphatase (Nebreda and Ferby 2000).

One of the early biochemical responses following progesterone stimulation is a transitory decrease in the level of cAMP, which is thought to translate into the inhibition of the cAMP-dependent protein kinase (PKA). In agreement with the idea that PKA plays an important role in oocyte meiotic arrest, the injection of PKA inhibitors produces maturation in the absence of progesterone; in contrast, the overactivation of PKA inhibits oocyte maturation (Palmer and Nebreda 2000).

The 39 kDa c-mos-phosphoprotein kinase is the mediator of the progesterone signal to reinitiate the meiotic process (Sagata 1997). An essential requirement for this event is the translation of the maternal c-mos mRNA stored in the oocytes.

The c-mos protein is present only during oocyte maturation and disappears upon fertilization, its main role being the release of the oocyte from the quiescent state. This protein also induces the mitogen activated protein kinase (MAPK) cascade, resulting in MPF activation. In addition to the mos-MAPK cascade, other signaling cascades are activated in parallel, resulting in a rapid MPF induction. The pole-like kinase cascade is thought to activate cdc25c phosphatase, which dephosphorylates and activates MPF (Qian *et al.* 2001).

The MPF phosphorylates several target proteins, including histones, the nuclear envelope-lamin proteins, and the regulatory subunit of cytoplasmic myosin. In consequence, several processes occur in the nucleus and in the cytoplasm of the oocytes including GVBD, which indicates the end of prophase I, the condensation of the chromatin, the formation of a mitotic spindle that will segregate homologous chromosomes, and the extrusion of the first polar body, which marks the completion of meiosis I.

The oocyte is arrested again in the metaphase of the second meiotic division by the combined actions of c-mos and of another protein, cyclin dependent kinase 2 (cdk2). Both proteins probably constitute the cytostatic factor (CSF) that is present in mature frog eggs blocking cell-cycle in metaphase II.

During fertilization the release of calcium ions activates a protease that specifically inactivates CSF, so that meiotic division can be completed.

3.6 OOCYTE-SOMATIC CELL INTERACTIONS DURING OOGENESIS

The ovary is a cyclically changing organ formed by two components, different in origin and function: a germinal cell line and a somatic cell line. These two components are involved in two main events: the endocrine functions and the ability to form the ova.

Germ cells interact with somatic cells not only during migration but also after settlement in the gonadal primordium. In the gonad, somatic cells envelope female germ cells to create an ensemble called a follicle. From the initial phases of follicle development both types of cells interact and influence each other.

Oocyte growth proceeds simultaneously with the growth and proliferation of follicle cells. Both undergo a highly coordinated sequence of changes essential for the production of a viable oocyte. Amphibians are an appropriate biological model to study the interactions that take place in ovarian follicles since they pass through different periods (previtellogenic, vitellogenic and maturation) during which various processes related to oocyte growth and maturation occur. The ultrastructural analysis of ovarian follicles during the different periods of oogenesis is important since structural features are likely to underlie follicle cells-oocyte intercellular signaling. The functional significance of the structural contacts between the amphibian oocyte and the associated follicle cells is still unclear. Follicle cells participate in a number of gonadotropin-regulated processes involved in amphibian oocyte growth and maturation such as steroidogenesis, initiation of yolk protein uptake, vitelline envelope formation and RNA transference.

3.6.1 Cell Interactions during Previtellogenesis

In general, similar structural features of previtellogenic oocytes have been found in *Xenopus laevis*, *Rana esculenta* and *R. pipiens*. In these species, oocytes are covered by a thin complete sheet of squamous follicle cells. These cells, closely apposed to the oolemma, establish adherent desmosomal junctions with each other, making contact with the underlying oocyte.

The study of the ultrastructural characteristics of oogenesis in *Ceratophrys cranwelli* shows that in early previtellogenesis, as in the case of other amphibian species, the membranes of the follicle cells are apposed to the oocyte membrane, without an interface. Follicle cells, which form the follicular epithelium, have weak electron-density and have a fair amount of ribosomes and a few cisternae of rough endoplasmic reticulum (clear follicle cells) (Fig. 3.15A).

As previtellogenesis progresses, the follicle plasm membrane begins to separate from the oolemma, forming an interface made up of small spaces with a homogeneous material. Only at certain points are both membranes in close apposition. An interesting characteristic is that the oolemma has an abundant glycocalyx, supporting the idea that several ligand-receptor interactions would be activated (Fig. 3.15A). The cytoplasmic cortex of these oocytes is devoid of organelles, except for a few scattered mitochondria.

During mid-previtellogenesis, follicle cells characterized by their electron-dense cytoplasm filled with rough endoplasmic reticulum, free ribosomes and glycogen (dark follicle cells) (Fig. 3.15 B, C) appear in the follicular epithelium together with clear follicle cells. Dark follicle cells, metabolically very active, are involved in the synthesis of nucleic acids. Autoradiographic studies demonstrate that follicle cells display strong radioactivity when incubated for 5 h with ³H adenosine as a precursor of nucleic acids synthesis, (Fig. 3.16). Up to now, the presence of both clear and dark follicle cells in the follicular epithelium of previtellogenic oocytes of *Ceratophrys* has not been observed in other amphibian species. However, a similar type of highly electron-dense metabolically active follicle cells has been described in Japanese quail (Callebaut 1991).

Inside the follicle cells, near the apical surface, there are certain vesicles containing two types of particles, one ranging from about 15 to 25 nm and the other from 25 to 50 nm, together with fibrillar material connecting the smaller particles. These vesicles are separated from the interface by a thin discontinuous cytoplasm (Fig. 3.15C). Extending from the margin of their apical surface, follicle cells exhibit long processes called macrovilli, many of which penetrate into the interface and make contact with the developing oocytes, often ending in morphologically specialized junctions such as gap junctions. Desmosomes are also observed between the macrovilli and the oocyte (Fig. 3.15D).

The oocyte surface displays only a few scattered short microvilli that differ from the macrovilli in that they are thinner and have a lower electron density. At the interface, both micro- and macrovilli delimit lacunae containing granular material. The oocyte surface also forms endocytic invaginations that incorporate the interface particles, forming peripherally arranged cortical vesicles ranging from about 2 μm to 3 μm in diameter (Fig. 3.15B). These cortical vesicles show a fibrillar and particulate content of the same nature as that observed in the vesicles of follicle cells and at the interface. The smaller particles, (15 to 25 nm), are positive staining for nucleic acids and disappear on treatment with RNAsa. RNA molecules has been detected inside the cortical vesicles at the ultrastructural level using the RNAsa-gold complex (Fig. 3.17A). Autoradiographs made of follicles of *Ceratophrys* incubated in the presence of ³H adenosine have shown an intense radioactive labeling in the follicle cells, in the germinal vesicle and in the cortical vesicles.

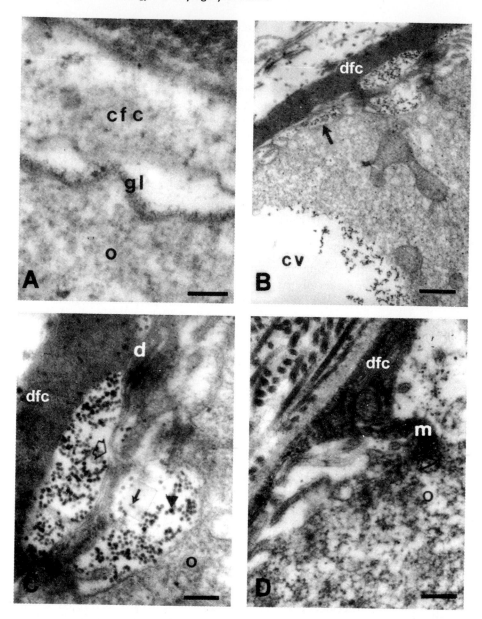

Fig. 3.15 Electron micrographs of *Ceratophrys cranwelli* early and mid-previtellogenic follicles showing the follicle cell-oocyte interface. **A.** The oocyte plasma membrane has a prominent glycocalyx (gl). cfc, clear follicle cell; o, oocyte cortex. Scale bar 0.16 µm. **B.** Dark follicle cells (dfc) with abundant free ribosomes and glycogen. At the interface (arrow) the micro and macrovilli delimit lacunae containing granular material. The oocyte cortex shows cortical vesicles (cv) with the same material as the interface. Scale bar 0.65 µm. **C.** Follicle cell with a vesicle containing two types of particles: one

Contd.

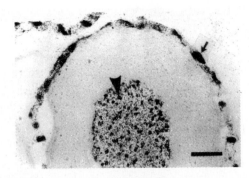

Fig. 3.16 Autoradiography of a section through a previtellogenic oocyte of *Ceratophrys cranwelli* incubated of 5 h in ³H adenosine. The follicle cells (arrow), the oocyte nucleoli (arrowhead) and the nucleoplasm display an intense radioactivity, Scale bar 20 μm. Villecco E.I. *et al.* 2002. Zygote 10: 163-173, Fig. 4.

Pulse-chase experiments have determined the migration of the radioactive granules from the follicular cells toward the oocyte cortex, suggesting a possible transference of these macromolecules (Fig. 3.17B). These results agree with the findings of Sánchez Riera *et al.* (1988 a, b), who observed a transport of radioactive molecules from the follicular envelopes to the fully grown *Bufo arenarum* oocytes.

The formation of cortical vesicles is a gradual process observed in previtellogenic and early vitellogenic *Ceratophrys cranwelli* oocytes, but not yet found in other species.

Throughout the previtellogenesis stage, oocytes accumulate RNAs for their subsequent use in early embryogenesis (Davidson 1986). Thus, an important characteristic of growing oocytes is the very active RNA synthesis as a consequence of rRNA being synthesized on amplified rDNA sequences and of heterogeneous RNA being synthesized on thousands of lampbrush chromosomes.

However, there are variations in the amount of RNAs present in eggs; these variations could be correlated with changes in the patterns of early development. The level of ribosomal RNA in the oocyte is correlated with the speed of embryonic development (Del Pino 1989 a, b). In *Xenopus*, a frog with aquatic eggs, rRNA is abundant because the ribosomal genes have been amplified, and the vulnerable embryo develops rapidly. Seven hours after fertilization it has already divided into 4,000 cells. In the marsupial frog

---

Fig. 3.15 *Contd.*
of 15-25 nm (arrow) and another of 25-50 nm (arrowhead). The open arrow shows discontinuous membranes. The follicle cells have desmosomes (d) between the macrovilli. The oocyte (o) surface has invaginations with the same particulate material. Scale bar 0.24 μm. **D.** The follicle cell (fc) displays a macrovillus (m) that passes through the interface and contacts the oocyte (o), forming a gap junction (open arrowhead). Scale bar 0.30 μm. Villecco E.I. *et al.* 2002. Zygote 10: 163-173, Figs. 2B and 3.

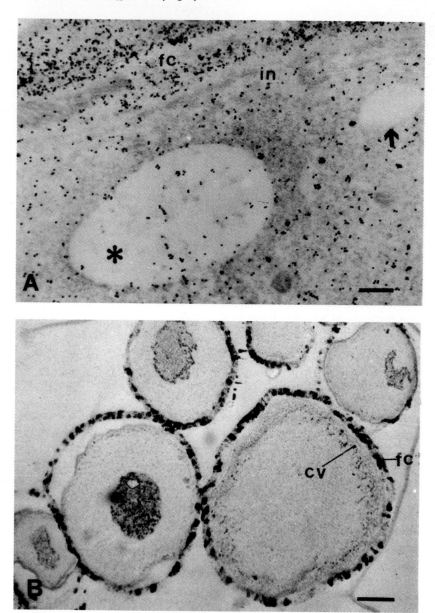

Fig. 3.17 Localization of RNA in *Ceratophrys cranwelli* previtellogenic oocytes. **A.** Immunoelectron localization of RNA with RNAsa-gold complex. Ultrathin section of the cortex of a previtellogenic oocyte created with RNAsa-gold complex. Gold particles appear inside the cortical vesicle (asterisk), in the follicle cells (fc) and in the interface (in) between the follicle cells and the oocyte. Note that the cortical vesicles devoid of their material do not contain gold particles (arrow). Scale bar 0.7 μm. **B.** Autoradiographyc study of sections of oocytes incubated 5 h ³H adenosine. Radioactivity is seen in the follicular epithelium, in the cortical vesicles (cv) and in the germinal vesicle. Scale bar 100 μm. Original.

Gastrotheca the level of rRNA is low as a consequence of the limited amplification of the ribosomal genes during oogenesis, so that in this frog, as in mammals (mice), development is slow. In both cases, however, the embryo is protected by the mother (Del Pino *et al.* 1989b). In another marsupial frog, *Flectonotus pygmaeus*, the oocyte has numerous nuclei, which add more rRNA and has a rapid development (Del Pino and Humphries 1978). This multinucleate oogenesis can be interpreted as the method that provides sufficient RNA to allow rapid development and can be considered as an adaptive mechanism to environmental conditions.

In the *Ceratophrys cranwelli* species, the oocyte contains high levels of RNAs as a consequence of a gene amplification that produces multiple nucleoli and of the RNAs from the follicle cells present in the cortical vesicles. This characteristic ensures a rapid development, taking into account that this species must become adapted to certain climatic conditions (storms) for egg deposition to occur (Fig. 3.18).

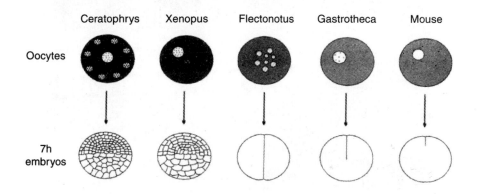

Fig. 3.18 Direct correlation between oocyte RNA levels and embryonic development rate. Shadowed area shows RNA content. Dots inside germinal vesicles are amplified nucleoli. In *Ceratophrys* oocytes spheres represent RNA inside cortical vesicles. Modified after Del Pino E.M. 1989b. Scientific American 260, 5: 110-118, Fig. 8.

3.6.1.1. Vitelline Envelope Formation

The vitelline envelope (VE) surrounding the amphibian oocyte begins to form during the previtellogenic stage, continues to differentiate throughout oocyte growth, and is complete in fully grown oocytes (Dumont 1972).

The origin of the components of the VE has long been controversial. In different amphibian species it has been attributed mainly to the oocyte with little, if any, contribution from follicle cells. Pinto *et al.* (1985), based on TEM visualization and biochemical analyses, have suggested that the oocyte is the site of synthesis of envelope glycoproteins in *Xenopus* oocytes, a result corroborated by Yamaguchi *et al.* (1989) by immunocytochemical methods. Wischnitzer (1964) and Bozzini and Pisanó (1987), on the other hand, pointed to the participation of follicle cells. Additionally, Cabada *et al.* (1996), studying the *Bufo arenarum* follicles by TEM and immunostaining with antibodies against VE, demonstrated that both the oocyte and the follicle cells are directly involved in the synthesis and secretion of the components of the VE. In previtellogenic *Bufo* oocytes the number of oocyte microvilli increases and numerous cross sections of them are found between the follicle cells and the oocyte. A filamentous material can be observed inside the follicle cells and between the follicle cells and the oocyte (Fig. 3.19). The participation of the oocyte in the process of VE formation is also revealed by the fact that large vesicles filled with an amorphous material of low and uniform electron density have been detected in the oocyte cortex, some of them in the process of releasing their content to the interface.

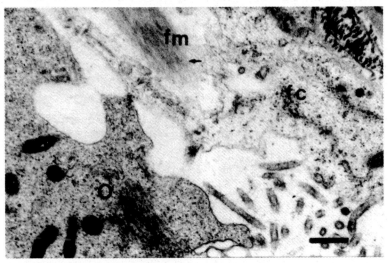

Fig. 3.19 Electron micrograph of *Bufo arenarum* primary vitellogenic follicle. Fibrous material (fm) is present inside the follicle cell (fc). o, oocyte. Scale bar 0.3 µm. From Cabada M.O. *et al.* 1996. BioCell 20, 1: 77-86, Fig. 5.

In previtellogenic *Ceratophrys oocytes*, Villecco *et al.* (2002) also observed fibrillar material in vesicles of the follicle cells and in the interface (Fig. 3.20A). This material is first evident as fine filaments in isolated patches that become continuous, forming a strip between the macro-and the microvilli sections. The interface is formed by a fibrillar pattern within which macro- and microvilli are arranged and which constitutes the precursor of the VE.

In the vitellogenic oocytes of this species the VE is made up of bundles of fibers of average electron density arranged in four-walled tunnels inside which sections of macro-and microvilli are observed (Fig. 3.20B).

In the fully grown oocyte in the interface the fibrillar network of the VE is divided in two zones: a loose one near the follicle cells and a more compact one near the oocyte (Fig. 3.21A). The VE shows a fibrillar electron-dense material that causes a decrease in the width of the tunnels (Fig. 3.21B) in which electron-dense particles can be seen (Fig. 3.21C).

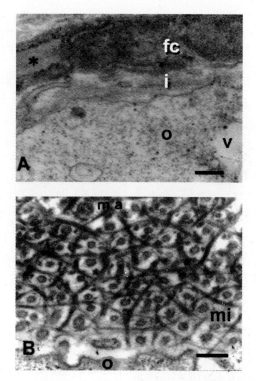

Fig. 3.20 Electron micrographs of *Ceratophrys cranwelli* late previtellogenic and vitellogenic follicles. **A.** Late previtellogenic follicle cells (fc) have abundant ribosomes, a rough endoplasmic reticulum and vesicles with fibrillar material (asterisk). The interface (i) shows the same material between the macro- and microvilli. The oocyte (o) cortex display a cortical vesicle (v). Scale bar 0.50 μm. **B.** Cross-section of the follicle cell-oocyte interface of vitellogenic oocyte showing the tunnel arranged in a net in the spaces of which macrovilli (ma) and microvilli (mi) can be seen. Scale bar 0.28 μm. From Villecco E.I. *et al.* 2002. Zygote 10: 163-173, Figs. 5 and 6.

From an ontogenetic viewpoint, the fibrillar material of the VE in amphibians may have originated, by selection, mainly for arranging the increasingly large number of oocyte microvilli so as to keep them perpendicular to the surface of oocyte, thus increasing the absorption area. The formation of VE in *Bufo* and *Ceratophrys* oocytes is the result of an interaction between follicle cells and the oocyte. In coelomic eggs, the VE changes its structural characteristics in order to prevent sperm penetration and consequent fertilization. In eggs that have passed through the initial segment of the oviduct the vitelline envelope changes again, increasing the probability of successful fertilization. At the time of fertilization the VE also changes in relation to the cortical reactions capable of blocking polyspermy.

3.6.2 Cell Interactions during Vitellogenesis

During vitellogenesis, in *Ceratophrys cranwelli* oocytes an intense activity related to the vitellogenic process is observed in the area of interaction between

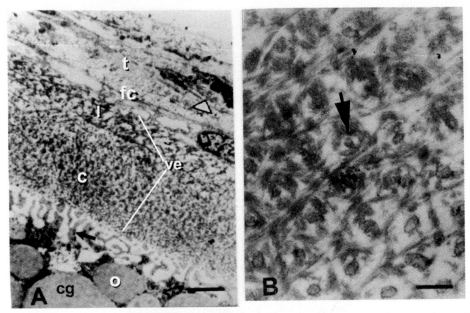

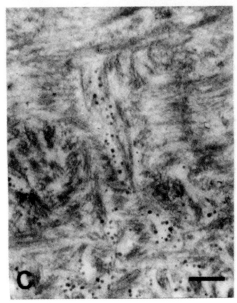

Fig. 3.21 Electron micrographs of *Ceratophrys cranwelli* fully grown follicles (stage VI) showing the follicle- oocyte interface. **A**. At the interface the fibrillar network of the vitelline envelope (ve) has two zones: a loose one (l) and a more compact one (c). The microvilli are shorter and thicker (arrow). Note the theca (t) with blood capillaries (open arrow head). The oocyte cortex also presents cortical granules (cg) arranged near the plasma membrane. Pigmented granules can also be seen. fc, follicle cell; o, oocyte. Scale bar 1.98 μm. **B**. Cross-section of the follicle cell-oocyte interface showing fibrillar material (arrow) that closes the tunnel of the vitelline envelope. Scale bar 0.29 μm. **C**. Perpendicular section of the interface showing a decrease in the width of the tunnels and particles inside them. Scale bar 0.29 μm. From Villecco E.I. *et al.* 2002. Zygote 10: 163-173, Fig. 8.

the follicular epithelium and the oocytes. At the apical surface of the follicle cells there is a significant increase in macrovilli, which penetrate the vitelline envelope and contact the outer surface of the oocyte, forming numerous gap junctions and desmosomes that are important interaction sites. At the same time, the oocyte surface exhibits numerous microvilli and deep crypts and protuberances with numerous coated pits and coated vesicles (Fig. 3.22). Macrovilli enter the oocyte crypts and make contact with them through gap junctions and desmosomes.

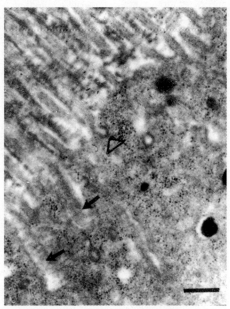

Fig. 3.22 Electron micrograph of *Ceratophrys cranwelli* vitellogenic follicle. Follicle cell oocyte interface in late vitellogenesis showing an increase in the number of macro- and macrovilli. At the oocyte surface, crypts into which the macrovilli (arrow) penetrate can be seen. Arrowhead, coated pits. Scale bar 0.50 μm. From Villecco E.I. *et al.* 2002. Zygote 10: 163-173, Fig. 6D.

Important interactions also take place at a distance by hormonal action. It is known that FSH-like gonadotropins promote vitellogenesis through the mediation of estradiol synthesized by follicle cells (Redshaw 1972; Wallace 1985). Moreover, gonadotropins promote the uptake of yolk proteins by vitellogenic oocytes, this effect being apparently mediated by follicle cells through an unknown mechanism (Wallace and Bergink 1974).

Considering that the vitellogenin uptake by the oocyte has proved to be a hormone-dependent process that requires the presence of the follicular epithelium, the possibility exists that the acquisition of endocytic competence by the oocyte could be transferred via the gap junction from the follicle cells.

3.6.3 Cell Interactions during Postvitellogenesis

During the postvitellogenic period, the follicular epithelium of the fully grown oocytes exhibits star-shaped follicle cells with lateral projections that contact the neighboring ones. The space between them constitutes a direct pathway from the blood capillaries along the tunnels of the vitelline envelope toward the oocyte.

In this stage, follicle cells possess numerous macrovilli contacting the oocyte surface, forming gap junctions (Fig. 3.23).

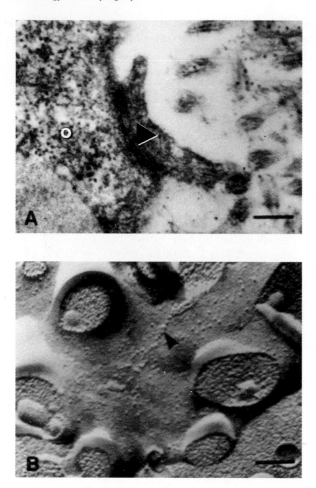

Fig. 3.23 A. Ultrathin section of the plasma membranes in the contact site between a follicle cell and the *Bufo arenarum* oocyte surface. o, oocyte; arrowhead, gap junctions. Scale bar 2 μm. **B**. Freeze-fracture replica of the oocyte plasma membrane. P-face showing a gap junction (arrowhead). Scale bar 0.5 μm. From Villecco E.I. *et al.* 1996. The Journal of Experimental Zoology 276: 76-85, Fig. 1.

The presence of these specialized junctions revealed by TEM and by fluorescent dye transfer is an important ultrastructural feature since it provides the pathway whereby one cell type can regulate the function of another (Lawrence *et al.* 1978).

The physiological significance of follicle cell-oocyte gap junction coupling in *Bufo* and *Ceratophrys* ovarian follicles has been investigated by Villecco *et al.* (1996) (Fig. 3.24). In these species, gap junction coupling varies during the annual sexual cycle. In winter there is either a marked decrease in the amount of heterologous gap-junction contacts or a total absence of them that coincides with the quiescent state of the oocyte (Legname *et al.* 1976).

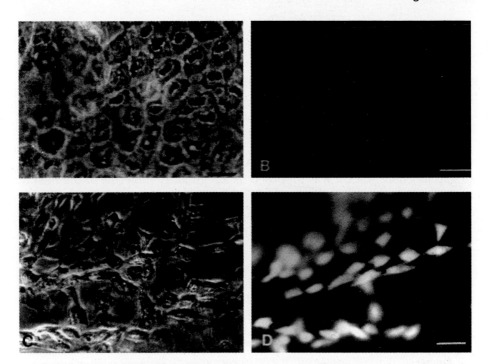

Fig. 3.24 Dye transfer from the *Bufo arenarum* oocyte to follicle cells **A,B**, in winter and **C,D**, summer. **A** and **C**, phase-contrast images of vitelline envelope/follicle cells preparations examined, and **B** and **D**, the same preparations examined for fluorescence. After injection of Lucifer Yellow into an oocyte, dye is normally transferred to the overlying follicles cells in summer (D), but dye transfer is not observed in winter (B). Scale bars 25 μm. From Villecco E.I. *et al.* 1996. The Journal of Experimental Zoology 276: 76-85, Fig. 2.

In contrast, during the breeding season, gap junctions are normally coupled in coincidence with high level of endogenous gonadotropin (FSH-like hormone) (Romero *et al.* 1998). This hormone induces, both in vivo and in vitro, the coupling of gap junctions, making possible the metabolic cooperation necessary for oocyte growth and maintenance. Thus, FSH plays an important role during vitellogenesis.

As regards nuclear maturation, a direct relation seems to exist between gap junction uncoupling and the interruption of the gap junctional flow of an oocyte maturation inhibition factor by the effect of an LH-like hormone or of progesterone.

Bufo arenarum follicles incubated with progesterone induce the gap junction uncoupling and nuclear maturation. Meiotic arrest in normal conditions requires gap junction coupling in order to ensure the transfer of signal molecules involved in the inhibition of meiosis (Villecco *et al.* 1996).

In *Bufo* fully grown oocytes, gap junctions are formed by connexin Cx43 and Cx32 (Villecco *et al.* 2000). This agrees with the fact that Cx43 protein and its respective mRNA are present in both the somatic and the germ cells of the *Xenopus* ovary (Gimlich *et al.* 1990).

The blocking of gap junctions by antibodies against Cx43 and Cx32 produces their uncoupling and the triggering of maturation, evidencing that gap junction uncoupling participates in the maturation process.

In *Bufo* oocytes, cAMP inhibits the maturation induced by the uncoupling of gap junctions. When *Bufo* oocytes are injected or pretreated with cAMP, the gap uncoupling induced maturation is inhibited. This suggests a possible connection between the intra-oocyte cAMP level and the coupling/uncoupling of gap junctions. cAMP was proposed as a regulatory molecule in the meiotic arrest of *Bufo* oocytes (Villecco *et al.* 2000).

Follicle cells might be an important source of cAMP and alterations in cAMP follicle cells levels could be reflected in cAMP oocyte concentrations (Tsafriri 1996).

During the breeding season, *Bufo arenarum* females exhibit high progesterone levels (Romero *et al.* 1998) and the effect of this hormone on the maturation process is probably inhibited by cAMP, which would pass through the gap junctions until environmental conditions (light, temperature and humidity) are those required for successful oocyte deposition.

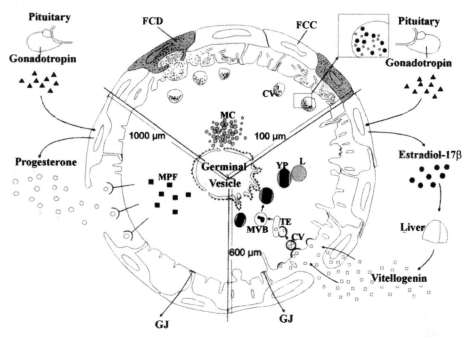

Fig. 3.25 Main processes in *Ceratophrys cranwelli* oogenesis in which follicle cells-oocyte interactions can be seen. MPF, maturation promoting factor; GJ, gap junction; MC, mitochondrial cloud; CV, cortical vesicle; DFC, dark follicle cell; CFC, clear follicle cell; TE, tubular endosome; cv, coated vesicle; L, lipid; YP, yolk platelet; MVB, multivesicular body. Original.

3.7 SUMMARY

Oogenesis is a complex process that leads to the formation and differentiation of the female gamete. Within the gonad, the development of each gamete is mainly mediated through the follicles which undergo progressive cellular and molecular interactions (Fig. 3.25).

One of the most relevant events in amphibian oogenesis is the marked gene expression and accumulation of transcripts during primary oocyte growth. The high pattern of accumulation of different RNAs is related to the requirement for stored components to be used in early development. Besides RNAs, oocytes store components to me*et al.*l nutritional requirements of the embryo until it is able to consume food. Thus, a large number of yolk platelets accumulate during the vitellogenic phase.

When the oocytes complete their growth in the ovary (full grown oocytes) they are arrested at the prophase of meiosis I. However, they are not able to be fertilized or support embryonic development unless they undergo oocyte maturation. This process corresponds to a complex differentiation program that transforms the oocyte into an egg that is fertilization competent.

3.8 ACKNOWLEDGEMENTS

Our research was supported by CONICET and CIUNT grants to S.S.S. We wish to thank Mrs Virginia Méndez for proofreading.

3.9 LITERATURE CITED

Anderson, D. M. and Smith, L. D. 1977. Synthesis of heterogeneous nuclear RNA in full grown oocytes of *Xenopus laevis* (Daudin). Cell 11: 663-671.

Anderson, D. M. and Smith, L. D. 1978. Patterns of synthesis and accumulation of heterogeneous RNA in lampbrush stage oocytes of *Xenopus laevis* (Daudin). Developmental Biology 67: 274-285.

Anderson, D. M., Richter, J. D., Chamberlin, M. E., Price, D. H., Britten, R. J., Smith, L. D. and Davidson, E. H. 1982. Sequence organization of the poly (A)$^+$ RNA synthesized and accumulated in lampbrush chromosome stage *Xenopus laevis* oocyte. Journal of Molecular Biology 155: 281-309.

Aybar, M. J., Genta, S. B., Villecco, E. I., Sánchez Riera, A. N. and Sánchez, S. S. 1996. Amphibian *Bufo arenarum* vitronectin-like protein: its localization during oogenesis. International Journal of Developmental Biology 40: 997-1008.

Bement, W. M. and Capco, D. G. 1990. Transformation of the amphibian oocyte into the egg: structural and biochemical events. Journal of Electron Microscopy Technology 16: 202-234.

Bossini, M. M. de and Pisanó, A. 1987. The development of the vitelline envelope during *Bufo arenarum* oogenesis. Microscopía Electrónica y Biología Celular 11: 36-45.

Brown, D. D. and Dawid, I. B. 1968. Specific gene amplification in oocytes. Science 160: 272-280.

Brummett, A. R. and Dumont, J. N. 1977. Intracellular transport of vitellogenin in *Xenopus* oocytes: an autoradiographic study. Developmental Biology 60: 482-486.

Buehr, M. L. and Blackler, A. W. 1970. Sterility and partial sterility in the South African clawed toad following the pricking of the egg. Journal of Embryology and Experimental Morphology 23: 375-384.

Cabada, M. O., Sánchez Riera, N. A., Genta, H. D., Sánchez, S. S. and Barisone, G. A. 1996. Vitelline envelope during oogenesis in *Bufo arenarum*. BioCell 20, 1: 77-86.

Callebaut, M. 1991. Pyriform-like and holding granulosa cells in the avian ovarian follicle wall. European Archives of Biology 102: 135-145.

Callen, J. C., Dennebouy, N. and Mounolou, J. C. 1980a. Development of the mitochondrial mass and accumulation of mtDNA in previtellogenic stages of *Xenopus laevis* oocytes. Journal of Cell Science 41: 307-320.

Callen, J. C., Tourte, M., Dennebouy, N. and Mounolou, J. C. 1980b. Mitochondrial development in oocytes of *Xenopus laevis*. Biology of the Cell 38: 13-18.

Capco, D. G. and Jeffery, W. R. 1982. Transient localizations of messenger RNA in *Xenopus laevis* oocytes. Developmental Biology 89: 1-12.

Chan, A. P., Kloc, M., Bilinski, S. and Etkin, L. D. 2001. The vegetally localized mRNA fatvg is associated with the germ plasm in the early embryo and is later expressed in the fat body. Mechanisms of Development 100: 137-140.

Coggins, L. W. 1973. An ultrastructural and radioautographic study of early oogenesis in the toad *Xenopus laevis*. Journal of Cell Science 12: 71-93.

Czolowska, R. 1969. Observations on the origin of the "germinal cytoplasm" in *Xenopus laevis*. Journal of Embryology and Experimental Morphology 22: 229-251.

Czolowska, R. 1972. The fine structure of the "germinal cytoplasm" in the egg of *Xenopus laevis*. Wilhelm Roux Archives Entwicklungsmechanik der Organismen 169: 335-344.

Davidson, E. H. 1986. *Gene Activity in Early Development*, 3 rd. Academic Press, Inc. Orlando, Florida.

Del Pino, E. M. 1989a. Modifications of oogenesis and development in marsupial frogs. Development 107: 169-187.

Del Pino, E. M. 1989b. Marsupial frogs. Scientific American 260 (5): 110-118.

Del Pino, E. M. and Humphries Jr., A. A. 1978. Multiple nuclei during early oogenesis in *Flectonotus pygmaeus* and other marsupial frogs. Biological Bulletin of Marine Laboratory, Wood Hole 154: 198-212.

Dixon, K. E. 1994. Evolutionary aspects of primordial germ cell formation. Ciba Foundation Symposium 182: 92-120.

Dumont, J. N. 1972. Oogenesis in *Xenopus laevis* (Daudin). I. Stages of oocyte development in laboratory maintained animals. Journal of Morphology 136: 153-180.

Eddy, E. M. 1975. Germ plasm and the differentiation of the germ cell line. International Review Cytology 43: 229-280.

Franke, W. and Scheer, U. 1970. The ultrastructure of the nuclear envelope of amphibian oocytes: A reinvestigation. Journal of Ultrastructural Research 30: 317-327.

Gard, D. L. and Klymkowsky, M. W. 1998. Intermediate filament organization during oogenesis and early development in the clawed frog, *Xenopus laevis*. Pp. 35-70. In Hermann and Harris (eds), *Subcellular Biochemistry, Vol. 31 Intermediate Filaments*. Plenum Press, New York, U.S.A.

Gard, D. L., Affleck, D. and Error, B. M. 1995. Microtubule organization, acetylation, and nucleation in *Xenopus laevis* oocytes: II. A developmental transition in microtubule organization during early diplotene. Developmental Biology 168: 189-201.

Gimlich, R. L., Kumar, N. M. and Gilula, N. B. 1990. Differential regulation of the level of three gap junctions mRNAs in *Xenopus* embryos. Journal of Cell Biology 110: 597-605.

Heasman, J., Hynes, R. D., Swan, A. P., Thomas, V. and Wylie, C. C. 1981. Primordial germ cell of *Xenopus* embryos: The role of fibronectin in their adhesion during migration. Cell 27: 437-447.

Heasman, J., Quarmby, J. and Wylie, C. C. 1984. The mitochondrial cloud of *Xenopus* oocytes: The source of germinal granule material. Developmental Biology 105: 458-469.

Ho, S. 1987. Endocrinology of vitellogenin. Pp. 145-169. In D. O. Norris and R. J. Jones (eds), *Hormones and Reproduction in Fishes, Amphibians and Reptiles*, Plenum Press, New York, U.S.A.

Houston, D. W. and King, M. L. 2000. A critical role for Xdazl, a germ plasm-localized RNA, in the differentiation of primordial germ cells in *Xenopus*. Development 127: 447-456.

Hudson, C. and Woodland, H. R. 1998. Xpat, a gene expressed specifically in germ plasm and primordial germ cells of *Xenopus laevis*. Mechanisms of Development 73: 159-168.

Ikenishi, K. 1998. Germ plasm in *Caenorhabditis elegans*, *Drosophila* and *Xenopus*. Development Growth and Differentiation 40: 1-10.

Ikenishi, K. and Kotani, M. 1975. Ultrastructure of the "germinal plasm" in *Xenopus* embryos after cleavage. Development Growth and Differentiation 17: 101-110.

Iwamatsu, T. and Chang, M. C. 1972. Sperm penetration in vitro of mouse oocytes at various tines during maturation. Journal of Reproduction and Fertility 31: 237-247.

Jorgensen, C. B. 1973. Mechanisms regulating ovarian function in amphibians (toads). Pp. 133-151. In H. Peters (ed.), *The Development and Maturation of the Ovary and its Functions*, Excerpta Medica, Amsterdam.

Kessel, R. G. 1968. Annulate lamellae. Journal of Ultrastructural Research (Suppl.) 10: 1-82.

King, M. L., Zhou, Y. and Bubunenko, M. 1999. Polarizing genetic information in the egg: RNA localization in the frog oocyte. BioEssays 21: 546-557.

Kloc, M., Bilinski, S., Chan, A. P. Y. and Etkin, L. D. 2000. The targeting of Xcat2 mRNA to the germinal granules depends on cis-acting germinal granule localization element within the 3'UTR. Developmental Biology 217: 221-229.

Kloc, M., Bilinski, S., Chan, A. P. Y. and Etkin, L. D. 2001b. Mitochondrial ribosomal RNA in the germinal granules in *Xenopus* embryos-revisited. Differentiation 67: 80-83.

Kloc, M., Bilinski, S., Chan, A. P. Y., Allen, L. H., Zearfoss, N. R. and Etkin, L. D. 2001a. RNA localization and germ cell determination in *Xenopus*. International Review Cytology 203: 63-91.

Kloc, M., Dougherty, M., Bilinski, S., Chan, A. P. Y., Brey, E., King, M. L., Patrick, C. W. and Etkin, L. D 2002. Three-dimensional ultrastructural analysis of RNA distribution within germinal granules of *Xenopus*. Developmental Biology 241: 79-93.

Kloc, M., Larabell, C., Chan, A. P. Y. and Etkin, L. D 1998. Contribution of METRO pathway localized molecules to the organization of the germ cell lineage. Mechanisms of Development 75: 81-93.

Kloc, M., Spohr, G. and Etkin , L. 1993. Translocation of repetitive RNA sequences with the germ plasm in *Xenopus* oocytes. Science 262: 1712-1714.

Ku, M. and Melton, D. A. 1993. Xwnt-11: A maternally expressed *Xenopus* wnt gene. Development 119: 1161-1173.

Lawrence, T. S., Beers, W. H. and Gilula, N. B. 1978. Transmission of hormonal stimulation by cell-to-cell communication. Nature 272: 501-506.

Legname, A. H., Salomón de Legname, H., Miceli, D. C, Sánchez, S. S. and Sánchez Riera, A. N. 1976. Endocrine control of amphibian oocyte metabolism. Acta Embryologiae Experimentalis 1: 37-49.

Legname, A. H., Salomón de Legname, H., Sánchez, S. S., Sánchez Riera, A. N. and Fernández, S. 1972. Metabolic changes in *Bufo arenarum* oocytes induced by oviducal secretions. Developmental Biology 29: 283-292.

MacGregor, H. C. 1972. The nucleolus and its genes in amphibian oogenesis. Biological Review 47: 177-210.

MacGregor, H. C. 1980. Recent developments in the study of lampbrush chromosomes. Heredity 44: 3-35.

Maller, J. L. and Krebs, E. G. 1980. Regulation of oocyte maturation. Current topics in cell regulation 16: 217-311.

Masui, Y. and Clarke, H. J. 1979. Oocyte maturation. International Review Cytology 57: 185-282.

Melton, D. A. 1987. Translocation of a localized maternal mRNA to the vegetal pole of *Xenopus* oocytes. Nature 328: 80-82.

Miller, O. L. and Hamkalo, B. A. 1972. Visualization of RNA synthesis on chromosomes. International Review Cytology 33: 1-25.

Mosquera, L., Forristall, C., Zhou, Y. and King, M. L. 1993. A mRNA localized to the vegetal cortex of *Xenopus* oocytes encodes a protein with a nanos-like zinc finger domain. Development 117: 377-386.

Mowry, K. L. and Cote, C. A. 1999. RNA sorting in *Xenopus* oocytes and embryos. Federation of American Societies for Experimental Biology Journal 13: 435-445.

Nebreda, A. R. and Ferby, I. 2000. Regulation of the meiotic cell cycle in oocytes. Current Opinion in Cell Biology 12 (6): 666-675.

Old, R. W., Callan, H. G. and Gross, K. W. 1977. Localization of histone gene transcripts in newt lampbrush chromosomes by in situ hybridization. Journal of Cell Science 27: 57-80.

Opresko, L. K. and Karft, R. A. 1987. Specific proteolysis regulates fusion between endocytic compartments in *Xenopus* oocytes. Cell 51: 557-568.

Opresko, L. K. and Wiley, H. S. 1987a. Receptor-mediated endocytosis in *Xenopus* oocytes. I Characterization of the vitellogenin receptor system. Journal of Biological Chemistry 262: 4109-4115.

Opresko, L. K. and Wiley, H. S. 1987b. Receptor-mediated endocytosis in *Xenopus* oocytes. II Evidence for two novel mechanisms of hormone regulation. Journal of Biological Chemistry 262: 4116-4123.

Palmer, A. and Nebreda, A. R. 2000. The activation of MAP kinase and p34cdc2/cycling B during the meiotic maturation of *Xenopus* oocytes. Progress in Cell Cycle Research 4: 131-143.

Pannese, M., Cagliani, R., Pardini, C. L. and Boncinelli, E. 2000. Xotx1 maternal transcripts are vegetally localized in *Xenopus laevis* oocytes. Mechanisms of Development 90: 111-114.

Pepling, M. E., de Cuevas, M. and Spradling, A. C. 1999. Germline cyst: A conserved phase of germ cell development? Trends in Cell Biology 9: 257-262.

Peters, H. 1978. Folliculogenesis in mammals. Pp. 121-144. In R. E. Jones (ed.), *The Vertebrate Ovary*, Plenum Press, New York, N.Y.

Pinto, M. A., Santella, A., Casazza, G., Rosati, F. and Monroy, A. 1985. The differentiation of the vitelline envelope of *Xenopus* oocytes. Development Growth and Differentiation 27: 189-200.

Qian, Y. W., Erikson, E., Taieb, F. E. and Maller, J. L. 2001. The polo kinase Plx1 is required for activation of the phosphatase Cdc25C and Cycling B-Cdc2 in Xenopus oocytes. Molecular Biology of the Cell 12: 1791-1799.

Redshaw, M. R. 1972. The hormonal control of the amphibian ovary. American Zoology 12. 289-306.

Ressom, R. E. and Dixon, K. L. 1988. Relocation and reorganization of germ plasm in *Xenopus* embryos after fertilization. Development 103: 507-518.

Ritcher, H. P. 1987. Membranes during yolk-platelet development in oocytes of the toad *Bufo marinus*. Roux´s Archives Developmental Biology 196: 367-371.

Ritcher, P. H. and Bauer, A. 1990. Ferromagnetic isolation of endosomes involved in vitellogenin transfer into *Xenopus* oocytes. European Journal of Cell Biology 51: 53-63.

Robb, D. L., Heasman, J., Raats, J. and Wylie, C. C. 1996. A kinesin-like protein is required for germ plasm aggregation in *Xenopus*. Cell 87: 823-831.

Romero, I. R., Atenor, M. B. and Legname, A. H. 1998. Nuclear maturation inhibitors in *Bufo arenarum* oocytes. BioCell 22, 1: 27-34.

Sagata, N. 1997. What does Mos do in oocytes and somatic cells? BioEssay 19: 13-21.

Salomón de Legname, H. 1969. Biochemical studies on the energetics of *Bufo arenarum* segmenting eggs. Archives Biologie 80: 471-490.

Salomón de Legname, H., Sánchez Riera, A. N. and Sánchez, S. S. 1971. Pathways of glucose breakdown during *Bufo arenarum* ontogenesis. Acta Embryologiae Experimentalis 3: 187-194.

Salomón de Legname, H., Sánchez Riera, A. N. and Sánchez, S. S. 1975. Source of precursors for nucleotides biosynthesis. Acta Embryologiae Experimentalis 2: 123-136.

Sánchez Riera A. N., Sánchez, S. S. and Cabada, M. O. 1983. RNA metabolism during progesterone-induced maturation of *Bufo arenarum* oocytes. Comunicaciones Biológicas1, 3: 273-288.

Sánchez Riera, A. N., Sánchez, S. S. and Cabada, M. O. 1988a. RNA metabolism in the follicle cells of *Bufo arenarum* oocytes. I Biochemical studies. Microscopía Electrónica y Biología Celular 12, 2:149-161.

Sánchez Riera, A. N., Sánchez, S. S. and Cabada, M. O. 1988b. RNA metabolism in the follicle cells of *Bufo arenarum* oocytes. II Autoradiographic studies. Microscopía Electrónica y Biología Celular 12, 2:163-175.

Savage, R. M. and Danilchik, M. V. 1993. Dynamics of germ plasm localization and its inhibition by ultraviolet irradiation in early cleavage *Xenopus* eggs. Developmental Biology 157: 371-382.

Scheer, U. 1973. Nuclear pore flow rate of ribosomal RNA and chain growth rate of its precursor during oogenesis of *Xenopus laevis*. Developmental Biology 30: 13-28

Scheer, U. and Dabauvalle, M. C. 1985. Functional organization of the amphibian oocyte nucleus. Pp. 385-430. In L. W. Browder (ed.), *Developmental Biology, Vol. 1 Oogenesis*, Plenum Press, New York, N.Y.

Smith, L. D. 1966. The role of a "germinal plasm" in the formation of primordial germ cells in *Rana pipiens*. Developmental Biology 14: 330-347.

Smith, L. D. 1989. The induction of oocyte maturation: transmembrane signaling events and regulation of the cell cycle. Development 107: 685-699.

Sommerville, J. 1977. Gene activity in the lampbrush chromosomes of amphibian oocytes. International Review Biochemistry 15: 79-156.

Stennard, F., Zorn, A. M., Ryan, K. Garrett, N. and Gurdon, J. B. 1999. Differential expression of Veg T and Antipodian protein isoforms in Xenopus. Mechanisms of Development 86: 87-98.

Tokarz, R. R. 1978. Oogonial proliferation, oogenesis and folliculogenesis in nonmammalian vertebrates. Pp. 145-179. In R. E. Jones (ed.), *The Vertebrate Ovary,* Plenum Press, New York, N.Y.

Tsafriri, A., Sang-Young, Ch., Ruobo, Z., Hsueh, A. J. W. and Conti, M. 1996. Oocyte maturation involves compartmentalization and opposing changes of camp levels in follicular somatic and germ cells: studies using selective phosphodiesterase inhibitors. Developmental Biology 178: 393-402.

Varriale, B., Pierantoni, R., Di Matteo, L., Minucci, S., Milone, M. and Chieffi, G. 1988. Relationship between estradiol-17β seasonal profile and annual vitellogenin content of liver, fat body, plasma and ovary in the frog (*Rana esculenta*). General Comparative Endocrinology 69: 328-334.

Villecco, E. I., Aybar, M. J., Genta, S. B., Sánchez, S. S. and Sánchez Riera, A. N. 2000. Effect of gap junction uncoupling in full *Bufo arenarum* ovarian follicles: participation of cAMP in meiotic arrest. Zygote 8: 171-179.

Villecco, E. I., Aybar, M. J., Sánchez Riera, A. N. and Sánchez, S. S. 1999. Comparative study of vitellogenesis in the anuran amphibians *Ceratophrys cranwelli* (Leptodactylidae) and *Bufo arenarum* (Bufonidae). Zygote 7: 11-19.

Villecco, E. I., Aybar, M. J., Sánchez, S. S. and Sánchez Riera, A. N. 1996. Heterologous gap junctions between oocyte and follicle cells in *Bufo arenarum*: Hormonal effects on their permeability and potential role in meiotic arrest. Journal Experimental Zoology 276: 76-85.

Villecco, E. I., Genta, S. B., Sánchez Riera, A. N. and Sánchez, S. S. 2002. Ultrastructural characteristics of the follicle cell-oocyte interface in the oogenesis of *Ceratophrys cranwelli*. Zygote 10: 163-173.

Wall, D. A. and Patel, S. 1987. Multivesicular bodies play a key role in vitellogenin endocytosis by *Xenopus* oocytes. Developmental Biology 119: 275-289.

Wallace, R. A. 1963. Studies on amphibian yolk. IV An analysis of the main-body components of the yolk platelets. Biochimica et Biophysica Acta 74: 505-518.

Wallace, R. A. 1985. Vitellogenesis and oocyte growth in nonmammalian vertebrates. Pp. 127-177. In L. W. Browder (ed.), *Developmental Biology, Vol. 1 Oogenesis,* Plenum Press, New York, N.Y.

Wallace, R. A. and Bergink, E. W. 1974. Amphibian vitellogenin: Properties, hormonal regulation of hepatic synthesis and ovarian uptake and conversion to yolk proteins. American Zoology 14: 1159-1175.

Wallace, R. A. and Dumont, J. N. 1968. The induced synthesis and transport of yolk proteins and their accumulation by the oocyte in *Xenopus laevis*. Journal of Cell Physiology (Suppl.) 72: 73-89.

Wessely, O. and De Robertis, E. M. 2000. The *Xenopus* homologue of Bicaudal-C is a localized maternal mRNA that can induce endoderm formation. Development 127: 2053-2062.

Whitington, P. M. and Dixon, K. E. 1975. Quantitative studies of germ plasm and germ cells during early embryogenesis of *Xenopus laevis*. Journal of Embryology and Experimental Morphology 33: 57-74.

Wiley, H. S. and Wallace, R. A. 1981. The structure of vitellogenin. Multiple vitellogenins in *Xenopus laevis* give rise to multiple forms of the yolk proteins. Journal of Biological Chemistry 256: 8626-8634.

Williams, M. A. and Smith, L. D. 1971. Ultrastructure of the "germinal plasm" during maturation and early cleavage in *Rana pipiens*. Developmental Biology 25: 568-580.

Wischnitzer, S. 1964. An electron microscope study on the formation of the zona pellucida in oocytes from *Triturus viridescens*. Zeitschrift für Zellforschung und Mikroskopische Anatomie 64: 196-2002.

Wylie, C. C. and Heasman, J. 1976. The formation of the gonadal ridge in *Xenopus laevis*. I. A light and transmission electron microscope study. Journal of Embryology and Experimental Morphology 35: 125-138.

Wylie, C. C. and Heasman, J. 1993. Migration, proliferation, and potency of primordial germ cells. Seminars in Developmental Biology 4: 161-170.

Wylie, C. C., Heasman, J., Swan, A. P. and Anderton, B. H. 1979. Evidence for substrate guidance of primordial germ cells. Experimental Cell Research 121: 315-324.

Yamaguchi, S., Hedrick, J. L. and Katagiri, C. 1989. The synthesis and localization of envelope glycoproteins in oocyte of *Xenopus laevis* using immunocytochemical methods. Development Growth and Differentiation 31: 85-94.

Yamashita, M., Mita, K., Yoshida, N. and Kondo, T. 2000. Molecular mechanisms of initiation of oocyte maturation: general and species-specific aspects. Progress in Cell Cycle Research 4: 115-129.

Yisraeli, J., Sokol, S. and Melton, D. A. 1990. A two-step model for the localization of maternal mRNA in *Xenopus* oocytes: Involvement of microtubules and microfilaments in translocation and anchoring of Vg1 RNA. Development 108: 289-298.

Züst, B. and Dixon, K. L. 1977. Events in the germ cell lineage after entry of the primordial germ cells into the genital ridges in normal and UV-irradiated *Xenopus laevis*. Journal of Embryology and Experimental Morphology 41: 33-46.

CHAPTER 4

Endocrinology of Reproduction

Silvia N. Fernández and Inés Ramos

4.1 INTRODUCTION

The reproductive activity of anuran amphibians is synchronized and adapted fundamentally to environmental conditions, the most important being temperature, photoperiods, humidity and food supply. Consequently, reproduction would be potentially limited to that period of the year in which conditions are optimal for the survival of the offspring. Although reproductive schemes vary considerably, most anuran amphibians exhibit a sexual cycle with a breeding period and a postreproductive one.

The breeding period comprises a preovulatory period in which the whole reproductive system is prepared for ovulation and an ovulatory period, often during the spring months, in which fully-grown mature oocytes are released into the female coelomic cavity, transported through the oviduct and finally released for external fertilization.

The postreproductive period, characterized by reproductive system recovery, also comprises two stages: the early postovulatory period, during the summer and fall seasons, signaled by follicular growth and differentiation (folliculogenesis), and the late postovulatory period, which includes hibernation, characterized by the completion of oogenesis and the later acquisition of the oocyte maturation capacity.

Different studies indicate that the reproductive activity in anurans is regulated by endocrine mechanisms acting through the hypothalamic-pituitary-gonadal axis (Whittier and Crews 1987). However, as these vertebrates are extremely sensitive to environmental influences, the pineal gland is also a significant regulator of the reproductive activity.

Departamento de Biología del Desarrollo, Instituto Superior de Investigaciones Biológicas (INSIBIO), Consejo Nacional de Investigaciones Científicas y Técnicas (CONICET) y Universidad Nacional de Tucumán. San Miguel de Tucumán 4000, Argentina

4.2. PINEAL GLAND

In animals with a seasonal modality of reproduction, the pineal gland is considered as a neuroendocrine transducer of photoperiod and temperature and seems to be involved in the coordination of the cyclic reproductive changes with the seasons of the year.

In anurans, the pineal gland is attached to the roof of the diencephalon and shows a simple structural organization based on a numerous epithelial cell population with a sensorial photoreceptor function similar to that of the retina. Glial cells are also found in smaller amounts.

The action of the gland is mediated by melatonin (N-Acetyl-5-methoxytriptamine), a rhythmically produced hormone. Circadian and seasonal differences in the melatonin content of the pineal gland, serum and retina have been found in some species (Delgado and Vivien-Roels 1989; d'Istria et al. 1994). Hormone biosynthesis is highest during the dark phase as a result of an increase in the activity of its rate-limiting enzyme, N-acetyltransferase (NAT), which converts serotonin into N-acetylserotonin. Methylation of the latter compound by hydroxyindole-O-methyltransferase (HIOMT) leads to the formation of melatonin.

The target sites of melatonin action involved in amphibian reproductive functions remain to be elucidated; however, using quantitative autoradiographic techniques, melatonin receptors have been found in the mesencephalon, hypothalamus and telencephalon (Tavolaro et al. 1995). Additional findings demonstrating that melatonin has a direct effect on *Bufo arenarum* ovaries (de Atenor et al. 1994) suggest that the hormone may act at several levels of the hypothalamic-pituitary-gonadal axis, thus regulating the development, growth and activity of the gonad.

Melatonin seems to have an inhibitory effect on the ovary of lower vertebrates. Melatonin treatment produces a decrease in the gonadosomatic index and in the vitellogenesis rate during gonadal recovery, an increase in the percentage of atretic follicles and a diminution in ovarian proteins and glycogen content (Kupwade and Saidapur 1986; Udaykumar and Joshi 1997). In *Xenopus laevis* (O'Connor 1969) and *Bufo arenarum* (de Atenor et al. 1994), both the pineal gland and melatonin produce an inhibitory effect in *in vitro* ovulation. The inhibition caused by the gland is higher than that of melatonin, probably because the gland contains not only melatonin but also 5-methoxytryptophol, a compound with an antigonadal effect (Skene et al. 1991). Different methoxyindols and neuropeptides, the physiological roles of which are still unknown, are also found in the gland.

In lower vertebrates the retina is also an important source of methoxyindols (Delgado et al. 1993), production being dependent on environmental factors.

4.3 HYPOTHALAMUS

Hypothalamus control on vertebrate reproduction is mediated by a decapeptide gonadotropin-releasing hormone (GnRH), the main regulator of the synthesis and of the release of gonadotropins from the anterior pituitary lobe.

In several anuran species, immunohistochemical methods have demonstrated the presence of immunoreactive GnRH in the perikarya of the prechiasmatic brain area, mainly in the medial septum and anterior diencephalon, from where the perikarya send their major projections through the ventral hypothalamus to the median eminence (Sotowska-Brochocka 1988; Muske and Moore 1990; Miranda *et al.* 1998). In *Xenopus laevis*, GnRH cells have also been localized in the ventral wall of the infundibulum. Microdissection-radioimmunoassay studies have shown that the GnRH cells in the hypothalamus are organized in two different regulatory centers involved in the control of gonadotropic functions. One of them, localized in the caudal hypothalamus, seems to be involved in the control of the seasonal reproductive cycle while the other, placed in the prechiasmatic region, is apparently involved in the control of ovulation. The finding of GnRH in extrahypothalamic areas of the central and peripheral nervous systems and nonneural tissues has led to the suggestion of possible new functions of this peptide such as behavior modulation, regulation of gonadal steroidogenesis and estradiol-17 β synthesis through prostaglandins mediation (Gobbetti and Zerani 1991).

The amount of hypothalamus GnRH changes according to the stage of development of amphibians, showing a significantly higher concentration when animals reach sexual maturity. Variations are also observed during the annual sexual cycle, with higher levels during the last hibernation period.

At least 11 GnRH variants have been found in vertebrates, most species exhibiting more than one. In anurans, three different variants have been identified: mammalian GnRH, chicken GnRH and a variant with chromatographic and immunological characteristics similar to those of salmon GnRH. However, the functional significance of these multiple forms within a single species is not clearly understood. In adult frogs such as *Rana catesbeiana*, *R. rugulosa* and *Bufo arenarum* the predominant form of brain GnRH shows the characteristics of mammalian GnRH (Miranda *et al.* 1998; Yuanyou and Haoran 2000). This variant may be the major active form involved in the direct regulation of pituitary gonadotropes while the other variants might have an extrapituitary role as neurotransmitters or neuromodulators at other brain sites (Yuanyou and Haoran 2000).

4.4 PITUITARY GLAND

The pituitary gland of anurans secretes two gonadotropins, a follicle stimulating hormone (FSH) and a luteinizing hormone (LH), both with biochemical and immunological properties similar to those of other vertebrates (Licht *et al.* 1983). Immunohistochemical studies have demonstrated that both gonadotropins are synthesized and secreted by gonadotrope cells distributed throughout the pars distalis. However, a small number of immunoreactive cells with only one or the other gonadotropin has also been found (Pinelli *et al.* 1996).

Anuran gonadotropins are glycoproteins made up of two subunits, α and β. The α-subunit, identical in both hormones, has 97 amino acid residues with one arginine insertion at position 29. Its molecular mass, not including the sugar chains, is 11,026 Da. The α-subunit has an approximately 70% sequence identity with the mammalian α-subunit (Hayashi *et al.* 1992a). The β-subunit, which is different in the two gonadotropins, gives them their specific activity. In bullfrog, for instance, the FSH β-subunit is composed of 107 amino acid residues with a molecular mass of 11,782 Da, while the LH β-subunit is composed of 112 amino acid residues with a molecular mass of 12,675 Da, taking into account the six cystine bridges and excepting the sugar chain (Hayashi *et al.* 1992b).

There is a close link between the period of the sexual cycle and the circulating levels of FSH and LH. Gonadotropins levels, which remain low during early hibernation, begin to rise progressively during the last third portion of the hibernation period and reach a peak just before ovulation (Polzonetti-Magni *et al.* 1998; Kim *et al.* 1998). This peak shows an LH concentration 4 to 6 times higher than that of FSH. A second peak of FSH occurs during the summer, at which time ovarian follicles grow rapidly, concurrently with the increase in the gonadosomatic index (Licht *et al.* 1983; Kim *et al.* 1998).

Although most of the data on the role of gonadotropins in the reproductive function of frogs have been obtained by indirect studies, several reports suggest that both hormones are involved in the control of ovarian steroid secretion (Kwon and Ahn 1994). A positive correlation between FSH and estradiol-17 β (Polzonetti-Magni *et al.* 1998) and between LH and progesterone secretions (Itoh and Ishii 1990) has been reported. The gonadotropic stimulation of steroidogenesis seems to depend on an increase in intracellular cAMP that would result in protein activation.

4.5 OVARIAN HORMONES

4.5.1 Gonadal Steroids

As regards the ovarian endocrine function, anuran female gonads show an enzymatic machine similar to that of mammals, which is necessary for the synthesis of steroid hormones. The follicular wall seems to be the primary site for ovarian steroidogenesis. The activity of enzymes such as 3β-hydroxysteroid dehydrogenase, 17α-hydroxylase, $C_{17,20}$-lyase and aromatase in the granulosa cells indicate that these cells are the main source of estradiol-17 β (E_2), progesterone (P), 17 α-hydroxyprogesterone and androstenedione. In contrast, the activity of 17 β-hydroxysteroid dehydrogenase, an enzyme involved in the biosynthesis of testosterone (T) from androstenedione, is greater in the theca layer than in the granulosa cells. In view of the above, a bi-directional cooperation between the two types of follicular components is obviously required for steroid hormonal synthesis to occur (Kwon and Ahn 1994) (Fig. 4.1).

The ovarian and circulating profiles of the gonadal steroids, examined by radioimmunoassays, reveal pronounced changes throughout the sexual cycle that correlate with the levels of circulating gonadotropins and with the morphological modifications in the sexual characteristics. These changes are also intimately associated with seasonal variations. Although in general an

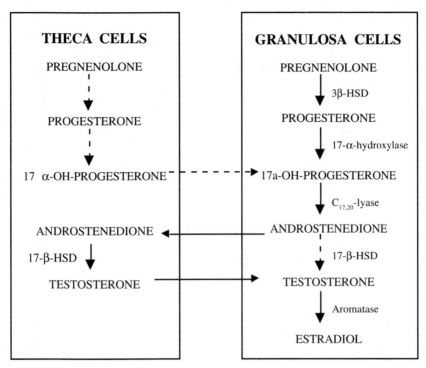

Fig. 4.1 Two-cell model for steroidogenesis in *Rana nigromaculata* ovarian follicles. Granulosa cells are responsible for synthesis of most ovarian steroids while theca cells are involved only in converting androstenedione to testosterone. 3β-HSD: 3β-hydroxysteroid dehydrogenase. Solid arrows indicate the major pathway and dotted arrows indicate the minor pathway. Modified after Kwon H.B. and Ahn R.S. 1994. General Comparative Endocrinology 94: 207-214, Fig. 6.

increase in hormone levels occurs during the breeding period, each steroid has its own profile that can differ between species.

The analysis of the seasonal rhythm of E_2 in *Bufo arenarum* shows its lowest concentration from the last phase of hibernation until the end of preovulatory period. A progressive increase in its levels occurs around the ovulatory and early postovulatory period concurrently with gonadal recovery (Fig. 4.2) and reaching, during the latter period, a peak of about 13–15 ng/ml (Fernández *et al.* 1984). A similar hormonal pattern appears in *Xenopus laevis* (Fortune 1983) and *Rana nigromaculata* (Kwon *et al.* 1993).

Throughout the sexual cycle the highest concentration in P circulating levels takes place during the preovulatory period, with a sharp decrease after ovulation (Fig. 4.2). Progesterone levels remain low during the initial phase of the early postovulatory period and then increase progressively until the values typical of the preovulatory period are reached.

A common feature in anuran females is the presence of high concentrations of the androgens T and dihydrotestosterone (DHT), higher even than those of E_2, both in the ovary and in blood (Fernández *et al.* 1984; Lutz *et al.* 2001). The

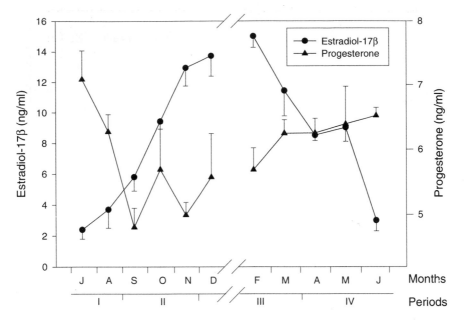

Fig. 4.2 Serum concentrations of estradiol-17β and progesterone of *Bufo arenarum* females during the reproductive cycle. Reproductive cycle periods: I- preovulatory; II- ovulatory; III- early postovulatory, IV- late postovulatory. Original.

patterns of these hormones (Fig. 4.3) reach a minimum concentration after ovulation and remain very low during the early postovulatory period, when previtellogenic oocytes are predominant in the ovary. T and DHT concentrations rise slightly during the late postovulatory period and then increase dramatically, reaching a peak during the preovulatory period. Although T and DHT present an almost identical profile during the sexual cycle, T concentration is always 4 or 5 fold higher than that of DHT. Although the exact physiological effects of androgens have not yet been completely elucidated, they probably play an important role in the reproductive process. The presence of aromatase activity provides evidence that T may be an important substrate for E_2 synthesis in different target organs such as ovary (Miyashita *et al.* 2000), oviduct (Kobayashi *et al.* 1996), brain (Guerriero *et al.* 2000) and liver (Di Fiore *et al.* 1998). In *Rana esculenta*, for instance, T may be involved in vitellogenin synthesis control, depending on its local conversion by an aromatase present in liver tissue (Di Fiore *et al.* 1998).

The profile of seric steroids is closely related to follicular growth. In fact, the smallest follicles present in the ovary immediately after ovulation secrete very low levels of the steroids but, during the early postovulatory period, the single layer of follicle cells that surrounds the growing oocytes synthesizes and secretes E_2 (Fortune 1983) , which is involved in the hepatic synthesis and secretion of vitellogenin, the yolk protein precursor. When the ovarian follicles reach an intermediate size, the steroidogenic potential for E_2 synthesis starts to

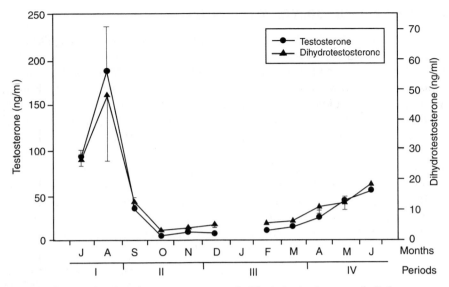

Fig. 4.3 Serum levels of testosterone and dihydrotestosterone of *Bufo arenarum* females during the reproductive cycle. Reproductive cycle periods: I- preovulatory; II- ovulatory; III- early postovulatory, IV- late postovulatory. Original.

decline. Concurrently, P and T secretion increases, reaching highest concentration when follicles attain maximum size (Sretarugsa and Wallace 1997). At the same time, oocytes acquire a progressive nuclear maturation competence. The peak in the levels of P and T, which takes place before ovulation, is related to their role in triggering oocyte maturation.

4.5.2 Gonadal Peptidic Hormones

Anuran gonads also secrete gonadopeptidic hormones such as inhibin- and activin-like proteins, whose role in reproductive functions has not yet been clearly established. These gonadopeptides, structurally related to the trans- forming growth factor-β family, are heterodimeric glycoproteins composed of two subunits, α and β. The active biological form of inhibin is constituted by an α subunit and one of two β subunits, $β_A$ and $β_B$, resulting in inhibin A and inhibin B. Activin is formed by any combination of the two β subunits.

These gonadopeptides act at different levels. For example, in bullfrog pituitary, activin B stimulates, in a dose dependent manner, the release not only of FSH but also of LH while inhibin B suppresses the activin-induced release of the gonadotropins without affecting the basal levels of FSH and LH (Uchiyama *et al.* 2000). On the other hand, in *Xenopus*, inhibin blocks, in a dose- dependent manner, oocyte maturation induced by pituitary homogenate and progesterone. Consequently, inhibin would be important not only as an endocrine modulator at the pituitary level but also as a local regulator of gonadal functions. As regards the ovary, inhibin seems to act both at the follicle cell level by blocking steroidogenesis and at the oocyte level by modifying the maturation process (Lin *et al.* 1999).

4.6 CONTROL OF OOCYTE MATURATION

Throughout the intraovarian growth phase, amphibian oocytes, like other vertebrates, remain arrested at the prophase of the first meiotic division (Masui 1985; Matten *et al.* 1996). The period of oocyte growth is followed by a process called oocyte maturation which is a prerequisite for successful fertilization and normal embryonic development. Maturation has long been associated with the events leading to meiotic resumption and with the progression of the cell cycle from prophase I to metaphase II (Sadler and Maller 1981; Liu and Patiño 1993). This process, morphologically characterized by the migration of the nucleus or germinal vesicle (GV) and the later dissolution of the nuclear envelope, is known as nuclear maturation. Meiotic resumption, however, is not the only significant aspect in gamete maturation, since fully-grown oocytes must undergo modifications not only at the nuclear but also at the cytoplasmic level (Legname and Bühler 1978) for normal embryonic development to occur. These facts have led to the concept of cytoplasmic maturation.

Up to the present time there is no single parameter to define and/or monitor cytoplasmic maturation. Some of the biological factors likely to be correlated with this process are: a) the capacity of the oocytes to generate the mRNA that codifies the synthesis of the maturation-promoting factor (MPF) which regulates meiotic resumption (Nguyen-Gia *et al.* 1986); b) the migration and redistribution of the cortical granules (Ramos *et al.* 1999) and c) the development of Ca^{2+}-releasing mechanisms from the intracellular deposits (Ramos *et al.* 1998). Other possible indicators of cytoplasmic maturation are the biochemical changes in the oxidative metabolism of carbohydrates found in *Bufo arenarum* oocytes (Legname and Salomón de Legname 1980). Fully-grown oocytes from the hibernation period present a predominantly energetic metabolism that utilizes glucose mainly through the Embden-Meyerhof glycolytic pathway followed by a classic tricarboxylic acid cycle. These are considered cytoplasmically immature oocytes since, although capable of ovulation and fertilization under experimental conditions, they do not continue their embryonic development (Legname and Bühler 1978). Oocytes from the breeding period, on the other hand, exhibit a predominantly anabolic metabolism with a significant utilization of glucose via the pentose phosphate pathway, followed by the variant of the tricarboxylic acid cycle known as the glutamic-aspartic cycle (Legname and Salomón de Legname 1980). Only these gametes, considered as cytoplasmically mature oocytes, are capable of aster development and of segmentation after fertilization (Bühler *et al.* 1987).

The maturation process, both at the cytoplasmic and nuclear levels, is under endocrine control to insure its sequentiation and coordination with ovulation.

Taking into account the fact that the pineal gland is active during the fall-winter season, it can be assumed that the characteristic immature stage of the fully-grown *Bufo arenarum* oocytes during the hibernation period is determined by the influence of this gland. In fact, both the gland and its hormone melatonin, acting through the follicular cells as well as directly on the oocytes, induce an oxidative metabolism of carbohydrates characteristic of the gametes from the hibernation period, which are considered as cytoplasmically immature oocytes (de Atenor *et al.* 1994) (Fig. 4.4). This points to the probable presence of melatonin receptors on the gamete, thus increasing the melatonin binding sites known at the moment in the reproductive system. Although melatonin seems to have a role in establishing the immature cytoplasmic stage, it does not by itself affect nuclear maturation.

Another hormone apparently involved in the determination of a cytoplasmically immature oocyte is noradrenaline. This catecholamine, in a way similar to melatonin, modifies the metabolic activity of oocytes by determining a type of oxidative metabolism typical of those from the hibernation period (Fig. 4.5) (Ramos *et al.* 2002). This effect is probably associated with the

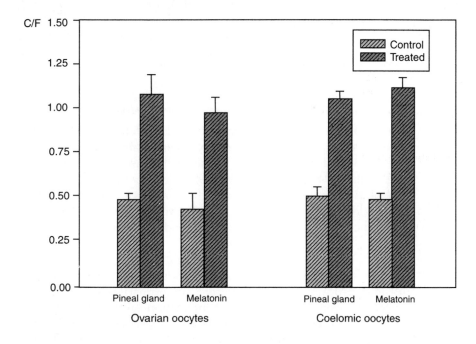

Fig. 4.4 Effect of the pineal gland and melatonin on the cytoplasmic maturation of *Bufo arenarum* ovarian and coelomic oocytes. Cytoplasmic maturation was estimated by assaying the capacity of isolated mitochondria to oxidize citrate and fumarate as intermediates of the Kreb's cycle. The metabolic behavior of the oocytes was expressed as the oxidation ratio of citrate/fumarate (C/F). C/F ratio close to or above 1 is considered an expression of cytoplasmically immature oocytes. C/F ratio whose values are between 0.3–0.6, is related to cytoplasmically mature oocytes. Original.

seasonal variations in the levels of noradrenaline, which reaches its maximum concentration both in the adrenal glands (Rapela and Gordon 1956) and in the plasma (Segura and D'Agostino 1964) of *Bufo arenarum* during the fall-winter months.

On the other hand, noradrenaline could act at two levels, directly on the oocyte or indirectly by regulating the activity of the pineal gland, a possibility supported by the presence of sympathetic nerve terminals in the pineal gland of *Bufo arenarum*. Consequently, the physiologically cytoplasmic immature stage in fully-grown oocytes from the hibernation period may result from either the individual or the joint effect of noradrenaline and melatonin.

At the end of the hibernation period, the progressive increase in the photoperiod causes a decrease in the activity of the pineal gland, with the consequent disappearance of the brake on the hypothalamic-pituitary-gonadal axis. This event permits the release of GnRH, whose action on the pituitary gland results in the increase in gonadotropins secretion (Kim *et al.* 1998).

As regards the effect of gonadotropins on oocyte maturation, FSH is one of the hormones involved in cytoplasmic maturation in *Bufo arenarum*. FSH changes the oocyte metabolic pattern to another characteristic of oocytes from the breeding period, that is, cytoplasmically mature ones. Similar metabolic changes occur when oocytes are incubated with dbcAMP or when the level of the second messenger is increased by the use of 3-isobutyl-1-methyl-xanthyne (IBMX), an inhibitor of phosphodiesterase (Budeguer de Atenor *et al.* 1989). This effect of FSH would involve the participation of the adenylate cyclase-cAMP system.

Adrenaline exerts an effect similar to that of FSH by modifying the metabolic behavior of oocytes (Fig. 4.5), making them cytoplasmically mature (Ramos *et al.* 2002). The above effect is probably cause by the amount of adrenaline present in the adrenal glands of *Bufo arenarum*, which fluctuates in accordance with the stage of the reproductive cycle, with high values during the period of sexual activity and an abrupt drop in the hibernation period (Rapela and Gordon 1956). Since adrenaline affects ovarian and coelomic oocytes, follicle cells seem not to be involved in its action.

The evidence obtained when using adrenergic agonists and antagonists such as ritodrine and xylometazoline (specific β_2 and α-adrenergic agonists respectively) and propranolol (β-adrenergic antagonist) suggests that the metabolic effect of adrenaline is caused by interaction with β_2-receptors. The capacity of β-adrenergic agonists to determine a cytoplasmically mature oocyte may be associated with the stimulatory effect of these agents on the activity of glucose 6-phosphate dehydrogenase (G-6-PDH), a key enzyme in glucose oxidation via the pentose phosphate pathway, which is predominant in oocytes during the breeding period.

Although in *Bufo arenarum* oocytes FSH and adrenaline are effective in inducing cytoplasmic maturation, they do not affect nuclear maturation by themselves. However, pretreatment of the follicles with each of the two hormones results in a significant inhibition of progesterone-induced nuclear maturation (Table 4.1). This inhibitory effect is possibly due to an increase in intracellular cAMP, taking into account that both FSH and adrenaline act through the adenylate cyclase-cAMP system. This hypothesis is strengthened when incubating follicles with dbcAMP or IBMX compounds, which determine high intracellular cAMP levels. Although these compounds have no direct

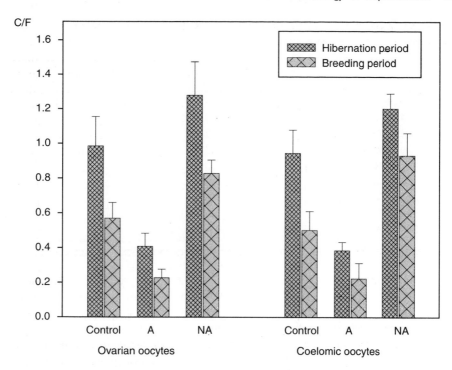

Fig. 4.5 Effect of catecholamines (A: 10 $^{-7}$ M adrenaline, NA: 10 $^{-7}$ M noradrenaline) on the cytoplasmic maturation of ovarian and coelomic oocytes obtained from *Bufo arenarum* females captured during the hibernation or the breeding period. C/F ratio close to or above 1 is considered an expression of cytoplasmically immature oocytes. C/F ratio between 0.3-0.6 is related to cytoplasmically mature oocytes. From Ramos I. *et al.* 2002. Zygote 10: 271-281, Fig. 3.

effect *per se* on meiotic resumption, they are effective in blocking progesterone-induced nuclear maturation (Kwon *et al.* 1989; de Romero *et al.* 1998).

The above data point to cAMP as a possible regulator and synchronizer of the maturation process through changes in its intracellular concentration. Thus, cytoplasmic maturation would occur first when cAMP reaches high levels; later, when cAMP levels decrease, nuclear maturation takes place (Kwon *et al.* 1989).

During the preovulatory period, gonadotropins cause follicular cells to secrete ovarian steroids such as progesterone, which plays an important role in oocyte nuclear maturation. This hormone, whose concentration reaches a maximum prior to ovulation, is considered as the natural inducer, whether *in vivo* or *in vitro*, of amphibian oocyte nuclear maturation (Sadler and Maller 1982). This process involves a series of morphological changes such as the migration of the germinal vesicle toward the surface of the animal hemisphere and the dissolution of the nuclear membrane. The arrival of the GV at the cortex causes a pigment displacement that originates a circular white spot, the first visible indication of oocyte maturation. Other morphological characteristics

Table 4.1 Effect of FSH, adrenaline, dbcAMP and IBMX on progesterone induce nuclear maturation in *Bufo arenarum* oocytes

Treatment	Incubation time (h)		
	8	16	24
Progesterone 1 µg/ml	95	100	100
FSH 0.1 µg/ml	0	0	0
FSH + Progesterone	0	28	65
Adrenaline 10^{-7}M	0	0	0
Adrenaline + Progesterone	0	0	15
dbcAMP 10^{-3} M	0	0	0
dbcAMP + Progesterone	0	0	0
IBMX 10^{-3} M	0	0	0
IBMX + Progesterone	0	0	0

The results are expressed as the mean of the percentage of germinal vesicle breakdown. Original

of nuclear maturation are chromosome condensation, spindle formation and the extrusion of the first polar body. After maturation is completed, oocyte meiosis once again becomes arrested at metaphase II until the mature egg is fertilized or parthenogenetically activated.

During the sexual cycle a close correlation exists between the sensitivity of oocytes to P and its circulating levels. *Bufo arenarum* oocytes in the breeding period (Fig. 4.6 A), for instance, exhibit a greater P sensitivity than in the hibernation period (Fig. 4.6 B), as revealed by the larger germinal vesicle breakdown (GVBD) percentages obtained after shorter incubation times and at lower P concentrations. Such differences may be caused by the high density of P receptors in fully- grown oocytes, which are characteristic of the preovulatory period.

The mechanism of P stimulation of oocyte nuclear maturation shows a sharp contrast with the well-known transcriptional regulation of gene expression by most other steroids, which bind to intracellular receptors within target cells. Progesterone fails to induce maturation when microinjected into amphibian oocytes but polymer-linked maturational steroids, which are unable to enter the cell across the plasma membrane, can induce maturation in *Xenopus* oocytes. These data point to the existence of a progesterone receptor in the plasma membrane, a fact confirmed by photoaffinity labeling with synthetic progestin and by radioreceptor binding assays (Liu and Patiño 1993; Bandyopadhyay *et al.* 1998).

Progesterone action seems to involve a series of transmembrane signs such as the inhibition of adenylate cyclase activity with a transient reduction in intracellular cAMP (Kwon *et al.* 1989) leading to a decrease in cAMP-dependent protein kinase A (PKA) activity. However, investigations suggest that oocytes may also contain a pathway independent of that involving cAMP. In fact, P reduces the activity of the phospholipase C (PLC) enzyme that hydrolyzes phosphatidylinositol 4,5-biphosphate (PIP_2) at the plasma membrane level. The products of PIP_2 hydrolysis are diacylglycerol (DAG), which is involved in

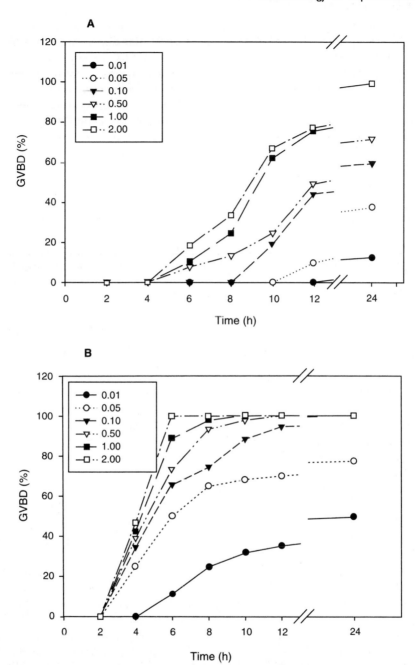

Fig. 4.6 Progesterone effect on nuclear maturation in *Bufo arenarum* fully-grown follicle oocytes. **A**. Follicles obtained from females captured during the hibernation period **B**. Follicles obtained from females captured during the breeding period. The results, expressed as percentage of germinal vesicle breakdown (GVBD), indicate a different oocyte sensitivity to the hormonal response between the two periods analyzed. From de Romero I.R. *et al.* 1998. BioCell 22: 27-34, Fig. 1A, 1B.

the regulation of protein kinase C (PKC) activity, and inositol 1,4,5 triphosphate (IP_3), which induces calcium release from intracellular deposits. Contradictory results have been reported concerning the role of PKC in amphibian oocyte maturation. In *Xenopus laevis* oocytes a decrease in PKC activity would be necessary for meiosis triggering (Varnold and Smith 1990) while in *Rana dybowskii* the activation of PKC increases the rate of progesterone-induced nuclear maturation (Kwon and Lee 1991). Although the second messenger IP_3 does not *per se* affect oocyte maturation, it increases P action, probably due to an increase in intracellular calcium ($[Ca^{2+}]_i$) levels. Now, while an increase in $[Ca^{2+}]_i$ caused either by progesterone exposure or by iontophoresis of Ca^{2+} into the oocyte cortex might be necessary and sufficient to induce oocyte maturation, other evidence has failed to show changes in $[Ca^{2+}]_i$ levels during oocyte maturation, thus suggesting that the cation may not be essential for progesterone action.

At least one additional protein kinase, the product of the c-mos proto-oncogene (Mos), seems to be involved in the initiation of oocyte maturation. Mos is a serine-threonine protein kinase that stimulates the mitogen-activated protein kinase (MAPK) pathway by direct phosphorylation and activation of the MAPK kinase (Matten *et al.* 1996). A positive feedback between MAPK and Mos apparently occurs during *Xenopus* oocyte maturation. The *de novo* synthesis of Mos and other proteins is required for the progression from meiosis I to the metaphase arrest at meiosis II. Therefore, one of the functions of MAPK during maturation may consist either in the stimulation of the synthesis or in the accumulation of Mos required for meiosis completion.

Whatever the signal ways involved in the induction of maturation, the process is finally triggered by the activation of a cytoplasmic factor, the maturation promoting factor (MPF), which phosphorylates targets responsible for GVBD and other events that mark the transition from meiotic prophase to metaphase. MPF, first purified from mature *Xenopus* oocytes (Lohka *et al.* 1988), has been identified as a protein kinase composed of a catalytic subunit p34^{cdc2} and a regulatory subunit, cyclin B. MPF activity is controlled by the phosphorylation and dephosphorylation of p34^{cdc2} after a complex formation with cyclin B. The MPF molecule and function are not species-specific, but the formation and activation of MPF differ between species. The mechanism can be divided in two types, the *Xenopus* and the goldfish type. The first involves an inactive MPF that is accumulated during terminal oocyte growth, forming a complex called preMPF. This preMPF is kept inactive by the inhibitory phosphorylation of p34^{cdc2} on threonine 14 and tyrosine 15. Under progesterone action the synthesis of Mos seems to be necessary for initiating oocyte maturation through thyroxine 15/threonine 14 dephosphorylation of preMPF. In the goldfish type, however, also suggested for other fishes and amphibians such as the frog *Rana japonica* and the toad *Bufo japonicus* (Tanaka and Yamashita 1995), preMPF is absent in immature oocytes. Gametes contain only monomeric cdc2 while cyclin B is not detectable. After hormonal induction of maturation, cyclin B is synthesized from its stored mRNA, binds to the pre-existing cdc2, is immediately activated through threonine 161 phosphorylation and activates MPF.

Consequently, with the exception of *Xenopus*, the Mos/MAPK kinase is not necessary for initiating maturation in amphibians (Yoshida *et al.* 2000).

During the preovulatory period high P levels together with a decrease in E_2 occur. E_2 diminution seems to be a requisite for meiotic resumption. *In vitro* experiments have shown that this hormone, although it does not by itself affect this process, is capable of producing an inhibition of progesterone-induced nuclear maturation (de Romero *et al.* 1998). The inhibitory effect of E_2 is probably due to its stabilizing effect on the viscoelastic properties of the ooplasm, thus blocking GV migration. E_2, then, seems to be more involved in the control than in the induction of maturation.

The nuclear maturation process can also be triggered by other hormones. Androgens, which exhibit high circulating levels during the preovulatory period, are important physiological mediators of oocyte nuclear maturation (Le Goascogne *et al.* 1985; Lutz *et al.* 2001; Ramos *et al.* 2001). In *Bufo arenarum*, DHT is effective in inducing meiotic resumption in a dose and time dependent manner. The oocyte hormonal response varies according to the phase of the sexual cycle, the highest sensitivity to DHT action occurring during the breeding period.

Androgen signaling seems to be mediated via a specific classical receptor isolated from the *Xenopus* oocyte cDNA library (Lutz *et al.* 2001) as well as through a transcription way known to be independent because its action is unaffected by actinomycin D. The way in which androgens mediate maturation is still open to further research. An unquestionably fact however, is that they produce cdc2 dephosphorylation and MAPK activation.

As regards the peptide hormones, it has been demonstrated that insulin plays an important role in the physiology of gonads, favoring the response of the ovary to gonadotropins and stimulating steroidogenesis. Conflicting data exist concerning insulin action on nuclear maturation. Studies carried out in oocytes from *Xenopus* and *Rana pipiens* (Lessman and Schuetz 1981) have proved that insulin can induce meiotic reinitiation by acting on specific low affinity receptors at the plasma membrane level. However, in *Xenopus* denuded oocytes, the data obtained indicate that insulin by itself does not affect nuclear maturation (Le Goascogne *et al.* 1985) but is able to increase P and T action. In fact, in both *Xenopus* denuded oocytes and in *Bufo arenarum* fully-grown follicle oocytes a synergistic effect of insulin and ovarian steroids (P, T or DHT) has been observed when they are present at subliminal doses (Fig. 4.7). It is important to note that in *Bufo arenarum* the association of progesterone or DHT with insulin is effective in inducing nuclear maturation in oocytes at the late vitellogenic stage, a period in which the steroids have no effect by themselves (Table 4.2). These results suggest that, although insulin potentiates the effect of P in oocytes in the last phases of oogenesis, at a previous stage in gonadal development it causes, by mechanisms as yet unknown, the acquisition of the capacity to respond to P or DHT (Ramos *et al.* 2001).

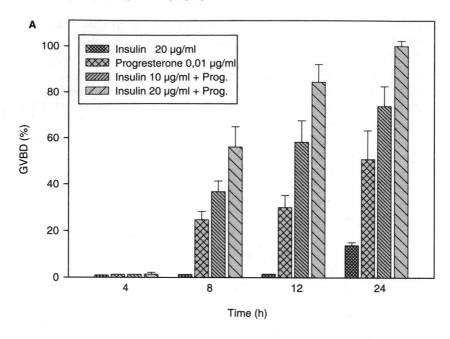

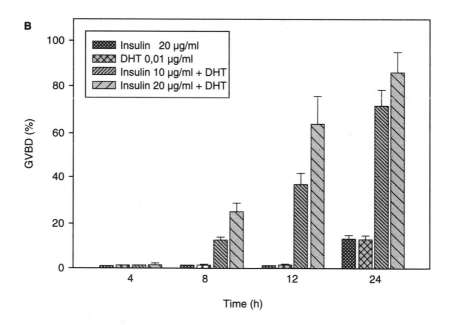

Fig. 4.7 Potentiating effect of insulin on DHT or progesterone induced nuclear maturation of *Bufo arenarum* fully-grown follicle oocytes. **A**. Insulin plus progesterone. **B**. Insulin plus DHT. Treatments were carried out using all hormones at subliminal doses. From Ramos I. *et al.* 2001. Zygote 9: 353-359, Fig. 2.

Table 4.2 Effect of the combination of insulin and progesterone on *Bufo arenarum* oocyte nuclear maturation at different folliculogenic stages.

	Progesterone (1 µg/ml)				Insulin (20 µg/ml) + progesterone (1 µg/ml)			
	4 h	8 h	12 h	24 h	4 h	8 h	12 h	24 h
Oocyte stage								
Late vitellogenic	0	0	0	0	0	0	40	80
Auxocytosis	0	75	100	100	57	100	100	100
Fully grown	28	97	100	100	85	100	100	100

The data are expressed as the percentage of germinal vesicle breakdown. From Ramos et al. 2001. Zygote 9: 353-359. Table 2.

4.7 OVULATION CONTROL

Ovulation is the physical process whereby the germ cell becomes separated from the follicle wall and, following its localized disintegration, the oocyte is extruded into the body cavity. Under physiological conditions, ovulation is temporally coupled to maturation; thus, it is restricted to those fully-grown oocytes that have undergone cytoplasmic and nuclear maturation. Ovulation shows a close relationship with seasonal variations and often occurs in spring, when environmental conditions are appropriate to insure normal embryonic development.

The hypothalamic-pituitary axis is involved in the control of ovulation. In fact, during the hibernation period, it has a low degree of activity, probably due to the inhibitory action of the pineal gland (de Atenor *et al.* 1994) or by a dopaminergic inhibition as reported for *Rana temporaria* (Sotowska-Brochocka *et al.* 1994). During the last third of hibernation a reduction in the central nervous system inhibition results in the release of GnRH; concurrently, pituitary LH contents and circulating LH levels gradually begin to rise. In *Rana catesbeiana* this LH surge is accompanied by an abrupt increase in progesterone levels followed by ovulation. In the gonadotropin control of ovulation the cAMP-dependent protein kinase and intracellular free calcium seem to be involved as second messengers (Skoblina 2000).

As ovulation requires the rupture of the ovarian wall in a proteolytic cascade, this process has been compared with the inflammatory response in which corticosteroids and prostaglandins (PGs) are involved. The ovary of anurans such as *Rana esculenta* releases $PGF_2\alpha$ and PGE_2, whose basal levels reveal seasonal changes, these PGs exhibiting an opposite effect on reproductive functions. During the preovulatory period PGE_2 reaches maximum concentration, exerting an inhibitory effect on ovulation, while $PGF_2\alpha$, with high levels during the ovulatory phase, can favor ovulation by increasing ovarian corticosteroids (Gobbetti and Zerani 1993).

During ovulation, anuran oocytes surrounded only by the vitelline envelope are released into the coelomic cavity. These oocytes cannot be fertilized under normal inseminating conditions. Only after their transit along the oviduct do they acquire the capacity to adequately fuse with the male gametes.

4.8 ANURAN OVIDUCT

4.8.1 Organization and Function

In several anuran species the success of fertilization depends on the products secreted around the eggs during their transit through the oviduct (Barbieri and Budeguer de Atenor 1973; Miceli *et al.* 1978; Fernández *et al.* 1984; Katagiri 1987; Omata 1993), which is structurally and functionally divided into three main zones: pars recta (PR), which collects the ovulated eggs through an open free ostium in the coelomic cavity, pars convoluta (PC), which represents the main body of the oviduct, and ovisac or uterus, where eggs accumulate before oviposition (See Tyler, chapter 2 of this volume, Fig. 2.2) (Moreno 1972; Lofts 1974).

In these vertebrates two different types of secretion, each produced by a particular oviductal zone and each having a different biological role, have been identified. The first one, secreted at the PR level, is a low viscosity product that contains a Ca^{2+} dependent trypsin-like serine protease enzyme named oviductin (Hardy and Hedrick 1992) which, through mild proteolysis, modifies the egg vitelline envelope at the structural and molecular levels, rendering it susceptible to sperm lysin and sperm penetration (Miceli and Fernández 1982; Katagiri *et al.* 1982; Bakos *et al.* 1990).

It has been demonstrated that a prerequisite for the acquisition of eggs fertilizability is either their passage through the PR, where they are bathed in a secretion containing the proteolytic enzyme, or their *in vitro* treatment with an extract of PR tissue or with its secretion fluid prior to insemination. This biological activity, observed for the first time in *Bufo arenarum* (Miceli *et al.* 1978), has also been confirmed in other anuran species including *Rana japonica* (Yoshizaki and Katagiri 1981), *Xenopus laevis* (Grey *et al.* 1977) and *Bufo japonicus* (Katagiri 1987).

The second secretion is jelly, a highly viscous material normally composed of multiple layers sequentially deposited on the eggs during their passage through the PC. This envelope, which is a component of the extracellular matrix, has been reported as a very important factor for successful gamete encounter (Del Pino 1973; Elinson 1974; Hedrick and Nishihara 1991). The proposed biological roles for jelly layers include the blocking of polysperm (Del Pino 1973), the prevention of species cross-fertilization (Elinson 1974; Katagiri 1987; Whitacre *et al.* 1996) and the induction of the sperm acrosome reaction (Raisman and Cabada 1977; Ishihara 1984; Miceli *et al.* 1987).

Despite the important biological functions attributed to oviductal secretions, the analyses of the cells involved in this process are scarce and limited either to the anterior third of the PC in *Rana japonica* (Yoshizaki and Katagiri 1981) and *Bufo japonicus* (Katagiri *et al.* 1982) or to the epithelial layer in *Xenopus laevis* (Yoshizaki 1985). The first exhaustive studies of the ultrastructural organization of the anuran oviduct were made in *Bufo arenarum* (Fernández *et al.* 1989a; 1997; Winik *et al.* 1999). The data obtained there demonstrate that the PR has a simple organization based only on a single epithelial layer that comprises an approximately equal number of both ciliated and secretory cells that follow a characteristic alternation pattern (Fig. 4.8 A).

Ciliated cells contain a large number of mitochondria in the apical area in close proximity to numerous glycogen granules, an organization that appears to respond predominantly to a regulatory role of cilia activity (Fig. 4.8 B). Secretory cells exhibit granules that differ from each other mainly in size and electron density (Fig. 4.8 C), indicating the presence of secretion products at various degrees of concentration that, when released into the lumen by exocytosis, give rise to the PR secretion responsible for biological activity (Fernández *et al.* 1989a). Confirming the nature of the material stored in the granules, in *Bufo japonicus* a protease with enzymatic properties similar to that of *Bufo arenarum* has been obtained from the isolated secretory granules of the

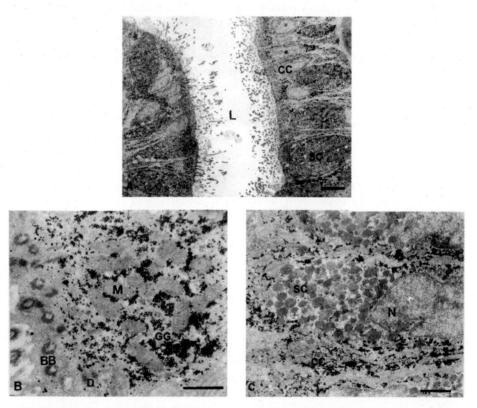

Fig 4.8 Electron micrographs of *Bufo arenarum* oviduct showing the PR zone during the preovulatory period. **A.** Low magnification view of epithelial layer showing secretory cells (SC) alternated with ciliated cells (CC). L, lumen. Scale bar 5 μm. **B.** Detail of a ciliated cell showing numerous mitochondria (M) and glycogen granules (GG). D, desmosomes. Basal bodies of the cilia (BB) appear in the apical area of the cell. Scale bar 2 μm. **C.** Detail of a secretory cell (SC) containing secretory granules which varies in size and electron density. CC, ciliated cell; N, nucleus. Scale bar 0.5 μm. From Fernández S.N. *et al.* 1989a. Microscopía Electrónica y Biología Celular 13: 211-220, Figs. 1, 3, 2.

oviductal PR (Takamune and Katagiri 1987). In disagreement with the data from *Rana japonica* (Yoshizaki and Katagiri 1981) suggesting that the only source of the material secreted in the PR is provided by the contents stored in the secretory granules, *Bufo arenarum* PR secretion is known to be composed not only of proteins synthesized and released by the epithelial secretory cells (Mansilla-Whitacre *et al.* 1992) but also of blood serum proteins, which also appear to be involved in the molecular events that lead to fertilization (Llanos *et al.* 1998).

In the PC, unlike those in the PR, the cells involved in the secretion of the different components that form the jelly coats are located at the epithelial as well as at the glandular level, those in each site showing distinctive characteristics (Winik *et al.* 1999). Moreover, as shown by ultrastructural analyses, the PC has secretory cell types different from those in the PR. A prominent feature of epithelial secretory cells lies in their secretory granules, which vary in their distribution, shape, size, arrangement of contents and electron density (Fig. 4.9 A), thus originating an ultrastructural mosaic along the whole length of the oviduct. The morphological heterogeneity of these granules, confirmed by histochemical light microscopic studies, can be ascribed to contents of a diverse nature . They are distributed following a characteristic pattern between epithelial and glandular secretory cells, with marked differences for each oviductal zone and period of the sexual cycle. Glandular secretory cells, on the contrary, exhibit granules of a homogeneous content characterized by the presence of one or more cores (Fig. 4.9 B).

These data demonstrate that the ultrastructural organization of *Bufo arenarum* oviductal secretory cells, their distribution pattern, and the diversity of the secretion products stored in secretory granules increase in complexity from the PR to the PC. This is in agreement with the provision of two secretions with a well defined different physicochemical composition: the simple, non structured PR secretion and the complex macromolecular composition and structural organization of the different jelly coats formed by the interaction of heterogenous products sequentially wrapped around the eggs as they traverse the PC (Hedrick and Nishihara 1991; Bonnell and Chandler 1996).

Once organized, jelly envelopes have a complex structure composed of a fibrillar structural stable matrix of high molecular weight glycoconjugates and a globular material consisting of low molecular weight proteins, some of which can diffuse from the jelly matrix into the surrounding medium during spawning (Barbieri and Del Pino, 1975; Bonnell *et al.* 1996).

Other important diffusing components present in the jelly are the divalent cations Ca^{2+} and Mg^{2+}, reported at this moment only for *Bufo japonicus* and *Bufo arenarum* (Ishihara *et al.* 1984; Medina *et al.* 2000). Ultracytochemical studies in *Bufo arenarum* have indicated for the first time that in the PC both epithelial (Fig. 4.10 A) and glandular (Fig. 4.10 B) secretory cells exhibit prominent Ca^{2+} deposits. This cation is stored in the cellular structures that, through different secretion mechanisms, release their products toward the oviductal lumen to form the jelly coats. Ca-ATPase activity outlining the granular and plasma membranes at both the apical and the basal levels (Figs. 4.10 C, D) is an

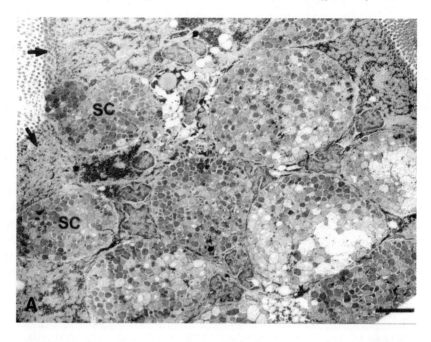

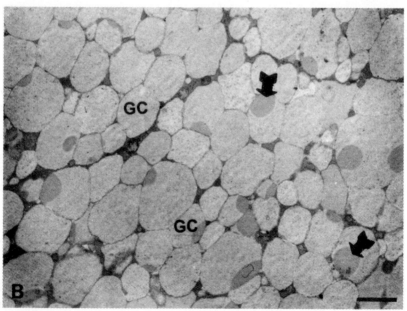

Fig 4.9 Electron micrographs of *Bufo arenarum* oviduct showing the PC zone during the preovulatory period. **A.** Panoramic view of the epithelial layer. Secretory cells (SC) are interspersed with ciliated cells (arrows). Scale bar 4 μm. **B.** Glandular cells (GC) containing large granules of moderate electron density, several of them with prominent cores (arrows). Scale bar 2 μm. From Winik B.C. *et al.* 1999. Journal of Morphology 239: 61-73, Figs. 2, 19.

indicator of Ca^{2+} transport from the extracellular fluids throughout the secretory cells toward the oviductal lumen (Medina *et al.* 2000), as is the case in avian oviducts (Wasserman *et al.* 1991). Immunohistochemical techniques using a monoclonal antibody (5F10) produced against the human erythrocyte calcium pump have confirmed that the Ca-ATPase activity present in *Bufo arenarum* oviducts corresponds to a calcium pump (Medina *et al.* 2000).

While PR secretion has been extensively studied, the exact role of the jelly components is still uncertain (Katagiri 1987; Hedrick and Nishihara 1991; Bonell and Chandler 1996; Bonnel *et al.* 1996). At present there is no conclusive evidence as to whether structural or diffusible molecules exert a biological effect on the sperm-egg interaction process. Studies in *Bufo arenarum* (Barbieri and del Pino 1975) and *Bufo japonicus* (Katagiri 1973) demonstrate that the loss of diffusible jelly components causes eggs to be refractory to fertilization. Additional findings have led to the conclusion that not only diffusible components but also the structural matrix are essential for fertilization (Barbieri and del Pino 1975). In *Xenopus laevis*, small diffusible proteins, present in both inner and outer jelly layers, also have a fertilization-promoting activity, while the macromolecules that form the structural matrix appear to be inactive (Olson and Chandler 1999). Experiments employing dejellied uterine eggs of the toad *Bufo japonicus* suggest that the glycoproteins in the jelly plays a role in fertilization through their single capacity to bind the divalent cations Ca^{2+} and/ or Mg^{2+}, which are essential for the induction of the sperm acrosome reaction (Ishihara *et al.* 1984). This hypothesis has been experimentally tested only in *Bufo arenarum*. Recent data obtained by means of fertilization and ultrastructural assays demonstrate that the Ca^{2+} present in the jelly actually triggers the acrosome reaction during the passage of the sperm through the jelly.

4.8.2 Annual Oviduct Cycle. Its Connection with the Ovarian Cycle and with the Circulating Sex Steroid Hormones

In annually reproducing anuran females, the oviduct undergoes a cycle which is secondary to the ovarian one (Fernández *et al.* 1984; Iela *et al.* 1986), showing pronounced seasonal changes that comprise structural and functional aspects (Licht *et al.* 1983; Fernández *et al.* 1984; 1989a; Winik *et al.* 1999).

Studies in *Bufo arenarum* have shown that the differentiation of secretory cells as well as the biological functions of oviductal secretions are closely correlated with the ovulatory process (Fernández *et al.* 1984; 1989a; 1997; Winik *et al.* 1999). Adult females with a mature ovary, captured during the preovulatory period, show a very well developed PR epithelium containing a large number of secretory cells (Fig. 4.8 A). They exhibit an optimum development of the organelles related to the biosynthetic activity, in agreement with the image of their cytoplasm filled with secretory granules. At this period, the secretory capacity of these cells is evinced by exocytosis processes together with the presence in the lumen of a secretion product consisting of flocculent material and membranous vesicles (Fernández *et al.* 1989a). Some similarities exist

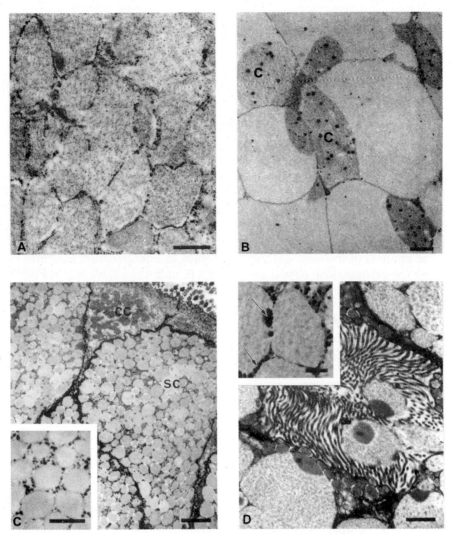

Fig. 4.10 Ultrastructural localization of calcium and active Ca-ATPase in *Bufo arenarum* PC during the preovulatory period. **A**. Presence of calcium in a secretory epithelial cell. Detail of the secretory granules showing electron-dense calcium deposits in their structural mesh and in the membranes that surround them. Scale bar 0.5 μm. **B**. Localization of calcium in the glandular cell. Conspicuous electron-dense deposits are located on the secretory granules, especially in the core (C). Scale bar 0.5 μm. **C**. Ca-ATPase activity in the secretory epithelial cells (SC) and ciliated cells (CC). Conspicuous electron-dense deposits indicate the presence of Ca-ATPase activity on the plasma and secretory granule membranes. Scale bar 2 μm. Inset: detail of secretory granules which exhibit a punctuate distribution of dense deposits on the limiting membranes. Scale bar 0.5 μm. **D**. Ca-ATPase activity in the glandular cells. Conspicuous electron-dense deposits indicate the presence of Ca-ATPase activity on the membranes of the secretory granules and also on the surface of the microvilli. Scale bar 1μm. Inset: arrows indicate the reaction products on the membranes that limit the secretory granules. Scale bar 0.5 μm. From Medina M.F. *et al.* 2000. Acta Histochemica and Cytochemica 33 (1) 49-58, Figs. 2, 7, 4, 8.

between these results and the ultrastructure reported for *Rana japonica* (Yoshizaki and Katagiri 1981) *Bufo japonicus* (Katagiri *et al*. 1982) and *Xenopus laevis* (Yoshizaki 1985).

In connection with this structural development, maximum PR activity (Fig. 4.11), that is, the capacity to induce the fertilizability of coelomic eggs, has been detected (Fernández *et al*. 1984).

In the PC, epithelial (Fig. 4.9 A) as well as glandular (Fig. 4.9 B) secretory cells show a fully differentiated state with secretory cells containing a large amount of secretory granules and a marked secretion at the lumen level. This secretion is composed of intact secretory granules released from the glandular cells and a flocculent material released from the epithelial cells by exocytosis (Fig. 4.12 A). At this time, before the passage of the eggs, both components are found in a disorganized pattern. Frequently, some apical cytoplasmic protrusions (Fig. 4.12 B) and entire cells (Fig. 4.12 C) are released through apocrine and holocrine mechanisms (Winik *et al*. 1999).

The analysis of the seasonal steroid profiles in *Bufo arenarum* indicates that during the preovulatory period serum T, DHT and P reach highest circulating levels while E_2 shows the lowest values detected during the cycle (Fernández *et al*. 1984). Under these hormonal conditions the whole oviduct achieves maximum development and differentiation of secretory cells and appropriate secretion into the lumen, thus assuring optimal conditions for eggs transport and acquisition of the capacity for later fertilization.

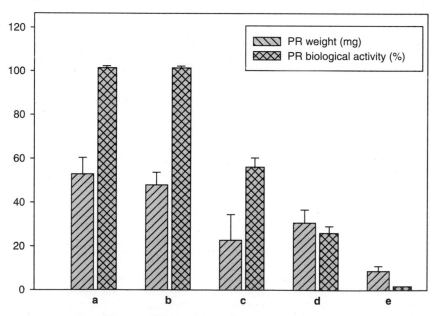

Fig. 4.11 Relation between PR weight of *Bufo arenarum* females and its biological activity during the ovarian cycle: a- preovulatory period; b- immediately after ovulation; c- early postovulatory period; d- late postovulatory period; e- eight weeks after ovariectomy. Original.

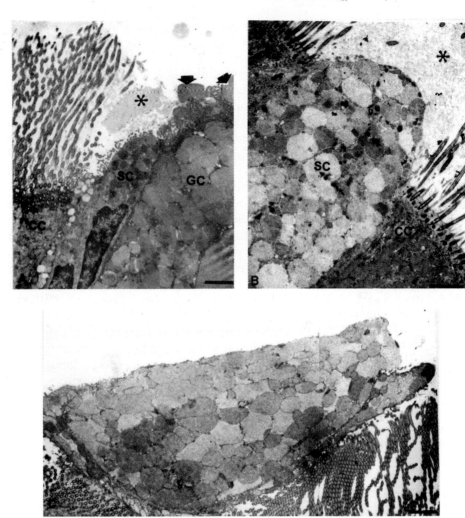

Fig. 4.12 Electron micrographs of *Bufo arenarum* oviduct. **A.** Glandular cell (GC) of the PC. Secretory granules (arrows) are released directly into the oviductal lumen. *, flocculent material; SC, secretory epithelial cell; CC, ciliated cell. Scale bar 2 μm. **B.** Apical protrusion of PC epithelial secretory cell (SC) during the preovulatory period. CC, ciliated cell. Flocculent material is present in the oviductal lumen (*). Scale bar 2 μm. **C.** Whole PC epithelial secretory cell released into the oviductal lumen through a holocrine mechanism. Scale bar 2 μm. From Winik B.C. *et al.* 1999. Journal of Morphology 239: 61-73, Figs. 4, 5, 21.

During the breeding period, characterized by a marked fall in androgens and P circulating levels and a progressive and steady increase in E_2, associated with the ovulation, differences between *Bufo arenarum* PR weight before and after spawning are not significant (Fig. 4.11). In this connection, although the presence of a larger surface of communication between these cells and the lumen is frequently observed, ultrastructural analyses indicate that only a slight amount of secretion is released to the lumen during the passage of eggs, as indicated by the small reduction in the number of granules within secretory cells (Fig. 4.13 A). This observation is in disagreement with the structural studies in *Rana japonica* and *Bufo japonicus*, since they reveal a severe depletion of secretory cell contents after ovulation, as opposed to the PR of preovulatory females (Yoshizaki and Katagiri 1981; Katagiri *et al.* 1982).

In *Bufo arenarum*, in accordance with the slight decrease in the amount of secretory granules, PR biological activity does not disappear immediately after the completion of spawning (Fig. 4.11). This activity persists without obvious changes during the period in which the largest serum levels of E_2 are detected. A slow ensuing reduction in PR activity with concomitant regressive morphological changes occurs at the end of the early postovulatory period, in association with a gradual decrease in E_2 concentrations. A maximum percentage of animals lacking an active PR and showing an atrophy of the epithelium, characterized by the presence of scattered secretory cells in a dedifferentiated state (Fig. 4.13 B), is usually reached approximately 6 months after the breeding period, together with a marked reduction in E_2 circulating concentrations. (Fernández *et al.* 1984).

In the PC, on the contrary, epithelial and glandular secretory cells undergo marked changes intimately connected with the ovulatory process. Such changes, which include a notable reduction in the number of granules and a concomitant diminution in cell volume (Fig. 4.13 C), finally determining an abrupt fall in oviductal weight, are likely to occur due to an additional abundant secretion into the lumen during the transit of the oocytes. The release of granular contents is probably caused by hormonal and mechanical stimuli (Winik *et al.* 1999). This secretion, which takes place when circulating androgens and P show high levels, completes the process started in the preovulatory period and ultimately leads to the formation of the jelly coats around the eggs through a still unknown mechanism.

The oviduct begins to increase its weight again during the late postovulatory period (Fig. 4.11), with a progressive proliferation and differentiation of secretory cells. The development and the biological activity of the oviduct reach a maximum prior to breeding, when E_2 begins to rise and androgens and P exhibit maximum concentrations.

4.8.3 Hormonal Regulation of the Functional Morphology of Anuran Oviduct

The seasonal morphological and functional changes exhibited by the oviduct, which depend on the ovarian cycle, point to the existence of steroid hormone modulation (Fernández *et al.* 1989a). This is supported by structural studies which demonstrate that ovariectomy, in accordance with a strong reduction in ovarian steroid levels, causes a regression in *Bufo arenarum* oviducts, which acquire the characteristics of a nonfunctional or quiescent state similar to that observed during the normal sexual cycle, approximately 6 months after ovulation (Fernández *et al.* 1984; 1989a).

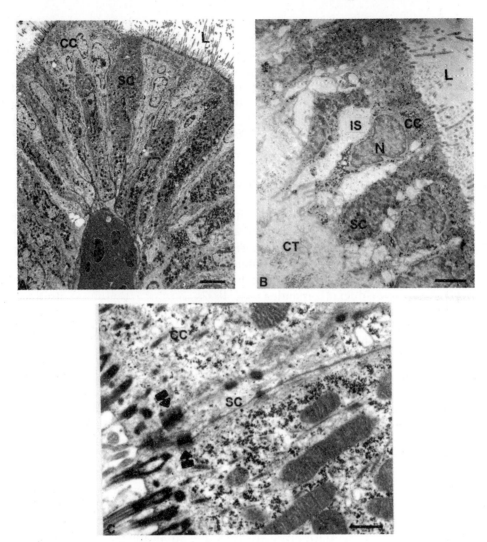

Fig. 4.13 Electron micrographs of *Bufo arenarum* oviduct. **A.** Low magnification view of PR epithelial layer during the early postovulatory period, showing epithelial secretory cells (SC) and ciliated cells (CC). L, lumen. Scale bar 5 μm. **B.** PR epithelium of an adult female in the late postovulatory period. Ciliated cells (CC) are severely reduced in height and scarce secretory cells (SC) are observed. L, lumen; N, nucleus; IS, Intercellular space; CT, connective tissue. Scale bar 3 μm. **C.** PC epithelial layer from the early postovulatory period showing a marked volume decrease in the secretory cells (SC) determined by the release of secretory granules. CC, ciliated cell; arrows, desmosome. Scale bar 0.5 μm.

A and B from Fernández S.N. *et al.* 1989a. Microscopia Electrónica y Biología Celular 13 (2): 211-220. C original.

The cytological changes that characterize this process have been exhaustively investigated in *Bufo arenarum* oviduct (Fernández *et al.*, 1989b). The most evident indications of regression are a reduction in the height of both epithelial and glandular layers and a marked loss of secretory cells, whose number is clearly lower (Fig. 4.14 A) than that of non-ovariectomized control females. The remaining secretory cells display evident signs of involution, with very few secretory granules in the cytoplasm (Fig. 4.14 B). The reduced secretion released to the oviductal lumen shows scattered particulate material and scanty flocculent products, while macrophages and cellular debris as well as the extrusion of whole cells are also frequently observed (Fig. 4.14 C).

The mechanism of regressive effects and loss of secretory cells is unknown. Although hormonal withdrawal induces programmed cell death in cycling tissues such as mammalian uterine epithelium (Rotello *et al.* 1992), this effect has not been reported in the amphibian reproductive system. Although no conclusive evidence exists concerning programmed cell death, signs of involution, chromatin condensation, extensive vacuolation and cell membrane blebbing suggest the possibility of induction of apoptosis by ovariectomy in *Bufo arenarum* oviducts (Fernández *et al.* 1997).

As regards PR, its biological activity declines at different rates according to the stage of the sexual cycle at which ovariectomy is performed. When this procedure is carried out during the fall-winter season, which comprises the hibernation period and the preovulatory period, the greater percentage of animals with inactive PR is obtained, concurrently with a sharp decrease in E_2 serum levels. Furthermore, PR activity disappears at a lower rate in normally spawned animals than in ovariectomized ones, which show only negligible circulating levels of E_2. Taken together, these data indicate that in *Bufo arenarum* the endogenous concentration of this hormone is an important regulator of PR activity.

The administration of androgens and E_2, but not P, induces the proliferation of PR secretory cells with the concomitant reappearance of dense secretory granules and the optimum development of the machinery required for the protein synthesis characteristic of active cells (Fernández *et al.* 1989a; b). Concomitantly, these hormones are effective in inducing PR activity in ovariectomized toads (Fernández *et al.* 1984). Thus, it is evident that both E_2 and androgens are required for the development and maintenance of a fully functional epithelium but, while DHT is more effective in evoking cellular proliferation and differentiation as well as ciliogenesis induction, E_2 is a more potent inducer of biological activity. In support of the above, in *Bufo arenarum* the effect of T and DHT on cellular differentiation is noticeable greater than in animals treated with E_2 (Fig. 4.15 A). Moreover, androgens, but not E_2, are capable of inducing the development of different secretory cell types (Fig. 4.15 B), giving a clear indication that a progressive specialization occurs in the epithelial layer as a result of androgen action. Androgens are the best indicators of oviductal stage as shown by the strong positive correlation found between the levels of these hormones and oviduct weight.

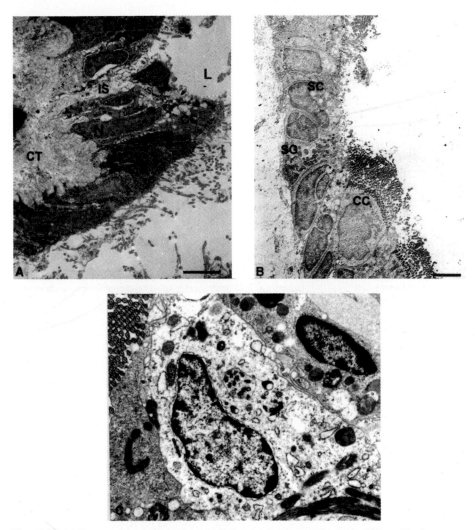

Fig. 4.14 Ultrastructural features of *Bufo arenarum* oviduct. **A.** PR epithelium two months after ovariectomy performed in the winter season. This layer, severely reduced in height, is constituted almost exclusively by ciliated cells (CC). N, nucleus; L, lumen; IS, intercellular spaces; CT, connective tissue. Scale bar 3 µm **B.** Panoramic view of the PC of an ovariectomized female showing a marked decrease in the height and development of the epithelial layer. Note the reduced number of secretory cells (SC) with a scarce amount of secretory granules (SG). Ciliated cells (CC). Scale bar 4 µm. **C.** PC section of an ovariectomized female showing whole cells released into the lumen. Scale bar 1 µm. Original.

These data agree with those reported for *Rana catesbeiana* (Licht *et al.* 1983), but differ from the ones found in *Pachymedusa dacnicolor*, in which oviductal weight shows a strong positive correlation with both T and E_2 (Iela *et al.* 1986) and in *Rana cyanophyctis* (Pancharatna *et al.* 2001) in which, although androgens are effective as inducers of oviductal growth, E_2 shows maximum effect.

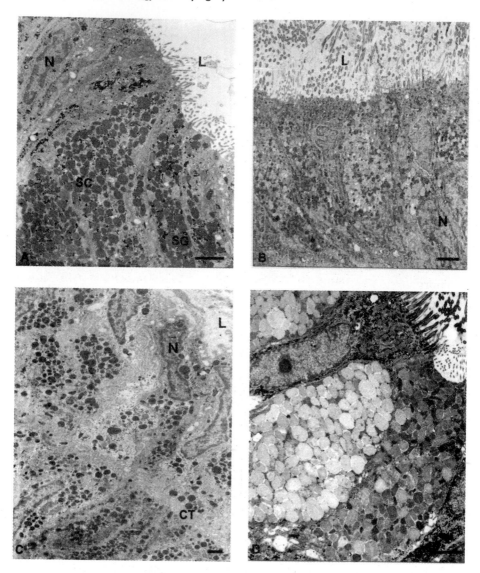

Fig. 4.15 Epithelium of *Bufo arenarum* PR after ovariectomy and hormonal treatments. **A.** Effect of estradiol-17β administration. Note the reappearance of secretory cells (SC) with secretory granules (SG). N, nucleus; L, lumen. Scale bar 2 μm. **B.** Under DHT treatment the epithelial layer has reverted the effect of ovariectomy, showing abundant secretory cells. L, lumen; N, nucleus. Scale bar 2 μm. **C.** PR epithelium after progesterone treatment. Its disorganized aspect evinces the inability of the hormone to revert the effect of ovariectomy. L, lumen; N, nucleus; CT, connective tissue. Scale bar 1 μm. **D.** PR epithelium from females treated with estradiol-17β plus progesterone evidencing different types of well-developed secretory cells. Scale bar 1.5 μm. From Fernández S.N. *et al.* 1989b. Microscopía Electrónica y Biología Celular 13: 201-210, Figs. 3, 7, 4, 9.

In spite of the important roles attributed to androgens in female reproductive physiology, there is little information at present about the basic mechanism that mediates hormonal events. Taking into account that T can stimulate oviduct growth in *Rana pipiens* whereas the non aromatizable androgen DHT cannot, it has been suggested that the promoting effect of androgens depends on their aromatization in the oviductal tissue to locally active estrogens (Dubowsky and Smalley 1993). *Rana pipiens* oviducts also contain an aromatase with low activity that shows characteristics and kinetic parameters similar to those of human placental aromatase (Kobayashi *et al.* 1996). However, it should be borne in mind that the possible aromatization of T may not be a significant physiological event in all anuran species. In *Bufo arenarum*, for instance, statistical analyses have shown a negative correlation between E_2 and oviductal weight. In agreement with that, E_2 induces only a moderate effect on PR growth and differentiation in ovariectomized animals (Fig. 4.15 A), far lower than that exerted by androgens. On the other hand, the results obtained with DHT, which cannot be aromatized to E_2, suggest a direct growth-promoting effect of this hormone on the oviduct.

As regards the main role of E_2 on the PR, that is, the induction of biological activity, the most remarkable effect of this hormone at the ultrastructural level is to promote the development of a functional endoplasmic reticulum and Golgi complexes that characteristically show the presence of intracisternal material and abundant vesicles (Figs. 4.16 A, B). In agreement with that, the induction of activity can be blocked by inhibitors of protein synthesis (Fernández *et al.* 1984). Additional findings have demonstrated that E_2 induces *de novo* synthesis of PR proteins from radioactive precursors both *in vivo* and *in vitro*, while ovariectomy causes a decrease in the process of synthesis (Mansilla-Whitacre *et al.* 1992). The newly synthesized proteins are released *in vivo* into the PR fluid and *in vitro* into a culture medium, both secretions showing the biological activity characteristic of the PR secretion fluid from the preovulatory period. These data indicate that protein synthesis is a key step in the action of estrogens on the PR.

Although results indicate that P fails to restore PR activity in ovariectomized animals, the hormone is effective in non-ovariectomized, normally lacking PR activity animals captured during the fall-winter season. Interestingly, the E_2 serum levels in these animals is approximately 3 ng/ml while that in ovariectomized animals is about 0.9 pg/ml (Fernández *et al.* 1984). Consequently, E_2 acts as a priming factor to sensitize the oviduct to P action.

Confirming the above, P is ineffective in counteracting the effect of ovariectomy at the structural level (Fig. 4.15 C), whereas pretreatment with E_2 before P administration causes a marked growth of the PR epithelium (Fig. 4.15 D). This acquires the characteristic appearance of epithelium in non-ovariectomized animals from the preovulatory period, with well developed secretory cells and a marked release to the lumen of the products stored in the secretory granules (Fernández *et al.* 1989b).

The hormonal regulation of the cytodifferentiation and of the secretion of jelly coat components shows a different pattern in the PC. In fact, though the

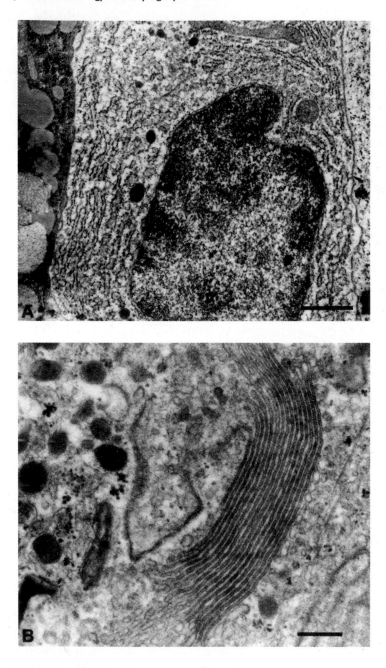

Fig. 4.16 *Bufo arenarum* PR epithelium from ovariectomized females after estradiol-17 β treatment. **A.** Epithelial secretory cell showing dilated profiles of the rough endoplasmic reticulum. The cisternae exhibit a fine granular material of moderate electron density. Scale bar 1 μm. **B.** A well-developed Golgi complex with numerous vesicles in close proximity. Scale bar 0.3 μm. Original.

administration of E_2, P or E_2 plus P provides the stimulus that triggers off the differentiation of secretory cells, the effectivity of these hormones is greater in the initial segment of the PC, responsible for the secretion of the inner jelly coats. Treatment with DHT is more effective than with other hormones, determining the maximum growth and differentiation in all PC zones.

An interesting finding is that, while PR is refractory to P alone, this hormone is an effective promoter of PC differentiation. Thus, the two zones of *Bufo arenarum* oviducts show a different reactivity to P.

Electron microscopy suggests that in the PC each hormone produces a distinctive effect. Thus, while E_2, as is the case in the PR, induces the development of the organelles involved in protein synthesis and the storage of products in secretory granules, P alone or combined with E_2 acts as a secretagogue, inducing the release of flocculent and particulate material from epithelial as well as glandular secretory cells. In a way similar to this effect determined by exogenous treatment, in the sexual cycle the secretion of jelly components occurs before and during egg transit through the oviduct, a period characterized by the highest circulating levels of P and androgens.

DHT stimulates the release of a secretion that can be observed before the passage of the eggs in a partially organized condition at the lumen level. DHT also induces an over proliferation of cilia (Fig. 4.17 A) (Fernández *et al.* 1997). In this respect, exogenous hormonal treatment supports the observations performed during the sexual cycle. Thus, maximum secretory activity and ciliogenesis occur during the preovulatory and ovulatory periods, when P and DHT reach highest levels.

Both processes are important features in anuran oviducts. In fact, ciliary movement is responsible for the propelling and rotation of the eggs along the oviduct, first facilitating their contact with PR secretion and then their sequential jelly coating, both secretions containing the components necessary for fertilization. Although secretion during the preovulatory period is only under hormonal control, the release of secretory products during the passage of the eggs probably involves an additional mechanical stimulus. The relative importance and contribution of each factor in controlling the discharge of the secretory granules during spawning is a matter for further research.

Two distinctive features of *Bufo arenarum* oviductal PC are the variation in the amount of calcium deposits stored in the secretory granules and the changes in the Ca-ATPase activity that occur during the sexual cycle (Medina *et al.* 2000).

Ultracytochemical observations clearly indicate that both parameters reach their maximum level at the preovulatory period, a marked reduction being observed after the eggs traverse the oviduct. This condition is progressively reverted in the late postovulatory period during hibernation, maximum calcium storage in co-localization with strong Ca-ATPase activity being reached immediately before ovulation. During the preovulatory period, the presence in the lumen of whole secretory cells, portions of them and flocculent material containing conspicuous Ca^{2+} deposits indicate that holocrine, apocrine and

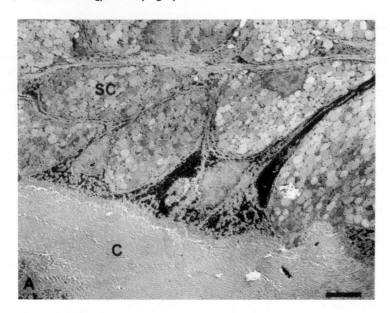

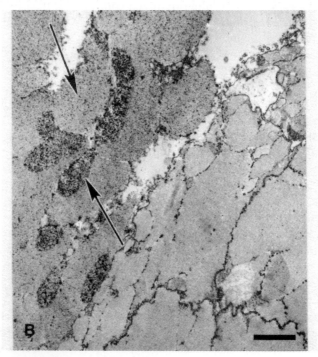

Fig. 4.17 **A.** *Bufo arenarum* oviductal PC from an ovariectomized and DHT treated female. Note the full development of the secretory cells (SC) and the marked ciliogenesis (C). Scale bar 4 μm. **B.** Localization of calcium in the *Bufo arenarum* PC lumen during the preovulatory period. Note the calcium deposits (arrows) in the glandular secretory granules released into the lumen. Scale bar 2 μm. Original.

exocytosis mechanisms are involved in the cation release (Fig. 4.17 B). Besides, the Ca^{2+} contents present in the jelly envelopes, which are higher than those at the serum level (Fig. 4.17), suggest that when eggs traverse the oviduct an active transport mechanism associated with the above secretion processes takes place between the serum compartment and the oviducal lumen.

Serum concentrations of Ca^{2+} show a profile strongly associated with the variations in the levels of circulating E_2, with minimum values in the preovulatory period, a progressive increase in the oviposition period and maximum levels during vitellogenesis. Similar seasonal fluctuations have been determined in *Rana pipiens* (Robertson 1977). In *Bufo arenarum*, the hormonal relation is further supported by the fact that ovariectomy diminishes the serum calcium levels (Fig. 4.18) (Medina *et al.* 2000). In agreement with the above, previous observations in *Rana temporaria* have shown that E_2 raises the Ca^{2+} contents in serum, this increase being connected with yolk formation (Pasanen and Koskela 1974).

From these results it is evident that there is a close relationship between oviductal calcium contents, E_2 serum concentrations and the stage of the sexual cycle which, associated with the annual cycle changes in circulating Ca^{2+} levels, indicate a possible hormonal regulation of this cation.

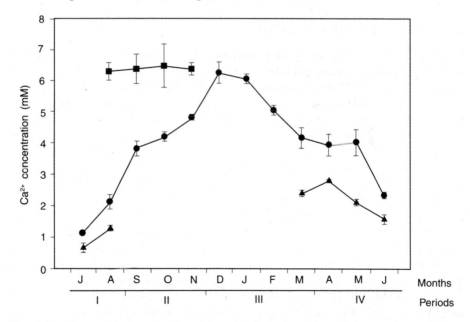

Fig. 4.18 Seasonal changes in calcium levels in serum and jelly envelopes (JE) of *Bufo arenarum* females during the sexual cycle. •, serum of adult control females obtained from each animal immediately after capture; ■, serum of ovariectomized adult females obtained 25 days after ovariectomy; ?, JE obtained from strings of eggs during the ovulatory period. Sexual cycle periods: I, preovulatory period; II, ovulatory period; III, postovulatory period. From Medina M.F. *et al.* 2000. Acta Histochemistry et Cytochemistry 33 49-58, Fig. 17.

Confirming the above, ovariectomy causes a clear reduction in Ca^{2+} deposits and Ca-ATPase activity in secretory cells from the epithelial and glandular layers. The effect of ovariectomy is partially counteracted by treatment with DHT, the administration of E_2, P or the combination of both hormones being more effective than DHT.

Although Ca^{2+} deposits can be observed throughout the oviduct, the detection of high amounts of this cation and the strong reaction indicating Ca-ATPase activity in the segments that secrete the products which form the inner jelly layers necessary for successful fertilization in *Bufo arenarum* (Barbieri and Budeguer de Atenor 1973) are interesting findings. In fact, according to the evidence collected up to the moment, it seems reasonable to suggest that the high concentrations of Ca^{2+} present in the secretory cells responsible for the formation of inner layers is released by different mechanism to the lumen together with the other jelly components. The sequential secretion determines that, once the jelly is structured, the innermost layers contain the greater concentration of Ca^{2+}. Thus, jelly envelopes could provide an appropriate calcium gradient during the passage of sperm. This gradient may act as an inducer of the acrosome reaction that is completed at or near the surface of the vitelline envelope which, sensitized to acrosomal enzymes by the PR oviductin, becomes penetrable by sperm, thus insuring fertilization.

4.9   SERUM BINDING PROTEIN

In some anuran species, sex steroid hormones are transported in the bloodstream bound to specific serum proteins. A high affinity and fair capacity steroid binding protein (SBP) that binds both C_{18} and C_{19} sex steroids (T, DHT and E_2), but not C_{21} steroids (progesterone and cortisol), is present in the β globulin fraction of *Bufo arenarum* serum (Ba SBP) (Fernández *et al.* 1994). A similar steroid binding protein showing a minor affinity has been determined in *Rana esculenta* (Paolucci and Di Fiore 1994). The characteristics of Ba SBP with regards to the rate of association and dissociation with its ligands as well as the sedimentation coefficient show evident similarities with mammalian SBP. Ba SBP specificity is comparable to human SBP, with the exception that it has a high affinity for the synthetic androgen methyltrienolone (R1881). The physicochemical characterization indicates a Stokes radius of 43.5 A and a molecular weight of 93,300. In relation to the steroid binding capacity this parameter, ranging from 3 to 6 10^{-7} M, is one order of magnitude higher than that reported for similar molecules partially characterized in other anuran amphibians (Martin and Ozon 1975; Smirnova *et al.* 1978), and two orders of magnitude higher than that reported for human serum (Burke and Anderson 1972).

The differences observed in the above mentioned SBP capacities may be largely explained by the fact that *Bufo arenarum* serum shows a high concentration of E_2, which can reach a maximum level during the amplexus period of about 14–16 ng/ml (10^{-7} M), this value being approximately 100 times higher than that found in human serum at the time of the preovulatory E_2 peak.

Furthermore, high levels of circulating androgens, 4–5 times higher than E_2, have been observed in *Bufo arenarum* females during the preovulatory period. These results agree on the one hand with the 100 fold concentration of Ba SBP with respect to the SBP of humans and, on the other, with the binding affinity of the *Bufo* protein that is slightly higher for T and DHT than for E_2 (Fernández *et al.* 1994). This lends some support to the assumption that, throughout vertebrates, serum sex steroid binding capacities are 10–50 times higher than those of the circulating sex steroids, thus assuring the binding of most of the circulating hormones.

The Ba SBP binding capacity during the sexual cycle shows a similar profile to that of serum E_2, higher concentrations having been found in females than in males (Fernández *et al.* 1994). Ovariectomy reduces in about 40% the Ba SBP level with respect to the concentrations found in control females. The administration of E_2 causes a substantial dose-dependent increase in this protein, indicating that its concentration, as in women, is under estrogenic control. This is not a common feature in anurans. In *Rana esculenta*, although SBP capacity to bind ^{3}H-E_2 and ^{3}H-T changes throughout the sexual cycle, neither gonadectomy nor the later steroid hormone treatment modifies this parameter (Paolucci and Di Fiore 1994).

The function of Ba SBP is still speculative, but the high capacity of this serum binder could explain why this toad is able to tolerate high levels of circulating steroids. The rapid association with its ligands renders this protein a suitable binder for steroids. The slow dissociation makes the complex formed stable for a long time, thus providing a good reservoir for circulating hormones. Consequently, SBP would be involved in the mechanism of transport and steroid hormone action by keeping high levels of circulating steroids protected from catabolism or excretion. The changes in its concentration, hormonally regulated, would control the availability of the active free steroid fraction.

The presence of a protein similar to the corticosteroid binding globulin (CBG) which binds C_{21} steroids has also been reported in amphibians (Martin and Ozon 1975).

4.10 STEROID RECEPTORS

Despite the regulatory effects of steroid hormones on the reproductive physiology of anuran females, studies on the mechanism of action of these hormones are scarce and limited to certain organs. In fact, an estrogen receptor related to vitellogenin synthesis has been identified and characterized in amphibian liver (Paolucci and Botte 1988). As regards P, a single class of binding site for this hormone has been reported at the oocyte plasma membrane, associated with the resumption of meiosis, whereas two P receptor forms similar to those in chickens and humans are present in the cytosol of *Rana pipiens* oocytes (Morril *et al.* 1997). However, at this time, there are no reports concerning steroid receptors in amphibian oviducts.

Studies in *Bufo arenarum* demonstrate the presence of E_2, DHT and T binding components in the cytosol fraction. The kinetics of association and dissociation, the specificity, affinity and binding capacity as well as the sedimentation analysis in equilibrium conditions show strong similarities with the characteristics previously determined for Ba SBP. Additionally, tissue culture experiments and affinity chromatography in DNA cellulose have demonstrated that, once this component of cytosol has been removed, there is no other component with the capacity to bind steroids, indicating the absence of specific receptors in this cellular compartment.

The analysis of the nuclear fraction by means of conventional techniques has not led to the detection of specific binding. Nevertheless, by using culture assays, a component with a high binding capacity and a specificity different from that reported for other target cells has been determined.

These results indicate that both classic cytosolic and nuclear receptors are absent in the female duct, the mechanism of action of steroid hormones in anuran oviducts being still uncertain. It seems reasonable to suggest that the "oviductal receptor" may be a different molecular entity from that of mammals or that SBP can function as a cytosolic receptor, both possibilities representing a promising field for future investigation.

4.11   CONCLUSIONS

The reproduction of most anuran amphibian species is characterized by a sexual cycle closely associated with the seasonal variations. The control of reproduction resides in the hypothalamus, which controls pituitary and ultimately gonadal functions. The analyses of the seasonal profiles of the circulating steroids demonstrate that different hormones, acting concurrently or sequentially, synchronize gonadal processes such as folliculogenesis and oocyte maturation, and oviductal processes involving growth and specific secretions. These significant events insure the release of a gamete capable of fertilization and successful embryonic development.

4.12   ACKNOWLEDGEMENTS

Our research was supported by CIUNT grants. We wish to thanks Dr Beatriz Winik for her assistance in the preparation of the micrographs and also to Mrs Virginia Méndez for proofreading.

4.13   LITERATURE CITED

Bakos, M., Kurosky, A. and Hedrick, J. L. 1990. Enzymatic and envelope converting activities of pars recta oviducal fluid from *Xenopus laevis*. Developmental Biology 138: 169-176.

Bandyopadhyay, A., Bandyopadhyay, J., Choi, H. H., Choi, H. S. and Kwon H. B. 1998. Plasma membrane mediated action of progesterone in amphibian (*Rana dybowskii*) oocyte maturation. General Comparative Endocrinology 109: 293-301.

Barbieri, F. D and del Pino, E. 1975. Jelly coats and diffusible factor in anuran fertilization. Archives de Biologie (Bruxelles) 86: 311-321.

Barbieri, F. D. and Budeguer de Atenor, M. S. 1973. Role of oviducal secretions in the fertilization of *Bufo arenarum* oocytes. Archives de Biologie Bruxelles 84: 501-511.

Bonnell, B. S. and Chandler, D. E. 1996. Egg jelly layers of *Xenopus laevis* are unique in ultrastructure and sugar distribution. Molecular Reproduction and Development 44: 212-220.

Bonnell, B. S., Reinhart, D. and Chandler, D. F. 1996. *Xenopus laevis* egg jelly coats consist of small diffusible proteins bound to a complex of structurally stable networks composed of high molecular weight glycoconjugates. Developmental Biology 174: 32-42.

Budeguer de Atenor, M. S., Salomon de Legname, H. and Legname, A. H. 1989. Effect of follicle-stimulating hormone on metabolism and maturation in *Bufo arenarum* oocytes. Gamete Research 23: 349-356.

Bühler, M. I., Petrino, T. and Legname, A. H. 1987. Sperm nuclear transformation and aster formation related to metabolic behavior in amphibian eggs. Development Growth and Differentiation 29 (2): 177-184.

Burke, C. W. and Anderson, D. C: 1972. Sex hormone-binding globulin is an estrogen amplifier. Nature 240: 38-40.

de Atenor, M. S. B., de Romero, I. R., Brauckmann, E., Pisanó, A. and Legname, A. H. 1994. Effects of the pineal gland and melatonin on the metabolism of oocytes in vitro and on ovulation in *Bufo arenarum*. Journal Experimental Zoology 268: 436-441.

de Romero, I. R., de Atenor, M. B. and Legname A. H. 1998. Nuclear maturation inhibitors in *Bufo arenarum* oocytes. BioCell 22: 27-34.

Delgado, M. J. and Vivien-Roels, B. 1989. Effect of environmental temperature and photoperiod on the melatonin levels in the pineal, lateral eye, and plasma of the frog *Rana perezi*: Importance of ocular melatonin. General Comparative Endocrinology 75: 46-53.

Delgado, M. J., Gutierrez, P. and Alonso-Bedate, M. 1993. Effect of daily melatonin injections on the photoperiodic gonadal response of the female frog *Rana ridibunda*. Comparative Biochemistry and Physiology 76: 389-392.

Del Pino, E. M. 1973. Interactions between gametes and environment in the toad *Xenopus laevis* (Daudin) and their relationship to fertilization. Journal Experimental Zoology 185: 121-131.

D'Istria, M., Monteleone, P., Serino, I. and Chieffi, G. 1994. Seasonal variations in the daily rhythm of melatonin and NAT activity in the harderian gland, retina, pineal gland, and serum of the green frog, *Rana esculenta*. General Comparative Endocrinology 96: 6-11.

Di Fiore, M. M., Assisi, L. and Botte, V. 1998. Aromatase and testosterone receptor in the liver of the female green frog, *Rana esculenta*. Life Sciences 62: 1949-1958.

Dubowsky, S. and Smalley, K. N. 1993. Testosterone induced growth of the oviduct in the frog *Rana pipiens*: evidence for local aromatization. General Comparative Endocrinology 89: 276-282.

Elinson, R. P. 1974. A block to cross-fertilization located in the egg jelly of the frog *Rana clamitans*. Journal Embryology and Experimental Morphology 32: 325-335.

Fernández, S. N., Mansilla, C. and Miceli, D. C. 1984. Hormonal regulation of an oviducal protein involved in *Bufo arenarum* fertilization. Comparative Biochemistry and Physiology 78 A (1): 147-152.

Fernández, S. N., Mansilla, C. and Miceli, D. C. 1989a. Correlation between the sexual cycle and ultrastructure of *Bufo arenarum* oviducal pars recta epithelium. Microscopia Electrónica y Biología Celular 13 (2): 211-220.

Fernández, S. N., Mansilla, Z. C. and Miceli, D. C. 1989b. Effect of ovariectomy and subsequent hormonal replacement on the proliferation of secretory cells of *Bufo arenarum* oviducal pars recta epithelium. Microscopía Electrónica y Biología Celular 13: 201-210.

Fernández, S. N., Mansilla-Whitacre, Z. C. and Miceli, D. C 1994. Characterization and properties of steroid binding protein in *Bufo arenarum* serum. Molecular Reproduction and Development 38: 364-372.

Fernández, S. N., Miceli, D. C. and Whitacre, C. M. 1997. Ultrastructural studies of the effect of steroid hormones on pars recta secretions in *Bufo arenarum*. Journal of Morphology 231: 1-10.

Fortune J. E. 1983. Steroid production by *Xenopus* ovarian follicles at different developmental stages. Developmental Biology 99: 502-509.

Gobbetti, A. and Zerani, M. 1991. Gonadotropin-releasing hormone stimulates biosynthesis of prostaglandin $F_2\alpha$ by the interregnal gland of the water frog, *Rana esculenta*, in vitro. General Comparative Endocrinology 84: 434-439.

Gobbetti, A. and Zerani, M. 1993. Prostaglandin E_2 and Prostaglandin $F_2\alpha$ involvement in the corticosterone and cortisol release by the female frog, *Rana esculenta*, during ovulation. Journal Experimental Zoology 267: 164-170.

Grey, R. D., Working, P. K. and Hedrick, J. L. 1977. Alteration of structure and penetrability of vitelline envelope after passage of eggs from coelom to oviduct in *Xenopus laevis*. Journal Experimental Zoology 201: 73-83.

Guerriero, G., Roselli, Ch. E., Paolucci, M., Botte, V. and Ciarcia, G. 2000. Estrogen receptors and aromatase activity in the hypothalamus of the female frog, *Rana esculenta*. Fluctuations throughout the reproductive cycle. Brain Research 880: 92-101.

Hardy, D. M. and Hedrick, J. L. 1992. Isolation and characterization of oviductin from *Xenopus laevis* pars recta oviduct. Biochemistry 31: 4466-4472.

Hayashi, H., Hayashi, T. and Hanaoka, Y. 1992a. Amphibian lutropin and follitropin from the bullfrog *Rana catesbeiana*. Complete amino acid sequence of the alpha subunit. European Journal of Biochemistry 203: 185-191.

Hayashi, T., Hanaoka, Y. and Hayashi, H. 1992b. The complete amino acid sequence of the follitropin beta-subunit of the bullfrog, *Rana catesbeiana*. General Comparative Endocrinology 88: 144-150.

Hedrick, J. L. and Nishihara, T. 1991. Structure and function of the extracellular matrix of anuran eggs. Journal of Electron Microscopy Technique 17: 319-335.

Iela, L., Rastogi, R. K., Del Rio, G and Bagnara, J. T. 1986. Reproduction in the Mexican leaf frog, *Pachymedusa dacnicolor*. III. The female. General Comparative Endocrinology 63: 381-392.

Ishihara, K., Hosono, J. Kanatani, H. and Katagiri, C. 1984. Toad egg-jelly as a source of divalent cations essential for fertilization. Developmental Biology 105: 435-442.

Itoh, M. and Ishii, S. 1990. Changes in plasma levels of gonadotropins and sex steroids in the toad, *Bufo japonicus*, in association with behavior during the breeding season. General Comparative Endocrinology 80: 451-464.

Katagiri, Ch. 1973. Chemical analysis of toad egg-jelly in relation to its "sperm-capacitating" activity. Developmental Growth and Differentiation 15:81-92.

Katagiri, Ch. 1987. Role of oviducal secretion in mediating gamete fusion in Anura amphibians. Zoological Science 1: 1-14.

Katagiri, Ch., Iwao, Y. and Yoshizaki, N. 1982. Participation of oviducal pars recta secretions in inducing the acrosome reaction and release of vitelline coat lysine in fertilization toad sperm. Developmental Biology 94: 1-10.

Kim, J. W., Im, W. B., Choi, H. H., Ishii, S. and Kwon, H. B. 1998. Seasonal fluctuations in pituitary gland and plasma levels of gonadotropic hormones in *Rana*. General Comparative Endocrinology 109: 13-23.

Kobayashi, F., Zimniski, S. J. and Smalley, K. N. 1996. Characterization of oviductal aromatase in the northern leopard frog, Rana pipiens. Comparative Biochemistry and Physiology B 113: 653-657.

Kupwade, V. A. and Saidapur, S. K. 1986. Effect of melatonin on oocyte growth and recruitment, hypophyseal gonadotrophs, and oviduct of the frog Rana cyanophyctis maintained under natural photoperiod during the prebreeding phase. General Comparative Endocrinology 64: 284-292.

Kwon, H. B. and Lee W. K. 1991. Involvement of protein kinase C in the regulation of oocyte maturation in amphibian (Rana dybowskii). Journal Experimental Zoology 257: 115-123.

Kwon, H. B. and Ahn, R. S. 1994. Relatives roles of theca and granulosa cells in ovarian follicular steroidogenesis in the amphibian, Rana nigromaculata. General Comparative Endocrinology 94: 207-214.

Kwon H. B., Lim, Y. K., Choi, M. J. and Ahn R. S. 1989. Spontaneous maturation of follicular oocytes in Rana dybowskii in vitro: seasonal influences, progesterone production and involvement of cAMP. Journal Experimental Zoology 252: 190-199.

Kwon, H. B., Ahn, R. S., Lee, W. K., Im, W.B., Lee, C. C. and Kim, K. 1993. Changes in the activities of steroidogenic enzymes during the development of ovarian follicles in Rana nigromaculata. General Comparative Endocrinology 92: 225-232.

Le Goascogne, C., Sananès, N., Gouézou, M. and Baulieu, E.-E. 1985. Testosterone-induced meiotic maturation of Xenopus laevis oocytes: evidence for an early effect in the synergistic action of insulin. Developmental Biology 109: 9-14.

Legname, A. H. and Bühler, M. I. 1978. Metabolic behavior and cleavage capacity in the amphibian egg. Journal Embryology Experimental Morphology 47:161-168.

Legname, A. H. and Salomón de Legname, H. 1980. Changes in the oxidative metabolism during maturation of amphibian oocytes. Journal Embryology Experimental Morphology 59: 175-186.

Lessman, C. A. and Schuetz, A. W. 1981. Role of follicle wall in meiosis reinitiation induced by insulin in Rana pipiens oocytes. American Journal Physiology 241: E51-E56.

Licht, P., McCreery, B. R., Barnes, R. and Pang, R. 1983. Seasonal and stress related changes in plasma gonadotropins, sex steroids, and corticosterone in the bullfrog, Rana catesbeiana. General Comparative Endocrinology 50: 124-145

Lin, Y-W. P., Petrino, T., Landin, A. M., Franco, S. and Simeus, I. 1999. Inhibitory action of the gonadopeptide inhibin on amphibian (Rana pipiens) steroidogenesis and oocyte maturation. Journal Experimental Zoology 284: 232-240.

Liu, Z. and Patiño, R. 1993. High-affinity binding of progesterone to the plasma membrane of Xenopus oocytes: characteristics of binding and hormonal and developmental control. Biology of Reproduction 49: 980-988.

Llanos, R, Whitacre, C. M., Medina, M., Fernández, S. N., Giunta, S. and Miceli, D. C. 1998. Involvement of blood serum in amphibian fertilization. BioCell 22: 67-72.

Lofts, B. 1974. Reproduction. Pp 185. In B. Lofts (ed.), The Physiology of the Amphibia. Academic Press, New York, NY.

Lohka, M. J., Hayes, M. K. and Maller, J. L. 1988. Purification of maturation-promoting factor, an intracellular regulator of early mitotic events. Proceedings of the National Academy of Sciences of the U.S.A. 85: 3009-3013.

Lutz, L. B., Cole, L. M., Grupta, K. W., Kwist, K. W., Auchus, R. J. and Hammes, S. R. 2001. Evidence that androgens are the primary steroids produced by Xenopus laevis ovaries and may signal through the classical androgen receptor to promote oocyte maturation. Proceedings of the National Academy of Sciences of the U.S.A. 98: 13728-13733.

Mansilla-Whitacre, Z. C., Fernández, S. N. and Miceli, D. C. 1992. In vivo and in vitro protein synthesis of oviductal pars recta by estradiol effect. Comparative Biochemistry and Physiology 102 A: 59-65.

Martin, B and Ozon, R. 1975. Steroid-protein interactions in nonmammalian vetertebrates. II. Steroid binding proteins in the serum of amphibians. A physiological approach. Biology of Reproduction 13: 371-80

Masui, Y. 1985. Meiotic arrest in animal oocytes. Pp. 189-219. In C. B. Metz and A. Monrroy (eds), Biology of Fertilization. Academic Press, Orlando.

Matten, W. T., Copeland, T. D., Ahn, N. G. and Vande Woude, G. F. 1996. Positive feedback between MAP kinase and Mos during Xenopus oocyte maturation. Developmental Biology 179: 485-492.

Medina, M. F., Winik, B. C., Crespo, C. A., Ramos, I., and Fernández, S. N. 2000. Subcellular localization of Ca-ATPase and calcium in Bufo arenarum oviducts. Acta Histochemistry et Cytochemistry 33 (1): 49-58.

Miceli, D. C. and Fernández, S. N. 1982. Properties of an oviducal protein involved in amphibian oocyte fertilization. Journal Experimental Zoology 221: 357-364.

Miceli, D. C., Fernández, S. N., Raisman, J. S. and Barbieri, F. D. 1978. A trypsin-like oviducal proteinase involved in Bufo arenarum fertilization. Journal Embryology and Experimental Morphology 48: 79-91.

Miceli, D. C., Fernández, S. N., Mansilla, Z. C. and Cabada, M. O. 1987. New evidence of anuran oviductal pars recta involvement on gamete interaction. Journal Experimental Zoology 244: 125-132.

Miranda, L. A., Paz, D. A., Affanni, J. M. and Somoza, G. M. 1998. Identification and neuroanatomical distribution of immunoreactivity for mammalian gonadotropin-releasing hormone (mGnRH) in the brain and neural hypophyseal lobe of the toad Bufo arenarum. Cell Tissue Research 293: 419-425.

Miyashita, K., Shimizu, N., Osanai, S. and Miyata, S. 2000. Sequence analysis and expression of the P450 aromatase and estrogen receptor genes in the Xenopus ovary. Journal Steroid Biochemistry Molecular Biology 75: 101-107.

Moreno, A. R. 1972. Histomorfología del oviducto de Bufo arenarum (Hensel). Revista Agronómica del Noroeste Argentino 9: 585-602.

Morril, G. A., Ma, G. Y. and Kostelow, A. 1997. Progesterone binding to plasma membrane and cytosol receptors in amphibian oocyte. Biochemical Biophysic Research Communication 6: 213-217.

Muske, L. E. and Moore, F. L. 1990. Ontogeny of immunoreactive gonadotropin-releasing hormone neuronal systems in amphibians. Brain Research 534: 177-187.

Nguyen-Gia, P., Bomsel, M., Labrousse, J. P., Gallien, C. L. and Weintraub, H. 1986. Partial purification of the maturationpromoting factor MPF from unfertilized eggs of Xenopus laevis. European Journal of Biochemistry. 161: 771-777.

O'Connors, J. H. 1969. Effect of melatonin on "in vitro" ovulation of frog oocytes. American Zoology 9: 577.

Olson, J. H. and Chandler, D. E. 1999. Xenopus laevis egg jelly contains small proteins that are essential to fertilization. Developmental Biology 210: 401-410.

Omata, S. 1993. Relative roles of jelly layers in successful fertilization of Bufo japonicus. Journal Experimental Zoology 265: 329-335.

Pancharatna, K., Rajapurohit, S. V., Hiregoudar, S. R. and Kumbar, S. M. 2001. Effect of androgens on oviductal growth in skipper frog Rana cyanophlyctis. Indian Journal Experimental Biology 39: 933-935.

Paolucci, M. and Botte, V. 1988. Estradiol-binding molecules in the hepatocytes of the female water frog, Rana esculenta, and plasma estradiol and vitellogenin levels during the reproductive cycle. General Comparative Endocrinology 70: 466-476.

Paolucci, M. and Di Fiore, M. M. 1994. Sex steroid binding proteins in the plasma of the green frog, *Rana esculenta*: changes during the reproductive cycle and dependence on pituitary gland and gonads. General Comparative Endocrinology 96: 401-411.

Pasanen, S. and Koskela, P. 1974. Seasonal changes in calcium, magnesium, copper and zinc content in the liver of the commom frog, *Rana temporaria* L. Comparative Biochemistry and Physiology 48 A 27-36.

Pinelli, C., Fiorentino, M., D'Aniello, B., Tanaka, S. and Rastogi, R. K. 1996. Immunohistochemical demonstration od FSH and LH in the pituitary of the developing frog, *Rana esculenta*. General Comparative Endocrinology 104: 189-196.

Polzonetti-Magni, A. M., Mosconi, G., Carnevali, O., Yamamoto, K., Hanaoka, Y. and Kikuyama, S. 1998. Gonadotropins and reproductive function in the anuran amphibian, *Rana esculenta*. Biology of Reproduction 58: 88-93.

Raisman, J. S. and Cabada, M. O. 1977. Acrosome reaction and proteolytic activity in the spermatozoa of an anuran amphibian, *Leptodactylus chaquensis*. Developmental Growth and Differentiation 19: 227-232.

Ramos I., Winik, B. C. and Cisint, S. 1998. Structural changes in the endoplasmic reticulum of *Bufo arenarum* oocytes during meiotic maturation. Electron Microscopy. Biological Sciences Vol IV: 185-186.

Ramos, I., Winik, B. C., Cisint, S., Crespo, C., Medina, M. and Fernández S. N. 1999. Ultrastructural changes during nuclear maturation in *Bufo arenarum* oocytes. Zygote. 7: 261-269.

Ramos, I., Cisint, S., Crespo, C. A., Medina, M. F. and Fernández, S. N. 2001. Nuclear maturation inducers in *Bufo arenarum* oocytes. Zygote 9: 353-359.

Ramos, I., Cisint, S., Crespo, C. A., Medina, M. F. and Fernández, S. N. 2002. Involvement of catecholamines in the regulation of oocyte maturation in frogs. Zygote 10: 271-281.

Rapela, C. E. and Gordon, M. F. 1956. Variación estacional del contenido de adrenalina y noradrenalina de la glándula suprarrenal del sapo. Revista Sociedad Argentina de Biología 32: 36-43.

Robertson, D. 1977. The annual pattern of plasma calcium in the frog and the seasonal effect of ultimobranchialectomy and parathyroidectomy. General Comparative Endocrinology 33: 336-343.

Rotello, R. J., Lieberman, R. C., Lepoff, R. B. and Gerschenson, L. E. 1992. Characterization of uterine epithelium apoptotic cell death kinetics and regulation by progesterone and RU486. American Journal Pathology 140: 449-456.

Sadler, S. E. and Maller, J. L. 1981. Progesterone inhibits adenylate cyclase in *Xenopus* oocytes: Action on the guanine nucleotide regulatory protein. Journal Biological Chemistry. 256 (12): 6368-6373.

Sadler, S. E., and Maller, J. L. 1982. Identification of a steroid receptor on the surface of *Xenopus* oocytes by photoaffinity labeling. Journal Biological Chemistry 257: 355-361.

Segura, E. T. and D'Agostino, S. A. 1964. Seasonal variations of blood pressure, vasomotor reactivity and plasmatic catecholamines in the toad. Acta Physiological Latinoamericana 14: 231-237.

Skene, D. J., Vivien-Roels, B. and Pevet, P. 1991. Day and nighttime concentrations of 5-methoxytryptophol and melatonin in the retina and pineal gland from different classes of vertebrates. General Comparative Endocrinology. 84: 405-411.

Skoblina, M. N. 2000. Role of protein kinase A and calcium ions in regulating ovulation of oocytes by gonadotropins in the grass frog (*Rana temporaria*) in vitro. Ontogenez 31: 382-387.

Smirnova, O. V., Smirnov, A. N., Shoshina, S. V. and Rozen, V. B. 1978. Some properties of sex hormone-binding protein from blood serum of the frog *Xenopus laevis* and the detection of estradiol-binding component differing from sex hormone-binding protein in frog liver cytosol. Biochemistry 43: 1133-1139.

Sotowska-Brochocka, J. 1988. The stimulatory and inhibitory role of the hypothalamus in the regulation of ovulation in grass frog, *Rana temporaria* L. General Comparative Endocrinology 70: 83-90.

Sotowska-Brochocka, J., Martynska, L. and Licht, P. 1994. Dopaminergic inhibition of gonadotropic release in hibernating frogs, *Rana temporaria*. General Comparative. Endocrinology 93: 192-196.

Sretarugsa, P. and Wallace, R. A. 1997. The developing *Xenopus* oocyte specifies the type of gonadotropin-stimulated steroidogenesis performed by its associated follicles cells. Developmental Growth and Differentiation 39: 87-97.

Takamune, K. and Katagiri, C. 1987. The properties of the oviductal pars recta protease which mediates gamete interaction by affecting the vitelline coat of a toad eggs. Developmental Growth and Differentiation 29: 193-203.

Tanaka, T. and Yamashita, M. 1995. Pre-MPF is absent in immature oocytes of fishes and amphibians except *Xenopus*. Development Growth and Differentiation 37: 387-393.

Tavolaro, R., Canonaco, M. and Franzoni, M. F. 1995. Comparison of melatonin-binding sites in the brain of two amphibians: an autoradiographic study. Cell Tissue Research 279 (3): 613-617.

Uchiyama, H., Koda, A., Komazaki, S., Oyama, M. and Kikuyama, S. 2000. Occurrence of immunoreactive activin/inhibin beta(B) in thyrotropes and gonadotropes in the bullfrog pituitary: possible paracrine/autocrine effects of activin B on gonadotropin secretion. General Comparative Endocrinology 118: 68-76.

Udaykumar, K. and Joshi, B. N. 1997. Effect of exposure to continuous light and melatonin on ovarian follicular kinetics in the skipper frog, *Rana cyanophlyctis*. Biology Signals 6 (2): 62-66.

Varnold, R. L. and Smith, L. D. 1990. Protein kinase C and progesterone-induced maturation in *Xenopus* oocytes. Development 109: 597-604.

Wasserman, R. H., Smith, C. A., Smith, C. M., Bruidak, M. E., Fullner, C. S., Krook, I., Penniston, J. T. and Kumar, K. 1991. Immunohistochemical localization of a calcium pump and calbindin-D28K in the oviduct of the laying hen. Histochemistry 96: 413-418.

Whitacre, C. M., Perdigón, G., Fernández, S. N. and Miceli, D. C. 1996. Role of oviductal pars recta extract on interspecific fertilization between anurans. Developmental Growth and Differentiation 38: 203-207.

Whittier, J. M. and Crews, D. 1987. Seasonal reproduction: Patterns and control. Pp. 385-409. In D. O. Norris and R.E. Jones (eds), *Hormone and reproduction in fishes, amphibians and reptiles*. Plenum Press, New York, N.Y.

Winik, B. C., Alcaide, M. F., Crespo, C. A., Medina M: F., Ramos, I. and Fernández, S. N. 1999. Ultrastructural changes in the oviductal mucosa throughout the sexual cycle in *Bufo arenarum*. Journal of Morphology 239: 61-73.

Yoshida, N., Mita, K. and Yamashita, M. 2000. Function of the Mos/MAPK pathway during oocyte maturation in the japanese brown frog *Rana japonica*. Molecular Reproduction and Development 57: 88-98.

Yoshizaki, N. 1985. Fine structure of oviducal epithelium of *Xenopus laevis* in relation to its role in secreting egg envelopes. Journal of Morphology 184: 155-169.

Yoshizaki, N. and Katagiri, Ch. 1981. Oviducal contribution to alteration of the vitelline coat in the frog *Rana japonica*. An electron microscopic study. Developmental Growth and Differentiation. 23: 495-506.

Yuanyou, L. and Haoran, L. 2000. Differences in mGnRH and cGnRH-II contents in pituitaries and discrete brain areas of *Rana rugulosa* W. according to age and stage of maturity. Comparative Biochemistry Physiology Part C 125: 179-188.

CHAPTER 5

Spermatogenesis and the Mature Spermatozoon: Form, Function and Phylogenetic Implications

David M. Scheltinga and Barrie G.M. Jamieson

5.1 INTRODUCTION

Of the 29 anuran families currently recognized (see Frost 2002), descriptions of the spermatozoa of 20 (Ascaphidae, Bombinatoridae, Bufonidae, Centrolenidae, Discoglossidae, Hylidae, Hyperoliidae, Leiopelmatidae, Leptodactylidae, Mantellidae, Megophryidae, Microhylidae, Myobatrachidae, Pelobatidae, Pelodytidae, Petropedetidae (here included in the Ranidae as a subfamily), Pipidae, Ranidae, Rhacophoridae, and Rhinodermatidae) have been given. This chapter collates the descriptions of anuran spermatozoa of various researchers, including much new information, for its interest *per se* and to allow an examination of anuran family interrelationships based on spermatozoal characters. It commences with a review of spermatogenesis and ends with a consideration of the function and reproductive biology of spermatozoa.

The taxonomy of the Anura is in a constant state of flux at all levels. We have here used the classifications of Frost (2002) and Duellman (Chapter 1 of this volume).

5.2 SPERMATOGENESIS

Cystic spermatogenesis. In the Lissamphibia sperm develop in testicular cysts. In such cystic spermatogenesis, each testicular cyst contains a clone of sperm cells that differentiate synchronously. This is confirmed as the basic anamniote condition, and has also been demonstrated for the caecilian *Chthonerpeton indistinctum* (Caeciliidae, Typhlonectinae) (de Sa and Berois 1986) and for the Urodela (see Uribe 2003).

Department of Zoology and Entomology, University of Queensland, Brisbane, Queensland 4072, Australia

The histological organization of the testis in Anura and the events of spermatogenesis were established for the toad *Bufo arenarum* by light (LM) and transmission electron microscopy (TEM) by Burgos and Fawcett (1956) (Fig. 5.1) but had been determined for several anurans by earlier workers.

Aspects of anuran spermatogenesis were described by LM for the discoglossoids *Bombina bombina, B. pachypus* (Champy 1913), and *B. variegata* (Obert 1976); *Discoglossus pictus* (Champy 1923; Favard 1955a; Sandoz 1970a,b,

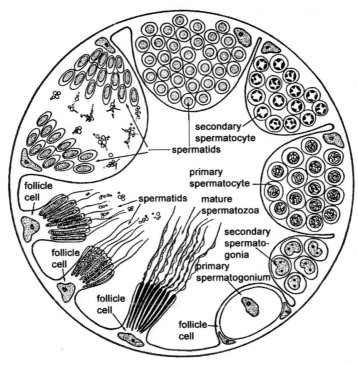

Fig. 5.1 *Bufo arenarum*. Diagrammatic representation of the arrangement of spermatocysts in the wall of a seminiferous tubule. Beginning at the bottom of the figure and reading counter clockwise, the stages in the spermatogenetic wave can be followed in temporal sequence. The follicle cells are Sertoli cells of current terminology. Modified after Burgos, M. H. and Fawcett, D. W. 1956. Journal of Biophysical and Biochemical Cytology 2: 223-240, Fig. 1.

1971, 1973, 1974, 1975); *Alytes obstetricans* (Champy 1913; Grassé 1929; Grassé 1986); the pipid *Xenopus laevis* (Kalt 1976, 1977; Gambino *et al.* 1981; Bernardini *et al.* 1990; Kobayashi *et al.* 1993; Takamune *et al.* 1995); the bufonids *Bufo vulgaris, B. calamita, B. pantherinus* (Champy 1913); *B. regularis* (Neyrand de Leffemberg and Exbrayat 1995); the hylid *Hyla arborea* (Champy 1913); the microhylid *Microhyla ornata*, in a study of testicular activity (Kanamadi and Hiremath 1993); the ranids *Rana temporaria* (Witschi 1915); *R. esculenta* (Champy 1913; Lofts 1974; Ogielska and Bartmanska 1999; and, in a demonstration of prothymsin α Aniello *et al.* 2002); *R. lessonae* (Neyrand de Leffemberg and

Exbrayat 1995); *R. pipiens* (Rugh 1939; Burgos 1955); *Hoplobatrachus tigerinus*, in a hormonal study (Srivastava and Gupta 1990); *Sphaeroteca breviceps*, in a study of testicular activity (Kanamadi and Jirankali 1991).

TEM studies on anuran spermatogenesis include: the discoglossoids *Discoglossus pictus* (Discoglossidae) (Sandoz 1971a,b, 1973, 1974, 1975); *Bombina bombina* (Bombinatoridae) (Folliot 1979); the pipid *Xenopus laevis* (Reed and Stanley 1972; Bernardini *et al.* 1986, 1990; Takamune *et al.* 1995); the leptodactylids *Telmatobius schreiteri* (Pisanó and Adler 1968); *Physalaemus* (Amaral *et al.* 1999); *Odontophrynus cultripes* (Báo *et al.* 1991; Fernandes and Báo 1998); *Batrachyla antartundica*, *B. taeniata* and *B. leptopus* (Garrido *et al.* 1989); the bufonids *Bufo arenarum* (Burgos and Fawcett 1956; Burgos and Vitale-Calpe 1967b; Cavicchia and Moviglia 1982, 1983); *Bufo bufo* (Nicander 1970); *Bufo gargarizans* (Kwon *et al.* 1993); *Bufo variegata* (Folliot 1979); *Melanophryniscus cambaraensis* (Báo *et al.* 2001); *Nimbaphrynoides occidentalis*, in a studies of the sexual cycle and factors affecting it (Gavaud 1976, 1977; Zuber-Vogeli and Xavier 1965); the hylids *Hyla japonica* (Lee and Kwon 1992); *H. chinensis* (Lin *et al.* 2000); *Scinax ranki* (Taboga and Dolder 1994b); *Pachymedusa dacnicolor* (Rastogi *et al.* 1988); *Litoria rheocola*, *L. latopalmata*, and *L. rubella* (present study, Fig. 5.2); the megophryid *Megophrys montana* (Asa and Phillips 1988); the ranids *Rana clamitans* (Poirier and Spink 1971, brief reference); *R. pipiens* (Poirier and Spink 1971, brief reference; Zirkin 1971a); *Rana catesbeiana* (Sprando and Russell 1987); *Rana* sp. (Pudney 1995); and the rhacophorid *Chiromantis xerampelina* (Mainoya 1981).

A brief review of anuran spermatogenesis by LM and TEM is given by Grassé (1986) to which the reader is referred for many earlier references and which partly reviews the exquisite work of Champy (1913-1923). Spermiogenesis in Anura and other Amphibia has been reviewed by Koch and Lambert (1990) and Pudney (1995).

Spermatogenetic wave. Spermatogenesis in the toad *Bufo arenarum* takes place within follicular structures termed spermatocysts that rest upon the basement membrane of the seminiferous tubules (Burgos and Fawcett 1956; Bernardini *et al.* 1986, 1990) (Fig. 5.1) or seminiferous lobules (Pudney 1995), see Fig. 5.2A for *Litoria rheocola*. At the beginning of what has loosely been recognized as a spermatogenic wave, the spermatocyst consists of a single large primary spermatogonium enclosed in a layer of Sertoli cells (also termed follicular, supporting, nurse or sustentacular cells) so that it is the only germ cell within the spermatocyst (Kalt 1976).

The stages of spermatogenesis reported by Burgos and Fawcett (1956) for *Bufo arenarum* (Fig. 5.1) have been demonstrated for other species. The stages in sequence are: diploid primary spermatogonia, secondary spermatogonia and primary spermatocytes, followed by haploid secondary spermatocytes, and spermatids transforming without division into spermatozoa. The process of production of these stages is spermatogenesis and transformation of spermatids to spermatozoa is the process of spermiogenesis.

Primary spermatogonia. The largest cells in the seminiferous tubules are the primary spermatogonia, also termed primordial germ cells or spermatoblasts.

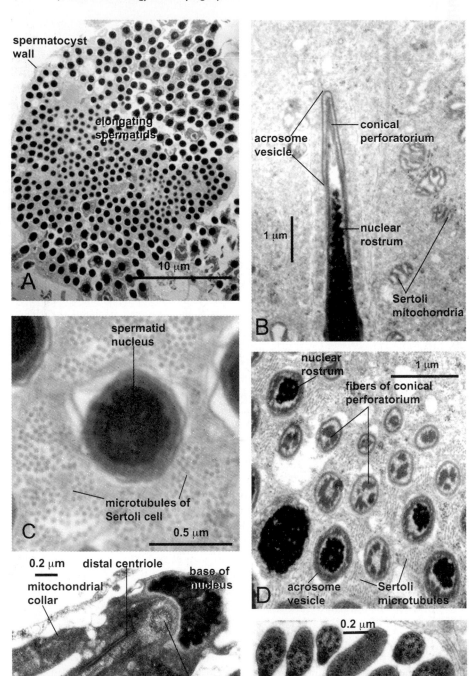

In the adult testis these have also been designated stem cell spermatogonia because in successive cell generations they proliferate from a residual population to produce the large number of gonia necessary to provide the secondary spermatogonia of the season (Rastogi *et al.* 1988). Primary spermatogonia are located at the periphery of the seminiferous tubules near the basal lamina, each supported by one or several Sertoli cells. They are distinguished by a lobate or highly polymorphic nucleus (Champy 1913; Cavicchia and Moviglia 1982; Rastogi *et al.* 1988) with one or more nucleolus-like structures. Pale and dark spermatogonia are distinguished on the basis of nuclear staining in *Pachymedusa dacnicolor* (Rastogi *et al.* 1988). The pale cells are hypothesized to be stem cell spermatogonia and the dark cells spermatogonia committed to spermatogenesis, respectively, as in *Rana esculenta* (Rastogi *et al.* 1985).

Secondary spermatogonia. As is normal for metazoan spermatogenesis, the primary spermatogonium gives rise by successive mitotic divisions to several generations of secondary spermatogonia. These finally differentiate into primary spermatocytes. All secondary spermatogonia divide synchronously but (Rastogi *et al.* 1988) no intercellular bridges have been observed between them. The secondary spermatogonia have pale nuclei but appear to arise from dark primary spermatogonia. The nuclei are multilobate. The nucleus of pale secondary spermatogonia tends to be situated at one pole of the cell, opposite a cytoplasmic area that is packed with as many as 50 or more mitochondria. Some mitochondria are modified as a lamellar body with several concentric layers. Poorly developed profiles of smooth endoplasmic reticulum (SER), some free ribosome-like bodies, and small Golgi complex are seen in pale and in dark spermatogonia. As the spermatogonia transform into primary spermatocytes the nucleus loses its lobate from, becoming spheroidal (Rastogi *et al.* 1988).

Tritiated thymidine autoradiography of *Xenopus laevis* germ testes has shown that the premeiotic DNA synthetic period occurs in late secondary spermatogonia, this S stage lasting six to seven days (Kalt 1976).

Primary spermatocytes. The number of primary spermatocytes in spermatocysts of *Xenopus laevis* at zygotene and pachytene has been estimated as 2^8, indicating 8 divisions from the primary spermatogonium (Takamune *et al.* 1995), in contrast to the seven mitotic divisions tentatively estimated by Rastogi *et al.* (1988) for *Pachymedusa dacnicolor*. As they differentiate into

---

Fig. 5.2 Some stages in spermatogenesis in hylids. **A**. Spermatocyst of *Litoria rheocola*. **B**. Anterior end of a mature spermatid of *Litoria rubella* embedded in a Sertoli cell. **C**. Transverse section (TS) of the nucleus of an elongating spermatid of *Litoria rubella* embedded in microtubule-rich cytoplasm of a Sertoli cell. **D**. TS of acrosomes and nuclei of advanced spermatids of *L. latopalmata*, embedded in microtubule-rich Sertoli cell cytoplasm. **E**. Longitudinal section of the centriolar region of a mature spermatid of *L. latopalmata* showing the proximal centriole, with 9 triplets visible, at right angles to the distal centriole which forms the basal body of the axoneme. The base of the nucleus, and the mitochondrial collar are also shown. **F**. TS of axonemes of *L. rubella*, anterior to the level where the juxta-axonemal fiber is continuous with an undulating membrane. Original.

zygotene/pachytene spermatocytes the nuclear volume of the preleptotene spermatocytes approximately doubles the 9-10 μm diameter of the oldest secondary spermatogonia. DNA synthesis, and meiosis, start before this increase in nuclear volume (Kalt 1977; Takamune *et al.* 1995) and we have seen that the cell with premeiotic DNA synthesis may be considered a terminal secondary spermatogonium. Preleptoene spermatocytes develop flattened vesicles with membranes which are more electron-dense than those of typical endoplasmic reticulum and are considered to be related to initiation of meiosis (Takamune *et al.* 1995). Flattened vesicles have also been reported for secondary spermatocytes of *Discoglossus pictus* (Sandoz 1970b), who relates them to forming Golgi dictyosomes, and in *Xenopus laevis*, in which they are cast off in residual cytoplasm of the spermatid (Reed and Stanley 1972). During the long meiotic prophase, SER increases and the Golgi complex becomes well developed; mitochondria are round and contain lamelliform cristae; a few microtubules are present, running parallel to the cell surface (Kalt 1977).

The primary spermatocytes have been shown in *Pachymedusa dacnicolor* to be interconnected by cytoplasmic bridges, which were not seen in earlier stages (Kalt 1977; Rastogi *et al.* 1988). Pachytene nuclei exhibit synaptonemal complexes between homologous chromosomes (Reed and Stanley 1972; Gambino *et al.* 1981; Rastogi *et al.* 1988; Takamune *et al.* 1995). The synaptonemal complexes terminate at the nuclear pore-lamina complex (NPL); and there is evidence that the NPL may be important in spatially organizing chromosomes during meiosis (Gambino *et al.* 1981). In *Xenopus laevis* it is at pachytene that the last vestiges of germ plasm disappear; they are visible earlier as dense granular substance usually closely associated with mitochondria and in small patches close to the nuclear membrane, often in nuclear pores (Kerr and Dixon 1974).

Giant chromatoid bodies have been demonstrated in the cytoplasm of primary spermatocytes, and in all stages from spermatogonia to spermatids; they are superficially similar to nucleoli but unlike these do not function in synthesis or accumulation of RNA (Kalt 1977).

Secondary spermatocytes. Secondary spermatocytes, produced by meiotic division of the primary spermatocytes, are readily distinguishable by their smaller size and rounded nuclei with dense clusters of chromatin. These clusters are suggestive of dyads [chromosomes consisting of paired chromatids] which at metaphase II become orientated at the equatorial plate (Rastogi *et al.* 1988). They are haploid cells and undergo the second, non-reductional division of meiosis to each give two spermatids.

Spermatids. As a result of cell multiplication and growth during spermatogenesis, the size of the spermatocyst increases and the enveloping Sertoli cells, which do not divide, become increasingly attenuated.

The youngest spermatids have round nuclei and differ from the secondary spermatocytes in that the nuclei are smaller and paler. Their cell diameters in *Pachymedusa dacnicolor* average 6-7 μm; the nuclear matrix is finely granular with a few darker, randomly distributed clumps (Rastogi *et al.* 1988).

The spermatids elongate, and become associated in clusters orientated perpendicularly to the wall of the spermatocyst with each cluster related to one

Sertoli cell. At this stage the anterior portions of the elongated spermatids are deeply embedded in the cytoplasm of the Sertoli cell with their tips situated near its nucleus, as here shown for *Litoria rubella* (Fig. 5.2B). The wall of the spermatocyst ultimately breaks down, leaving the Sertoli cells and their adherent clusters of spermatids still attached to the basement membrane of the seminiferous tubule. Thus neither the primary spermatogonia nor later stages of spermatogenesis are in direct contact with the basal lamina, being separated from it by Sertoli cell cytoplasm (Burgos and Fawcett 1956; Cavicchia and Moviglia 1982; Rastogi *et al.* 1988).

Annulate lamellae and nuclear pores. Annulate lamellae are abundant in the cytoplasm of advanced spermatids of *Bufo arenarum* (Cavicchia and Moviglia 1982) and *Telmatobius schreiteri* (Pisanó and Adler 1968) and are reported for the perinuclear cytoplasm of the early spermatid and in the midpiece of the later spermatid in *Bufo gargarizans* (Kwon *et al.* 1993). In elongating spermatids of *B. arenarum*, frequently in the posterior nuclear region, they form parallel arrays of double membranes joined by densities spaced at regular intervals. In transverse section or in freeze-fracture preparations these show a precise hexagonal pattern of circular fenestrae about 60 nm in diameter and with a center to center spacing of about 175 nm. The similarity of the fenestrations to nuclear pores supports the idea that the annulate lamellae may be fragments of nuclear envelope migrating to other locations in the cytoplasm. Nuclear pores are evenly distributed on Sertoli cell nuclei but in spermatogonia, spermatocytes and older round spermatids they are limited to only some areas of the nuclear envelope. Pore numbers are greatly reduced in round spermatids and they are usually absent from elongating spermatids. The clustering of nuclear pores in the spermatocytes has tentatively been attributed to the distribution of chromatin clusters attached to the nuclear envelope, the pores being absent where synaptonemal complexes attach to the envelope (Cavicchia and Moviglia 1982, and references therein).

Pores around the entire periphery of the nucleus in the young spermatid of *Discoglossus pictus* retract to its posterior pole while the nucleus is still spherical (Sandoz 1970a) (Fig. 5.3). This migration occurs before adhesion of the acrosome vesicle to the nucleus and is considered that the modification of the nuclear envelope determines the site of placement of the acrosome vesicle on the nucleus. The pores are said to persist in mature spermatozoa; as nucleo-cytoplasmic interchange of materials is unlikely, it is hypothesized that the annuli surrounding the pores may serve for attachment of chromatin fibers (Sandoz 1974).

Nuclear proteins. With the known exception of various crustaceans, during the haploid stages of spermatogenesis, histones are gradually replaced, first by transition proteins and later by sperm-specific proteins (Gavrila and Mircea 2001). There is a general trend in spermiogenesis towards the replacement of somatic histones by relatively simple basic arginine-rich proteins, resulting in a high arginine content. This change occurs late in spermiogenesis, during the absence of DNA replication. Whereas the histones are phylogenetically conservative, the genes coding for arginine-rich proteins in spermatozoa show

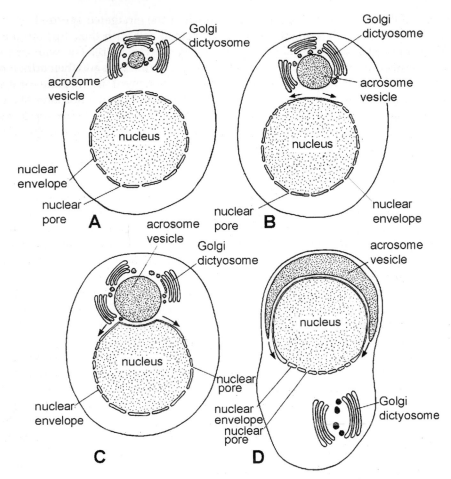

Fig. 5.3 *Discoglossus pictus*. Early development of the acrosome and modification of the nuclear envelope in the spermatid. **A**, **B**. The dictyosomes of the Golgi apparatus give rise via Golgi vesicles to proacrosomal vesicles which fuse to form the acrosome vesicle. **C**, **D**. The acrosome vesicle collapses onto the nucleus. Nuclear pores, retract to a posterior position with accompanying diminution in width of the perinuclear cisterna After Sandoz, D. 1970a. Pp. 93-113. In B. Baccetti (eds), *Comparative Spermatology*, Academic Press, New York and London, Schema 1.

exceptional variations, of unknown significance, even within small taxonomic groups and we will see that even the arginine content is not always high (*e.g.* Kasinsky 1989, 1995; Kasinsky *et al.* 1985). For example, *Rana* sperm proteins fall into Bloch's type 4 somatic-like histone category, while *Xenopus* and *Bufo* have type 3 intermediate sperm histones. The type 3 category is divisible into 2 groups: type 3B intermediate sperm histones of *Bufo* and intermediate type 3A sperm histones of *Xenopus*. *Rana* sperm histones are of the nucleosomal type, with a testis-specific, very lysine-rich H1 histone. The sperm protein in *Bufo* is richer in arginine than the proteins in *Xenopus*. Both of these genera contain lysine and histidine as well as arginine in their sperm proteins. Kasinsky *et al.* (1985) and Kasinsky (1989) indicate the types of sperm basic proteins in frogs on a phylogeny of the Anura. The existence of intrageneric variation in the composition of sperm basic protein in the Leptodactylidae has also been demonstrated for species of *Physalaemus* (Lopes *et al.* 1999), see also Taboga and Dolder (1994a) for *Scinax* (*=Hyla*) *ranki*.

In *Rana esculenta*, storage of Fos proteins in the cytoplasm and their phosphorylation in the nucleus of germ cells regulate progression of spermatogenesis during the seasonal breeding (Cobellis *et al.* 2002).

Connective tissue follicles. Spermatogenesis in *Bombina variegata* and *B. bombina* is said to differ from the cystic type in that the whole of spermiogenesis and sperm storage occurs in cellular connective tissue follicles, though the sperm ultimately reach the seminiferous tubules, and the sperm do not form bundles. It was not know whether this was characteristic of all discoglossids as *Alytes obstetricans, Discoglossus pictus* and *D. sardus* were not examined (Obert 1976). Favard (1955a) describes what appears to normal anuran testicular structure for *D. pictus*. In this species the testis is formed of long parallel, vertical tubes, inside which are packets of cells, the cysts, situated in contact with the walls of the tubes.

Sertoli cells. Sertoli cells are distinct from the Leydig cells that are found interposed between the basal lamina of the seminiferous tubules and the surrounding smooth muscle elements (*e.g.* Unsicker 1975; Kanamadi and Hiremath 1993). For a detailed description of Sertoli cells, see Rastogi *et al.* (1988) for *Pachymedusa dacnicolor*, and Burgos and Vitale-Calpe (1967a) and Cavicchia and Moviglia (1982, 1983) for a description referring particularly to their junctional specializations in *Bufo arenarum*. At spermiation, distension of the apical cytoplasm of the Sertoli cell containing the bundles of spermatozoa is caused by an influx of fluid and release of spermatozoa into the lumen and their subsequent flushing into the lumen of the seminiferous lobule (DeRobertis *et al.* 1946; Van Oordt *et al.* 1954; Burgos and Vitale-Calpe 1967a,b; Kobayashi and Iwasawa 1989; Pudney 1995).

The functions of Sertoli cells in general have been recognized as: maintenance of a permeability barrier to germinal cells during spermatogenesis; endocrine activity; the phagocytosis of degenerating germ cells and residual bodies; and the formation of specific antigens (Greer 1993). Most of these functions have been demonstrated, or suggested, or augmented for Anura.

For *Rana porosa* it has been shown that the lumina of the seminiferous tubules are not initially linked with the lumina of efferent ductules as the communication is blocked by adjacent Sertoli cells (Fig. 5.4). After experimental hormonal (hCG) treatment, dilatation of the Sertoli cells, but also of the seminiferous tubules, occurs; the lumina of the seminiferous tubules and of the efferent ductules become linked, allowing sperm to enter the latter ductules. The Sertoli cells were thus considered to have a valve-like function (Kobayashi and Iwasawa 1989). Finally, the swollen apical region of the Sertoli cell is sloughed into the lumen but further loss of Sertoli cytoplasm is prevented by apposition of residual, overlapping processes which preserve the integrity of

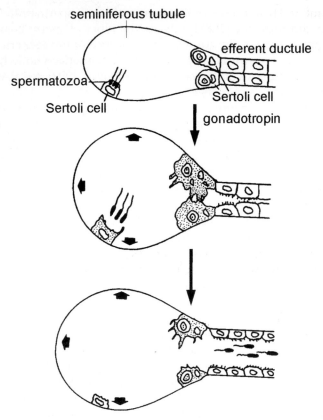

Fig. 5.4 Schematic representation of changes in the region adjoining the seminiferous tubules and the efferent ductules during sperm transport in *Rana porosa*. This process is divided into three stages. Stage 1: Seminiferous tubule and efferent ductule in the control are shown. Stage 2: After GTH treatment, expansion of the seminiferous tubule (large arrows) and dilatation of Sertoli cells are seen. Stage 3: Sertoli cells adjacent to the efferent ductule dilate and the lumina of the seminiferous tubule and efferent ductule are linked. As a result, sperm transport from the seminiferous tubule to the efferent ductule takes place. After Kobayashi, T. and Iwasawa, H. 1989. Zoological Science 6: 935-942, Fig. 6.

the remaining nucleated portion of the Sertoli cell attached to the basement membrane. This basal portion of the Sertoli cell, containing the nucleus, remains to develop the next generation of cysts. In anurans, therefore, a permanent germinal epithelium can be said to be present and this is considered to be a distinct difference from spermatogenesis in urodeles (see Pudney 1995).

In the Sertoli cells of the leptodactylid *Odontophrynus cultripes*, acid phosphatase was located in the lysosomes and glucose-6-phosphatase was observed in association with the endoplasmic reticulum and Golgi complex; they were also found in the spermatid and spermatozoon. Various phosphatases may play some role in spermatid differentiation and in the interactions of germ cells and Sertoli cells during spermiogenesis (Fernandes and Báo 1998).

As a result of the high degree of condensation and inactivation of spermatid and sperm chromatin, Sertoli cells are responsible for the nourishment of the spermatogenetic cells with ribosomal RNA and nutritive substances. In the Sertoli cell nucleus ribosomal gene amplification occurs by polytenisation (irreversible, transient development of polytene chromosomes), such cells being afterwards directed towards self-destruction (apoptosis). Cytoplasmic bridges between germ cells and Sertoli cells allow each germ cell to be supplied with the materials required for its growth and differentiation (Gavrila and Mircea 2001). The Sertoli cell processes were shown to be rich in glycogen in *Discoglossus pictus* (Sandoz 1970b). Those between adjacent Sertoli cells maintain a blood-testis barrier (see below).

As a further function of Sertoli cells, Báo *et al.* (1991) suggest that the microtubular arrangement in cytoplasmic processes of Sertoli cells possibly supplies structural support for germ cell differentiation. Kobayashi *et al.* (1993) found that, in vitro, nuclear elongation of *Xenopus laevis* spermatids did not occur in the absence of Sertoli cells. This suggests control of nuclear elongation by the Sertoli cells and endorses the view (Reed and Stanley 1972; Bernardini *et al.* 1990) that perinuclear microtubules are not responsible for elongation, while contradicting the suggestion (Reed and Stanley 1972) that internal reorganization of chromatin organizes elongation. Reed and Stanley (1972) suggest that the spermatid microtubules may be concerned with posterior migration of cytoplasm in the spermatid.

Bernardini *et al.* (1990) suggested that actin filaments, in addition to microtubules, in the Sertoli cell processes are involved in shaping the spermatids. They demonstrate, in Sertoli cell processes, a filamentous network, in which actin, localized using 10 nm gold particles, completely surrounds the spermatid head. A zone of filaments beneath the plasma membrane of Sertoli cells was already considered by Sprando and Russell (1987) to be actin. It is noteworthy that *Xenopus* spermatids do not themselves possess a microtubular manchette or actin system (Reed and Stanley 1972; Bernardini *et al.* 1990), further calling into question a function of perinuclear microtubules, or actin, in nuclear shaping.

Micrographs, or accounts, frequently show a rich array of microtubules in the Sertoli cells processes which surround the developing and mature spermatids, *e.g. Xenopus laevis* (Reed and Stanley 1972; Bernardini *et al.* 1990); *Pachymedusa dacnicolor* (Rastogi *et al.* 1988); *Batrachyla antartandica* (Garrido *et al.* 1989); *Odontophrynus cultripes* (Báo *et al.* 1991); *Hyla japonica* (Lee and Kwon 1992), *Bufo gargarizans* (Kwon *et al.* 1993), *Litoria rubella* and *L. latopalmata* (present study, Fig. 5.2.C, D), persisting at a time that microtubules are sparse or absent in the perinuclear cytoplasm of the spermatid.

As an additional function, lysosomes in the Sertoli cells of *Odontophrynus cultripes* and in *Pachymedusa dacnicolor* suggest a phagocytic role for these cells (Rastogi *et al.* 1988; Báo *et al.* 1991). In urodele testes abnormal spermatozoa are phagocytized by Sertoli cells (see Uribe 2003).

Cell junctions. Sertoli-Sertoli junctional specializations observed by freeze-fracture in *Bufo arenarum* are similar to those described in mammals though less elaborate (Cavicchia and Moviglia 1982, 1983). In *B. arenarum* they possess two of the components seen in mammalian junctions: (1) a series of focal contacts between the outer leaflets of both adjacent plasmalemmas which form a pentalaminar structure and (2) bundles of fine filaments located in the superficial cytoplasm and subjacent to each membrane fusion. However, they [usually] lack a third component: an extensive subsurface cisterna of the granular endoplasmic reticulum deep relative to the bundles of filaments and parallel to both Sertoli membranes (Cavicchia and Moviglia 1983). The filaments have been observed in *B. arenarum* (Cavicchia and Moviglia 1983; Lopez *et al.* 1983), *B. gargarizans* (Kwon *et al.* 1993), *Rana catesbeiana* (Sprando and Russell 1987) and *Hyla japonica* (Lee and Kwon 1992). Even an endoplasmic reticulum may sometimes be present, associated with the filaments in *Rana catesbeiana* (Sprando and Russell 1987).

After the leptotene stage of meiotic germ cells, these Sertoli-Sertoli junctions form a blood-testis barrier (Cavicchia and Moviglia 1983), as also demonstrated for urodele testes where it has been shown that all the steroid hormones must pass through Sertoli cells to reach the germ cells. Sertoli cells have all the ultrastructural and histochemical characteristics of steroid secretory activity (see Uribe Aranzábal 2003).

No junctions were observed between germ cells or between these and Sertoli cells in *Bufo arenarum* by Cavicchia and Moviglia (1982). However, desmosome-like junctions between spermatocytes and Sertoli cells have been reported in *Rana catesbeiana* (Sprando and Russell 1987) and for spermatocytes and spermatids in *Pachymedusa dacnicolor* in which no specialized contacts were present at the spermatogonial stage (Burgos and Vitale-Calpe 1967a,b; Rastogi *et al.* 1988). In the leptodactylid *Odontophrynus cultripes*, electron-dense plates resembling junctional structures appear where spermatids lie in close contact with the surface of microtubule-rich Sertoli cell processes (Báo *et al.* 1991).

Absence of Sertoli cells. *Bombina variegata* is reported to be unusual in that the germ cells develop in the absence of Sertoli cells and spermatozoa do not aggregate into bundles but appear as a mass of single cells (Obert 1976). However, the spermatids are embedded in Sertoli cell as far as their nuclei in *Discoglossus pictus* (Favard 1955a) and the observation for *B. variegata* requires confirmation.

Acrosome development. Although the acrosome vesicle (vacuole *sensu* Burgos and Fawcett 1956) and its precursors originate from the Golgi apparatus, there is no distinct acrosome granule in anuran spermiogenesis, unlike mammals and many other groups (*Bufo arenarum*, Burgos and Fawcett 1956; Cavicchia and Moviglia 1982; *Bufo gargarizans*, Kwon *et al.* 1993; *Xenopus laevis*, Reed and Stanley 1972; *Discoglossus pictus*, Sandoz 1970a,b; *Bombina bombina*, Folliot 1979; *Pachymedusa dacnicolor*, Rastogi *et al.* 1988; *Odontophrynus cultripes*, Báo *et al.* 1991; *Scinax ranki*, Taboga and Dolder 1994b). Golgi vesicles (dictyosomes) in the spermatid give rise to large numbers of vesicles which fuse to form the single acrosome vesicle (*Bufo arenarum*, Burgos and Fawcett 1956; Cavicchia and Moviglia 1982; *Xenopus laevis*, Reed and Stanley 1972; *Bombina bombina*, Folliot 1979; *Discoglossus pictus*, Sandoz 1970a,b; *Pachymedusa dacnicolor*, Rastogi *et al.* 1988; *Scinax ranki,* Taboga and Dolder 1994b). The vesicles arise on the opposite side of the Golgi complex to the location of the centrioles (Rastogi *et al.* 1988). A transitional stage is noted for *Discoglossus pictus* (Sandoz 1970b) in which Golgi vesicles first fuse to form the proacrosomal vesicles; a proacrosomal granule was observed at the center of each of these; the acrosome vesicle contains polysaccharides originating from the Golgi vesicles. In *Rana clamitans* the thick, short acrosome vesicle shows varying degrees of lobing which are attributed to incomplete fusion of the proacrosomal vesicles (Poirier and Spink 1971).

Whereas the acrosome vesicle in *R. clamitans* has inner and outer membranes these are said to be absent at maturity in *R. pipiens* (Poirier and Spink 1971). Grassé (1929; 1986), in a study of *Alytes obstetricans* by LM, recognized a single acrosome "granule" but from the evidence of other species, and as it was elaborated by more than one dictyosome which were incorporated into the acrosome, it seems likely that this also formed from proacrosomal vesicles. Development in *Discoglossus pictus* of the acrosome is said to be similar: in spermatocyte I, at pachytene, the dictyosomes, numbering about 40, are grouped in two or three complexes around the centrosome (cytoplasmic spheroid containing the centrioles). They disperse at metaphase and regroup in two lots at anaphase, one for each spermatocyte II, in each of which they form a group of 20. In the young spermatid they form the 'idiosome' with the centrosome at its center; shortly afterwards, the acrosome vesicle appears within the idiosome. Then, with the migration of the idiosome to the anterior pole of the nucleus, the acrosome vesicle greatly extends and covers the anterior hemisphere of the nucleus. A segregation of the acrosomal material then occurs: one part, against the nuclear surface, becomes the acrosome cap which extends for more than three quarters of the length of the nucleus; the remainder of the acrosome (an intra-acrosomal granule) elongates to become the helical axial perforatorium ('spirostyle'). As in *Alytes,* the dictyosomes become incorporated in the acrosome and are not seen in the residual cytoplasm (Favard 1955a; Grassé 1986).

The acrosome vesicle in mammalian spermatids becomes fixed to the nuclear membrane while the vesicle is still small but in anurans it attains a larger size before adhesion to this membrane (Folliot 1979; Cavicchia and Moviglia 1982; Rastogi *et al.* 1988; Taboga and Dolder 1994b; Lin *et al.* 2000), in *Xenopus laevis, Pachymedusa dacnicolor,* and *Scinax ranki* rivalling the nucleus in size (Reed and Stanley 1972; Rastogi *et al.* 1988; Taboga and Dolder 1994b). The acrosome vesicle then increases its area of contact with the nuclear membrane, extending further down the anterior pole of the nucleus, and at the same time becoming progressively narrower, appearing to collapse around the tip of the nucleus (Reed and Stanley 1972; Rastogi *et al.* 1988; Báo *et al.* 1991; Lee and Kwon 1992; Lin *et al.* 2000; Kwon *et al.* 1993). In the spermatid of *Discoglossus pictus,* the acrosome vesicle embraces the anterior half of the nucleus while the latter is still spherical (Favard 1955a; Sandoz 1970a) (Fig. 5.3). A double layered conical cap, the definitive acrosome vesicle, is thus formed (Burgos and Fawcett 1956; Rastogi *et al.* 1988; Taboga and Dolder 1994b) (see Fig. 5.2B for *Litoria rubella*). The vesicle is usually symmetrical but it may be longer on one side than the other as in *Discoglossus pictus* (Campanella *et al.* 1997) and *Hyla japonica* (Lee and Kwon 1992) (see section 5.3). In *Bufo*, homogeneous substance subsequently fills the lumen of the vesicle and, by maturity, the bounding membrane becomes difficult to discern from the contents (Burgos and Fawcett 1956), as in *Litoria rubella* (Fig. 5.2B) but the acrosome contents remain electron-lucent in some taxa, *e.g.*, *Scinax ranki* (Taboga and Dolder 1994b), *Pachymedusa dacnicolor* (Rastogi *et al.* 1988), and *Xenopus laevis* (Reed and Stanley 1972; Bernardini *et al.* 1990; Bernardini *et al.* 1986; Reed and Stanley 1972; Ueda *et al.* 2002).

A slender longitudinal bundle of microtubules which persist in *Pachymedusa dacnicolor* from the nuclear manchette at the apex of the subacrosomal space is regarded by Rastogi *et al.* (1988) as perforatorial.

Axial perforatorium. An axial fibrous rod, the axial perforatorium, lying in an endonuclear canal, is typical of the Discoglossoidea. Its development has been described for *Discoglossus pictus* (Favard 1955a; Sandoz 1970a,b, 1977) and *Bombina bombina* (Folliot 1979) (Fig. 5.6). It appears to develop from proteinaceous material initially segregated in the acrosome vesicle. In the spermatid of *D. pictus* a single granules is present posteriorly in the acrosome vesicle, soon becoming subacrosomal, and gives rise to the axial perforatorium ('baguette') which grows posteriad, simultaneously forming, around it, the endonuclear canal. The granule persists in the subacrosomal space at the head of the axial perforatorium at maturity. A terminal button of granular material develops anterior to the granule and persists within the tip of the mature acrosome vesicle (Sandoz 1970a) (Fig. 5.5).

In *Bombina bombina*, small dense masses within the acrosome vesicle aggregate in front of the forming axial rod (axial perforatorium). In that region the acrosome membrane is thickened, especially in two dark bands of the unit membrane. The clear cisterna of the unit membrane, approximately 3.5 nm wide, is crossed by very slender trabeculae. Later in morphogenesis the axial rod protrudes out of the nucleus at the apical end as a large subcylindrical blunt structure, with flat apex, capped by slender acrosome vesicle which extends down its sides (Folliot 1979).

Conical perforatorium. The perforatorium, here termed the conical perforatorium (also termed the subacrosomal cone by Lee and Jamieson 1992, Báo *et al.* 2001) to distinguish it from the axial perforatorium which may coexist with it as in *Ascaphus* and other discoglossoids, differentiates by fibrinogenesis in a narrow gap which develops between the acrosome vesicle and the nucleus. In bufonoids it consists of a number of coarse dense strands or fibers disposed

---

Fig. 5.5 *Discoglossus pictus*. Development of the axial perforatorium. **A**. A granule develops in the acrosome vesicle and becomes applied to the nuclear membrane. **B**. From it develops a rod (the axial perforatorium) which invaginates the nuclear membrane to form an endonuclear canal. **C**. The rod reaches the posterior end of the nucleus. Granular material around the granule is pressed forwards by the perforatorium and forms the terminal button in the acrosome vesicle. **D**. Granular material in the subacrosomal space persists as the conical perforatorium of current terminology. Relabelled after Sandoz, D. (1970a). In B. Baccetti (ed.), *Comparative Spermatology*, Academic Press, New York and London, Schema 2.

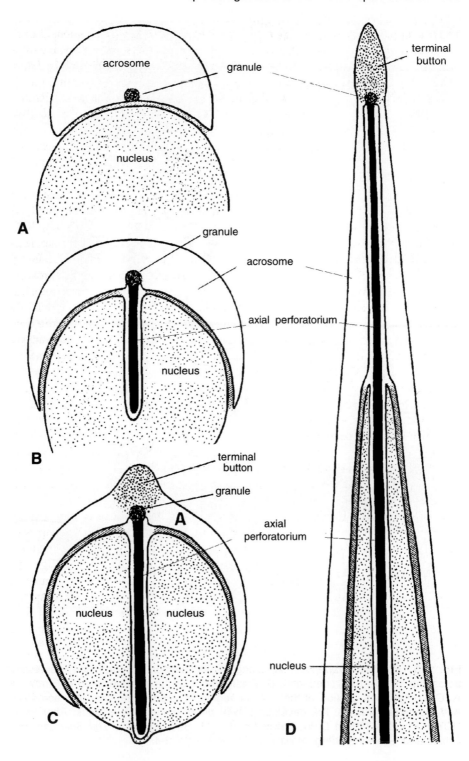

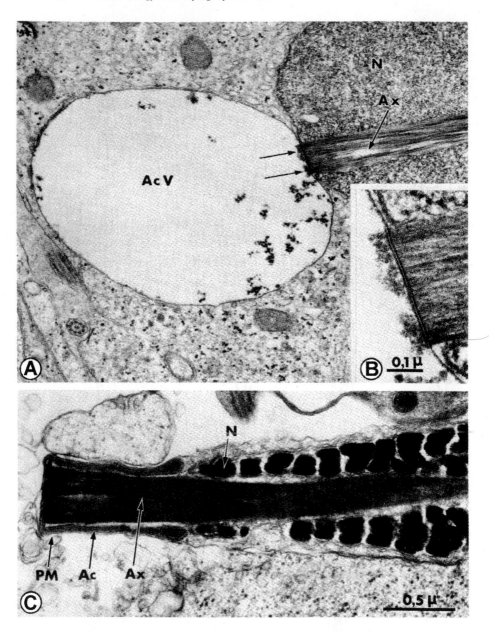

Fig. 5.6 *Bombina variegata*. **A**. Longitudinal section through the developing acrosome vesicle of a spermatid. **B**. Enlargement of the site indicated by arrows in A. C: Longitudinal section of the anterior region of a spermatozoon. Ac, acrosome. AcV, acrosome vesicle. Ax, developing axial perforatorium. N, nucleus. After Folliot (1979). Pp. 333-339. In D. W. Fawcett and J. M. Bedford (eds). *The Spermatozoon*, Urban and Schwarzenberg: Baltimore, Munich, Figs 3-4.

around the tapering end of the nucleus and converging anterior to it to form a dense, pointed structure; each of the coarse strands is a fascicle of closely packed, slender filaments (Burgos and Fawcett 1956), here shown in *Litoria rubella*, Fig. 5.2B, and *L. latopalmata*, Fig. 5.2D spermatids and for many mature spermatozoa (see section 5.3). The fibers seem to arise in very close relation to the nucleus but do not appear to be continuous with the nuclear membrane (Burgos and Fawcett 1956). In *Hyla japonica* these fibers have been reported to contain microtubules, as can be confirmed from a micrograph (Lee and Kwon 1992); microtubules are also said to assemble to form the subacrosomal cone (conical perforatorium) in *Hyla chinensis* (Lin *et al.* 2000).

In addition to profound differences from the rod-like perforatorium in shape and location, the bufonoid conical perforatorium develops very late in spermiogenesis (well after chromatin condensation and nuclear elongation) and in association with the nuclear membrane (Burgos and Fawcett 1956; Rastogi *et al.* 1988), whereas the endonuclear perforatorium develops very early in spermiogenesis and originates, as we have seen, from a granule at the base of the acrosome vesicle (*e.g.* Sandoz 1970a). In *Odontophrynus cultripes*, acid phosphatase was found in the conical perforatorium [labeled acrosome] and glucose-6-phosphatase in the acrosome vesicle [labeled acrosome membrane complex] (Fernandes and Báo 1998).

Fate of the Golgi complex. After the Golgi apparatus has produced the proacrosomal vesicles it moves to the posterior region of the spermatid where, in *Discoglossus pictus*, the dictyosomes produce polysaccharides before regressing (Sandoz 1970b).

Nuclear development. The nuclei of the spermatids are spherical in the early stages of differentiation but they later become ovoid and then progressively elongated until finally they are long slender rods, rounded at the base, which usually possesses a fossa, and tapering at the tip (Burgos and Fawcett 1956; Rastogi *et al.* 1988). In some rhacophorids, the nucleus, with the midpiece, then undergoes coiling (*Chiromantis xerampelina*, Mainoya 1981; *Rhacophorus arboreus* and *R. schlegelii*, Mizuhira *et al.* 1986; for mature sperm see section 5.3).

Nuclear condensation involves the progressive accretion of granules (*Bufo arenarum*, Burgos and Fawcett 1956; *B. gargarizans*, Kwon *et al.* 1993; *Bombina bombina*, Folliot 1979) or of fibrils (*Rana pipiens*, Zirkin 1971a; *Hyla japonica*, Lee and Kwon 1992; *H. chinensis*, Lin *et al.* 2000). In late spermiogenesis in *B. bombina* (Folliot 1979) preexisting membrane-bound nuclear vacuoles disappear and the chromatin granules condense into a number of almost prismatic equisized very large lumps regularly arranged along the nucleus (Folliot 1979). It seems possible that these are individual chromosomes. Lin *et al.* (2000) have shown, for *Hyla chinensis*, that nuclear material that is not involved in chromatin condensation is eliminated as a row of small vesicles in the cylindrical nucleus.

Nuclear development in *Megophrys montana* is unlike that reported for other species. The pattern of chromatin compaction does not follow nuclear membrane configuration. Instead, encased within the roughly spherical nuclear membrane, the developing nucleus forms a coil. When nuclear chromatin compaction is nearly complete, the cylindrical chromatin begins to uncoil. The nuclear membrane loses its spherical shape and starts to conform to the emerging form of its contents (Asa and Phillips 1988).

Nuclear manchette. An helical manchette of microtubules surrounding the spermatid nucleus has been demonstrated in the bufonid *Bufo arenarum*, where it is present only in the posterior region of the nucleus (Burgos and Fawcett 1956); the discoglossid *Discoglossus pictus* (Sandoz 1974, 1975) (Fig. 5.7); the hylids *Pachymedusa dacnicolor* (Rastogi *et al.* 1988) and *Hyla japonica* (Lee and Kwon 1992); the leptodactylids *Odontophrynus cultripes* (Báo *et al.* 1991) and *Physalaemus biligonigerus, P. fuscomaculatus*, and *P. gracilis* (Amaral *et al.* 1999). Abundant, fairly short, cytoplasmic microtubules are present in an irregular array in the spermatid of *Telmatobius schreiteri* (Pisanó and Adler 1968). In *D. pictus* the microtubules appear around the anterior region of the nucleus in which the perinuclear cisterna has diminished and nuclear pores have

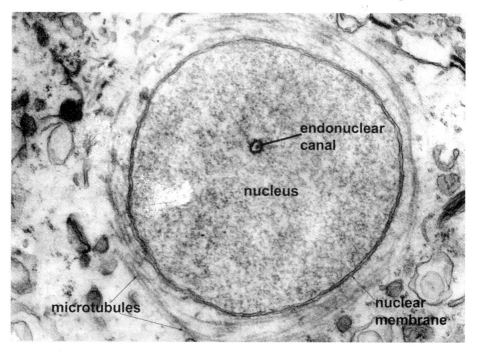

Fig. 5.7 *Discoglossus pictus*. Transverse section through the midregion of the nucleus, showing a helix of microtubules around the nucleus. Relabeled from Sandoz, D. 1974. Journal of Submicroscopic Cytology 6: 399-419, Fig. 16.

disappeared (Sandoz 1974). In *Xenopus laevis* a few microtubules are visible in the cytoplasm during formation of the acrosome but as nuclear elongation proceeds these are cast off, together with excess mitochondria, large vacuoles, and other cellular organelles such as endoplasmic reticulum and polysomes (Reed and Stanley 1972).

In the hylid *Pachymedusa dacnicolor*, nuclear shaping has been attributed to the array of perinuclear microtubules which disappear by the end of spermiogenesis and also, tentatively, to Sertoli microtubules (Rastogi *et al.* 1988; Báo *et al.* 1991). However, it is by no means certain that the manchette is concerned with nuclear morphogenesis, or at least that the presence of microtubules is an absolute requirement for this. Sandoz (1974) while noting that the presence of microtubules around the nucleus (Fig. 5.7) coincided with its elongation did not go so far as to suggest a causal relationship of one on the other. Furthermore, in the pipid *Xenopus laevis*, in contrast, there are a few microtubules, randomly distributed in the spermatid cytoplasm (Reed and Stanley 1972) and microtubules are not present around the nucleus in the spermatid of *Megophrys montana* (Asa and Phillips 1988), and, as shown above, the role of Sertoli microtubules in nuclear shaping has been emphasized by Kobayashi *et al.* (1993). However, a clear involvement of microtubules in chromatin condensation, or at least an intimate correlation of their presence with condensation and nuclear envelope modification, is seen in development of oligochaete sperm (see references in Jamieson 1981). Moreover, in mutants of *Drosophila melanogaster* in which microtubules (present around the nucleus in normal morphs) are absent, nuclear elongation does not occur (Wilkinson *et al.* 1974).

Nucleolus. The nucleoli originally present in the spermatids of *Bufo arenarum* consist of closely packed fine granules that are somewhat denser than the granules making up the rest of the karyoplasm. As nuclear differentiation proceeds, the nucleoli seem to break up into smaller aggregations of granules and these are no longer readily identifiable. Not uncommonly, in advanced spermatids, one or two sharply outlined spherical bodies 150 to 300 µm in diameter are seen among the coarse granules of chromatin. These may be related in some way to the nucleolus (Burgos and Fawcett 1956).

Cytoplasm. The cytoplasm of the spermatid contains numerous round mitochondria, abundant tubular and vesicular SER, and a prominent juxta-nuclear Golgi complex (Rastogi *et al.* 1988). Cytoplasm persists in the mature spermatozoon in many species, *e.g. Bombina bombina* (Folliot 1979).

Centrioles. Two centrioles are situated at the periphery of the spermatid of *Bufo arenarum* when the nucleus is still spherical (Burgos and Fawcett 1956). Initially, however, they are situated close to the Golgi complex (*e.g. Bombina bombina*, Folliot 1979; *Discoglossus pictus*, Sandoz 1975). Both proximal and distal centriole show the characteristic structure of nine triplets of microtubules (*e.g.* Rastogi *et al.* 1988; and *Litoria latopalmata*, Fig. 5.2E). The distal centriole organizes microtubules to form the flagellum ("template elongation") (Félix *et al.* 1994). During elongation of the spermatid the centrioles, and the base of the flagellum which is associated with one of these, move inwards followed by the cell surface and take up their definitive position at the caudal pole of the nucleus. One (Burgos and Fawcett 1956), as here shown in *Litoria latopalmata* (Fig. 5.2E), or both (Rastogi *et al.* 1988) come to be lodged within a basal nuclear fossa; they lie at approximately right angles to each other, with the distal centriole in the long axis of the spermatid (Burgos and Fawcett 1956; Rastogi *et al.* 1988). They are also mutually perpendicular in other species, *e.g. Telmatobius schreiteri* (Pisanó and Adler 1968); *Xenopus laevis* (Reed and Stanley 1972);

Bombina bombina (Folliot 1979); *Batrachyla antartandica, B. taeniata, B. leptopus* (Garrido *et al.* 1989); *Physalaemus biligonigerus, P. fuscomaculatus, P. gracilis* (Amaral *et al.* 1999); *Litoria latopalmata* (Fig. 5.2E); and many other cases may be seen in the account of mature sperm; they may also lie at an oblique angle or even parallel to each other (see section 5.3).

In *Bombina bombina* the centrioles are exceptional in lying in a small fossa near the anterior end of the nucleus. This indicates the presence of an acrosome and centriolar complex at opposite poles of the nucleus is not necessary for elongation of the spermatid as elongation occurs in *B. bombina*. Most of the peculiar features of the *Bombina* spermatozoon have been ascribed to the fact that the acrosome and centriolar complexes remain closely adjacent throughout the course of spermiogenesis (Folliot 1979).

In bufonoids, concomitant caudal displacement of cytoplasm from the anterior end of the cell results in the formation of a broad "cuff" (here termed a mitochondrial collar) of cytoplasm which surrounds the base of the nucleus for a distance of a few μm around the basal part of the tail from which it is separated by a gap, here termed the cytoplasmic canal. The collar contains numerous mitochondria with parallel or irregular cristae (Burgos and Fawcett 1956). Bufonoids with the collar were distinguished as eubufonoids by Lee and Jamieson (1992); in contrast, mitochondria border the axial fiber in the myobatrachids *Limnodynastes peronii* and *Neobatrachus pelobatoides* while in *Mixophyes fasciolatus* they are transient in a cytoplasmic droplet. The collar, described by Rastogi *et al.* (1988) as a 'mitochondrial sleeve'; is rich in SER and contains numerous mitochondria. These mitochondria apparently migrate from the cytoplasm surrounding the apical and middle portions of the elongating spermatid and become more numerous in the later stages of spermiogenesis of *Pachymedusa dacnicolor* (Rastogi *et al.* 1988). In *Telmatobius schreiteri* spermatids, the cytoplasmic canal appears to form by fusion of vesicles in the absence of a typical Golgi complex; mitochondria are seen in single file on each side in a collar in micrographs of late spermatids (Pisanó and Adler 1968). The sleeve is termed a 'cytoplasmic annular fold' for the leptodactylid *Batrachyla antartandica* by Garrido *et al.* (1989) who stress that the fold separates from the neck after the spermatozoon, freed from the Sertoli cells, is forced from the lumen of the seminiferous tubule into the collecting tubule and vasa efferentia. They suggested that the fold, with its mitochondria, has also detached in mature sperm of previously examined species: *Alsodes vittatus, Eupsophus roseus, Hylorina sylvatica, Bufo arenarum, B. bufo, B. variegatus, B. chilensis, Rhinoderma darwinii* and *R. rufum*, as also suggested for *Pachymedusa dacnicolor* by Rastogi *et al.* (1988). In *Pachymedusa dacnicolor*, an electron-dense mass develops in the immediate vicinity of the two centrioles, grows posteriorly and protrudes into the mitochondrial collar. A short distance from its origin, part of this mass passes towards the axoneme where it becomes orientated along one side of the axoneme within a common plasma membrane, thus forming an electron-dense rodlet (Rastogi *et al.* 1988), the juxta-axonemal fiber of the present chapter. The main part of the dense mass elongates parallel to the axoneme into the mitochondrial mass and forms the thicker axial rod (axial fiber). The plasma membrane around the axial fiber becomes continuous with that around the axoneme as the undulating membrane. The electron-dense material of the axial fiber passes through the undulating membrane and is continuous with the juxta-axonemal fiber. The three structures lie in the same plane as the two central singlets of the axoneme. The axial fiber is stiff whereas the axoneme, with its juxta-axonemal rod forms a spiral attached on one side by the undulating membrane to the axial rod (Rastogi *et al.* 1988).

Loss of the mitochondrial collar is also reported for *Hyla chinensis* but there the mitochondria move towards the proximal end of the tail and surround the distal centriole (Lin *et al.* 2000). However, the bufonid, leptodactylid (*sensu stricto*), and hylid families are considered by Lee and Jamieson (1993) to be united, and separated from myobatrachids, by the single synapomorphy of the thick collar-like cytoplasmic sheath containing the mitochondria. Scheltinga (2002; see also sections 5.3 and 5.4 below) also considers the collar in these families to persist at maturity.

Neck region. In *Bufo arenarum*, the proximal centriole is enveloped by a mantle of dense filamentous material continuous around the distal centriole with a thick band of dense material that is continuous distally with the undulating membrane complex. This pericentriolar material is covered by a filamentous outer layer or sheath. No annulus is said to be present, unlike urodele spermatids and spermatozoa (Burgos and Fawcett 1956). However, in the spermatid of *Discoglossus pictus* the dense proteinaceous material around the anterior end of the cytoplasmic canal was considered by Sandoz (1971) to be an annulus. It originates from laminae of rough endoplasmic reticulum (RER) that surround the anterior end of the tail (Sandoz 1971, 1975). It was referred to in the latter work merely as a 'ring' or 'muff' though it was ascribed a function in limiting the posterior migration of mitochondria (Sandoz 1975) as in a true annulus. After attachment of the centriolar complex to the nucleus, only the swollen anterior part of the 'muff 'persists and by the end of spermiogenesis it has become disymmetric, being widened at opposite sides in transverse section. The neck piece is inserted in the basal nuclear fossa. It is ovoid in form and consists of dense material which is continuous with the axial fiber of the tail. The pericentriolar material, as distinct, from the annulus, appears to form from Golgi vesicles (Sandoz 1975).

In *Bufo gargarizans*, a thick sheath forms around the proximal centriole. Posterior to this it divides into two columns, one large the other small. The large column connects to the axial fiber by three striated junctions. After the junction the columns extend tangentially, relative to the main axis, to the proximity of the midpiece whereas the small column ends at the level of the distal centriole (Kwon *et al.* 1993).

In *Physalaemus biligonigerus*, *P. fuscomaculatus*, and *P. gracilis* the 'connecting piece' located in the flagellum implantation zone of the spermatid also has transverse striations. These cross striations are reminiscent of the condition in the neck region of spermatozoa of mammals and of passerine birds and also ciliary roots (Amaral *et al.* 1999). In *Pachymedusa dacnicolor* there is also, at the base of the axial fiber ('axial rod'), a structure made up of cross striated fibers

(Rastogi *et al.* 1988). Cross striations are demonstrated for many mature sperm in this account.

A centriolar fossa is absent from the sperm of *Xenopus laevis*; the paired centrioles are associated with a pericentriolar apparatus consisting of nine radial fibers (see mature spermatozoa, section 5.3) (Bernardini *et al.* 1986).

Sperm tail and undulating membrane. A well developed flagellum, arising from the distal centriole, and consisting of the usual 9+2 arrangement of microtubules (Fig. 5.2F, for *Litoria rubella*), is visible before either acrosome formation or nuclear elongation begin (Rastogi *et al.* 1988). The centriole organizes development of the microtubules of the flagellum ("template elongation") in the spermatid but, when the sperm has entered the egg, pericentriolar material assembles and nucleates a microtubule aster ("astral nucleation"). There is experimental evidence that gamma-tubulin and certain phosphorylated epitopes appear in the centrosome only after entering the egg, the gamma-tubulin being recruited from the egg. This suggests that gamma-tubulin and/or associated molecules play a key role in centrosome formation and activity (Félix *et al.* 1994).

As described for *Bufo arenarum* by Burgos and Fawcett (1956), projecting perpendicularly from the side of the axoneme is a thin ribbon-like band of fibrous substance continuous with that surrounding the proximal centriole. This material forms the major part of the undulating membrane. In transverse sections through the base of the tail it appears as a thick band extending only a short distance from the side of the flagellum, as shown here for *Litoria rubella* (Fig. 5.2F). In sections farther caudad, the dense material of the undulating membrane is reduced to a thin sheet in its middle portion but remains thicker along its attached and its free margins. This entire structure is invested by an extension of the plasma membrane of the tail. The thickening along the attached border is roughly triangular in transverse section and is the juxta-axonemal fiber at doublet 3 in current terminology (Fig. 5.2F). The undulating membrane is always in the same plane as the two central singlets of the axoneme. The thickening in the free margin of the membrane is the axial fiber; it is larger than the juxta-axonemal fiber and is oval in transverse section.

A juxta-axonemal fiber (marginal fiber) at doublet 8, seen in urodele sperm, is absent in most anuran sperm but is seen as a transient structure in the spermatid of *Discoglossus pictus* (Sandoz 1975) and occasional spermatozoa of *Bufo marinus* (Swan *et al.* 1980).

Where an undulating membrane is present, an axial fiber usually develops at its extremity. An axial fiber is said to be absent, in the presence of an undulating membrane, in *Physalaemus biligonigerus*, *P. fuscomaculatus*, and *P. gracilis* (Amaral *et al.* 1999) but is here interpreted as possessing a paraxonemal rod (combined juxta-axonemal and axial fibers). An axial fiber and undulating membrane fail to develop in *Telmatobius schreiteri* (Pisanó and Adler 1968). For other conditions, including presence of an axial fiber in the absence of an undulating membrane, see mature sperm, section 5.3.

Spermatogenesis in a hybrid. The European water, or green frog, *Rana esculenta*, is a hybrid whose somatic genome is composed of haploid chromosome sets of its parental species *R. lessonae* and *R. ridibunda* (Vinogradov *et al.* 1988; Vinogradov *et al.* 1990; Ogielska and Bartmanska 1999). The DNA amount in the *ridibunda* genome is 16% greater than that of *lessonae*. The DNA content of *esculenta* somatic cells is exactly intermediate between those of both parental species. In contrast, the spermatozoa and the primary spermatocytes of *R. esculenta* have a DNA content which corresponds to the size of the *ridibunda* genome. Some hybridogenetic males have spermatogonia with only the *ridibunda* genome size, whereas the others have diploid cells with the *esculenta* (*i.e. ridibunda* + *lessonae*) genome size (Vinogradov *et al.* 1990). Thus, it is suggested that the selective elimination of the *lessonae* genome and compensatory doubling of the *ridibunda* one may occur in spermatogonia of *R. esculenta* males before premeiotic DNA synthesis occurs (Vinogradov *et al.* 1990; Ogielska and Bartmanska 1999). Meiosis then proceeds in the usual way on the basis of the diploid *ridibunda* genome (Vinogradov *et al.* 1990). In triploid males, the cytogenetic mechanism of hemiclonal inheritance is simpler than in diploids: after the elimination of a genome (always the genome in the minority in the triploid set; "homogenizing elimination"), no compensatory duplication of the remaining genetic material is necessary, unlike diploids (Vinogradov *et al.* 1990).

In *Rana esculenta*, gonad development is affected from the earliest stages: the gonads are smaller and composed of a reduced number of primary spermatogonia relative to the parent species; the phase of pale primary spermatogonia proliferation is prolonged up to the second year of life; and the structure of the gonads and of the germ cells is often abnormal (Ogielska and Bartmanska 1999).

The numbers of stem cell spermatogonia and differentiated spermatogonia vary according to the period of the year, as does the rate of turnover of stem cells, with nearly 60-90% of cells temporarily out of the cell cycle at any given time (Rastogi *et al.* 1985).

Centromere arrangement. Indirect immunofluorescence staining with human anti-kinetochore antibodies allows recognition of centromeres during vertebrate spermiogenesis. Many species of Amphibia have a low chromosome number and very large spermatids and spermatozoa. The number of kinetochore dots correlates exactly with the haploid chromosome number. This implies that kinetochore duplication occurs in the interval between meiosis I and meiosis II. The non-homologous centromeres are arranged in tandem during the entire course of spermiogenesis and in mature spermatozoa. A higher order centromere arrangement was found in spermiogenetic cells of anurans and urodeles. It is suggested that pair-wise association of centromeres is a universal principle of centromere arrangement at the postmeiotic stage (Haaf *et al.* 1990).

Control of the duration of spermatogenesis. Factors affecting spermatogenesis are beyond the scope of this chapter and details are given in Fernández and Ramos (Chapter 3 of this volume). However, we may note that it has recently been demonstrated that the normal annual testis cycle is controlled by different seasonal patterns of secretion of somewhat specifically acting FSH (follicle stimulating hormone) and LH (luteinizing hormone) gonadotropins (Jorgensen 1984). The kinetics of spermatogenesis is also linked to phyletic position and environmental factors. Tritiated thymidine injection in testis of gymnophionans, urodeles and anurans, has revealed close correlations between cycle type and climate, and between sexual maturity age and life duration. Meiosis duration is species-specific, and it is linked to genetic and ecological factors (Neyrand de Leffemberg and Exbrayat 1995).

Duration of spermatogenesis relative to the life span. We can take only some examples of estimations of the duration of spermatogenesis. It has been estimated using tritiated thymidine labeling in three tropical African anurans. At 30°C, there is a minimum duration of one week in *Phrynobatrachus calcaratus*, three weeks in *Bufo regularis* and an intermediate period in *Ptychadena bibroni*. This difference appears to depend on the longevity of each of the three species. Male *Ph. calcaratus* are only about 15 mm long, weigh on average 0.3 g, and live only about one year; *B. regularis* is up to 8 cm long, weighs 43 g, and lives longer than eight years; *Pt. bibroni* has a mean length of 4.5 cm, weighs 6.5 g, and lives three to four years (Gueydan-Baconnier *et al.* 1984).

In *Xenopus laevis*, at 18°C, the most rapidly maturing cells in the testis spend four days in leptotene, six days in zygotene, 12 days in pachytene, one day in diplotene, one day in meiotic division, and 12 days in spermiogenesis. The length of the premeitoic S stage was estimated as six to seven days. Spermatogenesis is continuously active and variations among males in numbers of individual stages are interpreted as the result of non-random entry of spermatogonial stem cells into the population of cells irreversibly committed to differentiation (Kalt 1976). However, there was variation between animals, and somewhat different figures have been given elsewhere, thus Kobayashi *et al.* (1993) observed that, at 22±1°C, primary spermatogonia were observed in newly metamorphosed toads; secondary spermatogonia three days after metamorphosis; and first leptotene and pachytene spermatocytes 17 and 20 days, respectively; the first diplotene spermatocytes, meiotic division and round spermatids 42 days; and the first mature spermatozoa 70 days, after metamorphosis.

In *Bufo arenarum* the testes become differentiated, during the 30 days after metamorphosis, directly from an undifferentiated pregonad. At the beginning of the third or fourth postmetamorphic month the fundamental anatomical testis organization is completed. During the first year, the testes show primary spermatogonia. The first spermatogenic wave (SW) commences during the first winter. The second SW shows an annual cycle initiated at the beginning of spring. It produces spermatocytes in the second spring to autumn and spermatids in winter. In the three-year old toads the third SW produces spermatozoa, but there is no interstitial tissue. During the fourth SW a large number of interstitial cells organized as a tissue appears, together with the nuptial pad on the polex. The testes complete maturation during the fourth spring (Echeverria and Maggese 1987).

The above estimates of the duration of different stages of spermatogenesis are a subsample only of the considerable literature (see, for instance, Neyrand de Leffemberg and Exbrayat (1995), for data on *Bufo regularis*, *Rana esculenta*, *Xenopus laevis*, and *Rana lessonae*; Gavaud (1977), for *Nimbaphrynoides occidentalis*).

5.3 MORPHOLOGY OF THE MATURE SPERMATOZOON

5.3.1 Introduction

The mature spermatozoa of the Anura are elongate, their length varying from approximately 25 to 180 µm, with a few exceptions (*Telmatobufo australis* and *Leiopelma hochstetteri* reach approximately 240-250µm, while *Discoglossus pictus* attains a length of 2,500µm (Furieri 1975a; Pugin-Rios 1980; Scheltinga *et al.* 2001; Scheltinga 2002). The spermatozoon consists of a head and a distinct tail which are in anteroposterior sequence excepting in the genus *Bombina*. The tail is generally longer than the head though in a few species from a variety of families it is equal, or very rarely shorter as in *D. pictus*.

The head is composed of an anterior acrosome complex (with or without a perforatorium) and nucleus. The form of the head may be cylindrical, curved cylindroconical, helical cylindroconical or fusiform, or may comprise apical whorls of a helical spermatozoon. The structure and organization of the tail is also variable. At the junction of the head and tail, termed the neck region, two centrioles are always present. Each centriole is composed of nine, circularly arranged, triplets of short microtubules. Mitochondria with well-developed linear cristae are usually located within the neck region and anterior region of the tail, though the number and disposition of the mitochondria varies greatly or they may even be absent. The 9+2 axoneme (with the plasma membrane, constituting the flagellum) may be simple, double, or connected to the axial fiber by an undulating membrane with or without the presence of a smaller juxta-axonemal fiber. A dense lamina within the undulating membrane connects the axial fiber and juxta-axonemal fiber. In some species a solitary paraxonemal rod (representing axial fiber, lamina, and juxta-axonemal fiber) is present, while in others the tail is composed of an axoneme and a cylindrical axial fiber only. The axial fiber, where present, is approximately in the longitudinal axis of the spermatozoon and the axoneme occupies a lateral position at the external, flexible margin of the undulating membrane.

5.3.2 Suborder "Archaeobatrachia"

The Archaeobatrachia is probably a paraphyletic group (see Duellman, Chapter 1 of this volume) both in terms of osteology and spermatology. It contains the Discoglossoidea (also probably paraphyletic), consisting of the Ascaphidae, Bombinatoridae, Discoglossidae, and Leiopelmatidae, which are the most basal anurans (Cannatella 1985; Hillis 1991; Scheltinga *et al.* 2001), the Pelobatoidea, and the Pipoidea. The Pelobatoidea and Pipoidea are often termed the Mesobatrachia. Sperm ultrastructure has been investigated in eight of the nine archaeobatrachian families, the exception being the pipoid family Rhinophrynidae.

5.3.3 Superfamily "Discoglossoidea"

This is here considered a paraphyletic group but the conventional complement of families will be included.

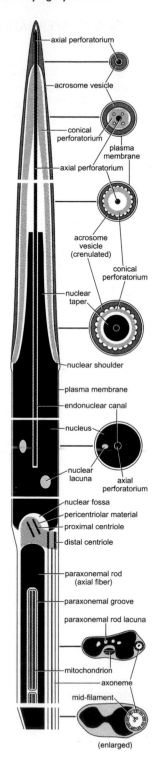

axial perforatorium

acrosome vesicle

conical
perforatorium

plasma
membrane

axial perforatorium

acrosome
vesicle
(crenulated)

conical
perforatorium

nuclear
taper

nuclear shoulder

plasma membrane

endonuclear canal

nucleus

nuclear
lacuna

axial
perforatorium

nuclear fossa

pericentriolar material

proximal centriole

distal centriole

paraxonemal rod
(axial fiber)

paraxonemal groove

paraxonemal rod lacuna

mitochondrion

axoneme

mid-filament

(enlarged)

5.3.4 Family Ascaphidae

The sperm of the sole living genus of the Ascaphidae, *Ascaphus*, has previously been very briefly described by light microscopy (Metter 1964). Its ultrastructure has been described by James (1970) and Jamieson *et al.* (1993). The ultrastructure of the spermatozoon of *Ascaphus truei* is shown diagrammatically in Figure 5.8.

The tip of the spermatozoon of *Ascaphus truei* is formed by the elongate, pointed acrosome complex, the base of which caps the tapered anterior end of the nucleus. Longitudinal and transverse sections through the basal region of the acrosome reveal (from center to periphery) an electron-dense, rod-shaped axial perforatorium; an endonuclear canal; the concavely tapering anterior region of the nucleus; an electron-lucent area (the "perinuclear space"); a conical periperforatorial sheath (subacrosomal cone) of moderate electron density (here recognized as a conical perforatorium); and another electron-lucent area (the subacrosomal space); all surrounded by a conical acrosome vesicle. The nucleus and the acrosome vesicle are each membrane-bound. The interior margin of the acrosome vesicle appears scalloped in transverse section. At the base of the acrosomal region, the nucleus, which does not penetrate the anterior region of the acrosome, flares out abruptly and the nuclear shoulders, conical perforatorium, and acrosome vesicle are contiguous. Behind this the nucleus forms a very elongate cylinder. The nucleus is approximately 25 μm long and 0.7 μm in diameter. The conical perforatorium extends posteriorly for a short distance beyond the base of the acrosome vesicle. The axial perforatorium protrudes anteriorly beyond the nucleus, penetrating the subacrosomal cone and extends into the nucleus within an endonuclear canal but does not penetrate a long basal region of the nucleus. The chromatin is electron-dense but contains large electron-lucent lacunae.

Two centrioles are present immediately behind the nucleus. Both are surrounded by diffuse, moderately electron-dense pericentriolar material. The proximal centriole is located in a fossa at the posterior end of the nucleus and is orientated obliquely with respect to the longitudinal nuclear axis. The distal centriole lies posterior to the proximal centriole with its length parallel to the longitudinal nuclear axis. It is continuous with the axoneme. The cytoplasm surrounding the centriolar region contains electron-lucent vesicles of varying size.

The spermatozoon tail consists of a 9 + 2 axoneme, lying directly adjacent to a large electron-dense longitudinal rod, termed the paraxonemal rod, which extends laterally for approximately four or five times the width of the axoneme. The midline of the paraxonemal rod is nearest doublet number 2 of the axoneme. The rod is confluent with the diffuse pericentriolar material and, in transverse section, has a distinct constriction so that it bears a longitudinal groove on one or both sides. The constricted region, forming a bridge between inner and outer regions of the paraxonemal rod, is considered to be the equivalent of the

Fig. 5.8 Diagrammatic representation of a spermatozoon of *Ascaphus truei* showing a longitudinal section with corresponding transverse sections as indicated. Modified from Jamieson, B. G. M. *et al.* 1993. Herpetologica 49: 52-65, Fig. 1.

undulating membrane of many anuran sperm. In the region of the constriction, there are small electron-lucent areas which form a longitudinal series. Proximally, the constriction is asymmetrical, giving a groove on one side only of the rod. It has been suggested by Jamieson *et al.* (1993) that in *Ascaphus*, the short, stout undulating membrane homologue is the result of paedomorphosis of the spermatozoon. In urodele spermatids (Barker and Biesele 1967), for instance, the structure of the tail is very similar to that in mature *Ascaphus* sperm: in both cases, the axoneme rests in a groove in the paraxonemal rod and both structures are surrounded by cytoplasm. Only in late urodele spermiogenesis does the undulating membrane develop and the cytoplasm disappear (Picheral 1972a). Deletion of this terminal stage should be a relatively simple process. It therefore seems likely that the derived features of the *Ascaphus* sperm tail (cytoplasmic coat, see below, reduction of the undulating membrane, electron-lucent areas in paraxonemal rod) can all be ascribed to a single process: spermatozoal paedomorphism.

Elongate mitochondria (numbering one to several in transverse section of the sperm tail) are present in the groove of the paraxonemal rod. Distally, the rod tapers gradually, the constriction becomes symmetrical, and mitochondria are finally absent. The axoneme has the usual 9 + 2 microtubular structure throughout its length but in addition, possesses a fine central "mid-filament" (*sensu* James 1970) which lies between the central singlets though displaced slightly towards doublet number 1. All tail structures (paraxonemal rod, axoneme, and mitochondria) are ensheathed by a thin layer of cytoplasm, the cytoplasmic coat.

5.3.5   Family Bombinatoridae

The sperm of *Bombina bombina*, *B. orientalis*, and *B. variegata* have previously been described by light microscopy (Retzius 1906), TEM (Furieri 1975a,b; Folliot 1979; Pugin-Rios 1980; Kwon and Lee 1992, 1995; Lee and Kwon 1996), and/or SEM (Kuramoto 1998). The ultrastructure of the spermatozoon of *B. variegata* is shown diagrammatically in Figure 5.9.

The spermatozoa of *Bombina* has an elongate (34-45 µm), slightly curved fusiform shape. The tail is juxtaposed to the convex side of the nucleus throughout the length of the head. The acrosome vesicle, 2.0-5.7 µm long, is cylindrical to cylindroconical, and truncated at the summit. The axial perforatorium, 20 µm long and 0.4-0.45 µm wide, is a cylindrical rod, consisting of longitudinal parallel coarse fibers. It has a flattened apical tip and continues posteriorly into the nucleus, within the endonuclear canal, for approximately half the length of the nucleus. The arrangement of fibers is less regular within the endonuclear canal in which the axial perforatorium progressively narrows before ending. The acrosome vesicle consists of a cylindrical part (sometimes absent) around the perforatorium with an external diameter of 0.5 µm and a more posterior region, 0.5-0.8 µm wide, which forms the base of a cone and surrounds the anterior end of the nucleus. The vesicle has a thickness of 0.03-0.04 µm, excepting a thicker basal annular dilatation. Its contents are moderately electron-dense and homogeneous.

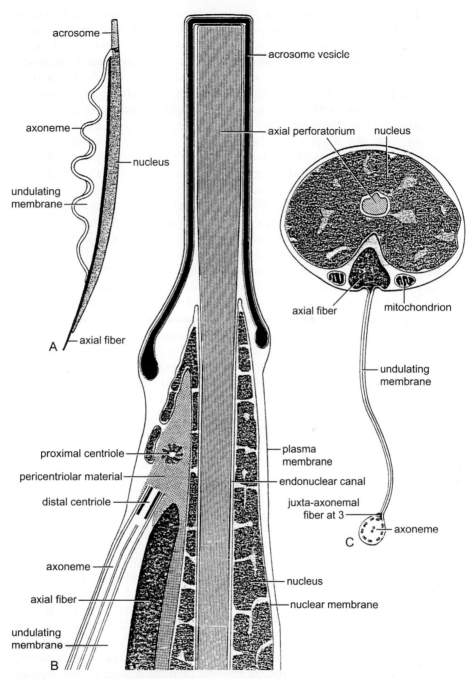

Fig. 5.9 Drawing of a spermatozoon of *Bombina variegata* showing; **A**. Whole spermatozoon. **B**. Longitudinal section through the anterior region. **C**. Transverse section through the mid region. From (relabelled) Pugin-Rios, E. 1980. Unpublished Ph.D. Thesis, L'Université de Rennes, France, Figs 3, 4, 5.

The nucleus is fusiform or clavate with the thin end anterior. In mature sperm the chromatin is seen to be condensed into medium-sized masses; scattered throughout the chromatin are lacunae which contain granules of varying size and opacity. A lateral implantation fossa (nuclear or centriolar fossa) is present near the anterior tip of the nucleus, which contains the proximal centriole at its anterior end and the distal centriole at its posterior end. The proximal centriole lies perpendicular to the long axis of the spermatozoon. The distal centriole forms the basal body of the axoneme, is separated from the nucleus by the anterior end of the axial fiber, and is orientated between 10° to 35° to the long axis of the spermatozoon. The fossa is continuous with a longitudinal groove which extends for a short distance along the nucleus and which contains the axial fiber. The two centrioles and the anterior end of the axial fiber are enveloped in a granular and fibrillar material which is in contact with the slightly thickened nuclear membrane of the fossa and the groove. The chromatin consists of dense masses, irregular in size and distribution, separated by clear, granular zones.

The axial fiber is 0.3-0.4 µm in diameter, though becoming thinner for a short distance at each end. Initially it is triangular in transverse section before becoming, in the posterior third, circular and finally irregular-shaped. The undulating membrane of the tail is strongly developed, being 3 µm wide, and has a central lamina which appears to envelope the axial fiber and ends at a thicker juxta-axonemal fiber adjacent to doublet 3 of the axoneme. An unusually large amount of cytoplasm persists, producing a groove with very thick walls, especially in the posterior two thirds, at least in testicular sperm, on each side of the axial fiber. Located within the cytoplasm, especially distally, are spherical mitochondria with distinct cristae which are especially numerous near the axial fiber, lamellar bodies, vacuoles, and glycogen granules. Details of spermiogenesis are given by Folliot (1979) (see section 5.2, above) who ascribes most of the peculiar features of the *Bombina* sperm to the fact that the acrosome and centriolar complexes remain closely adjacent throughout the course of spermiogenesis.

5.3.6 Family Discoglossidae

The sperm of the discoglossids *Discoglossus pictus* and *Alytes obstetricans* have previously been described and will be discussed separately below.

Discoglossus pictus. A variety of stages and processes of development of the spermatozoa of the Painted Frog, *Discoglossus pictus*, have been examined (see Favard 1955a,b; Sandoz 1970a,b, 1973, 1974, 1975; Furieri 1975a; Campanella and Gabbiani 1979; Pugin-Rios 1980). The ultrastructure of the spermatozoon of *D. pictus* is shown diagrammatically in Figure 5.10.

The sperm of *Discoglossus pictus* is the longest known in the Amphibia, measuring 2,300 µm to 2,500 µm. It has the form of a long helical filament composed of a head, 1,300 µm long and with a basal diameter of 1 µm, and a tail 1000 µm long. The nucleus, 700-800 µm long, is penetrated for almost its whole length by a narrow endonuclear canal. A perforatorium traverses the endonuclear canal in the late spermatid. This is said by Pugin-Rios (1980) to be

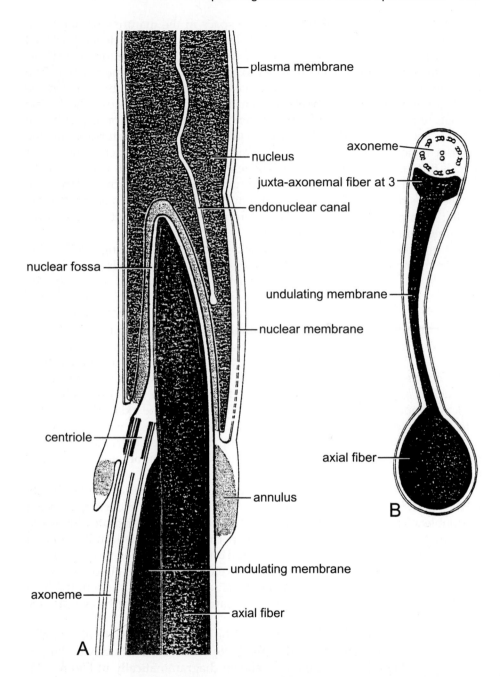

Fig. 5.10 Drawing of a spermatozoon of *Discoglossus pictus* showing; **A**. Longitudinal section through the neck region; and **B**. Transverse section through the principle piece of the tail. From (relabelled) Pugin-Rios, E. 1980. Unpublished Ph.D. Thesis, L'Université de Rennes, France, Figs 6, 7.

restricted to the prenuclear subacrosomal space in the mature spermatozoon. However, Campanella *et al.* (1997) have demonstrated a well developed axial perforatorium (acrosome rod), with long endonuclear and prenuclear regions, in mature sperm which were induced to undergo the acrosome reaction. The acrosome vesicle is in the form of very large elongate cone, almost as long as the head, covering all but the posterior 40 μm of the nucleus. The vesicle consists of a short thicker anterior segment and a long posterior segment. The anterior segment, about 0.6 μm wide, surrounds the perforatorium and the anterior region of the nucleus. The average thickness of the acrosome at this level is 0.18 μm. The long posterior segment has a width which varies from 0.4 to 0.5 μm and a constant thickness of 0.09 μm. Granular subacrosomal material lying between the acrosome vesicle and nucleus in *Discoglossus pictus* (Sandoz 1970a; Campanella *et al.* 1997) is tentatively recognized as a conical perforatorium in the present account.

The anterior region of the nucleus lacks chromatin, consisting solely of the nuclear membrane and that of the endonuclear canal. Further posteriorly it acquires chromatin which basally reaches a diameter of 1 μm. In the postacrosomal region the chromatin is rosette-shaped in transverse section owing to the presence of helical cortical bands. The nucleus has a deep basal implantation fossa which is almost entirely filled by the anterior end of the axial fiber. The two centrioles are situated very close together and lateral to the axial fiber. They and the anterior end of the axial fiber are surrounded by a dense fibrous bed or ring (identified with an annulus) which is connected to the nuclear membrane of the fossa by fine fibers. Sandoz (1974) states that the neck is composed of an ovoid mass, continuous with the axial fiber, inserted into the base of the nucleus. Sandoz (1974, 1975) equates it with the neck of urodeles and suggests that if may be compared to the centriolar annexe of insect sperm, the connecting piece of mammalian sperm, or ciliary rootlets of epithelia.

The short undulating membrane is continuous with the pericentriolar sheath. It has a thick internal lamina connecting the axial fiber to a thickening adjacent to doublets 2, 3 and 4 of the axoneme. This thickening constitutes the juxta-axonemal fiber at 3. In mature testicular sperm the small ovoid mitochondria around the base of nucleus have disappeared but a bed of nucleoplasm separates the dense chromatin from the nuclear envelope; at this level some nuclear pores are visible. The anterior region of the tail is partially surrounded by a short cytoplasmic collar from which it is separated by a space (the cytoplasmic canal).

Alytes obstetricans. The spermatozoa of the midwife toad, *Alytes obstetricans*, has been examined optically by Retzius (1906) and ultrastructurally by Furieri (1975a) and Pugin-Rios (1980) and is shown diagrammatically in Figure 5.11. It is 90 μm long and consists of a fusiform head (26 μm long and 1.2 μm wide) and a tail (64 μm long, consisting of a principal piece and endpiece). The acrosome vesicle, with fine granular moderately electron-dense homogeneous contents, is 3.6 μm long and has the form of a truncated cone which encloses the anterior portions of the axial perforatorium and nucleus. The perforatorium

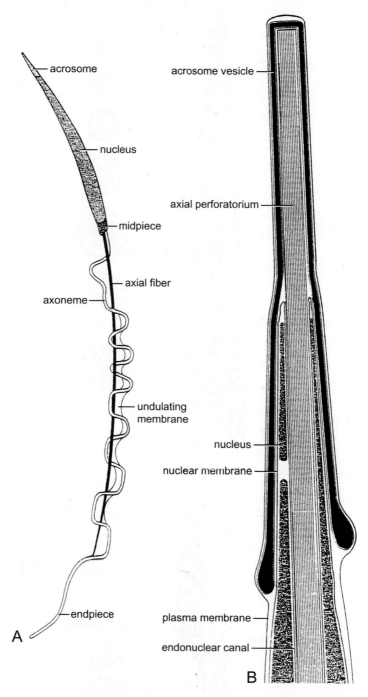

Fig. 5.11   Drawing of a spermatozoon of *Alytes obstetricans*. **A**. Whole spermatozoon. **B**. longitudinal section through the anterior region. From (relabelled) Pugin-Rios, E. 1980. Unpublished Ph.D. Thesis, L'Université de Rennes, France, Figs 1, 2.

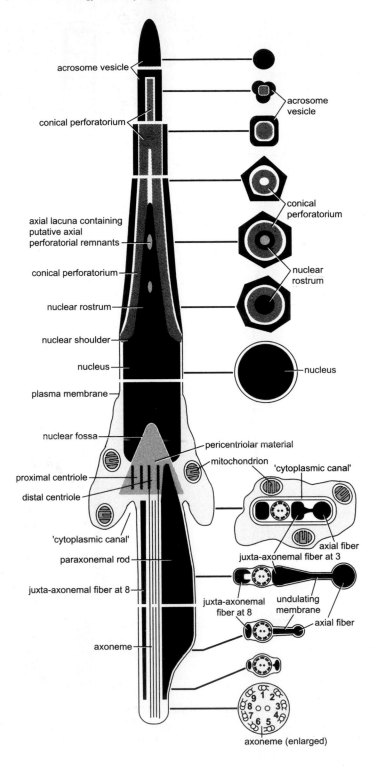

axoneme (enlarged)

is cylindrical, 26 µm long and 0.18 µm wide, and extends posteriorly within the nucleus for the entire length of an endonuclear canal. The canal is lined by the nuclear membrane. The acrosome vesicle is of uniform thickness (0.03 µm) but is thickened basally where it is turned outwards.

The tail is composed of a 9+2 axoneme, thin undulating membrane, and axial fiber. The axial fiber is similar to the axoneme in thickness but progressively narrows within the distal part of the principal piece and ends almost at the same level as the undulating membrane. It is circular in transverse section and consists of a homogeneous electron-dense central substance surrounded by a less dense layer. Doublets A and B of the 9+2 axoneme are of equal density. The endpiece consists of the axoneme only.

5.3.7 Family Leiopelmatidae

The ultrastructure of the spermatozoon of *Leiopelma hochstetteri* has previously been described by Scheltinga *et al.* (2001) and is illustrated diagrammatically in Figure 5.12.

The spermatozoa are filiform, being composed of a curved cylindroconical head (acrosome and nucleus) 66.8 µm long, a midpiece 2.3 µm long, and a tail 154 µm long and averaging 224 µm in total length.

The acrosome complex is composed of an elongate conical acrosome vesicle and an underlying putative conical perforatorium which caps the 3.24 µm long nuclear rostrum. The acrosome is a membrane bound vesicle and is filled with moderately electron-dense material. The conical perforatorium is not bound, lies free within the subacrosomal space, and is composed of diffuse material basally. Apically the material of the conical perforatorium compacts to form a dense rod which is flat-ended anteriorly. The conical perforatorium extends posteriorly for a short distance beyond the base of the acrosome vesicle. In transverse section the acrosome vesicle has a polygonal structure, being circular apically and becoming heptagonal basally.

The nucleus is elongate and cylindrical, is composed of electron-dense chromatin, and contains several axial lacunae which contain material which is possibly the homologue of an axial perforatorium. Although no continuous endonuclear canal is present, what appears to be the vestige of a short, more continuous endonuclear canal has been observed in the very tip of the nuclear rostrum in a late spermatid. Distinct nuclear shoulders are present at the base of the nuclear rostrum. At their level the nucleus is 0.41 µm in diameter; this increases throughout its length to a maximum diameter of 0.80 µm at the level of the basal nuclear fossa. The fossa is conical and 0.92 µm deep.

Two parallel centrioles, orientated in the long axis of the nucleus, lie at the base of the nuclear fossa. The two centrioles lie adjacent to each other embedded

---

Fig. 5.12 Diagrammatic representation of a spermatozoon of *Leiopelma hochstetteri* showing a longitudinal section with corresponding transverse sections as indicated. From (slightly relabelled) Scheltinga, D. M., Jamieson, B. G. M. *et al.* 2001. Zoosystema 23: 157-171, Fig. 1, ©Publications Scientifiques du Muséum National d'Histoire Naturelle, Paris.

in a common mass of electron-dense material, and do not extend into the nuclear fossa. One of the centrioles forms the basal body of the axoneme. The axial fiber of the tail extends through the neck region to the level of the base of the nucleus. A short cytoplasmic collar surrounding the tail contains a few, scattered mitochondria. A gap (termed the 'cytoplasmic canal' for many vertebrate sperm) separates the cytoplasmic collar from the tail.

The tail complex is composed of a 9+2 axoneme; an axial fiber, connected by a thickened undulating membrane to a juxta-axonemal fiber at doublet 3; and a further juxta-axonemal fiber at doublet 8. Anteriorly (in transverse section) the undulating membrane is shorter, and with the juxta-axonemal fiber at 3 and the axial fiber, forms an electron-dense, hourglass shaped paraxonemal rod. Posteriorly, the juxta-axonemal fiber at 3 becomes less distinct from the undulating membrane. The juxta-axonemal fiber at doublet 8 extends along much of the length of the axoneme, decreasing in size posteriorly. For some of its length this fiber has two thin extensions attaching it to the axoneme. A short portion of the axoneme extends alone beyond the juxta-axonemal and axial fibers as the endpiece.

5.3.8   Superfamily Pelobatoidea

The Megophryidae, Pelobatidae, and Pelodytidae comprise the Pelobatoidea and are considered basal anurans. The ultrastructure of spermatozoa of all three families has been examined and is discussed separately below.

5.3.9   Family Megophryidae

The sperm of the genera *Atympanophrys*, *Megophrys*, *Leptobrachium*, *Leptolalax*, and *Xenophrys* have previously been described by light microscopy and/or TEM (Asa and Phillips 1988; Zheng *et al.* 2000; Scheltinga 2002). The ultrastructure of the spermatozoon of *Leptolalax* sp. is shown diagrammatically in Figure 5.13.

Megophrys montana. Knowledge of ultrastructure of the male gamete in *Megophrys montana* is limited to a study of nuclear shaping in spermatids (Asa and Phillips 1988). There is an anterior acrosome, in the spermatid almost touching a Sertoli cell nucleus; a helical nucleus is followed by two parallel axonemes enclosed in a cytoplasmic sheath which contains mitochondria and electron-lucent vesicles. Throughout the length of the sperm tail the two axonemes are enclosed in a common membrane; a bridge connects doublets 3 and 8 of adjacent axonemes. Little cytoplasm and no vacuoles are present within the tail. The location of the mitochondria in *Megophrys* requires confirmation. The mitochondria appear to be in either a single linear series which occurs between the two axonemes and adjoining them or in two linear series, one adjacent to each axoneme anteriorly which changes to one linear series posteriorly.

Atympanophrys and *Xenophrys*. The mature spermatozoa of two species of *Atympanophrys* (=*Megophrys*) and 14 species of *Xenophrys* (=*Megophrys*) examined by Zheng *et al.* (2000) appear to be similar in general shape to those of *Leptobrachium* and *Leptolalax* given below. However, the spermatozoon head of

Atympanophrys and *Xenophrys* is more coiled (16.3-16.6 and 5.5-14.2 turns, respectively) and longer.

Leptobrachium and Leptolalax. The spermatozoa of the genera *Leptobrachium* and *Leptolalax* are filiform, being composed of a head (acrosome and nucleus) that is loosely coiled/spiral (approximately 3.5 turns) and a long tail (midpiece and principal piece). Under light microscopy no definite distinction can be made between the acrosome and nucleus, or between the midpiece and principal piece. Lengths of measurable components of the spermatozoa of the Pelobatoidea are given in Table 5.1.

Table 5.1   Dimensions (means in µm) of Pelobatoidea spermatozoa.

Species	Head length	Tail length	Total length	Author
Atympanophrys gigantica	91	75	167	Zheng *et al.* 2000
Atympanophrys shapingensis	107	79	187	Zheng *et al.* 2000
Leptobrachium aff. *hasseltii*	20	28	47	Scheltinga 2002
Leptolalax dringi	40	52	93	Scheltinga 2002
Leptolalax pictus	22	41	63	Scheltinga 2002
Pelobates syriacus	67	62	129	Scheltinga 2002
Pelodytes punctatus	25	44	68	Pugin-Rios 1980
Spea intermontana	14	25	39	Scheltinga 2002
Spea intermontana	13	22	35	Scheltinga 2002
Xenophrys boettgeri	46	60	106	Zheng *et al.* 2000
Xenophrys brachykolos	56	62	118	Zheng *et al.* 2000
Xenophrys glandulosa	41	61	103	Zheng *et al.* 2000
Xenophrys jingdongensis	57	65	122	Zheng *et al.* 2000
Xenophrys kuatunensis	40	54	94	Zheng *et al.* 2000
Xenophrys medogensis	38	53	91	Zheng *et al.* 2000
Xenophrys minor	64-76	71-87	135-164	Zheng *et al.* 2000
Xenophrys omeimontis	58-61	69-70	127-131	Zheng *et al.* 2000
Xenophrys parva	46	52	98	Zheng *et al.* 2000
Xenophrys spinata	47-56	65	111-121	Zheng *et al.* 2000
Xenophrys wuliangshanensis	58	61	119	Zheng *et al.* 2000
Xenophrys wushanensis	63	65	128	Zheng *et al.* 2000
Xenophrys zhangi	41	54	95	Zheng *et al.* 2000

The acrosome complex is loosely coiled and is composed of an acrosome vesicle which caps the conical perforatorium. The perforatorium is composed of coarse longitudinal fibers which appear as homogeneous separate fibers in transverse section, and overlies the attenuated nuclear tip. The acrosome complex of *Leptolalax* is longer and extends beyond the nuclear tip for a greater length than that of *Leptobrachium*.

The nucleus is loosely coiled, conical, and composed of condensed chromatin

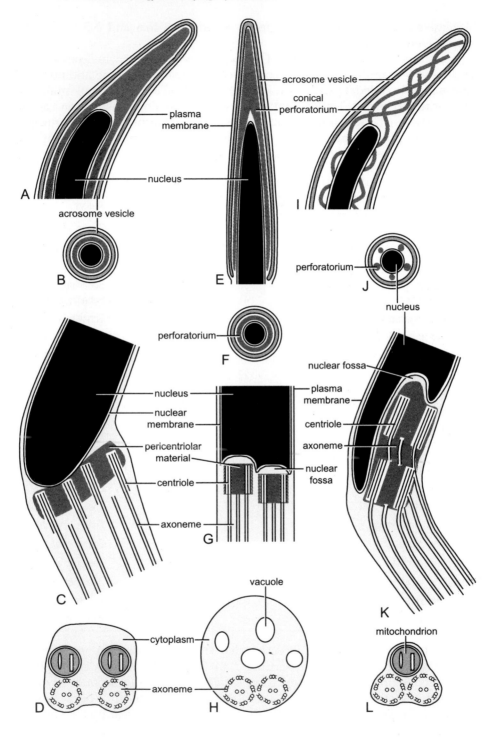

containing nuclear inclusions. Distinct nuclear shoulders are absent. The base of the nucleus is rounded with no fossa. Two centrioles are present which abut the base of the nucleus and each is continuous with a 9+2 axoneme of the tail. The centrioles are parallel and adjacent to each other and lie in the long axis of the spermatozoon. Pericentriolar material is present which also enters the centrioles. Spherical mitochondria are located in two longitudinal series, with each linear series adjacent to an axoneme. An axial fiber, undulating membrane, and juxta-axonemal fibers are absent. The two axonemes of the tail lie within a common plasma membrane and, unlike the axonemes of *Megophrys*, are separated from each other by a region of cytoplasm. Anteriorly, within the principal piece, a large amount of cytoplasm containing vacuoles of varying size is present.

5.3.10 Family Pelobatidae

The sperm of all three genera of the Pelobatidae, *Pelobates, Scaphiopus,* and *Spea*, have been previously examined (Retzius 1906; James 1970; Morrisett 1974; Scheltinga 2002). The ultrastructure of the spermatozoon of *Spea intermontana* and *Pelobates syriacus* is shown diagrammatically in Figure 5.13. The spermatozoa of the pelobatid genera *Pelobates* differ markedly in structure from those of *Scaphiopus* and *Spea* and are therefore described separately.

5.3.10.1 Spermatozoon of Pelobates

Pelobates syriacus **and** *P. fuscus*. The spermatozoa of *Pelobates fuscus* and *P. syriacus* are similar in general appearance. They are filiform, being composed of an elongate head (acrosome and nucleus) that is coiled and an elongate tail (midpiece and principal piece). Under light microscopy no definite distinction can be made between the acrosome and nucleus, or between the midpiece and principal piece.

Pelobates syriacus. The head of *P. syriacus* is composed of two regions, a tightly coiled anterior region with an average length of 14.6 µm and a loosely coiled 'wavy' posterior region with an average length of 52.3 µm.

The acrosome complex is composed of an extremely thin acrosome vesicle which caps both the perforatorium and the tapered end of the nucleus. The perforatorium is composed of one (apically) to five (basally) distinct longitudinal fibers that are homogeneous and which spiral around the nucleus. The nucleus is conical, circular in transverse section, composed of condensed chromatin and tapers to a fine point apically. Anteriorly, the nucleus is tightly coiled, the coils becoming looser posteriorly until the nucleus has a 'wavy' structure. Nuclear shoulders and inclusions are absent. An asymmetrical basal nuclear fossa is present in which the two centrioles lie.

The two centrioles are parallel and adjacent and lie in the long axis of the

Fig. 5.13 Diagrammatic representation of the spermatozoa of the Pelobatoidea. **A-D**. *Leptolalax pictus*. **E-H**. *Spea intermontana*. **I-L**. *Pelobates syriacus*. **A, E, I**. Longitudinal section (LS) through the anterior (acrosomal) region; **B, F, J**. Transverse section (TS) through the acrosome complex. **C, G, K**. LS through the centriolar region. **D, L**. TS through the midpiece. **H**. TS through the principal piece of the tail. Original.

spermatozoon with one centriole being more proximal than the other. Each centriole forms the basal body of an axoneme of the tail. The mitochondria are located in a single linear series. An axial fiber, undulating membrane, and juxta-axonemal fibers are absent. The two axonemes adjoin each other, with doublet 3 of one axoneme contacting doublet 8 of the other, and occur within a common plasma membrane. Little cytoplasm is present within the tail.

5.3.10.2 Spermatozoon of *Scaphiopus* and *Spea*

Scaphiopus **and** *Spea*. The spermatozoa of *Scaphiopus couchii* and *Sc. holbrookii* (James 1970; Morrisett 1974) and *Spea intermontana* and *Sp. intermontana* (Scheltinga 2002) are filiform, being composed of a short head (acrosome and nucleus) and a short tail (midpiece and principal piece). The spermatozoon head of *Spea* is slightly curved or bent but not coiled whereas that of *Scaphiopus* forms a single-turn helix (*i.e.* one coil). Under light microscopy no distinction can be made between the acrosome and nucleus, or between the midpiece and principal piece.

The acrosome complex is composed of an acrosome vesicle which caps a conical perforatorium composed of fine longitudinal fibers in *Spea* or the nucleus, in the absence of a perforatorium, in *Scaphiopus*. The nucleus is conical, tapering to a fine point anteriorly, circular in transverse section, and composed of condensed chromatin containing nuclear inclusions. Distinct nuclear shoulders are absent. The nucleus terminates basally in an asymmetrical basal nuclear fossa which has two distinct levels, each containing a centriole oriented parallel to the nuclear axis. One centriole is located more proximal than the other. The two centrioles are continuous with two parallel axonemes of the tail.

The exact position, number, and structure of the mitochondria have not been determined nor described with any accuracy in either genus but appear to be small-ovoid and located anteriorly for a short distance along the tail, starting just posterior to the centrioles. An axial fiber, undulating membrane, and juxta-axonemal fibers are absent. The two axonemes are closely associated with each other, with doublet 3 of one axoneme contacting doublet 8 of the other, and occur within a common plasma membrane. In *Scaphiopus* the axonemes may also be seen to be separated from each other by cytoplasm, particularly anteriorly. Within the anterior portion of the tail, a large amount of cytoplasm which contains numerous vacuoles of varying size is present. The cytoplasm decreases posteriorly until only the two axonemes, surrounded by the plasma membrane, are present.

5.3.11 Family Pelodytidae

The sperm of the monogeneric Pelodytidae has been examined in *Pelodytes punctatus* by Pugin-Rios (1980). They are filiform, being composed of a head (acrosome and nucleus) that is loosely coiled (1.5 to 2 turns), midpiece, and elongate tail (long principal piece and short endpiece). The ultrastructure of the spermatozoon of *P. punctatus* is shown diagrammatically in Figure 5.14.

The acrosome vesicle is moderately electron-dense and homogeneous, and surrounds a conical perforatorium consisting of longitudinal fibers. The nucleus is loosely coiled, circular in transverse section, and composed of condensed chromatin. A single symmetrical basal nuclear fossa is present and contains only the proximal centriole which is perpendicular to the long axis of the spermatozoon. The distal centriole lies at about 20° to the long axis and forms the basal body of the single axoneme of the tail. A moderately electron-dense and homogeneous axial fiber, circular in transverse section, is present. It is attached to the axoneme (near doublet 3) by a thin cytoplasmic lamina, the undulating membrane. Juxta-axonemal fibers are absent. The axoneme extends posteriorly beyond the limit of the axial fiber and undulating membrane as the endpiece.

5.3.12 Superfamily Pipoidea

The Pipoidea contains two families, the Pipidae and Rhinophrynidae. Only sperm of the Pipidae have been examined.

5.3.13 Family Pipidae

Only the sperm of *Xenopus laevis* (James 1970; Reed and Stanley 1972; Furieri 1972; van der Horst 1979; Pugin-Rios 1980; Bernardini *et al.* 1986; Yoshizaki 1987) has been examined ultrastructurally in the Pipidae. The spermatozoon is 47-62 μm long and composed of a loosely coiled (1-1.5 turns) head which is 17-22 μm long and a simple tail 30-40 μm long. The ultrastructure of the spermatozoon of *X. laevis* is shown diagrammatically in Figure 5.15.

The acrosome vesicle has the form of an apically rounded cone, covering the anterior part of the nucleus. The subacrosomal space is very small and a perforatorium is absent. The nucleus is surrounded posteriorly by a mitochondrial sheath 3-4 μm long. This sheath is formed by mitochondria interspersed with vesicles. The mitochondria are not fused, but are rounded, with a diameter of about 0.2 μm. The chromatin is electron-dense and homogeneous. Nuclear vacuoles of variable size and with granular contents are frequent. The double nuclear membrane is dense close to the inner acrosomal membrane and in the posterior nuclear region. Microtubules running for short distances along the nucleus and intra- and extra-nuclear glycogen packets have been observed. The nucleus tapers to a rounded end basally and thus a nuclear fossa is absent.

The tail is simple and consists of a short region surrounded by the mitochondrial sheath and a long simple axoneme. The number of mitochondria is reduced posteriorly within the nuclear region and they extend for only about 1 μm along, and around, the anterior portion of the axoneme. A very short cytoplasmic canal is occasionally evident. The proximal centriole, near the base of the nucleus is oblique relative to the long axis of the spermatozoon. The distal centriole is more elongate and forms the basal body of the axoneme, generally in a position lateral and slightly oblique relative to the long axis. The two centrioles lie at about 140° to each other. The centrioles are surrounded by dense pericentriolar material and give rise, via 9 radial fibrous bridges, to the system of cross striated fibrous bands of the centriolar complex. An axial fiber, undulating membrane, and juxta-axonemal fibers are absent.

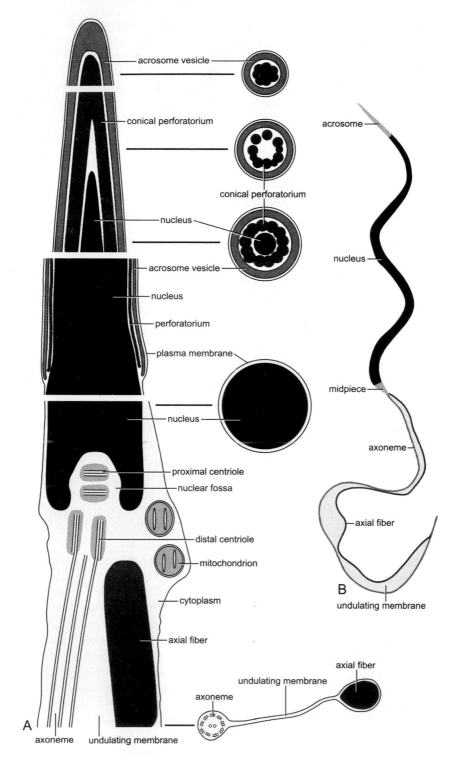

5.3.14 Suborder Neobatrachia

Whereas the Archaeobatrachia clearly comprise a paraphyletic assemblage, the suborder Neobatrachia appears on morphological and molecular grounds to be a monophyletic group (see Duellman, Chapter 1 of this volume). The Neobatrachia contain diverse and numerous groups of what are termed the advanced frogs. They are divided into two superfamilies, the Bufonoidea and Ranoidea, though some authors include a third superfamily, the Microhyloidea. Seven of the 11 bufonoid families and five (includes the Microhylidae) of the eight ranoid families have been investigated for sperm ultrastructure.

5.3.15 Superfamily Bufonoidea

The spermatozoa of the bufonoid families Allophrynidae, Brachycephalidae, Dendrobatidae, and Sooglossidae have not been described.

5.3.16 Family Bufonidae

The sperm of bufonid genera *Ansonia*, *Bufo*, and *Nimbaphrynoides* have previously been described (see Table 5.2). The ultrastructure of the spermatozoon of *Ansonia* sp. A and *Nimbaphrynoides occidentalis* is shown diagrammatically in Figures 5.16 and 5.17, respectively.

Table 5.2 Bufonid spermatozoa examined.

Species	Method Used	Author
Ansonia spp.	LM; TEM	Scheltinga 2002
Bufo arenarum	LM; TEM	Burgos and Fawcett 1956; Cavicchia and Moviglia 1982; Scheltinga 2002
Bufo arunco	LM; TEM	Pugin and Garrido 1981
Bufo boreas	LM; TEM	Scheltinga 2002
Bufo bufo	LM; TEM	Retzius 1906; Furieri 1961; Nicander 1970; Pugin-Rios 1980
Bufo calamita	LM; TEM	Pugin-Rios 1980
Bufo gargarizans	LM; TEM; SEM	Sze *et al.* 1981; Mo 1985; Kwon *et al.* 1993; Kwon and Lee 1995; Kuramoto 1998
Bufo japonicus	LM; TEM; SEM	Yoshizaki and Katagiri 1982; Takamune 1987; Kuramoto 1998
Bufo marinus	LM; TEM	Swan *et al.* 1980; Lee and Jamieson 1993; Scheltinga 2002
Bufo melanostictus	SEM	Kuramoto 1998
Bufo rangeri		van der Horst 1979; van der Horst *et al.* 1995
Bufo variegatus	LM; TEM	Pugin-Rios 1980; Pugin and Garrido 1981
Bufo woodhousii	TEM	Morrisett 1974
Nimbaphrynoides occidentalis	TEM	Pugin-Rios 1980

The spermatozoa of the bufonid genera *Ansonia* and *Bufo* (Fig. 5.17F) are filiform, being composed of a head (acrosome and nucleus), a short midpiece,

Fig. 5.14 Diagrammatic representation of a spermatozoon of *Pelodytes punctatus*. **A**. Longitudinal section with corresponding transverse sections as indicated. **B**. Whole spermatozoon. Drawn from light and TEM micrographs in Pugin-Rios, E. 1980. Unpublished Ph.D. Thesis, L'Université de Rennes, France.

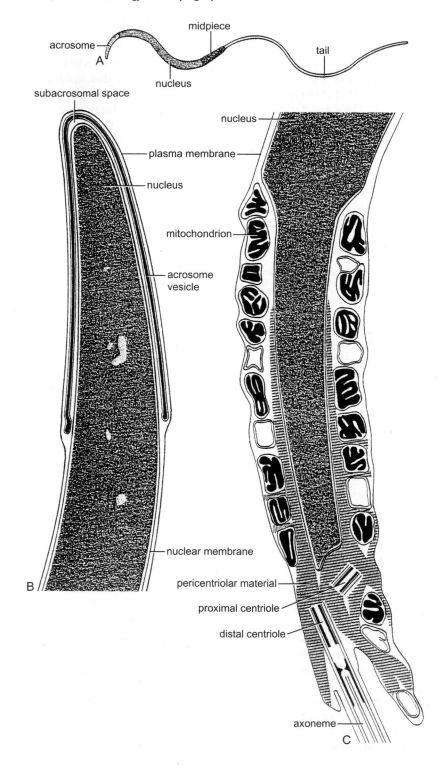

and a long tail (principal piece and endpiece) (Fig. 5.18I). Lengths of measurable components of the spermatozoa are given in Table 5.3.

Table 5.3 Dimensions (means in μm) of bufonid spermatozoa.

Species	Head length	Midpiece length	Tail length	Endpiece length	Total length	Author
Ansonia sp. A	27	2.8	57	8.8	87	Scheltinga 2002
Ansonia sp. B	26	2.7	61	9.9	90	Scheltinga 2002
Bufo arunco	16	?	35	?	≈ 51	Pugin and Garrido 1981
Bufo boreas	25	2.4	48	8	76	Scheltinga 2002
Bufo bufo	30-34	0	60-70	?	94-100	Furieri 1961; Pugin-Rios 1980
Bufo gargarizans	24	1.3	44	?	72	Kuramoto 1998
Bufo japonicus	27	1.4	58	?	89	Kuramoto 1998
Bufo marinus	18†-21	2.3-2.9	44-50	10	68-70	Lee and Jamieson 1993; Scheltinga 2002
Bufo melanostictus	19	1.4	42	?	64	Kuramoto 1998
Bufo variegatus	17	?	50	?	≈ 67	Pugin and Garrido 1981
Nimbaphrynoides occidentalis	40	7	60	?	100	Pugin-Rios 1980

†Includes midpiece.

The acrosome complex is composed of an acrosome vesicle that completely caps the conical perforatorium (Fig. 5.18A-C). The acrosome vesicle is membrane bound and filled with moderately electron-dense material. Anteriorly, the vesicle is relatively thick, becoming thin in the middle and posterior regions and terminating basally as a thickened bulb-shape (Fig. 5.18A, C). The perforatorium, situated within the subacrosomal space, is composed of coarse longitudinal fibers which appear as homogeneous separate sheaves in transverse section, and overlies the attenuated nuclear tip (Fig. 5.18A-C, E, F, J, K). The perforatorium is moderately electron-dense and is not membrane bound. The acrosome complex extends anteriorly beyond the nucleus, being symmetrically attached to the nucleus basally (Fig. 5.18A-C).

The nucleus is cylindro-conical, tapering to a rounded tip within the acrosome complex (Fig. 5.18A-C). It occupies most of the head length, is circular in transverse section, and composed of electron-dense condensed chromatin (Fig. 5.18G, L). Numerous small nuclear inclusions, which are electron-lucent or contain electron-pale material, are present (Fig. 5.18B, C, L). Distinct nuclear shoulders are absent. The nucleus increases in diameter throughout its length to a maximum at its base where it ends in a well-developed fossa (the basal nuclear, or centriolar, fossa) (Fig. 5.18D). The fossa is asymmetrical in one plane

Fig. 5.15 Drawing of a spermatozoon of *Xenopus laevis*. **A**. Whole spermatozoon. **B**. Longitudinal section (LS) through the anterior region. **C**. LS through the midpiece and centriolar region. From (relabelled) Pugin-Rios, E. 1980. Unpublished Ph.D. Thesis, L'Université de Rennes, France, Figs 8, 9, 10.

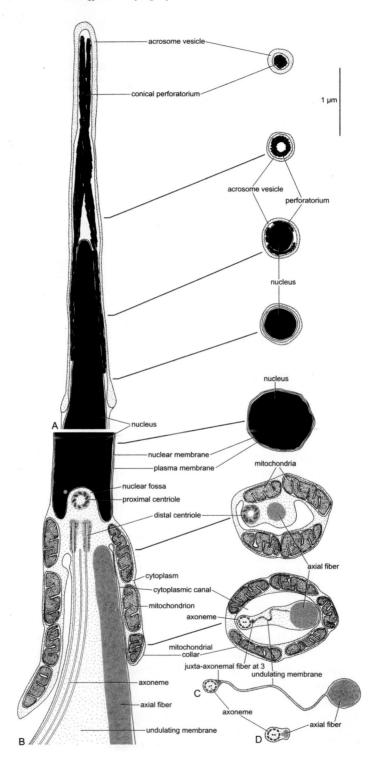

of longitudinal section, *i.e.* when the proximal centriole is also seen in longitudinal section.

Within the basal nuclear fossa lies the proximal centriole which is surrounded by a small amount of pericentriolar material that connects it to the nuclear fossa and distal centriole (Fig. 5.18D). The proximal centriole is orientated at approximately 70° to the long axis of the spermatozoon and at a right angle to the distal centriole which forms the basal body of the axoneme (Fig. 5.18D). Each centriole is composed of nine, circularly arranged, triplets of short microtubules. The axial fiber extends anteriorly into the neck region to the level of the distal centriole. A short mitochondrial collar, containing scattered mitochondria, cytoplasm, and the occasional vacuole, surrounds the anterior portion of the tail but is separated from it by a gap, the cytoplasmic canal (Fig. 5.18D, M, O, P). This collar is attached only to the centriolar/neck region of the spermatozoon and the cytoplasmic canal extends posteriorly from the level of the base of the distal centriole. The mitochondria have well-developed linear cristae and are of variable shape and size within a spermatozoon, being generally flattened and plate like in *Ansonia* (Figs 5.16, 5.18O) and spherical in *Bufo* (Fig. 5.18D, M, P).

The tail complex is composed of a 9+2 axoneme and an axial fiber, the fiber being connected to doublet 3 of the axoneme by a thin cytoplasmic lamina, the undulating membrane (Fig. 5.18I, N, Q). A juxta-axonemal fiber adjacent to doublet 3 is present but small, being less than the diameter of the axoneme (Fig. 5.18N, Q). The axial and juxta-axonemal fibers are continuous via a thin dense lamina sandwiched within the undulating membrane. The axial fiber is well developed and circular to ovoid in transverse section throughout its length. Anteriorly, it is slightly enlarged, the diameter being approximately 0.4 μm in *Ansonia* spp. and *Nimbaphrynoides occidentalis*, and approximately 0.15-0.22 μm in *Bufo* spp. (measured within the principal piece region) (Fig. 5.18I, N, Q). The axial fiber continues posteriorly, decreasing in diameter, for much of the tail length. The axoneme extends beyond the axial fiber and undulating membrane as the endpiece (Fig. 5.18H). Lengths of the endpiece are given in Table 5.3.

The minor differences in the spermatozoa of *Bufo woodhousii* from those of other *Bufo* (*e.g.* Golgi remnants, cytoplasm, and an unknown organ consisting of 'annulate lamellae' around the sperm nucleus) reported by Morrisett (1974), all appear to represent features of immature sperm. Lee and Jamieson (1993) reported the occasional asymmetry of the perforatorium which continues further posteriorly along one side of the nucleus than along the other in *B. marinus*. This condition was also observed in *Ansonia*, *B. arenarum*, and *B. boreas* by Scheltinga (2002) but its occasional nature, *i.e.* symmetrical perforatoria also

Fig. 5.16 Drawing of a spermatozoon of *Ansonia* sp. showing longitudinal sections through: **A.** Acrosome complex. **B.** Midpiece, with corresponding transverse sections (TSs) as indicated. **C.** TS through the main portion of the principal piece. **D.** TS through the distal portion of the principal piece just anterior to beginning of the endpiece. From Scheltinga, D. M. 2002. Unpublished Ph.D. Thesis, University of Queensland, Australia, Fig. 3.1.

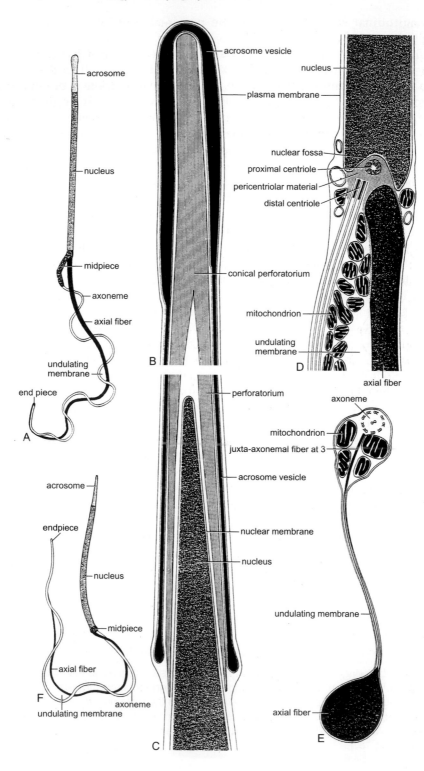

acrosome

nucleus

midpiece

axoneme

axial fiber

undulating
membrane

end piece

A

acrosome vesicle

plasma membrane

conical perforatorium

perforatorium

acrosome vesicle

nuclear membrane

nucleus

B

C

nucleus

nuclear fossa

proximal centriole

pericentriolar material

distal centriole

mitochondrion

undulating
membrane

axial fiber

D

axoneme

mitochondrion

juxta-axonemal fiber at 3

undulating membrane

axial fiber

E

acrosome

endpiece

nucleus

midpiece

axial fiber

F

undulating
membrane

axoneme

observed, limits its use as a phylogenetic character. In *B. bufo*, a cytoplasmic canal has been reported as present (Nicander 1970) or absent (Furieri 1961), both conditions were also observed in *B. japonicus*, *B. gargarizans*, and *B. melanostictus* by Kuramoto (1998). However, there is evidence that the mitochondrial collar may degenerate, with its mitochondria, in very late testicular spermatozoa (Pugin-Rios 1980; Garrido *et al.* 1989) or readily detach from the mature spermatozoa (Kuramoto 1998). In *B. marinus* occasionally spermatozoa with supernumerary (juxta-axonemal) accessory fibers are found. These supernumerary fibers are variable in position, being located adjacent to doublets 5-8 (Swan *et al.* 1980).

Nimbaphrynoides occidentalis is unusual in the Anura in having internal fertilization and in being viviparous (Xavier 1977; Duellman and Trueb 1986). However, its spermatozoon show no apparent modification for this specialized fertilization biology. The structure of the spermatozoon of *Nimbaphrynoides occidentalis* is similar to that given above for *Ansonia* and *Bufo*; however, two distinct differences are present. The acrosome complex differs from other bufonids in that the perforatorium ends slightly posterior to the acrosome vesicle. The distal centriole is unusual in being penetrated throughout its length by the central singlets of the axoneme. No mitochondrial collar is present, the mitochondria are located around the anterior axoneme, occupying two lateral compartments which are separated by the axoneme, juxta-axonemal fiber, and dense lamina of the undulating membrane.

5.3.17 Family Centrolenidae

The sperm of *Hyalinobatrachium fleischmanni* has previously been described by Scheltinga (2002) and is shown diagrammatically in Figure 5.19. The spermatozoa are filiform, being composed of a head (acrosome and nucleus) 19 µm long, a short midpiece 1.5 µm long, and a tail (principal piece and endpiece) of unknown length.

The acrosome complex is composed of an acrosome vesicle that completely caps the conical perforatorium. The acrosome vesicle is 4.6 µm long, membrane bound, relatively thin throughout its length, and filled with moderately electron-dense material. The perforatorium is composed of large longitudinal fibers, which appear as homogeneous separate sheaves in transverse section, and overlies the attenuated nuclear tip. The perforatorium is moderately electron-dense and is not membrane bound. The acrosome complex extends anteriorly beyond the nucleus, being symmetrically attached to the nucleus basally.

The nucleus is cylindro-conical, tapering to a pointed tip within the acrosome complex. It occupies most of the head length, is circular in transverse section, and composed of electron-dense condensed chromatin. Numerous small nuclear

Fig. 5.17 Drawing of a spermatozoon of **A-E**. *Nimbaphrynoides occidentalis* **F**. *Bufo bufo*. **A**. Whole spermatozoon. **B**. Longitudinal section (LS) through the apical acrosome complex. **C**. LS through the basal acrosome complex. **D**. LS through the centriolar region. **E**. Transverse section through the midpiece region. **F**. Whole spermatozoon. From (relabelled) Pugin-Rios, E. 1980. Unpublished Ph.D. Thesis, L'Université de Rennes, France, Figs 26, 27, 33, 34, 35, 36.

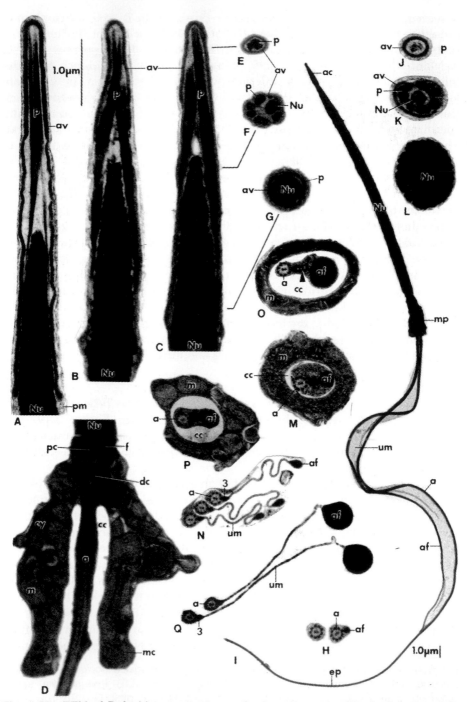

Fig. 5.18 TEM of Bufonidae spermatozoa. **A**. *Ansonia* sp. Longitudinal section (LS) through the acrosome complex. **B**. *Bufo boreas*. LS through the acrosome complex.

Contd.

inclusions, which are electron-lucent or contain electron-pale material, are present. Distinct nuclear shoulders are absent. The nucleus increases in diameter throughout its length to a maximum of 0.91 µm at its base where it ends in a well-developed basal nuclear fossa. The fossa is asymmetrical in one plane of longitudinal section, *i.e.* when the proximal centriole is also seen in longitudinal section.

Within the basal nuclear fossa lies the proximal centriole which is surrounded by a small amount of pericentriolar material that connects it to the nuclear fossa and distal centriole. The proximal centriole is orientated at approximately 85° to the long axis of the spermatozoon and at a right angle to the distal centriole which forms the basal body of the axoneme. The axial fiber extends anteriorly into the neck region to the basal most point of the nucleus. A short mitochondrial collar, containing scattered mitochondria, cytoplasm, electron dense granules, and the occasional vacuole, surrounds the anterior portion of the tail but is separated from it by a gap, the cytoplasmic canal. This collar is attached only to the centriolar/neck region of the spermatozoon and the cytoplasmic canal extends posteriorly from the level of the base of the distal centriole. The mitochondria have well-developed linear cristae and are of similar shape and size, being generally spherical.

The tail complex is composed of a 9+2 axoneme and an axial fiber, the fiber being connected to doublet 3 of the axoneme by a thin cytoplasmic lamina, the undulating membrane. A juxta-axonemal fiber adjacent to doublet 3 is present but small, being less than the diameter of the axoneme. The axial and juxta-axonemal fibers are continuous via a thin dense lamina sandwiched within the undulating membrane. The axial fiber is well developed and circular to ovoid in transverse section throughout its length. Anteriorly, it is slightly enlarged, the diameter being approximately 0.45 µm (measured within the principal piece region). The axial fiber continues posteriorly, decreasing in diameter, for much of the tail length. The axoneme extends beyond the axial fiber and undulating membrane as an endpiece.

Fig. 5.18 *Contd.*
C-I. *Bufo marinus*. **C**. LS through the acrosome complex. **D**. LS through the midpiece. **E**. Transverse section through the anterior acrosome complex. **F**. TS through the acrosome/nucleus junction. **G**. TS through the basal acrosome. **H**. TS through two different tails, one through the endpiece (left) and the other through the distal portion of the principal piece just before the beginning of the endpiece (right). **I**. A whole spermatozoon. **J-N**. *Bufo arenarum*: **J**. TS through the anterior acrosome complex. **K**. TS through the acrosome/nucleus junction. **L**. TS through the nucleus. **M**. TS through the midpiece. **N**. TS through the principal piece. **O**. *Ansonia* sp. B: TS through the midpiece. Note the thin lamina joining the axial fiber to doublet 3 of the axoneme (arrowhead). **P**. *B. boreas*. TS through the midpiece. **Q**. *Ansonia* sp. B. TS through the principal piece. A-H, J-Q to the same scale as indicated. I to the scale as indicated. Abbreviations: 3, juxta-axonemal fiber at doublet 3; a, axoneme; ac, acrosome complex; af, axial fiber; av, acrosome vesicle; cc, cytoplasmic canal; cy, cytoplasm; dc, distal centriole; ep, endpiece; f, basal nuclear fossa; m, mitochondrion; mc, mitochondrial collar; mp, midpiece; Nu, nucleus; p, conical perforatorium; pm, plasma membrane; um, undulating membrane. From Scheltinga, D. M. 2002. Unpublished Ph.D. Thesis, University of Queensland, Australia, Fig. 3.2.

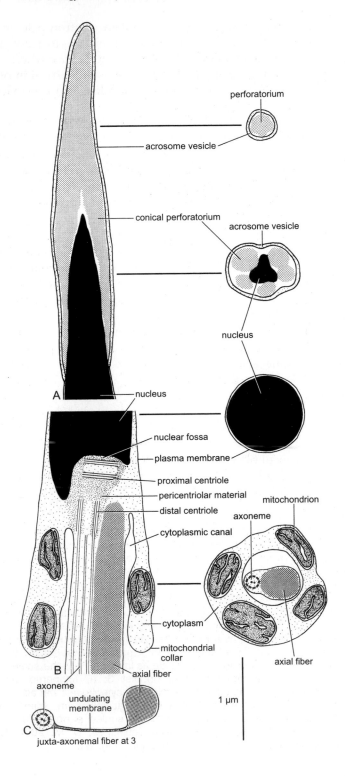

5.3.18 Family Heleophrynidae

The gross morphology of the sperm of *Heleophryne natalensis, H. purcelli,* and *H. orientalis* have been briefly described from SEM by Visser and van der Horst (1987). The tail of an unknown species of *Heleophryne* was briefly described from light microscopy by Fouquette and Delahoussaye (1977). The spermatozoa are filiform, being composed of a long, lance-like head (acrosome and nucleus), a short midpiece, and a tail. Lengths of measurable components of the spermatozoa are given in Table 5.4.

Table 5.4 Dimensions (means in μm) of heleophrynid spermatozoa.

Species	Head length	Midpiece length	Tail length	Total length	Author
Heleophryne natalensis	24	4.5	53	82	Visser and van der Horst 1987
Heleophryne purcelli	28	4.8	54	88	Visser and van der Horst 1987
Heleophryne orientalis	30	4.6	55	90	Visser and van der Horst 1987

The tail is composed of two filaments separated by an undulating membrane. The two filaments are said by Visser and van der Horst (1987) to be an axoneme and axial fiber but this can not be confirmed as no ultrastructural examinations have been performed.

5.3.19 Family Hylidae

The sperm of the hylid genera *Acris, Cyclorana, Hyla, Litoria, Nyctimystes, Pachymedusa, Phyllomedusa, Plectrohyla, Pseudacris, Pseudis,* and *Scinax* have been previously examined (see Table 5.5).

Table 5.5 Hylid spermatozoa previously examined.

Species	Method used	Author
Hylinae		
Acris crepitans	LM	Delahoussaye 1966
Acris gryllus	LM	Delahoussaye 1966
Hyla arborea	LM	Retzius 1906
Hyla arenicolor	TEM	Morrisett 1974
Hyla avivoca	LM	Delahoussaye 1966
Hyla chinensis	TEM; SEM	Kuramoto 1998; Lin *et al.* 1999
Hyla cinerea	LM	Delahoussaye 1966
Hyla femoralis	LM	Delahoussaye 1966
Hyla gratiosa	LM	Delahoussaye 1966
Hyla japonica	LM; TEM; SEM	Lee and Kwon 1992; Kwon and Lee 1995; Kuramoto 1998; Scheltinga 2002

Contd.

Fig. 5.19 Drawing of a spermatozoon of *Hyalinobatrachium fleischmanni* showing longitudinal sections through: **A**. Acrosome complex. **B**. midpiece, with corresponding transverse sections (TSs) as indicated. **C**. TS through the main portion of the principal piece. From Scheltinga, D. M. 2002. Unpublished Ph.D. Thesis, University of Queensland, Australia, Fig. 4.1.

Table 5.5 Contd.

Species	Method used	Author
Hyla squirella	LM	Delahoussaye 1966
Hyla meridionalis	TEM	Pugin-Rios 1980
Hyla versicolor	LM	Delahoussaye 1966
Plectrohyla matudai	LM; TEM	Scheltinga 2002
Pseudacris crucifer	LM	Delahoussaye 1966
Pseudacris regilla	LM; TEM	Scheltinga 2002
Pseudacris triseriata	LM	Delahoussaye 1966
Scinax acuminata	LM	Fouquette and Delahoussaye 1977
Scinax baumgardneri	LM	Fouquette and Delahoussaye 1977
Scinax blairi	LM	Fouquette and Delahoussaye 1977
Scinax boesemani	LM	Fouquette and Delahoussaye 1977
Scinax boulengeri	LM	Fouquette and Delahoussaye 1977
Scinax catharinae	LM	Fouquette and Delahoussaye 1977
Scinax crospedospila	LM	Fouquette and Delahoussaye 1977
Scinax cruentomma	LM	Fouquette and Delahoussaye 1977
Scinax cuspidatus	LM	Fouquette and Delahoussaye 1977
Scinax elaeochroa	LM	Fouquette and Delahoussaye 1977
Scinax fuscomarginata	LM	Fouquette and Delahoussaye 1977
Scinax fuscovaria	LM	Fouquette and Delahoussaye 1977; Gonzaga de Almeida and Cardoso 1985
Scinax garbei	LM	Fouquette and Delahoussaye 1977
Scinax hayii	LM	Fouquette and Delahoussaye 1977
Scinax megapodia	LM	Fouquette and Delahoussaye 1977
Scinax nasica	LM	Fouquette and Delahoussaye 1977
Scinax nebulosa	LM	Fouquette and Delahoussaye 1977
Scinax parkeri	LM	Fouquette and Delahoussaye 1977
Scinax perpusilla	LM	Fouquette and Delahoussaye 1977
Scinax proboscidea	LM	Fouquette and Delahoussaye 1977
Scinax quinquefasciata	LM	Fouquette and Delahoussaye 1977
Scinax ranki	TEM	Taboga and Dolder 1993, 1994b
Scinax rostrata	LM	Fouquette and Delahoussaye 1977
Scinax rubra	LM	Fouquette and Delahoussaye 1977
Scinax similis	LM	Fouquette and Delahoussaye 1977
Scinax squalirostris	LM	Fouquette and Delahoussaye 1977
Scinax staufferi	LM	Fouquette and Delahoussaye 1977
Scinax trachythorax	LM	Fouquette and Delahoussaye 1977
Scinax wandae	LM	Fouquette and Delahoussaye 1977
Scinax xsignata	LM	Fouquette and Delahoussaye 1977
Pelodryadinae		
Cyclorana alboguttata	LM; TEM	Meyer *et al.* 1997; Scheltinga 2002
Cyclorana brevipes	LM; TEM	Meyer *et al.* 1997; Scheltinga 2002
Cyclorana cryptotis	LM; TEM	Meyer *et al.* 1997; Scheltinga 2002
Cyclorana cultripes	LM; TEM	Scheltinga 2002
Cyclorana longipes	LM; TEM	Scheltinga 2002
Cyclorana maculosa	LM; TEM	Scheltinga 2002

Contd.

Table 5.5 Contd.

Species	Method used	Author
Cyclorana maini	LM; TEM	Scheltinga 2002
Cyclorana manya	LM; TEM	Scheltinga 2002
Cyclorana novaehollandiae	LM; TEM	Meyer *et al.* 1997; Scheltinga 2002
Cyclorana vagitus	LM; TEM	Scheltinga 2002
Litoria adelaidensis	LM; TEM	Scheltinga 2002
Litoria andiirrmalin	LM; TEM	Scheltinga 2002
Litoria angiana	LM; TEM	Scheltinga 2002
Litoria aurea	LM; TEM	Meyer *et al.* 1997; Scheltinga 2002
Litoria caerulea	LM; TEM; SEM	Lee and Jamieson 1992; Scheltinga 2002
Litoria cooloolensis	LM; TEM	Scheltinga 2002
Litoria cyclorhynchus	LM; TEM	Scheltinga 2002
Litoria dahlii	LM; TEM	Scheltinga 2002
Litoria electrica	LM; TEM	Scheltinga 2002
Litoria eucnemis	LM; TEM	Jamieson 1999; Scheltinga 2002
Litoria fallax	TEM; SEM	Lee and Jamieson 1992
Litoria freycineti	LM; TEM	Scheltinga 2002
Litoria genimaculata	LM; TEM	Scheltinga 2002
Litoria gracilenta	LM; TEM; SEM	Lee and Jamieson 1992; Scheltinga 2002
Litoria inermis	LM; TEM	Scheltinga 2002
Litoria infrafrenata	LM; TEM	Scheltinga 2002
Litoria latopalmata	LM; TEM	Scheltinga 2002
Litoria lesueuri	TEM; SEM	Lee and Jamieson 1992; Scheltinga 2002
Litoria longirostris	TEM; SEM	Scheltinga *et al.* 2002b
Litoria moorei	LM; TEM	Meyer *et al.* 1997; Scheltinga 2002
Litoria nannotis	LM; TEM	Scheltinga 2002
Litoria nasuta	LM; TEM	Scheltinga 2002
Litoria nigrofrenata	LM; TEM	Scheltinga 2002
Litoria pallida	LM; TEM	Scheltinga 2002
Litoria pearsoniana	LM; TEM	Scheltinga 2002
Litoria peronii	LM; TEM; SEM	Lee and Jamieson 1992; Scheltinga 2002
Litoria raniformis	LM; TEM	Scheltinga 2002
Litoria revelata	LM; TEM	Scheltinga 2002
Litoria rheocola	TEM	Jamieson 1999; Scheltinga 2002
Litoria rothii	LM; TEM	Scheltinga 2002
Litoria rubella	LM; TEM; SEM	Lee and Jamieson 1992; Scheltinga 2002
Litoria sp. aff. *modica*	LM; TEM	Scheltinga 2002
Litoria sp. 'iris'	LM; TEM	Scheltinga 2002
Litoria tyleri	LM; TEM	Scheltinga 2002
Litoria xanthomera	LM; TEM	Scheltinga 2002
Nyctimystes papua	LM; TEM	Scheltinga 2002
Phyllomedusinae		
Pachymedusa dacnicolor	TEM	Rastogi *et al.* 1988
Phyllomedusa vaillanti	LM; TEM	Scheltinga 2002
Pseudinae		
Pseudis paradoxa	LM; TEM	Scheltinga 2002

LM = light microscopy; TEM = transmission electron microscopy; SEM = scanning electron microscopy.

Thin undulating membrane

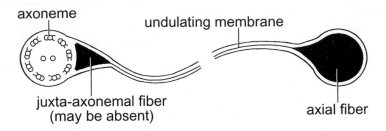

Cyclorana cryptotis, Litoria, Nyctimystes, Pachymedusa, Phyllomedusa, and Scinax.

Thick undulating membrane

All other Cyclorana spp. (in some species a distinct terminal axial fiber portion is present).

No undulating membrane

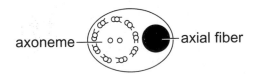

Acris, Cyclorana manya, Hyla, Plectrohyla, Pseudcris, and Pseudis (axial fiber absent).

Fig. 5.20 Diagrammatic representations of the different types of spermatozoon tails of hylid species. Original.

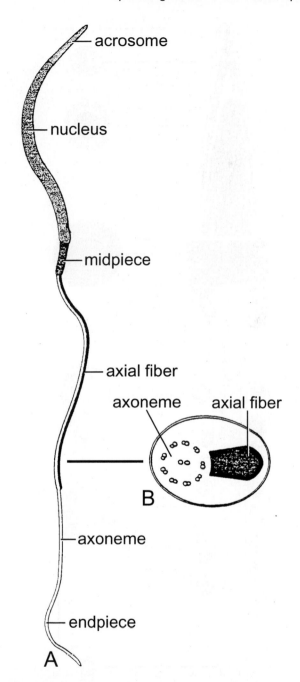

Fig. 5.21 Drawing of a spermatozoon of *Hyla meridionalis*. **A**. Whole spermatozoon. **B**. Transverse section through the principal piece of the tail as indicated. From (relabelled) Pugin-Rios, E. 1980. Unpublished Ph.D. Thesis, L'Université de Rennes, France, Figs 37, 38.

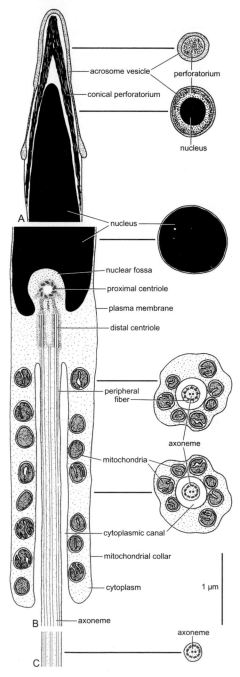

Fig. 5.22 Drawing of a spermatozoon of *Pseudis paradoxa* showing longitudinal sections through: **A**. Acrosome complex. **B**. Centriolar region and midpiece. **C**. Tail, with corresponding transverse sections as indicated. From Scheltinga, D. M. 2002. Unpublished Ph.D. Thesis, University of Queensland, Australia, Fig. 10.1.

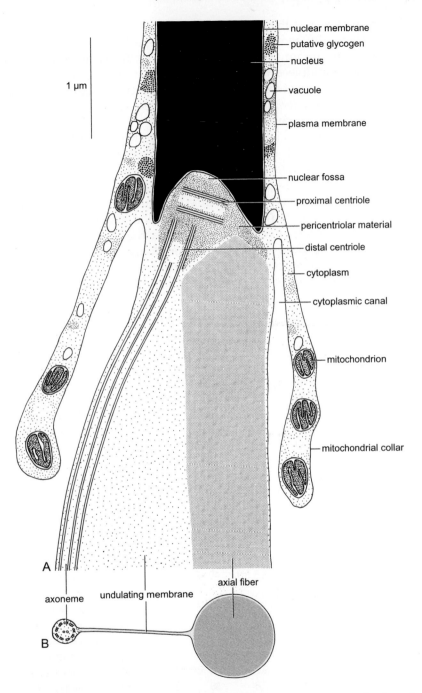

Fig. 5.23 Drawing of a spermatozoon of *Litoria longirostris*. **A**. Longitudinal sections through the centriolar region and midpiece. **B**. Transverse section through the principal piece of the tail. A, from Scheltinga D. M. *et al.* 2002. Memoirs of the Queensland Museum 48: 215-220, Fig. 2. B, original.

The spermatozoa of the Hylidae are uniform in many respects. However, there are several differences, particularly in respect to the tail (Fig. 5.20). The spermatozoa are filiform, being composed of a curved head (acrosome and nucleus), a short midpiece, and a long tail. Lengths of measurable components of the spermatozoa are given in Table 5.6. The spermatozoa of *Hyla meridionalis*, *Pseudis paradoxa*, and *Litoria longirostris* are depicted diagrammatically in Figures 5.21, 5.22, and 5.23, respectively.

The acrosome complex is composed of a moderately electron-dense acrosome vesicle which completely (or almost completely) caps the conical perforatorium (Fig. 5.24A-C). *Litoria dahlii, L. longirostris* and *Pseudis paradoxa* differ in that the perforatorium extends beyond the base of the vesicle for a significant distance. The perforatorium is composed of coarse longitudinal fibers which appear as homogeneous separate sheaves in transverse section, and overlies the attenuated nuclear tip (Fig. 5.24A-C), though, in some species the fibers of the perforatorium are tightly packed and appear as a homogeneous mass. The acrosome complex is symmetrically (asymmetrically in *L. longirostris*) attached to the nucleus and extends anteriorly beyond the nuclear tip (Fig. 5.24A).

Table 5.6 Dimensions (means in μm) of hylid spermatozoa.

Species	Head length	Midpiece length	Tail length[†]	Endpiece length	Total length	Author
Acris crepitans	13	12	-	-	-	Delahoussaye 1966
Acris gryllus	14	11	-	-	-	Delahoussaye 1966
Cyclorana alboguttata	23	5.5	52	-	80	Scheltinga 2002
Cyclorana brevipes	21-27	4.6-5	37-66	-	63-97	Meyer et al. 1997; Scheltinga 2002
Cyclorana cryptotis	14	3.3	39	5.8	56	Scheltinga 2002
Cyclorana manya	16	5.2	34	-	55	Scheltinga 2002
Cyclorana novaehollandiae	16	5.5	45	-	67	Scheltinga 2002
Hyla avivoca	12	7	-	-	-	Delahoussaye 1966
Hyla chinensis	18	6	47	-	73	Kuramoto 1998
Hyla cinerea	12	6.5	-	-	-	Delahoussaye 1966
Hyla femoralis	12	6.2	-	-	-	Delahoussaye 1966
Hyla gratiosa	14	6.9	-	-	-	Delahoussaye 1966
Hyla japonica	17	6.8-6.9	40-42	-	64-68	Kuramoto 1998; Scheltinga 2002
Hyla meridionalis	21	?	48		70	Pugin-Rios 1980
Hyla regilla	17	3.1	31	-	50	Scheltinga 2002
Hyla squirella	11	6.2	-	-	-	Delahoussaye 1966
Hyla versicolor	13	6.9	-	-	-	Delahoussaye 1966
Litoria adelaidensis	16	3.7	32	7	51	Scheltinga 2002
Litoria andiirrmalin	17	2.6	38	3	58	Scheltinga 2002
Litoria angiana	38	3.8	49	5.4	90	Scheltinga 2002
Litoria caerulea	16	3.4	29	7	48	Scheltinga 2002

Contd.

Table 5.6 Contd.

Species	Head length	Midpiece length	Tail length[†]	Endpiece length	Total length	Author
Litoria cooloolensis	16	3.4	46	17	63	Scheltinga 2002
Litoria dahlii	29	14	67	14	96	Scheltinga 2002
Litoria electrica	15	2.8	23	2	42	Scheltinga 2002
Litoria eucnemis	14	3.7	32	6	50	Scheltinga 2002
Litoria fallax	16	?	40	?	56	Lee and Jamieson 1993
Litoria freycineti	15	3	35	8.9	53	Scheltinga 2002
Litoria genimaculata	18	2.4	37	5.5	57	Scheltinga 2002
Litoria gracilenta	14	3.8	27	3.2	45	Scheltinga 2002
Litoria inermis	17	3.6	30	5	51	Scheltinga 2002
Litoria infrafrenata	14	3.1	32	2	50	Scheltinga 2002
Litoria lesueuri	15	2.8	32	4.3	50	Scheltinga 2002
Litoria moorei	16	2.3	35	6.7	53	Scheltinga 2002
Litoria nannotis	21	3.4	41	6.5	65	Scheltinga 2002
Litoria nasuta	13	2.8	29	5.7	45	Scheltinga 2002
Litoria nigrofrenata	14	3.5	32	7.3	50	Scheltinga 2002
Litoria pallida	14	2.7	30	3.9	46	Scheltinga 2002
Litoria pearsoniana	15	2.5	26	3.7	44	Scheltinga 2002
Litoria peronii	16	2.4	27	-	46	Scheltinga 2002
Litoria revelata	15	1.8	35	5.3	52	Scheltinga 2002
Litoria rheocola	20	2	40	3.9	62	Scheltinga 2002
Litoria rothii	16	3	27	2.7	46	Scheltinga 2002
Litoria rubella	14	3.3	30	2.9	47	Scheltinga 2002
Litoria sp. aff. modica	18	5.3	36	7.3	60	Scheltinga 2002
Litoria sp. 'iris'	22	3.3	34	5.4	58	Scheltinga 2002
Litoria xanthomera	15	2.8	31	4	48	Scheltinga 2002
Nyctimystes papua	23	2	51	7	76	Scheltinga 2002
Phyllomedusa vaillanti	36	-	55	5.6	91	Scheltinga 2002
Pseudacris crucifer	15	11	-	-	-	Delahoussaye 1966
Pseudacris triseriata	13	9.2	-	-	-	Delahoussaye 1966
Pseudis paradoxa	13	3.9	?	-	?	Scheltinga 2002
Scinax fuscovaria	31-32	?	59-64	-	90-96	Gonzaga de Almeida and Cardoso 1985

[†]Total tail length, i.e. includes endpiece

The nucleus is cylindro-conical, composed of electron-dense condensed chromatin and tapers to a rounded tip within the acrosome complex (Fig. 5.24A). Numerous small nuclear inclusions are present (Fig. 5.24E). Distinct nuclear shoulders are absent. The nucleus increases in diameter throughout its length to a maximum at its base where it ends in a well-developed asymmetrical fossa (Fig. 5.24E). The fossa of *Phyllomedusa vaillanti* differs in being symmetrical and poorly developed, appearing as little more than a slight concavity.

Within the basal nuclear fossa lies the proximal centriole which is orientated at between 70° to 80° to the long axis of the spermatozoon and at a right angle

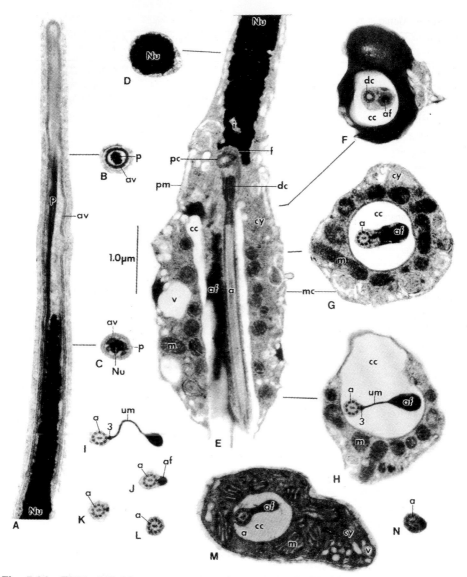

Fig. 5.24 TEM of Hylidae spermatozoa. **A-L**. *Litoria angiana*: **A**. Longitudinal section (LS) through the acrosome complex. **B-D**. Successive transverse sections (TSs) through the **B**, anterior acrosome complex, **C**. Acrosome nucleus junction, **D**. Nucleus. **E**. LS through the centriolar region and midpiece. **F-L**. Successive TSs through: **F**. Distal centriole. **G**. Anterior midpiece. **H**. Posterior midpiece. **I**. Principal piece. **J** and **K**. Distal end of principal piece. **L**. Endpiece. **M** and **N**. *Litoria rubella*. **M**. TS through the midpiece. **N**. Distal end of principal piece. All to the same scale as indicated. Abbreviations: 3, juxta-axonemal fiber at doublet 3; a, axoneme; af, axial fiber; av, acrosome vesicle; cc, cytoplasmic canal; cy, cytoplasm; dc, distal centriole; f, basal nuclear fossa; i, nuclear inclusion; m, mitochondrion; mc, mitochondrial collar; Nu, nucleus; p, conical perforatorium; pc, proximal centriole; pcm, pericentriolar material; pm, plasma membrane; um, undulating membrane; v, vacuole. From Scheltinga, D. M. 2002. Unpublished Ph.D. Thesis, University of Queensland, Australia, Fig. 5.9.

to the distal centriole (Fig. 5.24E, F). In *Phyllomedusa vaillanti* the proximal centriole is at an angle of 55° to the long axis of the sperm and the distal centriole. The axial fiber extends anteriorly into the neck region to the base of the distal centriole. In *Ph. vaillanti* the axial fiber extends anteriorly into the nuclear fossa and abuts the nucleus, whereas in *Hyla* and *Pseudacris* the axial fiber ends anteriorly well short of the distal centriole. In *Plectrohyla matudai* the axial fiber ends just short of the distal centriole. An axial fiber is absent from *Pseudis paradoxa*, however, for a short distance along the beginning of the axoneme (within the midpiece) nine thin dense fibers are present around the axoneme. One fiber associated with each doublet of the axoneme (Fig. 5.22).

A short mitochondrial collar, containing scattered mitochondria, granular material (putative glycogen), cytoplasm, and vacuoles, surrounds the anterior portion of the tail but is separated from it by the cytoplasmic canal (Figs 5.23A, 5.24E-H, M). This collar is attached only to the centriolar/neck region of the spermatozoon and the cytoplasmic canal extends posteriorly from the level of the base of the distal centriole. In *Hyla*, *Plectrohyla matudai* and *Pseudacris* the cytoplasmic canal ends anteriorly well short of the distal centriole. A mitochondrial collar and cytoplasmic canal are absent from the spermatozoon of *Phyllomedusa vaillanti*. In this species the mitochondria are located around the base of the nucleus, as also occurs in *Litoria dahlii*, though, a short mitochondrial collar is present in *L. dahlii*. The mitochondria have well-developed linear cristae and may be either spherical and constant or variable in size and shape (Fig. 5.24E, G, H, M).

The structure of the spermatozoon tail varies significantly between hylid genera, and in some cases between species (Fig. 5.20). The tail of *Pseudis paradoxa* is unique among examined hylids in that it is composed of an axoneme alone: an axial fiber, juxta-axonemal fiber, and undulating membrane are all absent (Fig. 5.22). The tail complex of all other hylids, though modified to varying degrees, is composed of a 9+2 axoneme and an axial fiber (near doublet 3 of the axoneme) surrounded by the plasma membrane (Fig. 5.24I-L, N). A juxta-axonemal fiber and/or undulating membrane may or may not be present. In *Acris*, *Cyclorana manya*, *Hyla*, *Plectrohyla*, and *Pseudacris* the axial fiber is in the form of a simple cylinder (circular in transverse section) and an undulating membrane and distinct juxta-axonemal fiber are absent. The axial fiber of other *Cyclorana* (excepting *C. cryptotis*) is in the form of a thick (laterally extended) paraxonemal rod, thus a thick undulating membrane is present. In *C. cryptotis*, *Litoria*, *Nyctimystes*, *Pachymedusa*, *Phyllomedusa*, and *Scinax* the axial fiber is well developed and circular to ovoid in transverse section. It is connected to the axoneme by a thin cytoplasmic lamina, the undulating membrane. A juxta-axonemal fiber adjacent to doublet 3 is either present or absent. The axial and juxta-axonemal fibers are continuous via a thin dense lamina sandwiched within the undulating membrane (Fig. 5.24H, I). This lamina is also present in those species lacking a juxta-axonemal fiber. The axoneme extends beyond the axial fiber (and undulating membrane) as the endpiece (Fig. 5.24L).

5.3.20 Family Leptodactylidae

Sperm structure has been examined in a wide variety of Leptodactylidae (24 species, 14 genera, 3 subfamilies). The species investigated are given in Table 5.7.

Table 5.7 Leptodactylid spermatozoa previously examined.

Species	Method used	Author
Ceratophryinae		
Ceratophrys cranwelli	LM; TEM	Scheltinga 2002
Lepidobatrachus laevis	LM; TEM	Waggener and Carroll 1998
Odontophrynus cultripes	LM; TEM	Báo et al. 1991; Fernandes and Báo 1998
Leptodactylinae		
Leptodactylus bufonius	LM	Barbieri 1950
Leptodactylus chaquensis	LM	Barbieri 1950; Raisman and Cabada 1977
Leptodactylus latinasus	LM	Barbieri 1950
Leptodactylus ocellatus	LM	Barbieri 1950
Leptodactylus wagneri	LM; TEM	Scheltinga 2002
Physalaemus biligonigerus	TEM	Amaral et al. 1999
Physalaemus fuscomaculatus	TEM	Amaral et al. 1999
Physalaemus gracilis	TEM	Amaral et al. 1999
Physalaemus pustulosus	LM; TEM	Scheltinga 2002
Pleurodema thaul	LM; TEM	Pugin-Rios 1980; Pugin and Garrido 1981
Pseudopaludicola falcipes	TEM	Amaral et al. 2000
Telmatobiinae		
Alsodes vittatus	TEM	Pugin-Rios 1980
Batrachyla antartandica	LM; TEM	Pugin-Rios 1980; Pugin and Garrido 1981; Garrido et al. 1989
Batrachyla leptopus	LM; TEM	Pugin-Rios 1980; Pugin and Garrido 1981; Garrido et al. 1989
Batrachyla taeniata	LM; TEM	Pugin-Rios 1980; Pugin and Garrido 1981; Garrido et al. 1989
Caudiverbera caudiverbera	LM; TEM	Pugin-Rios 1980; Pugin and Garrido 1981
Eupsophus roseus	LM; TEM	Pugin-Rios 1980; Pugin and Garrido 1981
Hylorina sylvatica	LM; TEM	Pugin-Rios 1980; Pugin and Garrido 1981
Telmatobius jelskii	LM; TEM	Scheltinga 2002
Telmatobius schreiteri	TEM	Pisanó and Adler 1968
Telmatobufo australis	LM; TEM	Pugin-Rios 1980; Pugin and Garrido 1981

LM = light microscopy; TEM = transmission electron microscopy.

The spermatozoa of the leptodactylid genera examined show several obvious differences, but also some similarities. The spermatozoa are filiform, being composed of a head (acrosome and nucleus), a short midpiece (occasionally absent), and a long tail (principal piece and endpiece). The sperm head of *Telmatobufo australis* differs from that of other leptodactylids in being helical, forming approximately 6 turns. The spermatozoa of *Pleurodema thaul, Telmatobufo australis, Alsodes vittatus,* and *Caudiverbera caudiverbera* is depicted diagrammatically in Figures 5.25, 5.26, 5.27, and 5.28, respectively. Lengths of measurable components of the spermatozoa are given in Table 5.8.

Table 5.8 Dimensions (means in μm) of leptodactylid spermatozoa.

Species	Head length	Midpiece length	Tail length#	Endpiece length	Total length	Author
Alsodes vittatus	40	0	70	?	110	Pugin-Rios 1980
Batrachyla antartandica	30	0	65	5	95	Pugin-Rios 1980; Pugin and Garrido 1981; Garrido *et al.* 1989
Batrachyla leptopus	19-25	0	35-60	?	80-85	Pugin-Rios 1980; Pugin and Garrido 1981; Garrido *et al.* 1989
Batrachyla taeniata	32	0	42-53	?	74-75	Pugin-Rios 1980; Pugin and Garrido 1981; Garrido *et al.* 1989
Caudiverbera caudiverbera	20	0	40-45	0	64	Pugin-Rios 1980; Pugin and Garrido 1981
Ceratophrys cranwelli	17[†]	1.6	45	?	62	Scheltinga 2002
Eupsophus roseus	24	0	62	?	86	Pugin-Rios 1980; Pugin and Garrido 1981
Hylorina sylvatica	18-20	0	35-37	?	55	Pugin-Rios 1980; Pugin and Garrido 1981
Lepidobatrachus laevis	17	0.87	55	?	73	Waggener and Carroll 1998
Leptodactylus wagneri	21[‡]	?	33[‡]	5.4	54	Scheltinga 2002
Physalaemus pustulosus	24	4.7	?	?	?	Scheltinga 2002
Pleurodema thaul	23	0	63-68	5	91	Pugin-Rios 1980; Pugin and Garrido 1981
Telmatobius jelskii	13	5.7	46	0	64	Scheltinga 2002
Telmatobufo australis	120	1-2	115-1200		240	Pugin-Rios 1980; Pugin and Garrido 1981

#Total tail length, *i.e.* includes endpiece. †Includes head and midpiece. ‡Both head and tail will include portions of the midpiece

Acrosome of *Alsodes, Batrachyla, Ceratophrys, Eupsophus, Hylorina, Lepidobatrachus, Odontophrynus, Pseudopaludicola,* **and** *Telmatobius*. The acrosome complex is composed of an acrosome vesicle that completely caps the conical perforatorium. The acrosome vesicle is membrane bound and filled with moderately electron-dense material. Anteriorly, the vesicle is relatively thick, becoming thin in the middle, and posteriorly appearing as little more than acrosome vesicle membranes. The perforatorium, situated within the subacrosomal space, is composed of coarse longitudinal fibers which appear as homogeneous separate sheaves in transverse section, and overlies the attenuated nuclear tip. The perforatorium is moderately electron-dense and is not membrane bound. The acrosome complex extends anteriorly beyond the nucleus, being symmetrically attached to the nucleus basally.

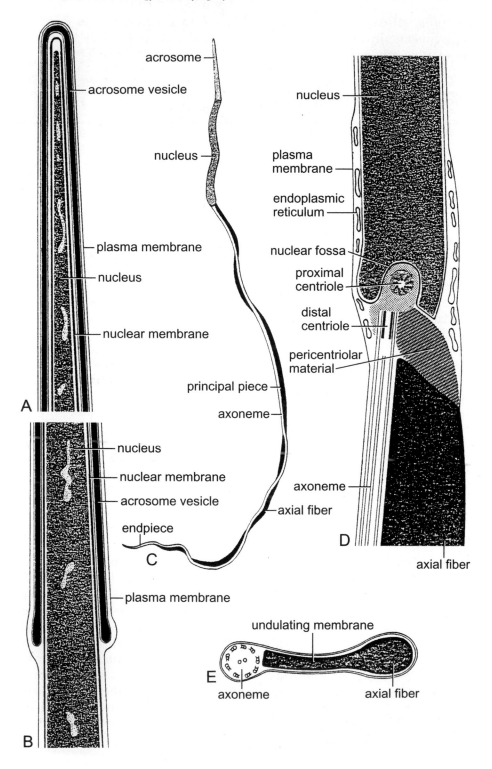

Acrosome of *Caudiverbera, Leptodactylus,* **and** *Pleurodema*. The acrosome complex is composed of an acrosome vesicle only. A perforatorium is absent. The acrosome vesicle is membrane bound and filled with moderately electron-dense, homogeneous material. It caps the tapered nuclear tip, extending anteriorly well beyond the nucleus. The acrosome vesicle is symmetrically attached to the nucleus.

Acrosome of *Physalaemus*. The acrosome complex of *Physalaemus pustulosus* is composed of an acrosome vesicle which caps the anterior portion of the conical perforatorium only. The vesicle ends just posterior to the tip of the nucleus, but the perforatorium extends beyond the base of the vesicle for a short, but significant, distance (Fig. 5.29A). The acrosome vesicle is relatively thick throughout its length, being membrane bound and filled with moderately electron-dense material (Fig. 5.29A, C, D). The perforatorium, situated within the subacrosomal space, is composed of fine longitudinal fibers which appear as a homogeneous mass in transverse section, and overlies the attenuated nuclear tip (Fig. 5.29A, C, D). The acrosome complex of *P. biligonigerus, P. fuscomaculatus,* and *P. gracilis* differ from that of *P. pustulosus* given above. In these three species the acrosome vesicle and perforatorium appear to be of the typical bufonoid-type, with the acrosome vesicles becoming thinner posteriorly and completely capping the perforatorium which is composed of sheaves. The acrosome complex extends anteriorly beyond the nucleus, being symmetrically attached to the nucleus basally.

Nucleus of leptodactylids. The nucleus is cylindro-conical, tapering to a pointed tip within the acrosome complex (Fig. 5.29A, H). It occupies most of the head length, is circular in transverse section, and composed of electron-dense condensed chromatin (Fig. 5.29E). Numerous nuclear inclusions, which are electron-lucent or contain electron-pale material, are present (Fig. 5.29B). Distinct nuclear shoulders are absent. The nucleus increases in diameter throughout its length to a maximum at its base where it ends in a well-developed fossa (the basal nuclear, or centriolar, fossa) (Fig. 5.29M). The fossa is asymmetrical in one plane of longitudinal section, *i.e.* when the proximal centriole is also seen in longitudinal section.

Neck of leptodactylids. Within the basal nuclear fossa lies the proximal centriole which is surrounded by a small amount of pericentriolar material that connects it to the nuclear fossa and distal centriole (Fig. 5.29B, F, M). The proximal centriole is orientated between 65° to 75° to the long axis of the spermatozoon and at a right angle to the distal centriole which forms the basal body of the axoneme (Fig. 5.29B).

Midpiece of *Alsodes, Batrachyla, Ceratophrys, Eupsophus, Hylorina, Lepidobatrachus, Odontophrynus, Physalaemus, Pseudopaludicola,*

Fig. 5.25 Drawing of a spermatozoon of *Pleurodema thaul.* **A.** Longitudinal section (LS) through the apical acrosome complex. **B.** LS through the basal acrosome complex. **C.** Whole spermatozoon. **D.** LS through the centriolar region. **E.** Transverse section through the principal piece of the tail. From (relabelled) Pugin-Rios, E. 1980. Unpublished Ph.D. Thesis, L'Université de Rennes, France, Figs 17, 18, 19, 20, 21.

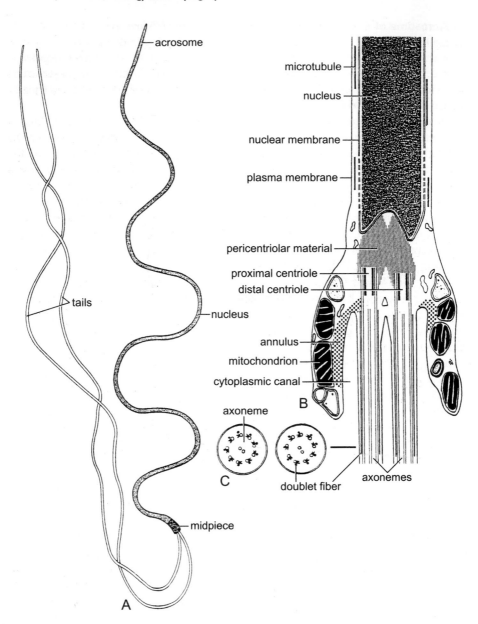

Fig. 5.26 Drawing of a spermatozoon of *Telmatobufo australis*. **A**. Whole spermatozoon. **B**. Longitudinal section through the centriolar region and midpiece. **C**. Transverse section through the anterior region of the tail as indicated. From (relabelled) Pugin-Rios, E. 1980. Unpublished Ph.D. Thesis, L'Université de Rennes, France, Figs 14, 15, 16.

Telmatobius, and *Telmatobufo.* A short mitochondrial collar, containing scattered mitochondria, cytoplasm, granular material, and vacuoles, surrounds the anterior portion of the tail but is separated from it by a gap, the cytoplasmic canal (Fig. 5.29H). This collar is attached only to the centriolar/neck region of the spermatozoon and the cytoplasmic canal extends posteriorly from the level of the base of the distal centriole. In *Physalaemus pustulosus* the canal extends posteriorly from a point well posterior of the distal centriole (Fig. 5.29B, H, M). Therefore, in *P. pustulosus* the mitochondria adjoin and surround the anteriorly portion of the axoneme and paraxonemal rod (Fig. 5.29G). *Physalaemus biligonigerus, P. fuscomaculatus,* and *P. gracilis* differ from that of *P. pustulosus* in the possession of a long cytoplasmic canal which extends posteriorly from the base of the distal centriole.

The mitochondria have well-developed linear cristae and are spherical to oval in shape, varying in size and number. In at least the leptodactylids *Alsodes, Batrachyla,* and *Eupsophus* the mitochondrial collar is lost at maturity.

Midpiece of *Caudiverbera* and *Pleurodema*. A mitochondrial collar is absent from both *Caudiverbera caudiverbera* and *Pleurodema thaul* spermatozoa. The mitochondria occur in a cytoplasmic droplet attached to the mid-nucleus in most testicular sperm of *C. caudiverbera,* but in the mature spermatozoon the droplet, and all mitochondria, are lost. In *P. thaul* the mitochondria occur in the neck region and in the cytoplasm of the basal quarter of the nucleus; there is no cytoplasmic collar.

Midpiece of *Leptodactylus*. The spermatozoal mitochondria of *Leptodactylus wagneri* are located within the posterior portion of the nucleus and anterior portion of the tail. They completely surround the nucleus but within the tail occur adjacent to the axoneme and within the adaxonemal region of the undulating membrane. No mitochondrial collar or cytoplasmic canal is present. The mitochondria have well-developed linear cristae and are small and spherical to oval in shape.

Tail of *Alsodes, Batrachyla, Ceratophrys, Eupsophus, Hylorina, Lepidobatrachus, Leptodactylus, Odontophrynus, Pseudopaludicola,* and *Telmatobius*. The tail complex is composed of a 9+2 axoneme and an axial fiber, the fiber being connected to doublet 3 of the axoneme by a thin cytoplasmic lamina, the undulating membrane. A juxta-axonemal fiber adjacent to doublet 3 is present and of moderate size (*i.e.* approximately the same size as the diameter of the axoneme) but is absent in *Leptodactylus wagneri* at least. The axial and juxta-axonemal fibers are continuous via a thin dense lamina sandwiched within the undulating membrane. This lamina is also present in *L. wagneri*. The axial fiber is well developed and circular or ovoid in transverse section throughout its length. The axial fiber extends anteriorly into the neck region to the level of the distal centriole. The axoneme extends beyond the axial fiber and undulating membrane as the endpiece. The spermatozoon of *Lepidobatrachus laevis* was described by Waggener and Carroll (1998) as having two axonemes and two juxta-axonemal fibers (termed axial fibers by them) connected to a central undulating membrane fiber by two

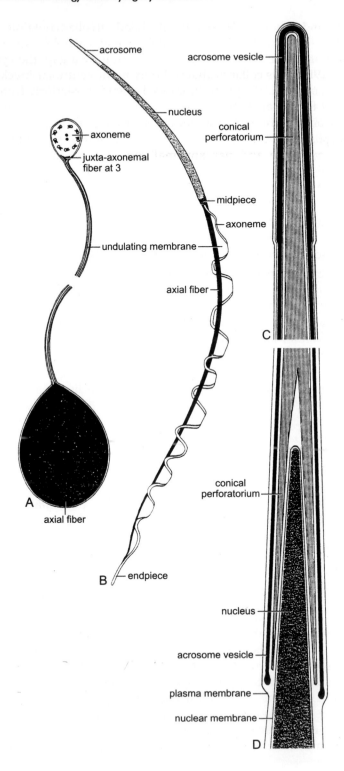

undulating membranes. However, the figures provided by them appear to be of a commonly observed artefact of sectioning through a typical tail with a single axoneme, juxta-axonemal fiber, axial fiber and thin undulating membrane. This is supported by their measured width of these double axonemes of 7 µm, which is approximately 5 times larger than the thickness of the tail in their drawing (figure 2) and light microscope pictures (figure 1a, b). A normal single axoneme, undulating membrane, and modified axial fiber tail section is provided (figure 5g) which approximately agrees in size with their drawing and light micrographs. Therefore, it appears unlikely that double axonemes are present in *Lepidobatrachus laevis* and it is here considered to have the normal bufonoid-type spermatozoon, though the structure of its midpiece remains unknown.

Tail of Physalaemus. In *Physalaemus pustulosus* a thin cytoplasmic collar surrounds the anterior portion of the tail, the two being separate by the cytoplasmic canal. The collar is composed of a thin layer of membrane bounded cytoplasm which extends posteriorly from the mitochondrial collar (Fig. 5.29B, I, M). The tail complex is composed of a 9+2 axoneme, an axial fiber, and an undulating membrane (Fig. 5.29J-L). The axial and juxta-axonemal fibers are continuous and in the form of a thick, electron-dense rod adjacent to doublet 3 of the axoneme and is termed the paraxonemal rod. A distinct juxta-axonemal fiber is absent. The paraxonemal rod has a distinct bulb at the abaxonemal end corresponding to an axial fiber (Fig. 5.29J). The anterior portion of the tail consists of the axoneme and paraxonemal rod only, however, posteriorly a thin cytoplasmic lamina, the undulating membrane, is present which extends from the abaxonemal end of the paraxonemal rod away from the axoneme (Fig. 5.29K). The paraxonemal rod and undulating membrane continue posteriorly, decreasing in size, for much of the tail length before the undulating membrane is lost, leaving only the axoneme and paraxonemal rod (Fig. 5.29J-L). The axoneme extends beyond the paraxonemal rod as the endpiece. *P. biligonigerus*, *P. fuscomaculatus*, and *P. gracilis* differ from that of *P. pustulosus* in the absence of a distinct abaxonemal bulb-like swelling on the paraxonemal rod and in the absence of a cytoplasmic collar.

Tail of Pleurodema. The tail is composed of an axoneme and axial fiber which is in the form of a paraxonemal rod with an abaxonemal bulb-like swelling.

Tail of Caudiverbera, Telmatobius, and Telmatobufo. The tail complex is composed of the 9+2 axoneme only. An axial fiber, juxta-axonemal fiber, and undulating membrane are absent. However, the tail of *Telmatobufo australis* is biflagellate, with two parallel centrioles and two axonemes being present. Anteriorly, within the midpiece region, distinct portions of electron-dense

Fig. 5.27 Drawing of a spermatozoon of *Alsodes vittatus*. **A**. transverse section through the principal piece of the tail. **B**. Whole spermatozoon. **C**. Longitudinal section (LS) through the apical acrosome complex. **D**. LS through the basal acrosome complex. From (relabelled) Pugin-Rios, E. 1980. Unpublished Ph.D. Thesis, L'Université de Rennes, France, Figs 22, 23, 24, 25.

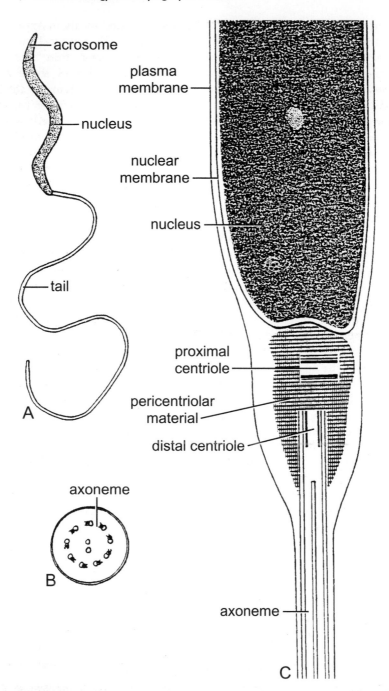

Fig. 5.28 Drawing of a spermatozoon of *Caudiverbera caudiverbera*. **A**. Whole spermatozoon. **B**. Transverse section through the tail. **C**. Longitudinal section through the centriolar region. From (relabelled) Pugin-Rios, E. 1980. Unpublished Ph.D. Thesis, L'Université de Rennes, France, Figs 11, 12, 13.

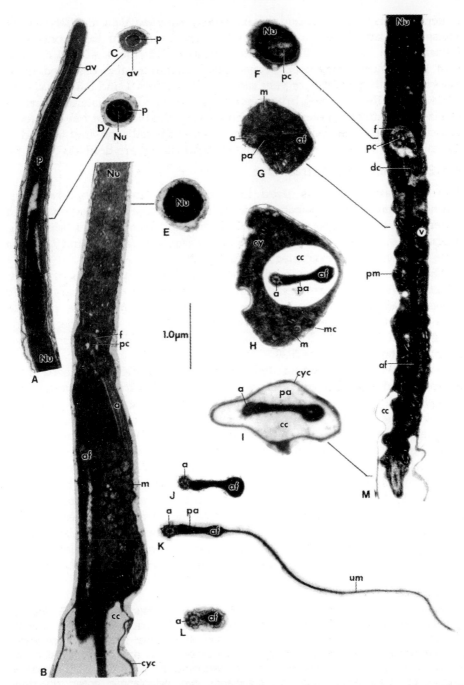

Fig. 5.29   TEM of *Physalaemus pustulosus* spermatozoa. **A**. Longitudinal section (LS) through the acrosome complex. **B**. LS through the basal portion of the nucleus and midpiece. **C-L**. Successive transverse sections (TSs) through: **C**. Anterior acrosome.

Contd.

material (fibers) occurs around the axoneme(s) of *Telmatobius* and *Telmatobufo*. These fibers are closely associated with each doublet of the axoneme in *Telmatobufo*. In *Telmatobius* the fibers are associated with doublets 2, 3, 4, 6, 7, and 8 of the axoneme. However, occasionally material is also observed to be associated with doublet 9, or very rarely with doublets 1 and 5 also (*i.e.* with all doublets of the axoneme). In both genera the electron-dense material decreases in size and number posteriorly (*i.e.* how many doublets that have material associated with them), being totally absent from the tail soon after the end of the midpiece. The last to lose their associated material are doublets 3 and 8.

5.3.21 Family Myobatrachidae

The Myobatrachidae as used here includes the two subfamilies, Limnodynastinae and Myobatrachinae. Sperm structure in the myobatrachid genera *Adelotus, Heleioporus, Kyarranus, Lechriodus, Limnodynastes, Mixophyes, Neobatrachus, Notaden, Pseudophryne, Rheobatrachus, Taudactylus,* and *Uperoleia* have been previously examined (see Table 5.9).

The spermatozoa of the myobatrachid genera examined are of similar structure. They are filiform, being composed of a straight, or only very slightly curved, head (acrosome and nucleus) and a long tail (midpiece, principal piece, and endpiece). Lengths of measurable components of the spermatozoa are given in Table 5.10. The spermatozoon of *Taudactylus liemi* is depicted diagrammatically in Figure 5.30.

The following ultrastructural description is based on the spermatozoa of the myobatrachid genera *Adelotus, Heleioporus, Kyarranus, Lechriodus, Limnodynastes, Mixophyes, Neobatrachus, Notaden, Pseudophryne, Rheobatrachus, Taudactylus,* and *Uperoleia*. The acrosome complex is composed of an acrosome vesicle that completely caps the conical perforatorium (Fig. 5.31A, C). The acrosome vesicle is membrane bound and filled with moderately electron-dense material. Anteriorly, the vesicle is relatively thick, becoming thin in the middle and posterior regions and terminating basally as a slightly thickened bulb-shape (Fig. 5.31A). The perforatorium occurs within the subacrosomal space and is composed of longitudinal fibers which overlie the attenuated nuclear tip. The longitudinal fibers are coarse and appear as homogeneous separate sheaves in at least some transverse sections, particularly those through the nuclear tip, but appear as a homogeneous mass anteriorly in *Heleioporus, Lechriodus,*

Fig. 5.29 *Contd.*

D. Mid-acrosome anterior to the nucleus. **E.** Nucleus. **F.** Nuclear fossa. **G.** Anterior midpiece. **H.** Mitochondrial collar of the midpiece. **I.** Cytoplasmic collar. **J.** Anterior portion of tail. **K.** Principal piece of tail showing abaxonemal undulating membrane. **L.** distal portion of tail. **M.** LS through the centriolar region and midpiece. All to the same scale as indicated. Abbreviations: a, axoneme; af, axial fiber; av, acrosome vesicle; cc, cytoplasmic canal; cy, cytoplasm; cyc, cytoplasmic collar; dc, distal centriole; f, basal nuclear fossa; m, mitochondrion; mc, mitochondrial collar; Nu, nucleus; p, conical perforatorium; pa, paraxonemal rod; pc, proximal centriole; pm, plasma membrane; um, undulating membrane; v, vacuole. From Scheltinga, D. M. 2002. Unpublished Ph.D. Thesis, University of Queensland, Australia, Fig. 7.2.

Table 5.9 Myobatrachid spermatozoa previously examined.

Species	Method used	Author
Adelotus brevis	TEM	Lee and Jamieson 1993
Crinia insignifera	LM	Scheltinga 2002
Crinia parinsignifera	LM	Scheltinga 2002
Heleioporus albopunctatus	LM; TEM	Scheltinga 2002
Heleioporus inornatus	LM	Scheltinga 2002
Kyarranus kundagungan	LM; TEM	Scheltinga 2002
Kyarranus loveridgei	LM; TEM	Scheltinga 2002
Lechriodus fletcheri	LM; TEM	Scheltinga 2002
Limnodynastes convexiusculus	LM; TEM	Scheltinga 2002
Limnodynastes ornatus	LM; TEM	Scheltinga 2002
Limnodynastes peronii	LM; TEM	Lee and Jamieson 1992; Scheltinga 2002
Limnodynastes salmini	LM	Scheltinga 2002
Limnodynastes tasmaniensis	LM; TEM	Scheltinga 2002
Limnodynastes terraereginae	LM; TEM	Scheltinga 2002
Mixophyes fasciolatus	LM; TEM	Lee and Jamieson 1992; Scheltinga 2002
Mixophyes iteratus	LM; TEM	Scheltinga 2002
Mixophyes schevilli	LM; TEM	Scheltinga 2002
Neobatrachus kunapalari	LM; TEM	Scheltinga 2002
Neobatrachus pelobatoides	LM; TEM	Lee and Jamieson 1992; Scheltinga 2002
Neobatrachus sudelli	LM; TEM	Scheltinga 2002
Notaden bennettii	LM; TEM	Scheltinga 2002
Notaden melanoscaphus	LM; TEM	Scheltinga 2002
Pseudophryne covacevichae	LM; TEM	Scheltinga 2002
Pseudophryne guentheri	LM; TEM	Scheltinga 2002
Pseudophryne major	LM; TEM	Scheltinga 2002
Pseudophryne raveni	LM; TEM	Scheltinga 2002
Rheobatrachus silus	LM; TEM	Scheltinga 2002
Rheobatrachus vitellinus	LM; TEM	Scheltinga 2002
Taudactylus acutirostris	LM; TEM	Scheltinga 2002
Taudactylus diurnus	LM; TEM	Scheltinga 2002
Taudactylus liemi	LM; TEM	Scheltinga 2002
Taudactylus rheophilus	LM; TEM	Scheltinga 2002
Uperoleia altissima	LM	Scheltinga 2002
Uperoleia fusca	LM	Scheltinga 2002
Uperoleia laevigata	LM	Scheltinga 2002
Uperoleia lithomoda	LM; TEM	Scheltinga 2002
Uperoleia littlejohni	LM; TEM	Scheltinga 2002
Uperoleia mimula	LM; TEM	Scheltinga 2002
Uperoleia rugosa	LM	Scheltinga 2002

LM = light microscopy; TEM = transmission electron microscopy.

Limnodynastes, Mixophyes, Neobatrachus (Fig. 5.31E), *Notaden,* and *Rheobatrachus.* In contrast, the fibers are fine and appears as a tightly packed homogeneous mass in all transverse sections in *Kyarranus, Pseudophryne, Taudactylus,* and *Uperoleia.* The perforatorium is moderately electron-dense and is not membrane bound. The acrosome complex is symmetrically attached to the nucleus. In *Pseudophryne, Taudactylus,* and *Uperoleia* the entire acrosome complex is closely associated with the tip of the nucleus, whereas, in *Heleioporus, Kyarranus,*

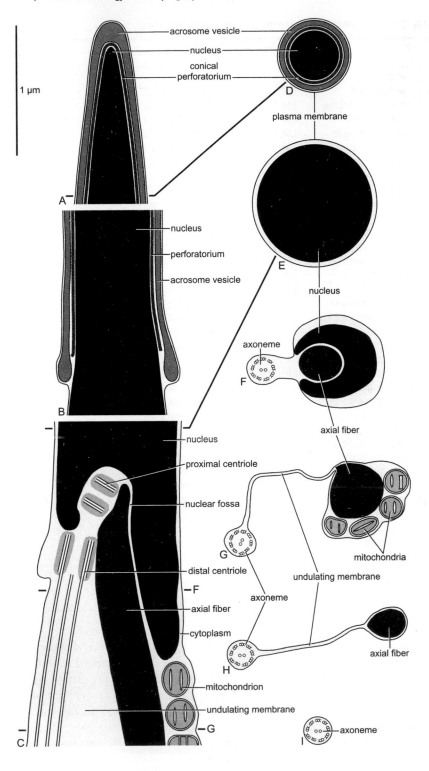

Table 5.10 Dimensions (means in μm) of myobatrachid spermatozoa.

Species	Head length	Axial fiber length[t]	Endpiece length	Total tail length	Total length	Author
Crinia insignifera	12	10	2.8	12	25	Scheltinga 2002
Crinia parinsignifera	11	11	2.3	13	24	Scheltinga 2002
Heleioporus albopunctatus	18	19	11	30	47	Scheltinga 2002
Heleioporus inornatus	22	24	9.8	34	56	Scheltinga 2002
Kyarranus kundagungan	25	28	0	28	53	Scheltinga 2002
Kyarranus loveridgei	22	26	0	26	48	Scheltinga 2002
Lechriodus fletcheri	16	14	9.5	24	39	Scheltinga 2002
Limnodynastes convexiusculus	18	18	11	29	46	Scheltinga 2002
Limnodynastes ornatus	10	10	17	27	37	Scheltinga 2002
Limnodynastes peronii	14	19	8.8	27-40	41-54	Lee and Jamieson 1992; Scheltinga 2002
Limnodynastes salmini	16	19	12	31	47	Scheltinga 2002
Limnodynastes tasmaniensis	16	17	8.5	26	41	Scheltinga 2002
Limnodynastes terraereginae	14	21	12	32	45	Scheltinga 2002
Mixophyes fasciolatus	14-21	20	3	23-30	44	Lee and Jamieson 1992; Scheltinga 2002
Mixophyes iteratus	20	26	5.2	32	53	Scheltinga 2002
Mixophyes schevilli	22	29	5.3	34	57	Scheltinga 2002
Neobatrachus kunapalari	18	22	11	33	50	Scheltinga 2002
Neobatrachus pelobatoides	26-30	38	20	58-60	84-90	Lee and Jamieson 1992; Scheltinga 2002
Neobatrachus sudelli	29	33	11	44	73	Scheltinga 2002
Notaden bennettii	12	18	2.6	20	32	Scheltinga 2002
Notaden melanoscaphus	11	10	3.6	14	24	Scheltinga 2002
Pseudophryne covacevichae	19	28	4.1	32	50	Scheltinga 2002
Pseudophryne guentheri	19	34	3.3	37	56	Scheltinga 2002
Pseudophryne major	20	28	7.2	35	54	Scheltinga 2002
Pseudophryne raveni	15	16	5.4	21	36	Scheltinga 2002
Rheobatrachus silus	19	19	2.7	22	41	Scheltinga 2002
Rheobatrachus vitellinus	24	22	2.7	25	49	Scheltinga 2002
Taudactylus acutirostris	12	13	2.7	17	30	Scheltinga 2002
Taudactylus diurnus	16	12	2.7	15	29	Scheltinga 2002
Taudactylus liemi	21	19	3.4	21	42	Scheltinga 2002
Taudactylus rheophilus	14	-	-	-	-	Scheltinga 2002
Uperoleia altissima	20	26	4.5	31	51	Scheltinga 2002
Uperoleia fusca	14	11	4.5	15	29	Scheltinga 2002
Uperoleia laevigata	15	11	5.5	17	32	Scheltinga 2002
Uperoleia lithomoda	20	30	4.5	34	54	Scheltinga 2002
Uperoleia mimula	14	-	-	-	-	Scheltinga 2002
Uperoleia rugosa	14	12	5.4	18	32	Scheltinga 2002

[t]Equivalent to undulating membrane length.

Fig. 5.30 A–C. Highly diagrammatic representation of the spermatozoon of *Taudactylus liemi* showing longitudinal sections. **A.** Apical acrosome complex and nucleus. **B.** Basal portion of the acrosome. **C.** Centriolar region, with corresponding transverse sections (TSs) as indicated by the corresponding letter on the LS. **H.** TS through the principal piece of the tail. **I.** TS through the endpiece. From Scheltinga, D. M. 2002. Unpublished Ph.D. Thesis, University of Queensland, Australia, Fig. 9.2.

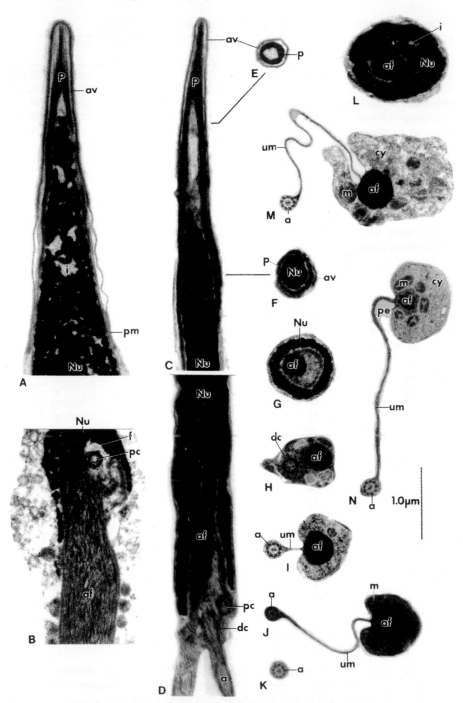

Fig. 5.31 TEM of Myobatrachidae spermatozoa. **A.** *Notaden bennettii*. Longitudinal section (LS) through the acrosome complex. **B.** *Notaden melanoscaphus*. LS through

Contd.

Lechriodus, Limnodynastes, Mixophyes, Neobatrachus (Fig. 5.31C), *Notaden* (Fig. 5.31A) and *Rheobatrachus* the acrosome complex extends anteriorly well beyond the nucleus.

The nucleus is cylindro-conical, tapering to a rounded tip within the acrosome complex. It occupies most of the head length, is circular in transverse section, and composed of electron-dense condensed chromatin. Numerous small to large nuclear inclusions, which are electron-lucent or contain electron-pale material, are present (Fig. 5.31A). Distinct nuclear shoulders are absent. The nucleus increases in diameter throughout its length to a maximum basally. The nucleus ends basally in an asymmetrical and well-developed fossa (Fig. 5.31B,D).

Within the neck or centriolar region of the spermatozoon are two centrioles, one proximal and one distal. Both centrioles are surrounded by a small amount of pericentriolar material. The proximal centriole is orientated perpendicular to the distal centriole in *Notaden* (Fig. 5.31B), *Pseudophryne, Taudactylus,* and *Uperoleia,* whereas, in *Kyarranus, Limnodynastes,* and *Neobatrachus* (Fig. 5.31D) the proximal and distal centrioles are parallel and lie at an oblique angle to the long axis of the spermatozoon. The orientation of the proximal centriole has not been determined in *Heleioporus, Lechriodus, Mixophyes,* and *Rheobatrachus.* In genera with parallel centrioles, as well as *Heleioporus, Lechriodus, Mixophyes,* and *Rheobatrachus,* the nuclear fossa is roughly symmetrical (in terms of the posterior extension of the nucleus) and the proximal centriole lies just outside or only just within the nuclear fossa (Fig. 5.31D). In contrast, in those genera with perpendicular centrioles, the nuclear fossa is strongly asymmetrical and the proximal centriole lies anteriorly within the fossa. *Notaden bennettii* differs slightly in that the perpendicular proximal centriole is not located anteriorly with the fossa but more within the mid region, and the nuclear fossa is not strongly asymmetrical. In all genera the distal centriole forms the basal body of the axoneme and lies at an oblique angle to the long axis of the spermatozoon. The axial fiber extends anteriorly through the neck region and into the nuclear fossa (Fig. 5.31B, D, G, L).

The midpiece is elongate and occurs along base of the nucleus and the anterior portion of the tail. The many small mitochondria are spherical to oval in shape, have linear cristae, lie adjacent to the outer surface of the axial fiber in a single layer, and are closely associated with the plasma membrane for most

---

Fig. 5.31 *Contd.*
the centriolar region. **C-K.** *Neobatrachus sudelli:* **C.** LS through the anterior acrosome complex. **D.** LS through the nuclear fossa. **E-K.** Successive transverse sections (TSs). **E.** Anterior acrosome. **F.** Acrosome/nucleus junction. **G.** Basal nuclear fossa. **H.** Distal centriole. **I and J.** Midpiece. **K.** Endpiece. **L-M.** *Neobatrachus kunapalari.* **L.** TS through the nuclear fossa. **M.** TS through the midpiece. **N.** *N. bennettii.* TS through the midpiece. All to the same scale as indicated. Abbreviations: a, axoneme; af, axial fiber; av, acrosome vesicle; cy, cytoplasm; dc, distal centriole; f, basal nuclear fossa; i, nuclear inclusion; m, mitochondrion; Nu, nucleus; p, conical perforatorium; pc, proximal centriole; pe, periaxial material; pm, plasma membrane; um, undulating membrane. From Scheltinga, D. M. 2002. Unpublished Ph.D. Thesis, University of Queensland, Australia, Fig. 9.4.

of the midpiece (Fig. 5.31I, J, M, N). Anteriorly, a large amount of cytoplasm is often observed and in which the mitochondria are scattered and separated from the axial fiber and plasma membrane. The mitochondria are relatively constant in shape and size within a spermatozoon. Occasionally, the anterior extremity of the axoneme is partially surrounded by a short cytoplasmic collar from which it is separated by a space (the cytoplasmic canal). The occurrence of a cytoplasmic collar/canal is sporadic within a species.

The midpiece of *Mixophyes* differs from that of the other myobatrachids studied here. In *Mixophyes* all the mitochondria are located within a cytoplasmic bead attached to the base of the nucleus and are not associated with the tail. The mitochondria have linear cristae, are variable in size, and distinctly oval in shape. Owing to inadequate fixation and alcohol storage the structure of the midpiece of *Rheobatrachus* was not determined with any certainty. However, it appears that the mitochondria are located within a cytoplasmic bead around the nucleus as in *Mixophyes*.

In *Heleioporus albopunctatus, Kyarranus kundagungan, Lechriodus fletcheri, Limnodynastes ornatus, Lim. peronii, Neobatrachus sudelli, Notaden bennettii* (Fig. 5.31N), *No. melanoscaphus, Pseudophryne raveni, Taudactylus diurnus, T. rheophilus*, and *Uperoleia mimula* the axial fiber, within the midpiece region, is almost completely surrounded by a thin layer of less electron-dense material, the periaxial sheath of Lee and Jamieson (1993). The periaxial sheath is lost posteriorly, leaving only the axial fiber to continue within the principal piece. Whether the periaxial material is actually part of the axial fiber or a densification of the cytoplasm surrounding it is uncertain. The two parts are more easily observed in formalin fixed material where the periaxial sheath is often seen separated from ('peeled off') the axial fiber, suggesting that it is cytoplasmic and not part of the axial fiber. The presence of periaxial material is variable within a species, suggesting that fixation and staining are important in its detection.

Within the midpiece region of *Pseudophryne, Taudactylus*, and *Uperoleia* the undulating membrane attaches to the side of the axial fiber, *i.e.* nearer the midpiece rather than in the middle of its 'free' surface, and extends perpendicular to a plane through the axial fiber and midpiece or nucleus (Fig. 5.30G). In all other myobatrachid genera the attachment of the undulating membrane, axial fiber, and midpiece or nucleus are in the same plane (*i.e.* parallel) (compare Figs 5.30G and 5.31J, M, N). Elongate tubular vesicles containing granular material are present within the midpiece of spermatozoa of *Pseudophryne* and *Uperoleia* but are absent from other myobatrachid genera.

The tail complex is composed of a 9+2 axoneme and an axial fiber, the fiber being connected to doublet 3 of the axoneme by a thin cytoplasmic lamina, the undulating membrane. A distinct juxta-axonemal fiber adjacent to doublet 3 is present but small, being less than the diameter of the axoneme in *Limnodynastes ornatus* and *Lim. peronii* but absent in other myobatrachids. A small juxta-axonemal fiber is occasionally seen in some *Mixophyes iteratus* tail sections. The axial and juxta-axonemal fibers are continuous via a thin dense lamina sandwiched within the undulating membrane. This lamina is also present in those species lacking a juxta-axonemal fiber. The axial fiber is well developed and circular to ovoid in transverse section throughout its length. The axial fiber continues posteriorly, decreasing in diameter, for much of the tail length. The axoneme extends beyond the axial fiber and undulating membrane as the endpiece (Fig. 5.31K). An endpiece was not observed in the spermatozoa of *Kyarranus*. However, an endpiece may be present but very short in *Kyarranus*. Lengths of the endpiece are given in Table 5.10.

5.3.22 Family Rhinodermatidae

The sperm of both species of *Rhinoderma* have previously been examined ultrastructurally by Pugin-Rios (1980), Pugin and Garrido (1981), and Scheltinga (2002). The spermatozoa of *Rhinoderma* are filiform, being composed of a head (acrosome and nucleus), a short midpiece, and a long tail (principal piece and endpiece). The sperm of *Rhinoderma rufum* is 80 µm long, consisting of a head 26-30 µm long, a 1 µm wide mitochondrial collar, and a tail 46 µm long. The spermatozoon of *R. darwinii* is 75 µm long of which the head is 23 µm long. The spermatozoon of *R. darwinii* is depicted diagrammatically in Figure 5.32.

The acrosome complex is composed of an acrosome vesicle that completely caps the conical perforatorium. The acrosome vesicle is membrane bound, filled with moderately electron-dense material. The thickness of the vesicle is constant throughout its length, being relatively thin. The perforatorium occurs within the subacrosomal space and is composed of fine longitudinal fibers which appear as homogeneous separate sheaves in transverse section through, and near, the nucleus. However, anteriorly the fibers are tightly packed and appear as a homogeneous mass and not separate sheaves. The base of the perforatorium overlies the attenuated nuclear tip. The perforatorium is moderately electron-dense and is not membrane bound. The acrosome complex extends anteriorly beyond the nucleus, being symmetrically attached to the nucleus basally.

The nucleus is cylindro-conical, tapering to a pointed tip within the acrosome complex. It occupies most of the head length, is circular in transverse section, and composed of electron-dense condensed chromatin. Numerous nuclear inclusions, which are electron-lucent or contain electron-pale material, are present. Distinct nuclear shoulders are absent. The nucleus increases in diameter throughout its length to a maximum at its base where it ends in a well-developed fossa. The fossa is asymmetrical in one plane of longitudinal section, *i.e.* when the proximal centriole is also seen in longitudinal section.

Within the basal nuclear fossa lies the proximal centriole which is surrounded by pericentriolar material that connects it to the nuclear fossa and distal centriole. The proximal centriole is orientated at a right angle to the distal centriole. The distal centriole lies in the long axis of the spermatozoon and forms the basal body of the axoneme. The axial fiber extends anteriorly into the neck region to the base of the distal centriole. A short mitochondrial collar, containing scattered mitochondria and cytoplasm, surrounds the anterior portion of the tail but is separated from it by a gap, the cytoplasmic canal. This collar is attached only to the centriolar/neck region of the spermatozoon and the cytoplasmic canal extends posteriorly from just below the base of the distal

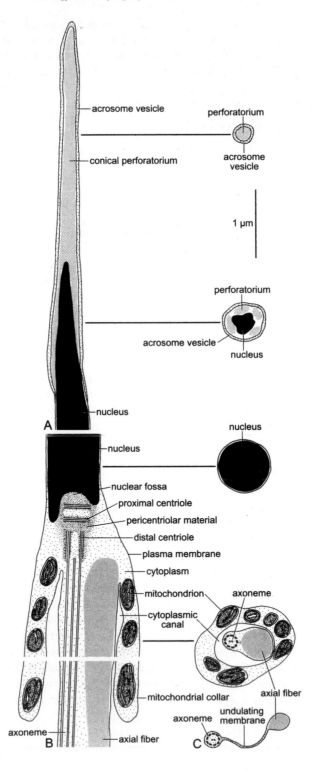

centriole. The mitochondria have well-developed linear cristae and are of variable shape (spherical to oval) and size within a spermatozoon.

The tail complex is composed of a 9+2 axoneme and an axial fiber, the fiber being connected to doublet 3 of the axoneme by a thin cytoplasmic lamina, the undulating membrane. A juxta-axonemal fiber adjacent to doublet 3 is absent. The axial fiber and axoneme are connected via a thin dense lamina sandwiched within the undulating membrane. The axial fiber is well developed and circular to ovoid in transverse section throughout its length. The axial fiber continues posteriorly, decreasing in diameter, for much of the tail length. The axoneme extends beyond the axial fiber and undulating membrane as the endpiece.

5.3.23 Superfamily Ranoidea

The structure of the spermatozoa of the ranoid families Arthroleptidae, Hemisotidae, and Scaphiophrynidae has not been described.

5.3.24 Family Hyperoliidae

Knowledge of spermatozoal structure in the Hyperoliidae is limited to SEM examination of the spermatozoa of *Leptopelis flavomaculatus* (Leptopelinae) and *Semnodactylus wealii* (Kassininae) by Wilson *et al.* (1994) and an ultrastructural study of the inadequately fixed spermatozoa of *Hyperolius puncticulatus* (Hyperoliinae) by Scheltinga (2002).

The spermatozoon of *Hyperolius puncticulatus* is described below. It is filiform and approximately 90 µm in total length, being composed of a head (acrosome and nucleus) approximately 45 µm long and a tail (midpiece and principal piece) approximately 45 µm long. Under light microscopy no definite distinction can be made between the acrosome and nucleus, or between the midpiece and principal piece. The spermatozoon of *H. puncticulatus* is depicted diagrammatically in Figure 5.33.

The acrosome complex is composed of an acrosome vesicle which caps the conical perforatorium. The acrosome vesicle appears to be membrane bound and filled with electron-pale material though its exact nature was difficult to determine owing to inadequate fixation. The perforatorium, situated within the subacrosomal space, is composed of fine longitudinal fibers which appear as homogeneous sheaves in transverse section, and overlies the attenuated nuclear tip. The perforatorium is moderately electron-dense, and is not membrane bound. The acrosome complex is relatively long (one incomplete longitudinal section measured 13 µm), extending anteriorly beyond the nucleus and being symmetrically attached to the nucleus basally.

The nucleus is cylindro-conical, tapering to a rounded tip within the acrosome complex. It is circular in transverse section, and is composed of condensed chromatin. Small nuclear inclusions, either electron-lucent or containing moderately electron-dense material, are present. Distinct nuclear shoulders are

Fig. 5.32 A, B. Drawing of a spermatozoon of *Rhinoderma darwinii* showing longitudinal sections. **A.** Acrosome complex. **B.** Centriolar region and midpiece, with corresponding transverse sections (TSs) as indicated. **C.** TS through the principal piece of the tail. From Scheltinga, D. M. 2002. Unpublished Ph.D. Thesis, University of Queensland, Australia, Fig. 13.1.

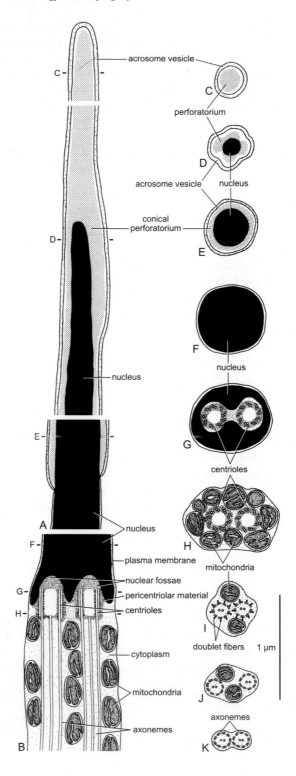

acrosome vesicle

perforatorium

acrosome vesicle   nucleus

conical
perforatorium

nucleus

nucleus

centrioles

mitochondria

doublet fibers   1 µm

axonemes

nucleus

plasma membrane
nuclear fossae
pericentriolar material
centrioles

cytoplasm

mitochondria

axonemes

absent. At the base of the nucleus there appear to be two distinct fossae but posteriorly these join medially. The nucleus increases in diameter throughout its length to a maximum near the nuclear fossae.

A single centriole lies within each nuclear fossa. The centrioles lie in the long axis of the spermatozoon and are parallel and adjacent to each other. Surrounding each centriole is pericentriolar material. Both centrioles are composed of nine, circularly arranged, triplets of short microtubules and each is continuous with a 9+2 axoneme of the flagellum. The mitochondria (midpiece) are located anteriorly along the tail, starting just posterior to the base of the nucleus. Initially, around the centrioles and anterior axonemes, the mitochondria are tightly packed and are located laterally around these structures. Posteriorly, the spherical to oval mitochondria with well-developed linear cristae, are in the form of two longitudinal series, of which a single linear series is adjacent to each axoneme. Initially, within the midpiece region the two axonemes are closely associated with each other, with doublet 3 of one axoneme adjacent to doublet 8 of the other. Posteriorly, the linear series of mitochondria comes to lie between the two axonemes before the axonemes again become closely associated within the principal piece. An axial fiber, undulating membrane, and juxta-axonemal fibers are absent. However, for a short distance along the anterior portion of each axoneme, nine small longitudinal fibers are present around the axoneme, one associated with each of the nine doublets. These fibers decrease in size posteriorly and disappear with the midpiece. A portion of the tail occurs in the absence of mitochondria as the principal piece and is composed of the two axonemes within, and closely associated with, a common plasma membrane. Little excess cytoplasm is present surrounding the axonemes.

The spermatozoa of *Leptopelis flavomaculatus* and *Semnodactylus wealii* clearly show a double tail complex. This differs from the single tail complex, containing two axonemes, observed for *Hyperolius puncticulatus*. Unfortunately the ultrastructure of the spermatozoon tail in *L. flavomaculatus* and *S. wealii*, i.e. if the double tail complex is composed of two axonemes or an axoneme and axial fiber, remains unknown.

5.3.25 Family Mantellidae

Structure of the spermatozoa of *Mantidactylus majori* has been described by Scheltinga (2002). The spermatozoa are filiform and 57 μm long, being composed of a relatively straight, 22 μm long head (acrosome and nucleus), a 2.2 μm long midpiece, and a 32 μm long tail. The spermatozoon of *M. majori* is depicted diagrammatically in Figure 5.34.

The acrosome complex is composed of an acrosome vesicle only. A perforatorium is absent. The acrosome vesicle is membrane bound and attaches

---

Fig. 5.33 A, B. Drawing of a spermatozoon of *Hyperolius puncticulatus* showing longitudinal sections (LSs). **A**. Acrosome complex. **B**. Nuclear fossae and anterior midpiece, with corresponding transverse sections (TSs) as indicated by the associated letter on LS. **I**. TS through the anterior portion of the midpiece. **J**. TS through the distal portion of the midpiece. **K**. TS through the tail. From Scheltinga, D. M. 2002. Unpublished Ph.D. Thesis, University of Queensland, Australia, Fig. 6.1.

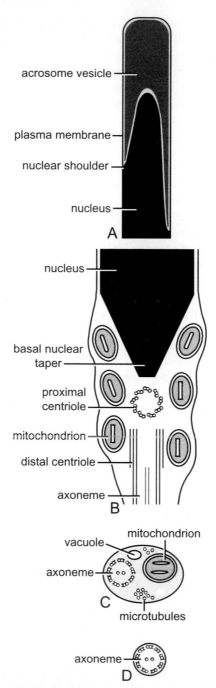

Fig. 5.34 Diagrammatic representation of the spermatozoa of *Mantidactylus majori*. **A**. Longitudinal section (LS) through the acrosome vesicle. **B**. LS through the centriolar region. **C**. Transverse section (TS) through the midpiece. **D**. TS through the tail. Original.

to (caps) the anterior portion of the nucleus asymmetrically, extending anteriorly well beyond the nucleus.

The nucleus is cylindrical with a rounded apical tip, composed of electron-dense condensed chromatin, and circular in transverse section. Occasional nuclear inclusions are present. Distinct nuclear shoulders are present. The nucleus increases in size posteriorly before basally tapering to a point within the midpiece.

The proximal centriole is located at the base of the nucleus and is orientated at a right angle to the distal centriole. The distal centriole lies in the long axis of the spermatozoon and forms the basal body of the axoneme. The mitochondria of the midpiece are located at the base of the nucleus, around the tapered point, and posteriorly to the level of the distal centriole. Numerous vacuoles and cytoplasm are also present within the midpiece. The relatively small and spherical to oval mitochondria are scattered throughout the midpiece. Auxiliary microtubules are present within the tail region. The numerous microtubules are the same diameter as those of the axoneme and occupy most of the cytoplasm surrounding the axoneme.

The tail is composed of a 9+2 axoneme, which for some of its length is surrounded by cytoplasm containing vacuoles. An axial fiber, juxta-axonemal fiber, and undulating membrane are absent.

5.3.26 Family Microhylidae

The structure of the spermatozoa of the microhylid genera *Calluella, Callulops, Cophixalus, Ctenophryne, Dermatonotus, Elachistocleis, Hylophorbus, Kaloula, Liophryne, Microhyla, Ramanella,* and *Sphenophryne* has previously been examined (see Table 5.11).

Table 5.11 Microhylid spermatozoa previously examined.

Species	Method used	Author
Asterophryinae		
Callulops sp.	LM; TEM	Scheltinga et al. 2002a
Hylophorbus rufescens	TEM	Scheltinga et al. 2002a
Dyscophinae		
Calluella guttulata	LM; TEM	Scheltinga et al. 2002a
Genyophryninae		
Cophixalus sp. A	LM; TEM	Scheltinga et al. 2002a
Cophixalus mcdonaldi	LM; TEM	Scheltinga et al. 2002a
Cophixalus ornatus	LM; TEM	Scheltinga et al. 2002a
Liophryne schlaginhaufeni	LM; TEM	Scheltinga et al. 2002a
Sphenophryne cornuta	LM; TEM	Scheltinga et al. 2002a
Microhylinae		
Ctenophryne geayi	LM; TEM	Scheltinga et al. 2002a
Dermatonotus muelleri	LM; TEM	Scheltinga et al. 2002a
Elachistocleis ovalis	LM; TEM	Scheltinga et al. 2002a
Kaloula pulchra	LM	Retzius 1906
Microhyla ornata	TEM; SEM	Kuramoto 1998; Kuramoto and Joshy 2001; Scheltinga et al. 2002a
Ramanella obscura	SEM	Kuramoto and Joshy 2001

LM = light microscopy; TEM = transmission electron microscopy; SEM = scanning electron microscopy.

The spermatozoa of all microhylids examined are of similar structure. They are filiform and lack an undulating membrane associated with the tail. The spermatozoa of *Cophixalus* spp., particularly *Cophixalus ornatus* and *C. mcdonaldi*, are extremely elongate, while those of *Liophryne* and *Sphenophryne* are moderately long and *Calluella, Callulops, Ctenophryne, Dermatonotus, Elachistocleis, Microhyla,* and *Ramanella* are relatively short. Measurements of head (acrosome complex and nucleus), midpiece (neck and mitochondrial sheath), tail (and/or tail+midpiece combined) and total length are given in Table 5.12. In some species it is impossible to distinguish between the end of the midpiece and the beginning of the tail.

Table 5.12   Dimensions (mean in μm) of microhylid spermatozoa.

Species	Head length	Midpiece length	Tail length	Tail and Midpiece length	Total length	Author
Callulops sp.	15	?	?	38	53	Scheltinga *et al.* 2002a
Calluella guttulata	15	2.3	29	31	47	Scheltinga *et al.* 2002a
Cophixalus mcdonaldi	66	14	58	67	133	Scheltinga *et al.* 2002a
Cophixalus ornatus	78	4.4	67	70	148	Scheltinga *et al.* 2002a
Cophixalus sp.	40	?	?	45	85	Scheltinga *et al.* 2002a
Ctenophryne geayi	11	2.1	≈ 37	39	50	Scheltinga *et al.* 2002a
Dermatonotus muelleri	11	1.8	≈ 25	27	38	Scheltinga *et al.* 2002a
Elachistocleis ovalis	13	1.5	≈ 31	33	45	Scheltinga *et al.* 2002a
Liophryne schlaginhaufeni	23	5	45	47	69	Scheltinga *et al.* 2002a
Microhyla ornata	8.5-11	1-1.5	40-41	35-43	48-54	Kuramoto 1998; Kuramoto and Joshy 2001; Scheltinga *et al.* 2002a
Ramanella obscura	9.8	?	?	27	37	Kuramoto and Joshy 2001
Sphenophryne cornuta	25	4.5	38	42	67	Scheltinga *et al.* 2002a

The spermatozoon of *Cophixalus ornatus* is depicted diagrammatically in Figure 5.35. The acrosome complex of *Cophixalus* differs from that found in the other microhylids examined. The acrosome complex is elongate, and composed of a relatively short cylindrical acrosome vesicle which caps only the apical portion of the conical perforatorium. The perforatorium is composed of fine longitudinal fibers which appear homogenous in transverse section and caps the nuclear tip. Posteriorly, the perforatorium extends beyond the base of the acrosome vesicle and is closely adpressed to one side of the nucleus. The acrosome complex of *Calluella, Callulops, Ctenophryne, Dermatonotus, Elachistocleis, Hylophorbus, Liophryne, Microhyla,* and *Sphenophryne* is composed of a conical

Fig. 5.35 Drawing of a spermatozoon of *Cophixalus ornatus* showing longitudinal sections with corresponding transverse sections as indicated. From Scheltinga, D. M., *et al.* 2002. Acta Zoologica (Stockholm) 83: 263-275, Fig. 2, ©The Royal Swedish Academy of Sciences; Blackwell Publishing.

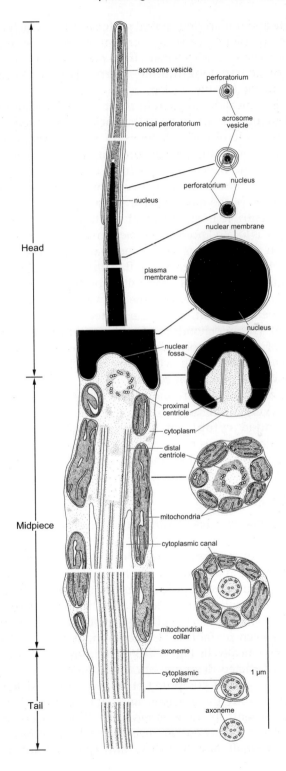

Head

Midpiece

Tail

acrosome vesicle

perforatorium

conical perforatorium

acrosome vesicle

nucleus

perforatorium

nucleus

nucleus

nuclear membrane

plasma membrane

nucleus

nuclear fossa

proximal centriole

cytoplasm

distal centriole

mitochondria

cytoplasmic canal

mitochondrial collar

axoneme

cytoplasmic collar

1 μm

axoneme

acrosome vesicle and underlying conical perforatorium which caps the nuclear tip. The vesicle completely caps the perforatorium. Posteriorly, the electron density of the acrosome vesicle changes from pale to lucent. The perforatorium is composed of coarse longitudinal fibers. The acrosome complex is symmetrically attached to the nucleus.

The nucleus is conical. It varies from short to elongate, and is composed of condensed chromatin. The apical tip of the nucleus is pointed in *Cophixalus* but rounded in other microhylids. Numerous nuclear inclusions containing moderately electron-dense material are present. At the base of the nucleus is an asymmetrical fossa in which the proximal centriole lies. Distinct nuclear shoulders are absent. The nucleus increases throughout its length to a maximum diameter at the level of the basal nuclear fossa.

The proximal centriole is orientated at 80° to the long axis of the nucleus. The distal centriole is located posterior to the proximal centriole and is orientated with its length in the long axis of the spermatozoon. The distal centriole is continuous with the 9+2 axoneme of the flagellum. Several mitochondria with prominent cristae form a sheath, the mitochondrial collar, surrounding the anterior portion of the axoneme. This sheath is attached only to the centriolar region of the spermatozoon, being separated from the axoneme by the cytoplasmic canal. The mitochondria of *Cophixalus* differ from the approximately spherical form of other microhylids in being elongate. This is particularly evident in *Cophixalus mcdonaldi* where the mitochondria extend along the length of the midpiece in a spiral pattern.

The axoneme of the tail is surrounded by a cytoplasmic collar, the two being separate by the cytoplasmic canal. The collar is composed of a thin layer of membrane bounded cytoplasm which extends posteriorly from the mitochondrial collar but does not contain mitochondria and is not considered part of the midpiece. An axial fiber, undulating membrane, and juxta-axonemal fibers are absent.

5.3.27 Family Ranidae

Spermatozoal structure has been examined in 11 genera of the Ranidae (Table 5.13).

The spermatozoa of the Ranidae are filiform, being composed of a relatively straight head (acrosome and nucleus), a short midpiece, and a long tail. Generally the shape of the spermatozoon head in ranids is straight, though it is loosely coiled in *Rana grayii* and *Nyctibatrachus major* or highly coiled in *Indirana beddomii* and *I. semipalmata*. Lengths of measurable components of the spermatozoa are given in Table 5.14. The spermatozoon of *Rana esculenta* is depicted diagrammatically in Figure 5.36, while that of *Rana temporaria* and *Rana dalmatina* is depicted diagrammatically in Figure 5.37.

The acrosome complex is composed of an acrosome vesicle alone, which is closely associated with the nucleus and does not extend anteriorly. A perforatorium is absent. In *Meristogenys*, and *Rana* (except *R. clamitans*, *R. dalmatina*, *R. japonica*, *R. pretiosa*, *R. rugosa*, and *R. temporaria*) the acrosome vesicle is relatively small, disc or sac-like, and apically located. The vesicle of

Table 5.13 Ranid spermatozoa previously examined.

Species	Method used	Author
Euphlyctis cyanophlyctis	LM; SEM	Kuramoto and Joshy 2000
Euphlyctis hexadactylus	SEM	Kuramoto and Joshy 2001
Hoplobatrachus rugulosus	SEM	Kuramoto 1998
Hoplobatrachus tigerinus	LM; TEM; SEM	Nath and Chand 1998; Kuramoto and Joshy 2000
Indirana beddomii	LM; SEM	Kuramoto and Joshy 2000
Indirana semipalmata	SEM	Kuramoto and Joshy 2001
Limnonectes greenii	SEM	Kuramoto and Joshy 2001
Limnonectes ingeri	LM; TEM	Scheltinga 2002
Limnonectes aff. keralensis	SEM	Kuramoto and Joshy 2001
Limnonectes keralensis	LM; SEM	Kuramoto and Joshy 2000
Limnonectes limnocharis	LM; SEM	Kuramoto 1998; Kuramoto and Joshy 2000; Scheltinga 2002
Limnonectes namiyei	SEM	Kuramoto 1998
Limnonectes rufescens	SEM	Kuramoto and Joshy 2001
Limnonectes syhadrensis	LM; SEM	Kuramoto and Joshy 2000
Meristogenys kinabaluensis	LM; TEM	Scheltinga 2002
Meristogenys orphnocnemis	LM; TEM	Scheltinga 2002
Nyctibatrachus major	LM; SEM	Kuramoto and Joshy 2000
Platymantis pelewensis	SEM	Kuramoto 1998
Phrynobatrachus natalensis	LM; TEM	Scheltinga 2002
Ptychadena mascareniensis	LM; TEM	Scheltinga 2002
Rana amamiensis	SEM	Kuramoto 1998
Rana arvalis	SEM	Herrmann 1987
Rana catesbeiana	TEM; SEM	Kuramoto 1998; Scheltinga 2002
Rana clamitans	TEM	Poirier and Spink 1971
Rana daemeli	LM; TEM	Scheltinga 2002
Rana dalmatina	TEM; SEM	Pugin-Rios 1980; Herrmann 1987
Rana dybowskii	TEM; SEM	Kwon and Lee 1995; Kuramoto 1998
Rana esculenta	LM; TEM	Retzius 1906; Pugin-Rios 1980
Rana fuscigula	TEM; SEM	Wilson et al. 1995
Rana grayii	TEM; SEM	Wilson et al. 1995
Rana grisea	LM; TEM	Scheltinga 2002
Rana ishikawae	SEM	Kuramoto 1998
Rana japonica	TEM; SEM	Yoshizaki 1987; Kuramoto 1998
Rana latouchii	SEM	Kuramoto 1998
Rana malabarica	SEM	Kuramoto and Joshy 2000
Rana narina	SEM	Kuramoto 1998
Rana nigromaculata	TEM; SEM	Sze et al. 1981; Mo 1985; Kwon and Lee 1995; Kuramoto 1998
Rana okinavana	SEM	Kuramoto 1998
Rana ornativentris	SEM	Kuramoto 1998
Rana perezi	LM	Serra and Vicente 1960
Rana pipiens	TEM	Zirkin 1971a,b; Poirier and Spink 1971
Rana pretiosa	LM; TEM	Scheltinga 2002
Rana ridibunda	TEM	Pugin-Rios 1980
Rana rugosa	TEM; SEM	Kwon and Lee 1995; Kuramoto 1998

Contd.

Table 5.13 Contd.

Species	Method used	Author
Rana supranarina	SEM	Kuramoto 1998
Rana supragrisea	LM; TEM	Scheltinga 2002
Rana tagoi	SEM	Kuramoto 1998
Rana temporalis	LM; SEM	Kuramoto and Joshy 2000
Rana temporaria	TEM; SEM	Nicander 1970; Pugin-Rios 1980; Herrmann 1987
Rana tsushimensis	SEM	Kuramoto 1998
Tomopterna cryptotis	TEM	Van der Horst 1979
Tomopterna delalandii	TEM; SEM	Wilson et al. 1995

LM = light microscopy; TEM = transmission electron microscopy; SEM = scanning electron microscopy.

Limnonectes, Phrynobatrachus, Ptychadena, R. clamitans, R. dalmatina, R. japonica, R. pretiosa, R. rugosa, and *R. temporaria* caps the anterior tip of the nucleus, extending posteriorly for a short distance around the nucleus. The vesicle of *R. dalmatina* differs further in extending anteriorly beyond the apical limit of the nucleus. The acrosome vesicle is symmetrically attached to the nucleus, or almost so, in all ranids examined except in *Ptychadena mascareniensis, R. clamitans, R. dalmatina,* and *R. temporaria* where it is distinctly asymmetrical.

The nucleus is cylindrical with a rounded apical tip and composed of electron-dense condensed chromatin. The nucleus is homogeneous throughout but occasional nuclear inclusions are present. Distinct nuclear shoulders are absent in *Meristogenys,* and *Rana* (except *Rana pretiosa*) but are present in *Limnonectes, Phrynobatrachus, Ptychadena,* and *R. pretiosa.* The nucleus is relatively constant in size throughout its length. In *Rana daemeli* the nucleus is flat basally but has a distinct small concavity present centrally, the nuclear fossa. A well developed basal nuclear fossa is present in *R. clamitans, R. pipiens,* and *R. temporaria,* though confirmation is required for *R. pipiens.* The proximal centriole is reported to lie within the fossa in *R. clamitans, R. pipiens,* and *R. temporaria* but not in *R. daemeli.* A fossa is absent from all other ranids. The base of the nucleus is flat (may be slightly concave or convex) in *Meristogenys, Ptychadena, Rana catesbeiana, R. grisea,* and *R. supragrisea.* Basally the nucleus tapers to a point within the midpiece in *Limnonectes, Phrynobatrachus,* and *R. pretiosa.*

At the base of the nucleus lies the proximal centriole which is orientated at a right angle to the distal centriole in all ranids except *Rana clamitans, R. pipiens, R. pretiosa,* and *R. temporaria.* In *R. pretiosa* the proximal centriole lies in the same plane, or almost so, as the distal centriole, whereas in *Rana clamitans, R. pipiens,* and *R. temporaria* it lies at an angle of 30º to the long axis of the spermatozoon. The proximal centriole of *R. japonica* is also obliquely orientated. In all ranids the distal centriole lies in the long axis of the spermatozoon and forms the basal body of the axoneme. The mitochondria of the midpiece are located at the base of the nucleus (around the tapered point in those possessing this) and posteriorly along the axoneme. Numerous vacuoles and cytoplasm are also present within the midpiece. The mitochondria are scattered throughout the midpiece, have well-developed linear cristae, are of spherical to oval in shape, and are relatively

Table 5.14   Dimensions (means in μm) of ranid spermatozoa.

Species	Head length	Midpiece length	Tail length	Total length	Author
Euphlyctis cyanophlyctis	17	?[†]	54	71	Kuramoto and Joshy 2000
Euphlyctis hexadactylus	13	?[†]	36	49	Kuramoto and Joshy 2001
Hoplobatrachus rugulosus	14	?[†]	35	48	Kuramoto 1998
Hoplobatrachus tigerinus	15	?[†]	47	62	Kuramoto and Joshy 2000
Indirana beddomii	38	?[†]	79	117	Kuramoto and Joshy 2000
Indirana semipalmata	31	?[†]	85	116	Kuramoto and Joshy 2001
Limnonectes greenii	18	?[†]	34	51	Kuramoto and Joshy 2001
Limnonectes ingeri	16	2	45	63	Scheltinga 2002
Limnonectes aff. *keralensis*	17	?[†]	35	52	Kuramoto and Joshy 2001
Limnonectes keralensis	17	?[†]	40	57	Kuramoto and Joshy 2000
Limnonectes limnocharis	14	3.8	38-46	51-57	Kuramoto 1998; Kuramoto and Joshy 2000; Scheltinga 2002
Limnonectes namiyei	16	?[†]	40	55	Kuramoto 1998
Limnonectes rufescens	14	?[†]	39	53	Kuramoto and Joshy 2001
Limnonectes syhadrensis	18	?[†]	37	55	Kuramoto and Joshy 2000
Meristogenys kinabaluensis	16	2	27	45	Scheltinga 2002
Meristogenys orphnocnemis	14	1.9	29	45	Scheltinga 2002
Nyctibatrachus major	24	?[†]	40	64	Kuramoto and Joshy 2000
Platymantis pelewensis	38	?[†]	57	95	Kuramoto 1998
Phrynobatrachus natalensis	15	1.9	30	46	Scheltinga 2002
Ptychadena mascareniensis	11	1.4	-	-	Scheltinga 2002
Rana amamiensis	32	?[†]	39	71	Kuramoto 1998
Rana catesbeiana	12	?[†]	30	42	Kuramoto 1998
Rana daemeli	15	1.9	24	41	Scheltinga 2002
Rana dalmatina	40	7	50	90	Pugin-Rios 1980
Rana dybowskii	14	?[†]	36	50	Kuramoto 1998
Rana esculenta	12-14	1.2-1.4	25	38-40	Pugin-Rios 1980
Rana fuscigula	12	2.4	35-45	49-59	Wilson *et al.* 1995
Rana grayii	16	?	37+	53+	Wilson *et al.* 1995
Rana grisea	19	1.9	32	53	Scheltinga 2002
Rana ishikawae	84	?[†]	56	139	Kuramoto 1998
Rana japonica	17	?[†]	35	52	Kuramoto 1998
Rana latouchii	15	?[†]	40	55	Kuramoto 1998
Rana malabarica	17	?[†]	28	45	Kuramoto and Joshy 2000
Rana narina	26	?[†]	40	67	Kuramoto 1998
Rana nigromaculata	12-15	?[†]	28-38	40-52	Kuramoto 1998
Rana okinavana	16	?[†]	35	51	Kuramoto 1998
Rana ornativentris	19	?[†]	40	59	Kuramoto 1998
Rana pretiosa	26	2.5	37	63	Scheltinga 2002
Rana rugosa	14	?[†]	41	55	Kuramoto 1998
Rana supranarina	24	?[†]	41	65	Kuramoto 1998
Rana supragrisea	19	1.8	32	52	Scheltinga 2002
Rana tagoi	18	?[†]	31	49	Kuramoto 1998
Rana temporalis	20	?[†]	45	66	Kuramoto and Joshy 2000
Rana temporaria	39	18	48	105	Pugin-Rios 1980
Rana tsushimensis	15	?[†]	25	40	Kuramoto 1998
Tomopterna delalandii	16	?	20-30	36-46	Wilson *et al.* 1995

[†]Midpiece length appears to be included within the head measurement.

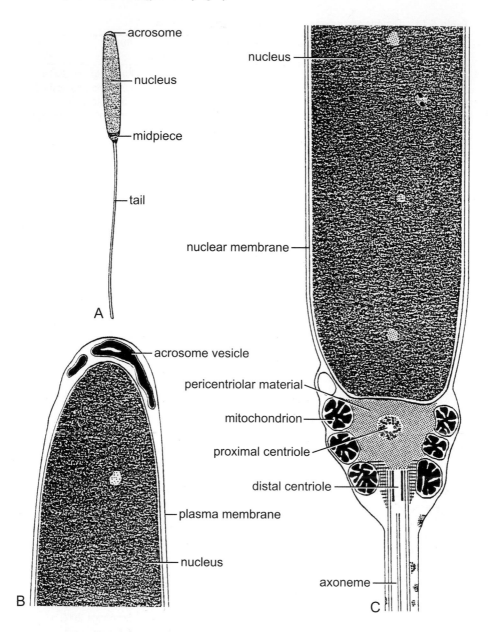

Fig. 5.36 Drawing of a spermatozoon of *Rana esculenta*. **A**. Whole spermatozoon. **B**. longitudinal section (LS) through the acrosome vesicle and anterior nucleus. **C**. LS through the centriolar region and midpiece. From (relabelled) Pugin-Rios, E. 1980. Unpublished Ph.D. Thesis, L'Université de Rennes, France, Figs 39, 42, 44.

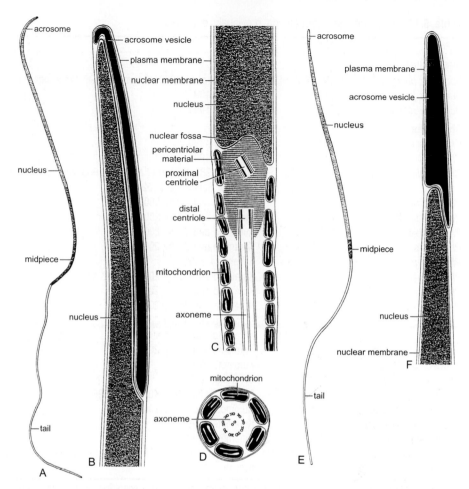

Fig. 5.37 Drawing of a spermatozoon of **A-D**, *Rana temporaria* and **E**, **F**, *Rana dalmatina*. **A**. Whole spermatozoon. **B**. Longitudinal section (LS) through the acrosome vesicle and anterior nucleus. **C**. LS through the centriolar region. **D**. Transverse section through the midpiece. **E**. Whole spermatozoon. **F**. LS through the acrosome vesicle and anterior nucleus. From (relabelled) Pugin-Rios, E. 1980. Unpublished Ph.D. Thesis, L'Université de Rennes, France, Figs 40, 41, 45, 46, 47, 48.

small. In *Limnonectes* auxiliary microtubules are present within the midpiece/tail which are absent from the other ranids. The microtubules are the same diameter as those of the axoneme and are few in number.

The tail is composed of a 9+2 axoneme, which for some of its length is surrounded by cytoplasm containing vacuoles. An axial fiber, juxta-axonemal fiber, and undulating membrane are absent.

5.3.28 Family Rhacophoridae

Spermatozoal structure has been examined in six genera from both subfamilies of the Rhacophoridae. The species investigated are given in Table 5.15.

Table 5.15 Rhacophorid spermatozoa previously examined.

Species	Method used	Author
Buergeriinae		
Buergeria buergeri	LM; TEM; SEM	Fukuyama et al. 1993; Kuramoto 1996; Scheltinga 2002
Buergeria japonica	SEM	Kuramoto 1996
Buergeria robusta	SEM	Kuramoto 1996
Rhacophorinae		
Chirixalus eiffingeri	SEM	Kuramoto 1996
Chirixalus idiootocus	SEM	Kuramoto 1996
Chiromantis xerampelina	TEM; SEM	Mainoya 1981; Wilson et al. 1991, 1995; van der Horst et al. 1995; Scheltinga 2002
Philautus sp. A	LM; SEM	Kuramoto and Joshy 2000
Philautus sp. B	LM; SEM	Kuramoto and Joshy 2000
Philautus sp. C	LM; SEM	Kuramoto and Joshy 2000
Philautus sp. D	LM; SEM	Kuramoto and Joshy 2000
Philautus bunitus	TEM	Scheltinga 2002
Philautus leucorhinus	SEM	Kuramoto and Joshy 2001
Philautus microtympanum	SEM	Kuramoto and Joshy 2001
Philautus variabilis	SEM	Kuramoto and Joshy 2001
Polypedates eques	SEM	Kuramoto and Joshy 2001
Polypedates leucomystax	LM; TEM; SEM	Kuramoto 1996; Scheltinga 2002
Polypedates maculatus	LM; SEM	Kuramoto and Joshy 2000
Polypedates megacephalus	SEM	Kuramoto 1996
Rhacophorus angulirostris	LM; TEM	Scheltinga 2002
Rhacophorus arboreus	LM; TEM; SEM	Mizuhira et al. 1986; Kuramoto 1996
Rhacophorus malabaricus	LM; SEM	Kuramoto and Joshy 2000
Rhacophorus moltrechti	SEM	Kuramoto 1996
Rhacophorus owstoni	SEM	Kuramoto 1996
Rhacophorus schlegelii	LM; TEM; SEM	Oka 1980; Mizuhira et al. 1986; Kuramoto 1996
Rhacophorus viridis	SEM	Kuramoto 1996

LM = light microscopy; TEM = transmission electron microscopy; SEM = scanning electron microscopy.

The spermatozoa of the rhacophorids are of similar general structure. However, distinctive differences, particularly in the neck, midpiece, and tail regions have been observed. They are filiform, being composed of a head (acrosome and nucleus), a short midpiece, and a long tail. Lengths of measurable components of the spermatozoa are given in Table 5.16. The head of the spermatozoa of *Chiromantis xerampelina* (10-12 coils) and some *Rhacophorus* spp. (9-20 coils) is distinct in being highly coiled/spiral. The spermatozoon of *Polypedates leucomystax* is depicted diagrammatically in Figure 5.38.

The acrosome complex is composed of a membrane bound acrosome vesicle only. A perforatorium is absent. In *Philautus bunitus* the acrosome vesicle is relatively small and apically located, capping the short, tapered nuclear tip, and does not extend anteriorly. A similar condition is observed in *Buergeria buergeri*, *Chiromantis xerampelina*, and *Rhacophorus angulirostris*, but in these species it extends further posteriorly around the nucleus. In *B. buergeri*, *C. xerampelina*, *Ph. bunitus*, and *R. angulirostris* the acrosome vesicle is of a constant thickness

Table 5.16 Dimensions (means in µm) of rhacophorid spermatozoa.

Species	Head length	Tail length	Total length	Author
Buergeria buergeri	44-45	44-66	89-110	Kuramoto 1996; Scheltinga 2002
Buergeria japonica	9.7	38	48	Kuramoto 1996
Buergeria robusta	9.2	29	38	Kuramoto 1996
Chirixalus eiffingeri	32-41	52-54	87-92	Kuramoto 1996
Chirixalus idiootocus	23	21	44	Kuramoto 1996
Chiromantis xerampelina	35	44	79	Wilson et al. 1995
Philautus sp. A	22	28	50	Kuramoto and Joshy 2000
Philautus sp. B	24	29	52	Kuramoto and Joshy 2000
Philautus sp. C	19	31	50	Kuramoto and Joshy 2000
Philautus sp. D	21	30	50	Kuramoto and Joshy 2000
Philautus leucorhinus	26	30	56	Kuramoto and Joshy 2001
Philautus microtympanum	33	43	76	Kuramoto and Joshy 2001
Philautus variabilis	20	29	49	Kuramoto and Joshy 2001
Polypedates eques	37	53	90	Kuramoto and Joshy 2001
Polypedates leucomystax	57	100	157-182	Kuramoto 1996; Scheltinga 2002
Polypedates maculatus	77	84	161	Kuramoto and Joshy 2000
Polypedates megacephalus	84	130	213	Kuramoto 1996
Rhacophorus angulirostris	28	49	77	Scheltinga 2002
Rhacophorus arboreus	76	56	132	Kuramoto 1996
Rhacophorus malabaricus	52	93	145	Kuramoto and Joshy 2000
Rhacophorus moltrechti	67	46	113	Kuramoto 1996
Rhacophorus owstoni	75	47	122	Kuramoto 1996
Rhacophorus schlegelii	56	57	113	Kuramoto 1996
Rhacophorus viridis	85	81	165	Kuramoto 1996

throughout and is closely associated with the nucleus. The acrosome vesicle of *Polypedates leucomystax* also caps the anterior portion of the nucleus but differs in that the vesicle extends anteriorly well beyond the nucleus. The acrosome vesicle is asymmetrically attached to the nucleus in *B. buergeri*, *Po. leucomystax* and *R. angulirostris* which differs from the condition in *Ph. bunitus* where attachment is symmetrical. The symmetry of attachment was not determined in *C. xerampelina* were the vesicle is thin and relatively poorly defined.

The nucleus is cylindrical and composed of electron-dense condensed chromatin. The nucleus is homogeneous throughout but occasional small nuclear inclusions are present. Distinct nuclear shoulders are absent in *B. buergeri*, *C. xerampelina* and *Po. leucomystax* but are present in *Ph. bunitus* and *R. angulirostris*. The non-tapered portion of the nucleus is relatively constant in size throughout its length, though it does increase posteriorly, particularly in *B. buergeri* and *Ph. bunitus*. Basally the nucleus tapers to a point within the midpiece in all rhacophorids examined excepting *B. buergeri*. In *B. buergeri* the base of the nucleus is mostly flat except for a lateral pointed extension, which is asymmetrical in longitudinal section. The nuclear taper in *Ph. bunitus* is asymmetrical, forming a flattened wedge of nuclear material with a flat base. This differs from the taper of *C. xerampelina*, *Po. leucomystax*, and *R. angulirostris* which is symmetrical, forming a cone of nuclear material with a rounded base.

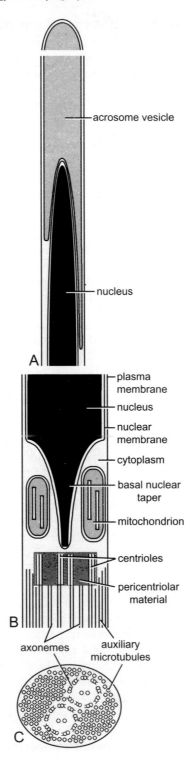

acrosome vesicle

nucleus

A

plasma membrane

nucleus

nuclear membrane

cytoplasm

basal nuclear taper

mitochondrion

centrioles

pericentriolar material

B

axonemes   auxiliary microtubules

C

The structure of the spermatozoon neck, midpiece, and tail varies significantly between rhacophorid genera.

Neck/midpiece and tail of *Buergeria* **and** *Philautus.* At the base of the nucleus lies the proximal centriole which is orientated at a right angle to the distal centriole. The distal centriole lies in the long axis of the spermatozoon and forms the basal body of the axoneme. The mitochondria of the midpiece are located at the base of the nucleus (around the nuclear taper in *Philautus bunitus*) and posteriorly along the axoneme. Numerous vacuoles and cytoplasm are also present within the midpiece. Several auxiliary microtubules are present within the midpiece/tail region. The microtubules are the same diameter as those of the axoneme and are scattered throughout the cytoplasm.

The tail is composed of a 9+2 axoneme, which for some of its length is surrounded by cytoplasm containing vacuoles. An axial fiber, juxta-axonemal fiber, and undulating membrane are absent.

Neck/midpiece and tail of *Chiromantis, Polypedates,* **and** *Rhacophorus.* The mitochondria (midpiece) are limited to the region around the tapered base of the nucleus and the centrioles. No mitochondria are associated with the axonemes. The mitochondria occur in a single layer around the nucleus and are oval in shape, being elongate in *Polypedates leucomystax* and *Rhacophorus angulirostris*. The mitochondria of *Po. leucomystax* differ in that they spiral around the nucleus. At the basal tip of the nucleus are two parallel and adjacent centrioles which lie in the long axis of the spermatozoon, each is continuous with a 9+2 axoneme of the flagellum.

Surrounding the two 9+2 axonemes of the tail are numerous microtubules. These auxiliary microtubules are the same diameter as those of the axoneme and occupy most of the cytoplasm surrounding the axonemes. Both axonemes are aligned in a single plane. An axial fiber, juxta-axonemal fiber, and undulating membrane are absent.

5.4 DISCUSSION OF ANURAN SPERMATOZOAL CHARACTERS

5.4.1 Acrosome Vesicle

An anterior acrosome vesicle is present in all investigated lissamphibian spermatozoa. The shape of the vesicle varies in two main ways within the Anura, being either closely associated with or well separated (a conical perforatorium intervening) from the apical tip of the nucleus. The vesicle of Ascaphidae, Bombinatoridae, Discoglossidae, Leiopelmatidae, Megophryidae, Pelobatidae (except *Scaphiopus*), Pelodytidae, Bufonidae, Centrolenidae, Hylidae, Hyperoliidae, Leptodactylidae (except *Caudiverbera, Leptodactylus,* and *Pleurodema*), Microhylidae, Myobatrachidae (except Myobatrachinae), and Rhinodermatidae is of a relatively constant thickness and is well separated from the apical tip the nucleus. In the pelobatid *Scaphiopus*, Pipidae, the

Fig. 5.38 Diagrammatic representation of the spermatozoa of *Polypedates leucomystax*. **A**. Longitudinal section (LS) through the acrosome vesicle. **B**. LS through the centriolar region and midpiece. **C**. Transverse section through the principal piece of the tail. Original.

leptodactylids *Caudiverbera, Leptodactylus* and *Pleurodema,* Myobatrachinae myobatrachids, Ranidae, and Rhacophoridae the vesicle is closely associated with the nucleus throughout its length. The closely associated vesicle of *Leptodactylus* and some Rhacophoridae is not of a constant thickness, extending anteriorly beyond the nuclear tip, whereas in *Scaphiopus,* Pipidae, *Caudiverbera* and *Pleurodema,* Myobatrachinae, some Ranidae, and other Rhacophoridae it is of a constant thickness throughout. In other Ranidae the vesicle is apically located on the nucleus and disc or sac-like in shape. The absence of such separation in the Myobatrachinae appears to be homoplastic with other anurans possessing a closely associated vesicle of constant thickness as a perforatorium is present only in the myobatrachines (see below). Close association of the vesicle (of constant thickness) with the nucleus is here deduced to be the result of spermatozoal paedomorphism. During spermiogenesis the acrosome vesicle of anurans is closely associated with the nucleus, separating from the nucleus tip late in development with the development of the perforatorium. The presence of an acrosome vesicle which is well separated from the apical tip of the nucleus appears to be the plesiomorphic condition as it occurs in urodeles, gymnophionans, and the most primitive anurans.

5.4.2   Axial Perforatorium and Endonuclear Canal

A rod-like axial perforatorium and endonuclear canal are characteristic of spermatozoa of basal tetrapods. They occur in urodeles (Fawcett 1970; Picheral 1979; Selmi *et al.* 1997; Scheltinga 2002) and gymnophionans, though, there lodged posteriorly in a greatly shortened endonuclear canal (Scheltinga *et al.* 2003). They also characterise the more basal anurans, *Ascaphus* (Jamieson *et al.* 1993) and the bombinatorid *Bombina* and discoglossids *Alytes* and *Discoglossus* (Sandoz 1970a,b; Furieri 1975a,b; Folliot 1979; Pugin-Rios 1980; Kwon and Lee 1992; Campanella *et al.* 1997). Furthermore one or both are present in sarcopterygian fish (Dipnoi and *Latimeria chalumnae*) (Mattei *et al.* 1988; see reviews in Jamieson 1991). It is thus reasonable to deduce that an endonuclear canal containing a long rod-like axial perforatorium is plesiomorphic for the Lissamphibia. This is endorsed by persistence of this condition in lower amniotes, the Chelonia, Sphenodontida and Crocodylia, and in palaeognath birds (Jamieson and Healy 1992; Jamieson 1995, 1999; Jamieson *et al.* 1997). Although *Leiopelma* lacks a definite endonuclear canal, there are what appear to be remnants of an endonuclear canal and axial perforatorium in the form of axial lacunae, containing electron-pale material, within the nuclear rostrum (Scheltinga *et al.* 2001). An axial perforatorium traverses the endonuclear canal in the late spermatid of *Discoglossus pictus* (Sandoz 1970b) but is restricted to the prenuclear subacrosomal space in the mature spermatozoon, though the endonuclear canal persists (Pugin-Rios 1980). An axial perforatorium and endonuclear canal are apomorphically absent from all other known Anura.

5.4.3   Conical perforatorium

Homology of the conical perforatorium in Megophryidae, Pelobatidae (absent in *Scaphiopus*), Pelodytidae, Bufonidae, Centrolenidae, Hylidae, Hyperoliidae,

Leptodactylidae (absent in *Caudiverbera*, *Leptodactylus*, and *Pleurodema*), Microhylidae, Myobatrachidae, and Rhinodermatidae with either the axial or conical perforatorium (also termed the subacrosomal cone or periperforatorial material) found in urodeles appears unlikely. The conical perforatorium of at least the hylids *Hyla japonica* and *Pachymedusa dacnicolor*, appears to develop much later in spermiogenesis, in a different position, and with a different ontogeny and final form than that of urodeles (Folliot 1979; Lee and Kwon 1992; Picheral 1972a; Rastogi *et al.* 1988), and is probably a unique development of those anurans possessing it and therefore, a synapomorphy. Homology of the conical perforatorium (also termed subacrosomal cone) of the primitive frogs *Ascaphus* and *Leiopelma* with that of urodeles or other Anura remains to be determined from spermatogenic stages. However, the presence in the mesobatrachians Megophryidae, Pelobatidae (Scheltinga 2002), and Pelodytidae (Pugin-Rios 1980) of a distinct conical perforatorium in the absence of an axial perforatorium suggests that this condition developed relatively early in anuran evolution. A conical perforatorium is absent from the Bombinatoridae, the pelobatid *Scaphiopus*, Pipidae, the leptodactylids *Caudiverbera*, *Leptodactylus*, and *Pleurodema*, Ranidae, and Rhacophoridae. As noted in section 5.3, granular subacrosomal material lying between the acrosome vesicle and nucleus in *Discoglossus pictus* (Sandoz 1970a; Campanella *et al.* 1997) is here tentatively recognized as a conical perforatorium. The absence of all perforatorial material from *Scaphiopus*, Pipidae, *Caudiverbera*, *Leptodactylus*, and *Pleurodema*, Ranidae, and Rhacophoridae is apomorphic. Its absence in *Scaphiopus* is questionable as Morrisett (1974) reported the presence of microtubules within the subacrosomal space which are also present in the spermatids of *Hyla japonica* in which they form the perforatorium late in spermiogenesis (Lee and Kwon 1992).

5.4.4 Nucleus

At the base of the acrosome complex, the nucleus flares out abruptly to form distinct nuclear shoulders (tetrapod synapomorphy) in basal amniotes (Healy and Jamieson 1992; Jamieson 1995), urodeles (Picheral 1979; Selmi *et al.* 1997; Scheltinga 2002), *Ascaphus* (Jamieson *et al.* 1993) and *Leiopelma* (Scheltinga *et al.* 2001). Behind this the nucleus forms a very elongate cylinder. Distinct nuclear shoulders are absent from all other Anura, though shoulders occur homoplastically in the ranid *Rana clamitans* and the rhacophorids *Philautus bunitus* and *Rhacophorus angulirostris*.

The chromatin is electron dense but the type of condensation is variable. The chromatin is condensed into large, almost prismatic masses in *Bombina*, or in smaller masses to rather homogeneous in other Anura. Numerous small to large nuclear inclusions, which are electron-lucent or contain electron-pale material, are present in all anuran spermatozoa.

5.4.5 Nuclear Fossa

Within the Anura a nuclear (centriolar) fossa containing no or one centriole is poorly developed in species with a simple flagellum, well developed in those with an undulating membrane and especially well developed in *Discoglossus pictus* in which the tail is very long (1000 µm). The nuclear fossa occurs in Ascaphidae, Bombinatoridae, Discoglossidae, Leiopelmatidae, Pelobatidae, Pelodytidae, Bufonidae, Centrolenidae, Hylidae, Hyperoliidae, Leptodactylidae (little more than a concavity in *Caudiverbera*), Microhylidae, Myobatrachidae, the ranids *Rana clamitans* (questionably), *R. daemeli* (greatly reduced) and *R. temporaria*, and Rhinodermatidae. It is basal except in the Bombinatoridae where it is located far anteriorly within the nucleus. Double fossae are present in Pelobatidae and Hyperoliidae. A nuclear fossa is absent from Megophryidae, Pipidae, other Ranidae, and Rhacophoridae. The base of the nucleus is rounded in Megophryidae, tapered in Pipidae, some Ranidae, and some Rhacophoridae, or flat (or only very slightly convex or concave) in other Ranidae, and other Rhacophoridae. A single basal nuclear fossa is present in most urodeles (Picheral 1979; Scheltinga 2002), caecilians (Seshachar 1945; van der Horst *et al.* 1991; Scheltinga *et al.* 2003), amniotes (Fawcett 1975; Jamieson 1995, 1999), and fishes (Jamieson 1991) and is therefore plesiomorphic for the Anura.

The spermatozoa of Discoglossidae, Leiopelmatidae, the ranids *Rana clamitans* and *R. daemeli* differ from those of other Anura possessing a basal nuclear fossa in that no centriole is present within the nuclear fossa. The centriole resides within the fossa of urodeles and gymnophionans (Scheltinga 2002; Scheltinga *et al.* 2003) and this is therefore the plesiomorphic state.

5.4.6 Centrioles

Within the Lissamphibia both centrioles are always present and have the classic structure of nine triplets of short microtubules. The proximal (anterior) centriole lies either parallel (Leiopelmatidae, Megophryidae, Pelobatidae, Hyperoliidae, the leptodactylid *Telmatobufo*, the myobatrachids *Kyarranus*, *Limnodynastes* and *Neobatrachus*, some Rhacophoridae), oblique (Ascaphidae, the discoglossid *Discoglossus*, Pipidae, and the ranids *Rana clamitans*, *R. japonica*, *R. pipiens*, *R. pretiosa*, *R. rugosa* and *R. temporaria*) or mutually perpendicular (or almost so) (Bombinatoridae, the discoglossid *Alytes*, Pelodytidae, Bufonidae, Centrolenidae, Hylidae, Leptodactylidae (except *Telmatobufo*), Microhylidae, other Myobatrachidae, other Ranidae, other Rhacophoridae, and Rhinodermatidae), to the distal (posterior) centriole. The distal centriole is situated at the base of the axoneme, for which it forms the basal body, and is orientated in the long axis, but in sperm with an undulating membrane the distal centriole is often slightly oblique. Where the spermatozoon has two axonemes, as in Megophryidae (see below), there is no differentiation into proximal and distal centrioles but each centriole forms the basal body of an axoneme. In *Leiopelma* both centrioles lie adjacent and parallel, and in the long axis of the spermatozoon, conditions which are unique to *Leiopelma* among lissamphibians with only one axoneme (Scheltinga *et al.* 2001). The myobatrachids *Kyarranus*, *Limnodynastes* and *Neobatrachus* also have parallel centrioles with only one axoneme but the centrioles differ from those in *Leiopelma* in being laterally located and at an oblique angle to the long axis of the spermatozoon. The centrioles of gymnophionans and urodeles (except ambystomatids and dicamptodontids where they are oblique), lie perpendicular to each other (Scheltinga 2002; Scheltinga *et al.* 2003) and this appears to be the plesiomorphic state. However, parallel and obliquely orientated centrioles appear early in anuran evolution.

5.4.7 Midpiece

The number and disposition of the mitochondria in lissamphibian sperm are variable and they are often absent at maturity, a fact which has interesting/unknown implications for motility. In the primitive cryptobranchid and hynobiid urodeles (Baker 1963; Picheral 1979; Kuramoto 1995, 1997; Scheltinga 2002) the mitochondria are located in a protoplasmic bead around the nucleus and not around the tail, even in the mature sperm. In salamandrids, ambystomatids and plethodontids small, ovoid mitochondria are present in cytoplasm around a long anterior region of the axial fiber where, in transverse section of the sperm, they form an arc (Baker 1966; Picheral 1967, 1979; Fawcett 1970; Wortham *et al.* 1977; Jamieson *et al.* 1993; Scheltinga 2002). The urodele type of midpiece was reported to be absent in anurans (Pugin-Rios 1980) but in Myobatrachidae (except *Mixophyes* and *Rheobatrachus* see below) an incomplete ring of mitochondria surrounding the axial fiber has been demonstrated, much as in salamandrids (Lee and Jamieson 1992; Scheltinga 2002). A somewhat similar arrangement is also seen in *Bombina* and the discoglossid *Alytes* (Folliot 1979; Furieri 1975b; Pugin-Rios 1980). The mitochondria lie in a cytoplasmic mass around the base of the nucleus in the myobatrachids *Mixophyes* and *Rheobatrachus*. A similar condition occurs in the discoglossid *Discoglossus* though in it there is no cytoplasmic mass and the mitochondria are closely associated with the nucleus before being lost late in spermiogenesis (Sandoz 1975). Mitochondria occur within a cytoplasmic mass in the leptodactylids *Caudiverbera* and *Pleurodema* and are also lost late in development (Pugin-Rios 1980).

Mitochondria occur in one or more longitudinal series in a groove in the paraxonemal fiber in the ascaphid *Ascaphus*, a condition which does not appear to be plesiomorphic (arguments having been presented for a paedomorphic origin) despite the primitive (basal) status of this species (Jamieson *et al.* 1993). Mitochondria are located in either one (the pelobatid *Pelobates*) or two (Megophryidae and Hyperoliidae) linear series which are closely associated with the axoneme(s). In the pelobatids *Spea* and *Scaphiopus* the mitochondria appear to be scattered along the midpiece. In Pelodytidae, Pipidae, Ranidae, and Rhacophoridae the mitochondria occur within the neck region of the spermatozoa and are closely associated with the tapered base of the nucleus (in those possessing this condition), the centrioles and anterior axoneme, though in the rhacophorids *Chiromantis*, *Mantidactylus*, *Polypedates*, and *Rhacophorus* the mitochondria do not extend posteriorly along the axoneme. The pipid *Xenopus* differs slightly in possessing a very short cytoplasmic canal posteriorly which separates the more posterior of the mitochondria from the tail. In contrast, in Bufonidae (except *Nimbaphrynoides*), Centrolenidae, Hylidae, Leptodactylidae (except *Caudiverbera*, *Leptodactylus*, and *Pleurodema*), Microhylidae, and Rhinodermatidae the mitochondria are distributed around the anterior region of the sperm tail in a well-developed and distinct collar, which may be lost at maturity in many of these species (Garrido *et al.* 1989; Lee and Jamieson 1992; Meyer *et al.* 1997; Pugin-Rios 1980). The cytoplasmic canal of the hylid *Plectrohyla* and the leptodactylid *Physalaemus* is, like that of pipids, much reduced. In *Leptodactylus* and *Nimbaphrynoides* a mitochondrial collar is absent and the mitochondria are locate along the tail, adjacent to the axoneme. The conditions seen in *Leiopelma* of slight development of a cytoplasmic collar and rudimentary cytoplasmic canal is similar to that seen in the lungfish *Neoceratodus forsteri* (Jespersen 1971; Jamieson 1999). Until the examination of *Leiopelma* spermatozoa no extant species were known which presented an arrangement of mitochondria which could confidently be regarded as plesiomorphic for the Anura. It seems reasonable to deduce that the *Neoceratodus*-like condition seen in *Leiopelma* is plesiomorphic for anurans. The well-developed, but often transient, mitochondrial collar in many anurans, resemble a condition seen in many acanthopterygian fish (Jamieson 1991) and therefore on first analysis appear plesiomorphic, but may well be a reversal to a pre-lissamphibian condition as a collar is absent from urodeles, gymnophionans, and most of the basal Anura.

5.4.8 Tail Complex

The structure of the sperm tail is highly distinctive of the Lissamphibia. The generalized lissamphibian sperm has two longitudinal fibers, one on each side of the axoneme adjacent to doublets 3 and 8. Swan *et al.* (1980) showed, as did James (1970) and Pugin-Rios (1980), that the axial fiber and the juxta-axonemal fiber at 3 are interconnected by a dense lamina within the undulating membrane, an observation later confirmed for *Litoria* (Jamieson *et al.* 1993; Lee and Jamieson 1993) and for other genera here. It therefore seems likely that the two fibers are subdivisions of a single accessory axonemal fiber. The two widely separated, axial and juxta-axonemal, fibers occurring in many anurans are here considered to be homologous with the paraxonemal rod as suggested by Jamieson *et al.* (1993) for *Ascaphus*, or with the simple unmodified axial fiber of other anurans. The axial fiber, which is associated with doublet 3 is usually separated from it by an undulating membrane. The juxta-axonemal fiber at 8, which is closely associated with the axoneme and characteristic of urodeles, is absent in gymnophionans and was previously thought to be absent in anurans. The plesiomorphic condition seen in *Leiopelma* of the presence of juxta-axonemal fibers at 3 and 8 is unique among the Anura. However, a temporary fiber is present in the spermatids of *Discoglossus pictus* (Sandoz 1975) and also in occasional *Bufo marinus* sperm (Swan *et al.* 1980). Fibers associated with additional doublets, or with all doublets of the axoneme have also been observed in spermatids of the hylid *Hyla regilla* and the leptodactylid *Telmatobius jelskii*, and in the mature spermatozoa of the leptodactylid *Telmatobufo australis*, the hyperoliid *Hyperolius puncticulatus*, and the hylid *Pseudis paradoxa*. The presence of nine external longitudinal fibers, one adjacent to each doublet, in the absence of separate axial and juxta-axonemal fibers and of an undulating membrane, is a noteworthy link with the similar condition typical of amniote sperm. Its presence may indicate retention of a genetic system for a 9+9+2 arrangement retained from an ancestral tetrapod which is not activated in most anurans. It seems less likely that it is an entirely independent, homoplastic development.

The phylogenetic origins and, therefore, homologies of the lissamphibian flagella complex have been discussed by Jamieson (1995, 1999) who suggested dipnoan affinities. Thus, in *Neoceratodus forsteri* where the anterior region of the sperm axoneme has a large fiber on each side, at doublets 3 and 8, and each fiber is continuous with a 'fin' which terminates with a further, smaller lateral fiber (Jespersen 1971; Jamieson 1999). Jamieson (1995, 1999) has tentatively recognized homology between each fin and an amphibian sperm undulating membrane, between the fibers at doublets 3 and 8 and the amphibian juxta-axonemal fibers, and between the terminal lateral fibers and the axial fiber. He considered that this, if valid, would suggest that lissamphibians have retained, from an ancestor shared with dipnoans (and with other sarcopterygian fish?), only one of two former, bilateral, undulating membranes, and only one of a former pair of axial fibers. He further argued that the two juxta-axonemal fibers of urodeles (and *Leiopelma*, Scheltinga *et al.* 2001) are a persistence of the paired ancestral condition, the fiber at doublet 8 normally being lost in the Anura. Thus the unilateral location of the undulating membrane and its axial fiber, rather than presence of undulating membranes *per se*, constitutes the synapomorphic condition for the Lissamphibia (Jamieson *et al.* 1993; Jamieson 1995, 1999). Sarcopterygian fish (for definition see Jamieson 1991), and particularly dipnoans, appear to be the nearest extant non-tetrapod relatives of amphibians.

5.4.9 Axial Fiber

An axial fiber associated with the axoneme is present in Ascaphidae, Bombinatoridae, Discoglossidae, Leiopelmatidae, Pelodytidae, Bufonidae, Centrolenidae, Hylidae, Leptodactylidae (absent from *Caudiverbera, Telmatobius,* and *Telmatobufo*), Myobatrachidae, and Rhinodermatidae. The axial fiber is situated further posteriorly than, or adjacent to the distal centriole. However, in *Discoglossus* the base of the axial fiber protrudes deeply into the basal nuclear fossa and the centrioles occupy a lateral position at the caudal end of the fossa (Pugin-Rios 1980). This condition is also seen in the Myobatrachidae where it is considered a distinctive synapomorphy (Lee and Jamieson 1992) presumably homoplastic relative to *Discoglossus*. In urodeles the axial fiber reaches the base of the fossa or extends only shortly into the fossa (Werner 1970; Picheral 1979, Selmi *et al.* 1997; Scheltinga 2002). In gymnophionans the axial fiber reaches the distal centriole and into which it apomorphically penetrates (Scheltinga *et al.* 2003). The axial fiber can be either simple and circular in transverse section (hylids *Cyclorana manya, Plectrohyla* and *Hyla*), modified in being elongate and thick in transverse section as a paraxonemal rod (Ascaphidae, the discoglossid *Discoglossus*, Leiopelmatidae, the hylid *Cyclorana* (except *C. cryptotis* and *C. manya*), and the leptodactylids *Physalaemus* and *Pleurodema*), or modified in being divided into a juxta-axonemal fiber (occasionally much reduced or absent) joined by a thin lamina to a bulb-shaped axial fiber (Bombinatoridae, the discoglossid *Alytes*, Pelodytidae, Bufonidae,

Centrolenidae, other Hylidae (including *Cyclorana cryptotis*), other Leptodactylidae, Myobatrachidae, and Rhinodermatidae). A simple axial fiber is present in primitive urodeles and all gymnophionans suggesting that modification of the axial fiber is apomorphic. In urodele it is divided into distinct cortex and medulla regions which has also been reported in *Discoglossus* (Sandoz 1975; Campanella and Gabbiani 1979) and the leptodactylids *Eupsophus vittatus* and *Hylorina sylvatica* (Pugin-Rios 1980). The presence of a periaxial sheath around the axial fiber reported in some myobatrachids (Lee and Jamieson 1993) appears similar on initial inspection to cortex and medulla, but this sheath is here considered to be cytoplasmic and not part of the axial fiber.

Absence of an axial fiber from Megophryidae, Pelobatidae, Pipidae, Hyperoliidae, the leptodactylids *Caudiverbera*, *Telmatobius* and *Telmatobufo*, the hylid *Pseudis paradoxa*, Microhylidae, Ranidae, and Rhacophoridae is apomorphic.

5.4.10   Undulating Membrane

The narrow unilateral undulating membrane occurs in the mature spermatozoa of Bombinatoridae, the discoglossid *Alytes*, Pelodytidae, Bufonidae, Centrolenidae, Hylidae (except *Cyclorana*, *Hyla*, *Plectrohyla* and *Scinax*; but including *Cyclorana cryptotis*), Leptodactylidae (except *Caudiverbera*, *Pleurodema*, *Telmatobius* and *Telmatobufo*), Myobatrachidae, and Rhinodermatidae. A narrow undulating membrane occurs in urodeles and the dipnoan *Neoceratodus* and is thus presumed to be the plesiomorphic state. In the leptodactylid *Physalaemus* the narrow undulating membrane is apomorphic in its location, extending from the abaxonemal end of the axial fiber. A short thick equivalent of the undulating membrane is present in Ascaphidae, the discoglossid *Discoglossus*, Leiopelmatidae, the hylid *Cyclorana* (except *C. cryptotis* and *C. manya*), and the leptodactylid *Pleurodema*. The tail may lack an undulating membrane in Megophryidae, Pelobatidae, Pelodytidae, Pipidae, the hylids *C. manya*, *Hyla*, *Plectrohyla*, *Pseudis*, and *Scinax*, Hyperoliidae, the leptodactylids *Caudiverbera*, *Telmatobius* and *Telmatobufo*), Microhylidae, Ranidae, and Rhacophoridae. With the exception of three of the four hylids (*Hyla*, *Plectrohyla*, and *Scinax*), all taxa which lack an undulating membrane also lack an axial fiber.

5.4.11   Axoneme

One or two axonemes can be present within the anuran tail. A single axoneme is present in Ascaphidae, Bombinatoridae, Discoglossidae, Leiopelmatidae, Pelodytidae, Pipidae, Bufonidae, Centrolenidae, Hylidae, Leptodactylidae (except *Telmatobufo*), Microhylidae, Myobatrachidae, Ranidae, Rhacophoridae (except *Chiromantis*, *Polypedates* and *Rhacophorus*), and Rhinodermatidae, whereas two are present in Megophryidae, Pelobatidae, Hyperoliidae, the leptodactylid *Telmatobufo*, and the rhacophorids *Chiromantis*, *Polypedates* and *Rhacophorus*. Where two axonemes are present, the spermatozoa of Megophryidae and rhacophorids have no nuclear fossa, Pelobatidae and Hyperoliidae have two nuclear fossae, and *Telmatobufo* has a single fossa. The biflagellate condition has clearly evolved independently more than once. In the

Sarcopterygii biflagellarity of spermatozoa is normal in the lungfish *Protopterus* and *Polypterus* in addition to those frog sperm displaying this condition. It has doubtful phylogenetic value, although it may characterize some fish clades, as it occurs sporadically both in teleosts and in non-chordate phyla (Jamieson 1991).

5.4.12 Conclusion

Spermatozoa of the Anura share with those of urodeles and gymnophionans the apomorphic character of the presence of a unilateral undulating membrane which connects the axial fiber to doublet 3 of the axoneme. This character is an autapomorphy of the Lissamphibia as suggested by Jamieson (1995, 1999).

The Anura were previously defined spermatologically on a single, negative autapomorphy, the loss of a longitudinal fiber adjacent to axonemal doublet 8, a fiber which occurs in urodeles. However, as noted by Jamieson (1999), this fiber is also absent from gymnophionans and is occasionally found in other Anura (see above) and therefore, the Anura cannot be defined spermatologically on the negative apomorphy of loss of this fiber. Jamieson (1999) therefore proposed that the autapomorphy would be better described as the retention of the juxta-axonemal fiber at doublet 3. A juxta-axonemal fiber at doublet 8 and 3 is a feature of the anuran ground plan, as it is present in the basal frog *Leiopelma* and is very rarely retained in other anurans. A juxta-axonemal fiber at 3 is absent in urodeles and caecilians but cannot be considered an autapomorphy of the Anura as it appears to have a dipnoan precursor. Also, the juxta-axonemal fiber of anurans appears to be a modification of the axial fiber which is present in both urodeles and gymnophionans but in these latter taxa it is well separated from the axoneme. Therefore, Scheltinga (2002) proposed that the Anura be defined spermatologically on a single autapomorphy, the presence of an axial fiber which is closely associated with doublet 3, being modified into juxta-axonemal and axial portions. An axial fiber being apomorphically lost in some Anura.

There appear to be four major trends in the evolution of the structure of anuran spermatozoa. 1) Modification of the perforatorium (loss of the endonuclear canal and axial perforatorium which are present in a few basal taxa, and development of a conical perforatorium or loss of a perforatorium altogether). 2) Modification of the base of the nucleus (development of two fossae or loss of a nuclear fossa and the base becoming tapered). 3) Modification of the axial fiber (dividing into distinct juxta-axonemal and axial parts or loss of the axial fiber altogether). 4) Tail becoming biflagellate from the plesiomorphic uniflagellate condition seen in most taxa. However, there appear to be no spermatozoal characters that can be mapped onto currently accepted trees without resulting in much homoplasy. In general the spermatozoa of the archaeobatrachians Ascaphidae, Bombinatoridae, Discoglossidae, and Leiopelmatidae appear the most basal, which is in agreement with most phylogenetic analyses. In contrast, the mesobatrachians (Megophryidae, Pelobatidae, Pelodytidae, and Pipidae) do not appear to group together spermatologically and group with differing neobatrachian families. The

Neobatrachia are generally accepted as being a monophyletic group but this is not confirmed from spermatozoal ultrastructure if mesobatrachian taxa are excluded. The spermatozoa of Megophryidae and Pelobatidae are similar to those of Hyperoliidae, whereas the spermatozoa of the Pelodytidae are similar to those of the Bufonidae, Centrolenidae, Hylidae, Leptodactylidae, and Rhinodermatidae. The spermatozoa of the Microhylidae and the hylid *Pseudis paradoxa* are similar in structure, whereas those of the Myobatrachidae are distinctive but all three families appear to share more characters with the pelodytids. However, the Microhylidae and the hylid *P. paradoxa* are apomorphic in the absence of an axial fiber as are the pipids and ranoids. Several phylogenetic analyses have grouped the Microhylidae with the Ranoidea, whereas the hylid *P. paradoxa* is grouped with the Bufonoidea suggesting that the absence of an axial fiber may be homoplastic between these two groups. The spermatozoa of the Pipidae are similar in many respects to those of the Ranidae and basal Rhacophoridae.

5.5 FERTILIZATION BIOLOGY

5.5.1 Vocalizaton and Courtship

Particularly in the warmer parts of the world, vocalizaton (calling) by anurans is often the dominant sound at night. Considerations of space do not permit detailed treatment here of vocalizaton but, as it is an essential component of the process of reproduction in most species, some mention follows.

A successful advertisement call results in mating, involving amplexus (see next section).

Anuran vocalization has been classified into the following categories (Bogert 1960; Littlejohn 1977; Wells 1977; for details see Duellman and Trueb 1986, 1994):

1. **Advertisement call.** This is emitted by males and has the function of (1) attraction of conspecific females and (2) signalling of occupied territory to males of the same or different species. Both functions may be subserved by the same call or a different call may be used for each. Three kinds of advertisement call have been recognized:
    A. *Courtship call.* Produced by males to attract conspecific females.
    B. *Territorial call.* Produced by resident male, often in response to the advertisement call of another male.
    C. *Encounter call.* Evoked during close range, agonistic interactions between males.
2. **Reciprocation call.** Emitted by a receptive female in some species, usually in response to advertisement calls of conspecific males.
3. **Release call.** Produced agonistically by a male or an unreceptive female in response to amplexus. Release calls are well exemplified by *Bufo* species but are unknown, for instance, in hylids.
4. **Distress (alarm) call.** A loud vocalization by either sex in response to disturbance, for instance by a predator. This call differs from all of the above in being given with the mouth open (Duellman and Trueb 1986, 1994; Stebbins and Cohen 1995).

A successful advertisement call results in mating, involving amplexus (see next section).

5.5.2 Mating (Amplexus)

Whether the species is externally or internally fertilizing, mating in the Anura occurs by the male holding or otherwise approximating itself to the female in the process termed amplexus. In the more primitive frogs, including all archaeobatrachians, most myobatrachids, some telmatobiine leptodactylids and sooglossids, amplexus is inguinal. Amplectic positions have been classified, *inter alia*, as inguinal (with the male's forelimbs around the waist of the female); axillary (around the armpit or neck); cephalic (around the head); straddle (male dorsally straddling the female); glued (male adherent by secretions to the female); and independent (vent to vent, in opposite directions). The male may change position from one to the other of these during mating (see Duellman and Trueb 1986, 1994). Inguinal amplexus is exemplified by the myobatrachid *Limnodynastes ornatus* (Fig. 5.39) but is axillary in the cofamilial *Heleioporus australiacus* (Fig. 5.40) (Littlejohn *et al.* 1993).

In externally fertilizing anurans (the vast majority of the approximately 5000 species) sperm are shed externally onto the eggs during amplexus. Internal fertilization is rare, though probably more widespread than presently known. It occurs in *Ascaphus truei* and *A. montanus* (Ascaphidae); four genera of African Bufonidae; and in the leptodactylids *Eleutherodactylus jasperi* and *E. coqui*. *Ascaphus* achieves insemination by insertion of the penis into the cloaca of the female. Sever *et al.* (2001), introduced the term copulexus for this penial amplexus. The few internally fertilizing bufonids and leptodactylids probably inseminate by cloacal apposition (references in Sever, Chapter 7 of this volume).

Nuptial pads (as in *Heleioporus australiacus*, Fig. 5.40) and breeding glands, involved in amplexus, are discussed by Brizzi *et al.* (Chapter 6 of this volume).

5.5.3 Reproductive Modes

The sperm are shed in the immediate vicinity of the eggs or directly onto them in all external fertilizers, or into the female cloaca in the few internally fertilizing species (see also section 5.5.4). The male often minimizes dispersal of the eggs during spawning by trapping the eggs with the hind legs. It appears that the sperm swim only short distances, if they swim at all outside the egg layers, even in aquatic fertilizers. For instance, in *Discoglossus pictus* sperm motility seems to be limited to movement of the sperm from the egg surface, on which they are deposited, to the oolemma (Campanella *et al.* 1997) despite the fact that fertilization is aquatic.

Duellman (Table 1 in Chapter 1 of this volume) has given a comprehensive classification of reproductive modes which need not be repeated here except to indicate the chief modes recognized (for details of sub-modes and taxa, see that chapter). Where eggs are aquatic they are deposited in water, in foam or bubble nests or are imbedded in the dorsum of the aquatic female. If eggs are terrestrial they are deposited on the ground, in burrows, or otherwise concealed, or in arboreal situations, or in foam nests (the latter also occur in some aquatic breeders, Fig. 5.39), or are carried by either the male or the female (see Lehtinen and Nussbaum, Chapter 8 of this volume). Exceptionally, but probably more

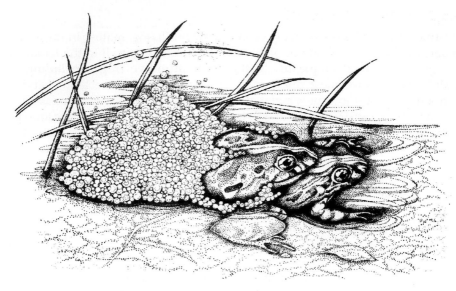

Fig. 5.39 A pair of *Limnodynastes ornatus* in inguinal amplexus depositing foam nest at the water surface. From Roberts, J. D. 1993. Fauna of Australia, Vol. 2A, Pp. 28-34, Fig. 5.3.

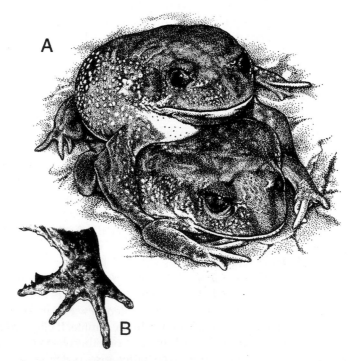

Fig. 5.40 A. Amplexus in *Heleioporus australiacus*. **B**. Nuptial excrescences on the hand of the male. From Littlejohn, M. J., Roberts, D. J., Watson, G. F. and Davies, M. 1993. Fauna of Australia, Vol. 2A, Pp. 41-57, Fig. 7.3.

commonly than has been recorded, the eggs are retained in the oviducts and internal fertilization occurs accompanied by ovoviparity or viviparity (see Sever, Chapter 7 of this volume).

5.5.4 Sperm Motility

5.5.4.1 General

The only rigorous comparative analysis of anuran sperm motility and the morphological features underlying this is that of van der Horst *et al.* (1995) on 13 species in seven families of South African anurans (Table 5.17). Their account is summarized here.

Table 5.17   Anura for which sperm motility was studied by van der Horst *et al.* 1995.

Family	Species	Mode
Bufonidae	*Bufo rangeri*	AF
Hemisotidae	*Hemisus marmoratus*	TF
Hyperoliidae	*Hyperolius horstockii*	AF
	Semnodactylus wealii	AF
	Leptopelis flavomaculatus	TF
Microhylidae	*Breviceps gibbosus*	TF
Pipidae	*Xenopus laevis*	AF
Ranidae	*Arthroleptella lightfooti*	TF
	Pyxicephalus adspersus	AF
	Rana fuscigula	AF
	Strongylopus grayii	AF
	Tomopterna delalandii	AF
Rhacophoridae	*Chiromantis xerampelina*	TF

AF, aquatic fertilizers. TF, terrestrial fertilizers

Among vertebrates, amphibians represent the widest range of "fertilization environments". Internal fertilization occurs in most salamanders and in all caecilians. In anurans such as *Ascaphus* internal fertilization takes place and their sperm are classified as introsperm in the system of Jamieson and Rouse (1989). External fertilization is the rule for most anurans and their sperm can be classified as ect-aquasperm in that terminology. The mode of external fertilization varies among anurans. In the South African species studied, direct sperm deposition on the eggs occurs in *Arthroleptella lightfooti* and *Breviceps gibbosus* and in which the sperm only swim through the mucous/egg jelly surrounding of the egg but are not substantially in contact with external fluids or fluids that do not originate from the adults themselves. Species exhibiting this pattern are referred to by van der Horst *et al.* (1995) as terrestrial fertilizers (TF). In *Bufo* spp. and *Xenopus laevis*, in contrast, sperm are ejaculated into fresh water and sperm swim through the external medium to reach the eggs. Species exhibiting this pattern are referred to as aquatic fertilizers (AF).

van der Horst *et al.* (1995) proposed that sperm shape (particularly the head) and flagellar beat should largely determine the swimming behaviour of sperm. Sperm motility represents a summation of form, mitochondrial function, membrane integrity, exchange of important ions such as calcium and unmasking of receptors. They contrasted fertilization biology with features that are important in establishing phylogenetic relationships and investigated quantitatively sperm motility patterns as a function of fertilization biology. Quantitative sperm motility was studied in representative examples of frogs exhibiting either the terrestrial or aquatic mode of fertilization and it was ascertained whether sperm motion was species specific in anurans. The parameters measured are shown in Fig. 5.41.

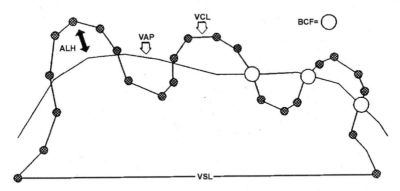

Fig. 5.41   Sperm track indicating relevant terminology. VAP, average path velocity; VCL, curvilinear velocity; VSL, Straight line velocity; ALH, Amplitude of lateral head displacement; BCF, Beat cross frequency. LIN = Linearity = VSL/VCL. From van der Horst, G., Wilson B., and Channing, A. 1995. Pp. 333-342. In Jamieson, B. G. M., Ausio, J. and Justine, J.-L. (eds). 1995. *Advances in Spermatozoal Phylogeny and Taxonomy*. Mémoires du Muséum National d' Histoire Naturelle Paris 16, Fig. 1.

5.5.4.2   Structural considerations of anuran sperm

Aquatic fertilizers. Of the sperm studied the simplest [by apomorphic loss] is that of *Rana fuscigula* which has a straight, short symmetrical head, a short midpiece and a single flagellum with no undulating membrane or associated fibers. The acrosomal cap is short. The sperm of *Semnodactylus wealii* have similar features except that the flagellum has in addition an undulating membrane and axial fiber. *Xenopus laevis* has a single flagellum but the head is partly corkscrew-shaped with one and a half coils. *X. laevis* sperm furthermore differ from those of most anurans in that the first few rows of mitochondria surround the posterior part of the nucleus. *Bufo rangeri* sperm have a spear-shaped head, distinct but short midpiece, and tail with undulating membrane and axial fiber (*i.e.*, basic features of anuran sperm). The head length of aquatic fertilizers ranges from 10.3 to 20.5 µm (Table 5.18).

Terrestrial fertilizers. In contrast the sperm of South African anurans which exhibit the terrestrial mode of fertilization have very long and slender heads (*Hemisus marmoratus, Leptopelis flavomaculatus, Breviceps gibbosus, Arthroleptella lightfooti*) or long heavily coiled heads (*Chiromantis xerampelina*) varying from 21.4 to 45 µm in length (Table 5.18). Both the head length and acrosome length of the terrestrial fertilizers were significantly (p<0.05) longer than those of the aquatic fertilizers. However, the flagella of the TF represent either a single axoneme or an axoneme with an undulating membrane and axial fiber like the AF. The exception was *C. xerampelina* which has a single flagellar complex containing two axonemes surrounded by a multitude of microtubules. The term ect-terrasperm was suggested by van der Horst *et al.* (1995) for amphibians which exhibit the terrestrial mode of fertilization. A clear morphological and functional distinction between ect-aquasperm and ect-terrasperm is therefore evident.

5.5.4.3. Sperm motility of South African anurans

Sperm of all aquatic fertilizers exhibited forward progression and the curvilinear velocity (VCL) varied from 20 to 31 µm/s, straight line velocity (VSL) from 8 to 23 µm/s, Linearity (LIN = VSL/VCL) from 40 to 72%, beat cross frequency (BCF) from 3 to 6Hz and amplitude of lateral head displacement (ALH) from 4.7 to 7.5 µm. In contrast, the sperm of two species of terrestrial fertilizers (*Breviceps* and *Arthroleptella*) were immotile in a wide range of physiological media and osmotic concentrations varying from 10 to 300 mOsm/kg. In *Chiromantis xerampelina*, sperm only exhibited an initial rapid "starspin" movement followed by uncoiling of the head and became immotile within a minute. Star symbol plots were used to visually express sperm motility as a pattern of movement (Fig. 5.42). Distinct motility patterns could be constructed that were clearly species specific. Sperm with partly coiled heads such as *Xenopus* and *Strongylopus* swim in a corkscrew fashion and the linearity of the sperm is almost similar (41.3% and 42.4% respectively) whereas *Bufo* sperm have straight heads and a LIN of 67.6%. Furthermore, more closely related species (*Strongylopus grayii* and *Pyxicephalus adspersus*) had different sperm motility patterns but also shared many similarities for several motion parameters. This can also be seen in Figure 5.42, indicating the similarity in star symbol plots.

Table 5.18   Sperm dimensions of aquatic and terrestrial South African anuran fertilizers.

Fertilization grouping	Head length (µm)	Head width (µm)	Acrosome length (µm)
Aquatic			
Mean (± SD)	16.0 (2.2)	0.7 (0.4)	2.2 (1.2)
Range	10.3 - 20.5	0.1 - 1.6	0.7 - 4.6
Terrestrial			
Mean (± SD)	29.6 (6.0)	0.6 (0.3)	5.9 (1.6)
Range	21.4 - 45.0	0.1 - 1.1	3.2 - 7.3

Fertilization mode and the sperm tail. It has been hypothesized that primitive lissamphibians (and possibly the Lepospondyli and Labyrithodontia) were internally fertilizing and that external fertilization was a secondary reversion in the anurans (Jamieson *et al.* 1993). The 9+2 axoneme-undulating membrane-axial fiber (AUA) is basic to lissamphibian sperm. As noted by van der Horst *et al.* (1995), in reptiles, birds and mammals, where internal fertilization is the rule, the sperm flagellum is associated with a 9+9+2 pattern where the

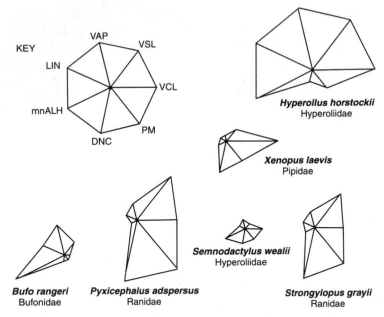

FIG. 5.42   Star Symbol Plots (seven motion parameters used - see abbreviations in Fig. 5.8 - PM = percent motile sperm) depicting patterns of sperm motion among six aquatically fertilizing anuran species. For any set of stars (above) the smallest observed value for a parameter is plotted as an arm one-tenth the length of the arm representing the largest observed value. From van der Horst, G., Wilson B., and Channing, A. 1995. Pp. 333-342. In Jamieson, B. G. M., Ausio, J. and Justine, J.-L. (eds). 1995. *Advances in Spermatozoal Phylogeny and Taxonomy*. Mémoires du Muséum National d' Histoire Naturelle Paris 16, Fig. 9. ©Publications Scientifiques du Muséum National d'Histoire Naturelle, Paris.

nine outer coarse fibers seem to function as a strengthening device for sperm swimming in a viscous environment. The juxta-axonemal fiber as well as the axial fiber in amphibians seem to be homologous with the coarse outer fibers 8 and 3 of amniotes and insects (Picheral 1972b). van der Horst *et al.* (1995) consider it possible that the AUA of the early internal fertilizing amphibians represents an intermediate structure from external to internal fertilization and that this has been retained on reversion to external fertilization in many anurans. Alternatively, they consider, the AUA of early internal fertilizing Lissamphibia may represent a simplification of an ancestor that had a 9+9+2 flagellar arrangement. They consider their study to support the view that presence of the AUA in anurans is a plesiomorphic character within a particular group or family (Fouquette and Delahoussaye 1977; Jamieson and Lee 1993), as considered by Scheltinga and Jamieson (present account). In accordance with this view, Jamieson (1995, 1999) considered the undulating membrane of lissamphibian sperm to be homologous with one of the lateral fins of the sperm tail in dipnoan fish which are externally fertilizing. A comparison of three species of *Heleophryne* was also considered to support the view that the AUA is basic (Visser and van der Horst 1987). The investigation of van der Horst *et al.* (1995) accords with the view expressed here (section 5.4) that a flagellum without an axial fiber represents a derived condition from the plesiomorphic AUA.

Head length correlation. A distinct relationship between sperm head length and mode of fertilization among externally fertilizing anurans has been demonstrated (van der Horst *et al.* 1995). The sperm heads and acrosomes of aquatic fertilizing anurans are significantly shorter than those of the TF. It is also known that the egg coverings of TF are generally thicker than those of AF, presumably to protect the eggs against dehydration (Passmore and Carruthers 1979). A longer sperm head and a larger acrosome, which may contain more digestive enzymes, may have been an advantage in penetrating these thick egg coats (Wilson 1994). The information on head and acrosome length may also be of predictive value in other anuran species in estimating whether they belong to the TF or AF grouping. Three South African *Heleophryne* species posses sperm heads with lengths varying from 23 to 28 μm (Visser and van der Horst 1987). They are predominantly TF and further support the findings of van der Horst *et al.* (1995). It was therefore reasonably concluded from their data on South African anurans that sperm head and acrosome dimensions reflect the fertilization biology rather than the phylogeny and are independent of the type of flagellum (whether only an axoneme or with undulating membrane and axial fiber). Observations on types of acrosomes and location of fibers associated with the axoneme are examined in sections 5.3 and 5.4.

Motility and fertilization mode. The study of quantitative sperm motility in South African anurans indicates that sperm from all aquatic fertilizers (AF) possess motile sperm that swim progressively forward in a physiological medium of 30 mOsm/kg. In contrast sperm from terrestrial fertilizers (TF) are immotile in a wide range of physiological and culture media ranging from 10 to 300 mOsm/kg. In AF, sperm swim in a low osmolality environment even if the male and female cloacae are in close proximity as in *Xenopus*. Sperm of *Chiromantis xerampelina* (TF) only exhibited a brief spurt (seconds) of hyperactivated motility. In TF the sperm are deposited directly on the eggs and the need for vigorous and longer term motility seems less than in the AF group. It was conceded that sperm activation in TF might be dependent on specific substances associated with the egg coat/surface and that this might explain the immotile status of TF sperm in the experiments.

Taxon specificity. Sperm motion is highly species specific among the South African anurans. The pattern of sperm motion appears to be related to the form of the sperm head, the type of tail (presence or absence of undulating membrane) and the type of flagellar beat. The AF anurans with only an axoneme seem to exhibit greater values for most motion parameters than those with an AUA and the AF accordingly have larger star symbol plots (Fig. 5.42).

Several sperm structural features in the Lissamphibia were found by van der Horst *et al.* (1995) to be diagnostic in separating the three main orders, families (urodeles), and even closely related species (*Heleophryne*). Sperm morphology also assists in phylogenetic inferences and is predictive in terms of fertilization biology. Quantitative sperm motility analysis of representative examples of seven of the nine South African anuran families suggest that sperm motility is species specific and also relates to fertilization biology (van der Horst *et al.* 1995).

Motility in *Xenopus laevis*. A further study has been made of sperm motility in *Xenopus laevis*, which has become an experimental animal and therefore a model for anuran biology. However, it must be stressed that *Xenopus* sperm do not represent a plesiomorphic anuran condition as the tail lacks an axial fiber, undulating membrane, and juxta-axonemal fibers.

Xenopus sperm are motile in water and media of low osmolarity but not in media of osmolarities higher than 200 mOsm/l, regardless of the ionic composition. External calcium and pH are not involved in the inhibition of motility which occurs at high osmolarities. The sperm are motile for less than 10 minutes and both flagellar beat frequencies and sperm velocities decline progressively. The flagellum propagates a three dimensional wave which imparts a spinning motion to the spermatozoon. This pattern of movement is not dependent on the shape of the head as isolated reactivated flagella exhibit three dimensional waves (Bernardini *et al.* 1988).

5.5.5 Function of the Undulating Membrane

It is clear that the term undulating membrane is appropriate for the typical structure of this name in anuran sperm insofar as the axoneme undulates on the axial fiber to which it is attached by the membrane, much as the trypanosome undulating membrane (axoneme and membrane) undulates on the body of the protozoon. This is especially evident in higher urodeles where the axial fiber has a trifoliate transverse section, resembling the structure of a girder, a structure which must provide a rigid axis for undulation of the axoneme. The presence of an undulating membrane and such appurtenances as are present in addition to the axoneme (juxta-axonemal fibers, axial fiber) cannot be related to the mode of fertilization as it is present in internal fertilizers (most urodeles, caecilians and some frogs) and external fertilizers (some urodeles and most frogs).

It has been shown above that aquatic fertilizing anurans with only an axoneme seem to exhibit greater values for most motion parameters than those with an axoneme-undulating membrane-axial fiber (AUA). This is in agreement with the view of the present authors that the undulating membrane complex may have been retained by selection in externally fertilizing anurans to lessen sperm dispersal. This does not, however, explain its presence in internal fertilizers. The same dichotomy exists between the externally fertilizing Dipnoi and the internal fertilizing *Latimeria*, both of which have paired homologues of the undulating membrane (see Jamieson 1991).

Within the Anura, a spermatozoon near the generalized condition for the group is seen in *Bufo marinus* (see whole spermatozoon in Fig. 5.18I) and many other neobatrachians. The well developed undulating membrane and moderately developed axial fiber must be considered retentions of plesiomorphic lissamphibian features as they are seen in all three orders. It is remarkable therefore that the most plesiomorphic anurans, the admittedly paraphyletic Discoglossoidea, show highly modified sperm tails in which rigidification of the tail, with or without involvement of the undulating membrane, appears to be a trend. In what way this trend relates to fertilization biology is unknown. An extreme example is seen in *Bombina variegata* (Fig. 5.9) in which not only the axial fiber but also the elongate nucleus lies parallel to the axoneme, but in which the undulating membrane remains well developed. Presumed rigidity of the undulating membrane complex, where juxta-axonemal fiber, undulating membrane and axial fiber are fused as a solid structure is seen in *Ascaphus* (Fig. 5.8) *Discoglossus* (Fig. 5.10) and *Leiopelma* (Fig. 5.12). However, *Alytes obstetricans* (Fig. 5.11) is more typical of the Anura in having an albeit rigid axial fiber supporting a well developed undulating membrane and its axoneme. Solidification of the AUA (though contractility of the axoneme appears always to be retained) is also seen in most species of the hylid genus *Cyclorana* (Fig. 5.20) and in the leptodactylid *Pleurodema* (Fig. 5.25). Further trends in modification of the AUA, including loss of the undulating membrane and/or of the axial fiber, are apparent in section 5.4, above, but it remains clear that the typical AUA does not augment locomotion of the spermatozoon.

5.5.6 Insemination, Fertilization and Early Embryogenesis

5.5.6.1 Sperm penetration

Fertilization is achieved only if the sperm penetrates the animal hemisphere of the egg. Penetration by the spermatozoon activates the egg which completes the second meiosis, restores the diploid chromosome complement on fusion of the sperm and egg nuclei (karyogamy) and initiates cleavage. The egg at spawning is a secondary oocyte arrested in metaphase of meiosis II (see Sanchez and Villeco, Chapter 3 of this volume). Entry of the sperm induces completion of meiosis: the two chromatids of each dyad separate on the spindle and pass to the poles of the nucleus in anaphase and telophase. One of the resulting nuclei is extruded with the second polar body, the other is the female pronucleus. The haploid chromosome set of this pronucleus mingles with that of the male pronucleus from the spermatozoon to produce the diploid nucleus of the zygote.

At gamete interaction, sperm swim in the jelly coats which surround the amphibian oocyte (collectively the egg), pass through the vitelline coat (VC), and arrive in minutes at the egg plasma membrane (oolemma) where gametes fuse. Presence of the jelly coat around the egg is essential for fertilization. A fundamental role for the molecules of the jelly coat is the ability to retain calcium and/or magnesium ions (Campanella *et al.* 1997, and references therein). The site of the acrosome reaction varies in different species (see section 5.5.6.2).

It is commonly stated that in anurans, as opposed to urodeles, if polyspermy occurs the embryo is abnormal. However, in *Discoglossus pictus*, at least, polyspermy appears to be normal. Insemination in this species occurs at a predetermined site for sperm entrance, the dimple, seen also in urodeles. The sperm have a total motility of about 14-15 sec. Upon egg insemination (Fig. 5.43), sperm bundles stick to the surface of the outermost jelly layer (J3); they are ejected from the bundle in about 5 sec, and penetrate into the egg investments and the dimple contents (DC) to the egg surface (oolemma) in about 10 sec, becoming embedded along their whole length in the jelly layers. The apical subacrosomal rod (axial perforatorium) is seen next to the oolemma 15 sec postinsemination. Electrophysiological measurements of *D. pictus* eggs record fertilization potentials from 15-40 sec after insemination (Campanella *et al.* 1997, and references therein).

5.5.6.2 Acrosome reaction

The most recent paper on the acrosome reaction in Anura is that of Ueda *et al.* (2002) for *Xenopus laevis* from whose review this account is largely drawn.

The acrosome reaction involves penetration of egg envelopes and therefore some mention of these is necessary. In amphibians, the unfertilized egg is surrounded by a vitelline envelope (VE) and several jelly layers, which play an important role in both protecting the embryo and assuring fertilization (Iwao 2000). In *Xenopus laevis*, eggs ovulated from the ovary into the coelom are surrounded by coelomic envelopes before they enter the oviducal funnel (ostium). The physicochemical characterization of the coelomic envelopes changes after passing through the pars recta region of the oviducts, and the envelopes acquire a sperm-binding ability (Grey *et al.* 1977; Yoshizaki and Katagiri 1984; Takamune and Katigiri 1987; Bakos *et al.* 1990; Tian *et al.* 1997a,b; Katagiri *et al.* 1999). The eggs acquire three jelly layers as they pass through the pars convoluta region (Chapter 2, Fig. 2.2) in the posterior region of the oviducts (Freeman 1968; Yurewicz *et al.* 1975). The sperm must pass through these layers if fertilization is to be achieved. The acrosome possesses a VE lysin and, experimentally, sperm have an ability to lyse proteinaceous material in the form of gelatin in the frogs *Rana pipiens* (Elinson 1971; Penn and Gledhill 1972), *Leptodactylus chaquensis* (Raisman and Cabada 1977), *Bufo japonicus* (Iwao and Katagiri 1982), and the urodele *Cynops pyrrhogaster* (Iwao *et al.* 1994). The proteolytic activity of lysins is suppressed by trypsin inhibitors (Raisman and Cabada 1977; Cabada *et al.* 1978; Iwao and Katagiri 1982). The VE lysin appears crucial in allowing sperm to pass through the VE following an acrosome reaction.

The conical perforatorium, in the terminology of this chapter, when present (see section 5.3) is located between the nucleus and the acrosomal vesicle. An acrosome reaction induced under hypotonic conditions was observed by light microscopy in the sperm of *Leptodactylus chaquensis* (Raisman *et al.* 1980), and acrosome-reacted sperm are detectable with peanut agglutinin (PNA) staining in *Bufo japonicus* (Takamune 1987). The acrosome reaction appears to occur in the jelly layers (specifically the outermost jelly coat, J3) of the egg of the frog, *Discoglossus pictus* (Campanella *et al.* 1997, see below), and of the urodeles *Pleurodeles waltl* (Picheral 1977) and *Cynops pyrrhogaster* (Nakai *et al.* 1999; Onitake *et al.* 2000), whereas it occurs on the VE in *B. japonicus* (Yoshizaki and Katagiri 1982).

Several acrosome reaction-inducing substances have been identified in echinoderms and chordates (references in Ueda *et al.* 2002). Treatment of the

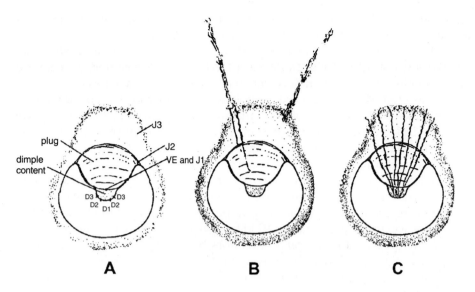

Fig. 5.43 Schematic drawing of *Discoglossus pictus* egg at sperm penetration. **A.** A dimple is located at the center of a large concavity in the animal half. Dimple regions D1, D2, and D3 are indicated. Jelly layers J1, J2, and J3 surround the whole egg. J3 is about threefold thicker at the animal half, in correspondence with the jelly plug, than in the rest of the egg. The gelatinous plug is located uniquely in the egg concavity. VC and J1 are separated from the dimple surface by the dimple content, consisting of finely granular material. **B.** At insemination, spermatozoa are ejected from their bundle into the jelly. **C.** Sperm become embedded along their whole length in the jelly layers. After Campanella C. *et al.* 1997. Molecular Reproduction and Development 47: 323-333, Fig. 1.

sperm with pars recta extract (PRE) induces a strong lytic activity against the VE in *Bufo japonicus* (Katagiri *et al.* 1982) and the acrosome reaction has been confirmed by electron microscopy (Katagiri *et al.* 1982; Yoshizaki and Katagiri 1982) and by LysoSensor Green fluorescence by confocal laser scanning microscopy (Ueda *et al.* 2002).

An acrosome has been demonstrated ultrastructurally in the sperm of *Xenopus laevis* (James, 1970; Reed and Stanley, 1972; Furieri, 1972; van der Horst, 1979; Pugin-Rios, 1980; Bernardini *et al.*, 1986; Yoshizaki, 1987; see section 5.3.13). Sperm (>1000 sperm/egg) bind to the VE of unfertilized *Xenopus* eggs before fusion of the sperm with the egg membrane (Tian *et al.* 1997a; Lindsay and Hedrick 1998; Lindsay *et al.* 1999). Sperm-egg binding appears to require oligosaccharide chains on the VE (Tian *et al.* 1997a; Vo and Hedrick 2000), of which one (gp 69/64) is a homologue of the mammalian sperm receptor ZP2 (Tian *et al.* 1999). The coelomic envelopes of eggs ovulated into the coelom lack sperm-binding ability as this is acquired only in passage through the oviduct (Tian *et al.* 1997b; Lindsay and Hedrick 1998; Katagiri *et al.* 1999) concomitantly with the conversion of some glycoproteins on the coelomic envelopes. This is the so-called coelomic/vitelline envelope transition (Gerton and Hedrick 1986;

Hardy and Hedrick 1992; Tian *et al.* 1997b). However, the state of the acrosomes in the sperm that bind the VE is not known (Ueda *et al.* 2002).

Ueda *et al.* 2002 found that about 40% of *Xenopus laevis* sperm in which an acrosome was detected with LysoSensor Green fluorescence under a confocal laser scanning microscope underwent an acrosome reaction in response to Ca^{2+} ionophore A23187, as evidenced by a loss of LysoSensor Green stainability, accompanied by breakdown of the acrosomal vesicle. About 53% of the sperm bound to isolated vitelline envelopes underwent an acrosome reaction, whereas both jelly water and solubilized vitelline envelopes induced only a weak acrosome reaction. About 40% of sperm treated with an oviductal extract obtained from the pars recta underwent an acrosome reaction. Extracts from the pars convoluta did not invoke a reaction. The pars recta substance with acrosome reaction-inducing activity appeared to be heat-unstable with a molecular weight of greater than 10 kDa. The activity is not proteolytic or at least, is not inhibited by protease inhibitors but it requires extracellular Ca^{2+} ions. It was thus concluded that the acrosome reaction occurs on the vitelline envelopes in response to the substance deposited from the pars recta during the passage of the oocytes (Ueda *et al.* 2002).

The ultrastructure of sperm changes and penetration in the egg of *Discoglossus pictus* has been described by Campanella *et al.* (1997) and is of particular interest in view of the basal status of the Discoglossoidea and the probability that its acrosome reaction is of a type basal to the Anura. The spermatozoon has an acrosome cap with an apical rod (axial perforatorium); a conical perforatorium is absent (Figs 5.3, 5.5, 5.44). The first stage of the sperm axial perforatorium and acrosome reaction (AR) (Fig. 5.44) consists in vesiculation between the sperm plasma membrane and the outer acrosome membrane. The two components of the acrosome cap are released in sequence. The innermost component (component B) is dispersed first. The next acrosome change is the dispersal of the outermost acrosome content (component A). At 30 sec post-insemination, when the loss of component B is first observed, holes are seen in the innermost jelly coat (J1), surrounding the penetrating sperm. Therefore, this acrosome constituent might be related to penetration through the innermost egg investments (Campanella *et al.* 1997). It is concluded that the subacrosomal cone (periperforatorial sheath of *Ascaphus*) of anuran sperm is perforatorial and should be termed the conical perforatorium.

At 1 min post-insemination, during sperm penetration into the egg, a halo of finely granular material is observed around the inner acrosome membrane of the spermatozoon, suggesting a role for component A at this stage of penetration. Gamete-binding and fusion take place between DI (the egg-specific site for sperm interaction) and the perpendicularly oriented sperm. Spermatozoa visualized at their initial interaction (15 sec post-insemination) with the oolemma are undergoing vesiculation. The first interaction is likely to occur between the DI glycocalyx and the plasma membrane of the hybrid vesicles surrounding the apical rod (axial perforatorium). Fusion is observed between the internal acrosome membrane and the oolemma. It was postulated that gametic interaction might be followed by fusion of the oolemma "with the apical rod

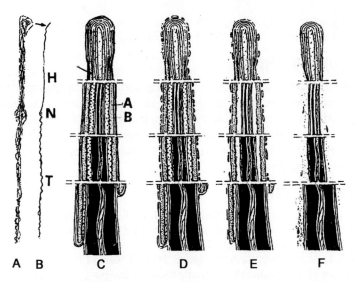

Fig. 5.44 The acrosome reaction in *Discoglossus pictus*. **A**. Sperm bundle. **B**. A single spermatozoon. Arrow, apical rod (axial perforatorium) and surrounding membranes here considered to be acrosomal. H, head. N, neck. T, tail. **C**. Intact sperm head. Components in the terminology of this chapter are: A, acrosome vesicle. B, conical perforatorium. In the narrow nucleus chromatin and an endonuclear canal are present. The small arrow indicates the inner acrosome membrane. **D**. Vesiculation of the acrosome vesicle. **E**. Loss of component B (the conical perforatorium). **F**. Loss of component A (the acrosome vesicle). After Campanella C. *et al.* 1997. Molecular Reproduction and Development 47: 323-333, Fig. 3.

internal membrane that extends posteriorly into the inner acrosome membrane". In the present chapter the axial perforatorium (apical rod *sensu* Campanella *et al.* 1997) is defined as a non-membrane-bound structure and the overlying membranes are clearly the inner and outer membranes of the acrosome vesicle.

Insemination of the outermost jelly layer (J3) dissected out of the egg, and observations of the ultrastructural changes of spermatozoa in this coat, indicate that J3 rather than the vitelline coat (VC) induces the AR. At the late post-insemination stage, VC fibrils are seen cross-linking the inner acrosome membrane (Campanella *et al.* 1997).

5.5.6.3 Cleavage

Cleavage of the zygote is normally dependent on the presence of a centriole from the sperm (Maller *et al.* 1976), though parthenogenetic development of a normal blastula from the egg can occur. It is in fact pericentriolar material which assembles and nucleates a microtubule aster ("astral nucleation"). The CTR2611 antigen is present in the sperm neck in the pericentriolar region and there is evidence that gamma-tubulin is recruited from the region of the egg around the centrioles (Félix *et al.* 1994). The grey crescent, which determines the dorsal surface of the embryo and bilateral symmetry, develops on the opposite side to the point of sperm entry. The events of early embryogenesis are summarized by Altig (Chapter **9** of this volume) and molecular aspects of development by Key (Chapter **10** of this volume).

5.6 ACKNOWLEDGEMENTS

We thank Keith McDonald for his unstinting aid in procuring specimens and for sharing his knowledge of frogs with us. Bjorn Afzelius is thanked for drawing our attention to, and donating to us, the thesis of Pugin-Rios. Professor P. Jego, Université de Rennes kindly gave permission for reproduction of illustrations from the thesis. An Australian Research Council grant to BGMJ made our work possible.

5.7 LITERATURE CITED

Amaral, M. J. L. V., Fernandes, A. P., Báo, S. N. and Recco-Pimentel, S. M. 1999. An ultrastructural study of spermatogenesis in three species of *Physalaemus* (Anura, Leptodactylidae). Biocell 23: 211-221.

Amaral, M. J. L. V., Fernandes, A. P., Báo, S. N. and Recco-Pimentel, S. M. 2000. The ultrastructure of spermatozoa in *Pseudopaludicola falcipes* (Anura, Leptodactylidae). Amphibia-Reptilia 21: 498-502.

Aniello, F., Branno, M., De Rienzo, G., Ferrara, D., Palmiero, C. and Minucci, S. 2002. First evidence of prothymosin α in a non-mammalian vertebrate and its involvement in the spermatogenesis of the frog *Rana esculenta*. Mechanisms of Development 110: 213-271.

Asa, C. S. and Phillips, D. M. 1988. Nuclear shaping in spermatids of the Thai Leaf Frog *Megophrys montana*. The Anatomical Record 220: 287-290.

Baker, C. L. 1963. Spermatozoa and spermateleosis in *Cryptobranchus* and *Necturus*. Journal of the Tennessee Academy of Science 38: 1-11.

Baker, C. L. 1966. Spermatozoa and spermateleosis in the Salamandridae with electron microscopy of *Diemictylus*. Journal of the Tennessee Academy of Science 41: 2-25.

Bakos, M. A., Kurosky, A. and Hedrick, J. L. 1990. Physicochemical characterization of progressive changes in the *Xenopus laevis* egg envelope following oviductal transport and fertilization. Biochemistry 29: 609-615.

Báo, S. N., Dalton, G. C. and De Oliveira, S. F. 1991. Spermiogenesis in *Odontophrynus cultripes* (Amphibia, Anura, Leptodactylidae): ultrastructure and cytochemical studies of proteins using E-PTA. Journal of Morphology 207: 303-314.

Báo, S. N., Vieira, G. H. C. and de Paiva, F. A. 2001. Spermiogenesis in *Melanophryniscus cambaraensis* (Amphibia, Anura, Bufonidae): Ultrastructural and cytochemical studies of carbohydrates using lectins. Cytobios 106 (suppl. 2): 203-216.

Barbieri, F. D. 1950. Observaciones sobre los cromosomas y espermatozoides en algunos batracios del genero *Leptodactylus*. Acta Zoologica Lilloana 9: 455-462.

Barker, K. R. and Biesele, J. J. 1967. Spermateleosis of a salamander *Amphiuma tridactylum* Cuvier; a correlated light and electron microscope study. La Cellule 67: 91-118.

Bernardini, G., Andrietti, F., Camatini, M., Cosson, M.-P. 1988. Xenopus spermatozoon: correlation between shape and motility. Gamete Research 21 : 165-175.

Bernardini, G., Podini, P., Maci, R. and Camatini, M. 1990. Spermiogenesis in *Xenopus laevis*: from late spermatids to spermatozoa. Molecular Reproduction and Development 26: 347-355.

Bernardini, G., Stipani, R. and Melone, G. 1986. The ultrastructure of *Xenopus* spermatozoon. Journal of Ultrastructure and Molecular Structure Research 94: 188-194.

Bogert, C. M. 1960. The influence of sound on the behavior of amphibians and reptiles. Pp. 137-320. In W. E. Lanyon and W. N. Tavolga (eds), *Animal Sounds and Communication*. American Institute of Biology Scientific Publications 7.

Burgos, M. H. 1955. Histochemistry of the testis in normal and experimentally treated frogs (*Rana pipiens*). Journal of Morphology 96: 283-299.

Burgos, M. H. and Fawcett, D. W. 1956. An electron microscope study of spermatid differentiation in the toad, *Bufo arenarum* Hensel. Journal of Biophysical and Biochemical Cytology 2: 223-239.

Burgos, M. H. and Vitale-Calpe, R. 1967a. The fine structure of the Sertoli cell-spermatozoan relationship in the toad *Bufo arenarum*. Journal of Ultrastructure Research 19: 221-237.

Burgos, M. H. and Vitale-Calpe, R. 1967b. The mecahanism of spermiation in the toad. American Journal of Anatomy 120: 227-252.

Cabada, M. O., Mariaono, M. I. and Raisman, J. S. 1978. Effect of trypsin inhibitors and concanavalin A on the fertilization of *Bufo arenarum* coelomic oocytes. Journal of Experimental Zoology 204: 409-416.

Campanella, C. and Gabbiani, G. 1979. Motile properties and localization of contractile proteins in the spermatozoon of *Discoglossus pictus*. Gamete Research 2: 163-175.

Campanella, C., Carotenuto, R., Infante, V., Maturi, G. and Atripaldi, U. 1997. Sperm-egg interactions in the painted frog (*Discoglossus pictus*): an ultrstructural study. Molecular Reproduction and Development 47: 323-333.

Cannatella, D. C. 1985. *A Phylogeny of Primitive Frogs (Archaeobatrachians)*. 404 pp. Ph. D. Thesis, The University of Kansas, USA.

Cavicchia, J. C. and Moviglia, G. A. 1982. Fine structure of the testis in the toad (*Bufo arenarum* Hensel): a freeze-fracture study. The Anatomical Record 203: 463-474.

Cavicchia, J. C. and Moviglia, G. A. 1983. The blood-testis barrier in the toad (*Bufo arenerum* Hensel): a freeze-fracture and lanthanum tracer study. Anatomical Record 205: 387-396.

Champy, C. 1913. Recherches sur la spermatogénèse des batraciens et les éléments accessoires du testicule. Archives de Zoologie Expérimentale et Générale 52: 13-304.

Champy, C. 1923. La spermatogenèse chez *Discoglossus pictus*. Comparaison avec celle des autres discoglossides et des Vertébrés en général. Archives de Zoologie Expérimentales et Général 62: 1-52 (Fide Grassé 1986).

Cobellis, G., Meccariello, R., Fienga, G., Pierantoni, R. and Fasano, S. 2002. Cytoplasmic and nuclear Fos protein forms regulate resumption of spermatogenesis in the frog, *Rana esculenta*. Endocrinology 143: 163-170.

Delahoussaye, A. J. 1966. The comparative sperm morphology of the Louisiana Hylidae (Amphibia: Anura). Proceedings of the Louisana Academy of Science 29: 140-152.

DeRobertis, E., Burgos, M. H. and Breyeter, E. 1946. Action of anterior pituitary on Sertoli cells and release of toad spermatozoa. Proceedings of the Society for Experimental Biology and Medicine 61: 20-22.

de Sa, R. and Berois, N. 1986. Spermatogenesis and histology of the testes of the caecilian, *Chthonerpeton indistinctum*. Journal of Herpetology 20: 510-514.

Duellman, W. E. and Trueb, L. 1986. *Biology of Amphibians*. McGraw-Hill, New York. 670 pp (reprint, Johns Hopkins University Press, 1994).

Echeverria, D. D. and Maggese, M. C. 1987. Development of the testis of *Bufo arenarum* (Anura, Bufonidae): Spermatogenic cycles and seasonal variations in young toads. Revista Del Museo Argentino De Ciencias Naturales "Bernardino Rivadavia" E Instituto Nacional De Investigacion De Las Ciencias Naturales Zoologia 14: 125-138.

Elinson, R. P. 1971. Sperm lytic activity and its relation to fertilization in the frog *Rana pipiens*. Journal of Experimental Zoology 177: 207-217.

Favard, P. 1955a. Spermatogenèse de *Discoglossus pictus* Otth. Étude cytologique. Maturation du spermatozoïde. Annales des Sciences Naturelles, Zoologie 11: 369-394.

Favard, P. 1955b. Mise en évidence d'une sécrétion acrosomique avant la fécondation chez les spermatozoïdes de *Discoglossus pictus* Otth. et de *Rana temporaria* L. Comptes Rendus des Académie des Sciences (Paris) 240: 2563-2565.

Fawcett, D. W. 1970. A comparative view of sperm ultrastructure. Biology of Reproduction Supplement 2: 90-127.

Fawcett, D.W. 1975. The mammalian spermatozoon. Developmental Biology 44: 394-436.

Félix, M. A., Antony, C., Wright, M. and Maro, B. 1994. Centrosome assembly in vitro: Role of gamma-tubulin recruitment in *Xenopus* sperm aster formation. Journal of Cell Biology 124: 19-31.

Fernandes, A. P. and Báo, S. N. 1998. Cytochemical localization of phosphatases in the germ- and sertoli cells of *Odontophrynus cultripes* (Amphibia, Anura, Leptodactylidae). Biocell 22: 93-101.

Folliot, R. 1979. Ultrastructural study of spermiogenesis of the anuran amphibian *Bombina variegata*. Pp. 333-339. In D. W. Fawcett and J. M. Bedford (eds), *The Spermatozoon*, Urban and Schwarzenberg, Baltimore.

Fouquette, M. J. Jr. and Delahoussaye, A. J. 1977. Sperm morphology in the *Hyla rubra* group (Amphibia, Anura, Hylidae), and its bearing on generic status. Journal of Herpetology 11: 387-396.

Freeman, S. B. 1968. A study of the jelly envelopes surrounding the egg of the amphibian *Xenopus laevis*. Biological Bulletin 135: 341-350.

Frost, D. R. 2002. *Amphibian Species of the World: an online reference*. V2. 21 (15 July 2002). The American Museum of Natural History. http://research.amnh.org/herpetology/amphibia/index.html

Fukuyama, K., Miyazaki, K. and Kusano, T. 1993. Spermatozoa and breeding systems in Japanese anuran species with special reference to the spiral shape of sperm in foam-nesting rhacophorid species. Abstracts. Second World Congress of Herpetology. Adelaide, South Australia. 29th December 1993-6th January 1994, pp. 92-93.

Furieri, P. 1961. Caratteri ultrastrutturali di spermi flagellati di Anfibi e Uccelli. Studio al microscopio elettronico. Archivio Zoologico Italiano (Napoli) 46: 123-147.

Furieri, P. 1972. La morfologia degli spermi di alcuni Anfibi Anuri. Bollettino di Zoologia 39: 618-619.

Furieri, P. 1975a. La morfologia comparata degli spermi di *Discoglossus pictus* Otth., *Bombina variegata* (L.) e *Alytes obstetricans* (Laurenti). Bollettino di Zoologia 42: 458-459.

Furieri, P. 1975b. The peculiar morphology of the spermatozoon of *Bombina variegata* (L.). Monitore Zoologico Italiano 9: 185-201.

Gambino, J., Eckhardt, R. A. and Risley, M. S. 1981. Nuclear matrices containing synaptonemal complexes from *Xenopus laevis*. The Journal of Cell Biology 91: 63a.

Garrido, O., Pugin, E. and Jorquera, B. 1989. Sperm morphology of *Batrachyla* (Anura: Leptodactylidae). Amphibia-Reptilia 10: 141-149.

Gavaud, J. 1976. La gamétogenèse du mâle de *Nectophrynoides occidentalis* Angel (amphibien anoure vivipare). Annales de Biologie Animale Biochimie Biophysique 16 (1): 1-12.

Gavaud, J. 1977. La gamétogenèse du mâle de *Nectophrynoidesoccidentalis* Angel (amphibien anoure vivipare). II. - Etude expérimentale du rôle des facteurs externes sur la spermatogenèse de l'adulte, au cours du cycle annuel. Annales de Biologie Animale Biochimie Biophysique 17 (5A): 679-694.

Gavrila, L. and Mircea, L. 2001. Chromatin and chromosomal fine structure in spermatogenesis of some species of amphibians. Zygote 9: 183-192.

Gerton, G. L. and Hedrick, J. L. 1986. The coelomic envelope to vitelline envelope conversion in eggs of *Xenopus laevis*. Journal of Cell Biochemistry 30: 341-350.

Gonzaga de Almeida, C. and Cardoso, A. J. 1985. Variabilidade em medidas dos espermatozóides de *Hyla fuscovaria* (Amphibia, Anura) e seu significado taxonômico. Revista Brasileira de Biologia 45: 387-391.

Grassé, P.-P. 1929. Les constituants cytoplasmiques des éléments mâles du crapaud acchoucheur (*Alytes obstetricans* Laur.). Comptes Rendues de la Société Biologique 101: 79-82.

Grassé, P.-P. 1986. Appareil urogénital des amphibiens. La spermatogenèse. Pp. –20. In P.-P. Grassé and M. Delsol (eds), 'Traité de Zoologie. Anatomie, Systématique, Biologie', Vol.14 (I-B). Masson, Paris).

Greer, H. J. 1993. Comparative organization of Sertoli cells including the Sertoli cell barrier. Pp. 704-739. In L. D. Russell and M. D. Griswold (eds), *The Sertoli Cell*. Cache River Press, Clearwater, FA.

Grey, R. D., Working, P. K. and Hedrick, J. L. 1977. Alteration of structure and penetrability of the vitelline envelope after passage of eggs through ceolom to oviduct in *Xenopus laevis*. Journal of Experimental Zoology 201 : 73-83.

Gueydan-Bacconier, M., Neyrand de Leffemberg, F., Pujol, P., Delsol, M. and Flatin, J. 1984. Étude comparative du dynamisme de la spermatogenèse chez trois batraciens tropicaux par autohistoradiographie. Annales Des Sciences Naturelles, Zoologie, Paris: 191-196.

Haaf, T., Grunenberg, H. and Schmid, M. 1990. Paired arrangement of nonhomologous centromeres during vertebrate spermiogenesis. Experimental Cell Research 187: 157-161.

Hardy, D. M. and Hedrick, J. L. 1992. Oviductin. Purification and properties of the oviductal protease that processes the molecular weight 43,000 glycoprotein of the *Xenopus laevis* egg envelope. Biochemistry 31: 4466-4472.

Healy, J. M. and Jamieson, B. G. M. 1992. Ultrastructure of the spermatozoon of the tuatara (*Sphenodon punctatus*) and its relevance to the relationships of the Sphenodontida. Philosophical Transactions of the Royal Society of London B 335: 193-205.

Herrmann, H. -J. 1987. Rasterelektronenmikroskopische aufnahmen von spermien der mitteleuropäischen braunfrösche *Rana arvalis*, *R. dalmatina* und *R. temporaria* (Amphibia, Anura, Randiae). Mauritiana (Altenburg) 12: 163-165.

Hillis, D. M. 1991. The phylogeny of amphibians: current knowledge and the role of cytogenetics. Pp. 7-31 In D. M. Green and S. K. Sessions (eds), *Amphibian Cytogenetics and Evolution*, Academic Press, New York, NY.

Iwao, Y. 2000. Fertilization in amphibians. Pp. 147-191. In J. J. Tarin and A. Cano (eds), *Fertilization in Protozoa and Metazoan Animals*, Springer-Verlag, Heidelberg.

Iwao, Y. and Katigiri, C. 1982. Properties of the vitellin coat lysin from toad sperm. Journal of Experimental Zoology 219: 87-95.

Iwao, Y., Miki, A., Kobayashi, M. and Onitake, K. 1994. Activation of *Xenopus* eggs by an extract of *Cynops* sperm. Development, Growth and Differentiation 36: 469-479.

James, W. S. 1970. *The Ultrastructure of Anuran Spermatids and Spermatozoa.* 121 pp. Ph. D. Thesis, University of Tennessee, USA.

Jamieson, B. G. M. 1981. *The Ultrastructure of the Oligochaeta.* Academic Press, London, 462 pp.

Jamieson, B. G. M. 1991. *Fish Evolution and Systematics: Evidence from Spermatozoa.* Cambridge University Press, Cambridge. 319 pp.

Jamieson, B. G. M. 1995. Evolution of tetrapod spermatozoa with particular reference to amniotes. Pp. 343-358. In B. G. M. Jamieson, J. Ausio, J. and J.-L. Justine (eds), *Advances in Spermatozoal Phylogeny and Taxonomy.* Mémoires du Muséum National d'Histoire Naturelle, vol. 166, Paris.

Jamieson, B. G. M. 1999. Spermatozoal phylogeny of the Vertebrata. Chapter 29, Pp. 303-331. In C. Gagnon (ed.), *The Male Gamete: From Basic Science to Clinical Applications.* Cache River Press, St. Louis, Missouri.

Jamieson, B. G. M. and Healy, J. M. 1992. The phylogenetic position of the tuatara, *Sphenodon* (Sphenodontida, Amniota), as indicated by cladistic analysis of the ultrastructure of spermatozoa. Philosophical Transactions of the Royal Society of London B 335: 207-219.

Jamieson, B. G. M. and Rouse, G. W. 1989. The spermatozoa of the Polychaeta (Annelida): An ultrastructural review. Biological Reviews 64: 93-157.

Jamieson, B. G. M., Lee, M. S. Y. and Long, K. 1993. Ultrastructure of the spermatozoon of the internally fertilizing frog *Ascaphus truei* (Ascaphidae: Anura: Amphibia) with phylogenetic considerations. Herpetologica 49: 52-65.

Jamieson, B. G. M., Scheltinga, D. M. and Tucker, A. D. 1997. The ultrastructure of spermatozoa of the Australian Freshwater Crocodile, *Crocodylus johnstoni* Krefft, 1873 (Crocodylidae, Reptilia). Journal of Submicroscopic Cytology and Pathology 29: 265-274.

Jespersen, Å. 1971. Fine structure of the spermatozoon of the Australian lungfish *Neoceratodus forsteri* (Krefft). Journal of Ultrastructure Research 37: 178-185.

Jorgensen, C. B. 1984. Testis function in the toad *Bufo bufo* (Amphibia, Anura) at organ and subunit levels. Videnskabelige Meddelelser Fra Dansk Naturhistorisk Forening 0 (145): 117-130.

Kalt, M. R. 1976. Morphology and kinetics of spermatogenesis in *Xenopus laevis.* Journal of Experimental Zoology 195: 393-408.

Kalt, M. R. 1977. Cytoplasmic inclusions in *Xenopus* spermatogenic cells. Ultrastructural and cytochemical analysis of the action of antimitotic agents on subcellular elements. Journal of Cell Science 28: 15-28.

Kanamadi, R. D. and Jirankali, C. S. 1991. Testicular activity in the burrowing frog *Tomopterna breviceps.* Zoologischer Anzeiger 227: 80-92.

Kanamadi, R. D. and Hiremath, C. R. 1993. Testicular activity in the frog *Microhyla ornata* (Microhylidae). Proceedings of the Indian National Science Academy Part B, Biological Sciences 59: 489-500.

Kasinsky, H. E. 1989. Specificity and distribution of sperm basic proteins. Pp. 73-163. In L. S. Hnilica, G. S. Stein and J.L. Stein (eds), *Histones and other basic nuclear proteins.* CRC Press, Inc., Boca Raton, Florida.

Kasinsky, H. E. 1995. Evolution and origins of sperm nuclear basic proteins. Pp. 463-473. In B. G. M. Jamieson, J. Ausio and J.-L. Justine (eds), *Advances in spermatozoal phylogeny and taxonomy,* Vol. 166. Mémoires du Muséum National d'Histoire Naturelle, Paris.

Kasinsky, H. E., Huang, S. Y., Mann, M., Roca, J. and Subirana, J. A. 1985. On the diversity of sperm histones in the vertebrates: 4. Cytochemical and amino acid analysis in Anura. Journal of Experimental Zoology 234: 33-46.

Katigiri, C., Iwao, Y. and Yoshizaki, N. 1982. Participation of oviducal pars recta secretions in inducing the acrosome reaction and release of vitelline coat lysin in fertilizing toad sperm. Developmental Biology 94: 1-10.

Katigiri, C., Yoshizaki, Y., Kotani, M. and Kubo, H. 1999. Analyses of oviductal pars recta-induced fertilizability of coelomic eggs in *Xenopus laevis*. Developmental Biology 210: 269-276.

Kerr, J. B. and Dixon, K. E. 1974. An ultrastructural study of germ plasm in spermatogenesis of *Xenopus laevis*. Journal of Embryology and Experimental Morphology 32: 573-592.

Kobayashi, T. and Iwasawa, H. 1989. The role of Sertoli cells adjacent to efferent ductules in sperm transport in *Rana porosa porosa*. Zoological Science 6: 935-942.

Kobayashi, T., Asakawa, Y. and Iwasawa, H. 1993. Kinetics of first spermtaogenesis in young toads of *Xenopus laevis*. Zoological Science 10: 189-193.

Koch, R. A. and Lambert, C. C. 1990. Ultrastructure of sperm, spermiogenesis, and sperm-egg interactions in selected invertebrates and lower vertebrates which use external fertilization. Journal of Electron Microscopy Technique 16: 115-154.

Kuramoto, M. 1995. Scanning electron microscopic studies on the spermatozoa of some Japanese salamanders (Hynobiidae, Cryptobranchidae, Salamandridae). Japanese Journal of Herpetology 16: 49-58.

Kuramoto, M. 1996. Generic differentiation of sperm morphology in treefrogs from Japan and Taiwan. Journal of Herpetology 30: 437-443.

Kuramoto, M. 1997. Further studies on sperm morphology of Japanese salamanders, with special reference to geographic and individual variation in sperm size. Japanese Journal of Herpetology 17: 1-10.

Kuramoto, M. 1998. Spermatozoa of several frog species from Japan and adjacent regions. Japanese Journal of Herpetology 17: 107-116.

Kuramoto, M. and Joshy, H. S. 2001. Scanning electron microscopic studies on spermatozoa of anurans from Indian and Sri Lanka. Amphibia-Reptilia 22: 303-308.

Kuramoto, M. and Joshy, S. H. 2000. Sperm morphology of some Indian frogs as revealed by SEM. Current Herpetology 19: 63-70.

Kwon, A. S. and Lee, Y. H. 1992. The fine structure of spermatozoa in *Bombina orientalis* (Anura, Amphibia). Nature and Life 22: 15-22.

Kwon, A. S. and Lee, Y. H. 1995. Comparative spermatology of anurans with special references to phylogeny. Pp. 321-332. In B. G. M. Jamieson, J. Ausio and J. -L. Justine, (eds), *Advances in Spermatozoal Phylogeny and Taxonomy*, vol. 166, Mémoires du Muséum National d'Histoire Naturelle, Paris.

Kwon, A. S., Kim, H. J. and Lee, Y. H. 1993. Fine structure of the neck of spermatozoa and spermiogenesis in *Bufo bufo gargarizans* (Amphibia, Anura). Nature and Life 23: 95-105.

Lee, M. S. Y. and Jamieson, B. G. M. 1992. The ultrastructure of the spermatozoa of three species of myobatrachid frogs (Anura, Amphibia) with phylogenetic considerations. Acta Zoologica (Stockholm) 73: 213-222.

Lee, M. S. Y. and Jamieson, B. G. M. 1993. The ultrastructure of the spermatozoa of bufonid and hylid frogs (Anura, Amphibia): implications for phylogeny and fertilization biology. Zoologica Scripta 22: 309-323.

Lee, Y. H. and Kwon, A. S. 1992. Ultrastructure of spermiogenesis in *Hyla japonica* (Anura, Amphibia). Acta Zoologica (Stockholm) 73: 49-55.

Lee, Y. H. and Kwon, A. S. 1996. Ultrastructure of spermatozoa in Urodela and primitive Anura (Amphibia) with phylogenetic considerations. The Korean Journal of Systematic Zoology 12: 253-264.

Lin, D. J. You, Y. L. and Zhong, X. 1999. The structure of the spermatozoon of *Hyla chinensis* and its bearing on phylogeny. Zoological Research 20: 161-167.

Lin, D. J. You, Y. L. and Zummo, G. 2000. Spermiogenesis in *Hyla chinensis*. Acta Zoologica Sinica 46: 376-384.

Lindsay, L. L. and Hedrick, J. L. 1998. Treatment of *Xenopus laevis* coelomic eggs with trypsin mimics pars recta oviductal transit by selectively hydroloyzing envelope glycoprotein gp43, increasing sperm binding to the envelope, and rendering the eggs fertilizable. Journal of Experimental Zoology 281: 132-138.

Lindsay, L. L., Wieduwilt, M. J. and Hedrick, J. L. 1999. Oviductin, the *Xenopus laevis* oviductal protease that processes egg envelope glycoprotein gp43, increases sperm binding to envelope, and is translated as part of an unusual mosaic protein composed of two proteases and several CUB domains. Biology of Reproduction 60: 989-995.

Littlejohn, M. J. 1977. Long range acoustic communication in anurans: an integrated and evolutionary approach. Pp. 263-294. In D. H. Taylor and S. I. Guttman (eds), *The Reproductive Biology of Amphibians*, Plenum Press, New York.

Littlejohn, M. J., Roberts, D. J., Watson, G. F. and Davies, M. 1993. Family Myobatrachidae. Pp. 41-57. In C. J. Glasby, G. J. B. Ross and P. L. Beesley (eds), *Fauna of Australia, Vol. 2A, Amphibia and Reptilia*, Australian Government Publishing Service: Canberra.

Lofts, B. 1974. Reproduction. Pp. 107-218. In B. Lofts (ed.), *Physiology of the Amphibia*, vol. 2, Academic Press, New York.

Lopes, V. d. A. M. J., Recco, P. S. M. and Cardoso, A. J. 1999. A comparison of the sperm nucleoprotein composition in the genus *Physalaemus* (Amphibia, Anura). Cytobios 100: 147-157.

Lopez, L. A., Vincenti, A. and Burgos, M. H. 1983. Microtubular stability and cytoplasmic vacuolization in the toad Sertoli cell during spermiation. Microscopia electronica y biologia cellular 7: 73-85 (Fide Kwon *et al.* 1993).

Mainoya, J. R. 1981. Observations on the ultrastructure of spermatids in the testis of *Chiromantis xerampelina* (Anura: Rhacophoridae). African Journal of Ecology 19: 365-368.

Maller, J. D., Poccia, D., Nishioka, D., Kidd, P., Gerhardt, J. and Hartman, M. 1976. Spindle formation and cleavage in *Xenopus* eggs injected with centriole-containing fractions from sperm. Experimental Cell Research 99: 285-294.

Mattei, X., Siau, Y. and Seret, B. 1988. Etude ultrastructurale du spermatozoïde du coelacanthe: *Latimeria chalumnae*. Journal of Ultrastructure and Molecular Structure Research 101: 243-251.

Metter, D. E. 1964. On breeding and sperm retention in *Ascaphus*. Copeia 1964: 710-711.

Meyer, E., Jamieson, B. G. M. and Scheltinga, D. M. 1997. Sperm ultrastructure of six Australian hylid frogs from two genera (*Litoria* and *Cyclorana*): phylogenetic implications. Journal of Submicroscopic Cytology and Pathology 29: 443-451.

Mizuhira, V., Futaesaku, Y., Ono, M., Ueno, M., Yokofujita, J. and Oka, T. 1986. The fine structure of the spermatozoa of two species of *Rhacophorus* (*arboreus, schlegelii*). I. Phase-contrast microscope, scanning electron microscope, and cytochemical observations of the head piece. Journal of Ultrastructure and Molecular Structure Research 96: 41-53.

Mo, H. 1985. Ultrastructural studies on the spermatozoa of the frog *Rana nigromaculata* and the toad *Bufo bufo asiaticus*. Zoological Research 6: 381-390.

Morrisett, F. W. 1974. *Comparative Ultrastructure of Sperm in Three Families of Anura (Amphibia)*. 56 pp. M. Sc. Thesis, Arizona State University, USA.

Nakai, S., Watanabe, A. and Onitake, K. 1999. Sperm surface heparin/heparin sulfate is responsible for sperm binding to the uterine envelope of the newt *Cynops pyrrhogaster*. Development, Growth and Differentiation 41: 101-107.

Nath, A. and Chand, G. B. 1998. Ultrastructure of spermatozoa correlated with phylogenetic relationship between *Heteropneustes fossilis* and *Rana tigrina*. Cytobios 95: 161-165.

Neyrand de Leffemberg, F. and Exbrayat, J.-M. 1995. Étude comparative du dynamisme de la spermatogenèse chez les amphibiens par la méthode histoautoradiographique à la thymidine tritée. Bulletin Mensuel de la Société Linnéenne de Lyon 64 (8): 356-372.

Nicander, L. 1970. Comparative studies on the fine structure of vertebrate spermatozoa. Pp. 47-55. In B. Baccetti (ed.), *Comparative Spermatology*. Accademia Nazionale dei Lincei, Rome.

Obert, H. J. 1976. Die spermatogenese bei der Gelbauchunke (*Bombina variegata variegata* L.) im Verlauf der jahrlichen Aktivitatsperiode und die Korrelation sur Paarungsrufaktivitat (Discoglossidae, Anura). Zeitshcrift für Mikroskopische Anatomie Forschung, Leipzig 90: 908-924.

Ogielska, M. and Bartmanska, J. 1999. Development of testes and differentiation of germ cells in water frogs of the *Rana esculenta*-complex (Amphibia, Anura). Amphibia-Reptilia 20: 251-263.

Oka, T. 1980. Ultrastructural observations on the sperm in a frog, *Rhacophorus schlegelii*. Japanese Journal of Herpetology 8: 137.

Onitake, K., Takai, H., Ukita, M., Mizuno, J., Sasaki, T. and Watanabe, A. 2000. Significance of egg-jelly substances in the internal fertilization of the newt *Cynops pyrrohogaster*. Comparative Biochemistry, Physiology B. Biochemistry and Molecular Biology 126: 121-128.

Passmore, N. I. and Carruthers, V. C. 1979. *South African Frogs*. Johannesburg, Witwatersrand University Press: 1-270.

Penn, A. and Gledhill, B. L. 1972. Acrosomal protelolytic activity of amhibian sperm: A direct demonstration. Experimental Cell Research 74: 285-288.

Picheral, B. 1967. Structure et organisation du spermatozoïde de *Pleurodeles waltlii* Michah. (Amphibien Urodèle). Archives de Biologie, (Liege) 78: 193-221.

Picheral, B. 1972a. Les éléments cytoplasmiques au cours de la spermiogenèse du Triton *Plurodeles waltlii* Michah. I. La genèse de l'acrosome. Zeitschrift für Zellforschung und Mikroskopische Anatomie 131: 347-370.

Picheral, B. 1972b. Les éléments cytoplasmiques au cours de la spermiogenèse du triton *Pleurodeles waltlii* Michah. III. L'évolution des formations caudales. Zeitschrift für Zellforschung und Mikroskopische Anatomie 131: 399-416.

Picheral, B. 1977. La fécondation chez le triton pleurodèle. II. La pénétration des spermatozoïdes et la réaction locale de l'oeuf. Journal of Ultrastructure Research 60: 181-202.

Picheral, B. 1979. Structural, comparative, and functional aspects of spermatozoa in urodeles. Pp. 267-287. In D. W. Fawcett and J. M. Bedford (eds), *The Spermatozoon*, Urban and Schwarzenberg, Baltimore.

Pisanó, A. and Adler, R. 1968. Submicroscopical aspects of *Telmatobius hauthali schreiteri* spermatids. Zeitschrift für Zellforschung 87: 345-349.

Poirier, G. R. and Spink, G. C. 1971. The ultrastructure of testicular spermatozoa in two species of *Rana*. Journal of Ultrastructure Research 36: 455-465.

Pudney, J. 1995. Spermatogenesis in nonmammalian vertebrates. Microscopy Research and Techniques 32: 459-497.

Pugin, E. and Garrido, O. 1981. Morfologia espermatica en algunas especies de anuros pertenecientes al bosque temperado del sur de Chile. Ultraestructura comparada. Medio Ambiente 5: 45-57.

Pugin-Rios, E. 1980. *Étude Comparative sur la Structure du Spermatozoïde des Amphibiens Anoures. Comportement des Gamètes lors de la Fécondation*. 114 pp. Ph. D. Thèse, L'Université de Rennes, France.

Raisman, J. S. and Cabada, M. O. 1977. Acrosomic reaction and proteolytic activity in the spermatozoa of an anuran amphibian, *Leptodactylus chaquensis* (Cei). Development, Growth and Differentiation 19: 227-232.

Raisman, J. S., de, C. R. W., Cabada, M. O., del, P. E. J. and Mariano, M. I. 1980. Acrosome breakdown in Leptodactylus chaquensis (Amphibia: Anura). Development Growth and Differentiation 22 (3): 289-297.

Rastogi, R. K., Bagnara, J. T., Iela, L. and Krasovich, M. A. 1988. Reproduction in the Mexican Leaf Frog, *Pachymedusa dacnicolor*. IV. Spermatogenesis: a light and ultrasonic study. Journal of Morphology 197: 277-302.

Rastogi, R. K., Di, M. M., Di, M. L., Minucci, S. and Iela, L. 1985. Morphology and cell population kinetics of primary spermatogonia in the frog (*Rana esculenta*) (Amphibia: Anura). Journal of Zoology Series A 207: 319-330.

Reed, S. C. and Stanley, H. P. 1972. Fine structure of spermatogenesis in the South African clawed toad *Xenopus laevis* Daudin. Journal of Ultrastructure Research 41: 277-295.

Retzius, G. 1906. Die spermien der amphibien. *Biologische Untersuchungen, Neue Folge* 13: 49-70.

Roberts, J. D. 1993. Natural History of the Anura. Pp. 28-34. In C. J. Glasby, G. J. B. Ross and P. L. Beesley (eds), *Fauna of Australia, Vol. 2A, Amphibia and Reptilia*. Australian Government Publishing Service, Canberra.

Rugh, R. 1939. The reproductive processes of the male frog *Rana pipiens*. Journal of Experimental Zoology 80 : 81-105.

Sandoz, D. 1970a. Étude ultrastructurale et cytochimique de la formation de l'acrosome du discoglosse (Amphibien Anoure). Pp. 93-113. In B. Baccetti (ed.), *Comparative Spermatology*, Accademia Nazionale dei Lincei, Rome.

Sandoz, D. 1970b. Étude cytochimique des polysaccharides au cours de la spermatogénese d'un Amphibien Anoure: le discoglosse *Discoglossus pictus* (Otth.). Journal de Microscopie 9: 243-262.

Sandoz, D. 1971. Formation du cou et de l'anneau dans les spermatides du discoglosse (amphibien anoure). Journal de Microscopie, Paris 11: 91 (Fide Grassé 1986).

Sandoz, D. 1973. Participation du réticulum endoplasmique a l'élaboration de l'anneau dans les spermatides du discoglosse (Amphibien Anoure). Journal de Microscopie 17: 185-198.

Sandoz, D. 1974. Modifications in the nuclear envelope during spermiogenesis of *Discoglossus pictus* (Anuran Amphibia). Journal of Submicroscopic Cytology 6: 399-419.

Sandoz, D. 1975. Development of the neck region and the ring during spermiogenesis of *Discoglossus pictus* (Anuran Amphibia). Pp. 237-247. In B. A. Afzelius (ed.), *The Functional Anatomy of the Spermatozoon*. Pergamon Press, Oxford.

Sandoz, D. 1977. Étude comparative des phénomenes sécrétoires dans l'appareil génital de quelques vertébrés. Paris VI. Thèse Doctorat d'État. (Fide Folliot 1979).

Scheltinga, D. M. 2002. Ultrastructure of Spermatozoa of the Amphibia: Phylogenetic and Taxonomic Implications. 320 pp. Ph.D. Thesis, University of Queensland, Australia.

Scheltinga, D. M., Jamieson, B. G. M., Bickford, D. P., Garda, A. A., Báo, S. N. and McDonald, K. R. 2002a. Morphology of the spermatozoa of the Microhylidae (Anura, Amphibia). Acta Zoologica (Stockholm) 83: 263-275.

Scheltinga, D. M., Jamieson, B. G. M. and McDonald, K. R. 2002b. Ultrastructure of the spermatozoa of *Litoria longirostris* (Hylidae, Anura, Amphibia): modifications for penetration of a gelatinous layer surrounding the arboreal egg clutch. Memoirs of the Queensland Museum 48: 215-220.

Scheltinga, D. M., Jamieson, B. G. M., Eggers, K. E. and Green, D. M. 2001. Ultrastructure of the spermatozoon of *Leiopelma hochstetteri* (Leiopelmatidae, Anura, Amphibia). Zoosystema 23: 157-171.

Scheltinga, D. M., Wilkinson, M., Jamieson, B. G. M. and Oommen, O. V. 2003. Ultrastructure of the mature spermatozoa of caecilians (Amphibia: Gymnophiona). Journal of Morphology : (In Press).

Selmi, M. G., Brizzi, R. and Bigliardi, E. 1997. Sperm morphology of salamandrids (Amphibia, Urodela): implications for phylogeny and fertilization biology. Tissue and Cell 29: 651-664.

Serra, J. A. and Vicente, M. J. 1960. New structures of spermatozoa of *Rana* in relation to lipid localization. Revista Portuguesa de Zoologia e Biologia Geral 2: 223-242.

Seshachar, B. R. 1945. Spermateleosis in *Uraeotyphlus narayani* Seshachar and *Gegenophis carnosus* Beddome (Apoda). Proceedings of the National Institute of Sciences, India 11: 336-340.

Sever, D. M., Moriarty, E. C., Rania, L. C. and Hamlett, W. C. 2001. Sperm storage in the oviduct of the internal fertilizing frog, *Ascaphus truei*. Journal of Morphology 248: 1-21.

Sprando, R. L. and Russell, L. D. 1987. A comparative study of Sertoli cell ectoplasmic specializations in selected non-mammalian vertebrates. Tissue and Cell 19: 479-493.

Srivastava, R. K. and Gupta, M.I. 1990. Comparative effect of FSH, LH and homoplastic pars distalis homogenate on the spermatogenesis of common Indian frog *Rana tigrina* during spermatogenetic recrudescence phase. European Archives of Biology 101: 89-99.

Stebbins, R. C. and Cohen, N. W. 1995. *A Natural History of Amphibians*, Princeton University Press, Princeton, New Jersey, 316 pp.

Swan, M. A., Linck, R. W., Ito, S. and Fawcett, D. W. 1980. Structure and function of the undulating membrane in spermatozoan propulsion in the toad *Bufo marinus*. Journal of Cell Biology 85: 866-880.

Sze, L. C., Zhang, J. -Z., Yan, Y. -C. and Mao, Z. -C. 1981. A comparative study on the basic proteins and ultrastructure of the sperm chromatins from *Rana nigromaculata* and *Bufo asiaticus*. Acta Biologiae Experimentalis Sinica 14: 89-94.

Taboga, S. R. and Dolder, H. 1993. Ultrastructural analysis of the *Hyla ranki* spermatozoan tail (Amphibia, Anura, Hylidae). Cytobios 75: 85-92.

Taboga, S. R. and Dolder, H. 1994a. Basic nuclear protein substitution during spermiogenesis in *Hyla ranki* (Amphibia, Anura, Hylidae). Ciencia e Cultura Sao Paulo 46: 161-163.

Taboga, S. R. and Dolder, H. 1994b. Ultrastructural study of acrosome formation in *Hyla ranki* (Amphibia, Anura, Hylidae). Cytobios 77: 247-252.

Takamune, K. 1987. Detection of acrosome-reacted toad sperm based on specific lectin binding to the inner acrosomal membrane. Gamete Research 18: 215-223.

Takamune, K. and Katigiri, C. 1987. The properties of the oviducal pars recta protease which mediates gamete interaction by affecting the vitelline coat of a toad egg. Development, Growth and Differentiation 29: 193-203.

Takamune, K., Teshima, K., Maeda, M., and Abé, S. 1995. Characteristic features of preleptotene spermatocytes in *Xenopus laevis*: increase in the nuclear volume and first appearance of flattened vesicles in these cells. The Journal of Experimental Zoology 273: 264-270.

Tian, J., Gong, H. and Lennarz, W. J. 1999. *Xenopus laevis* sperm receptor gp69/64 glycoprotein is a homologue of the mammalian sperm receptor ZP2. Proceedings of the National Academy of Sciences of the USA 96: 829-834.

Tian, J., Gong, H., Thomsen, G. H. and Lennarz, W. J. 1997a. *Xenopus laevis* sperm-egg adhesion is regulated by modifications in the sperm receptor and the egg vitelline envelope. Developmental Biology 187: 143-153.

Tian, J., Gong, H., Thomsen, G. H. and Lennarz, W. J. 1997b. Gamete interactions in *Xenopus laevis*: identification of sperm binding glycoproteins in the egg vitelline envelope. Journal of Cell Biology 136: 1099-110.

Ueda, Y., Yoshizaki, N. and Iwao, Y. 2002. Acrosome reaction in sperm of the frog *Xenopus laevis*: its detection and induction by oviductal pars recta secretion. Developmental Biology 243: 55-64.

Unsicker, K. 1975. Fine structure of the male genital tract and kidney in the Anura *Xenopus laevis* Daudin, Rana temporaria L. and Bufo bufo L. under normal and experimental conditions. Cell and Tissue Research 158: 215-240.

Uribe Aranzábal, M. C. 2003. The Testes, Spermatogenesis and Male Reproductive Ducts. Pp. 183-202. In D. M. Sever (ed.), *Reproductive Biology and Phylogeny of Urodela*. B. G. M. Jamieson (Series ed.), Vol. 1. Science Publishers, Inc., Enfield, New Hamsphire.

van der Horst, G. 1979. Spermatozoon structure of three anuran (Amphibia) species. Proceedings of the Electron Microscope Society of Southern Africa 9: 153-154.

van der Horst, G. and van der Merwe, L. 1991. Late spermatid-sperm/sertoli cell association in the caecilian, *Typhlonectes natans* (Amphibia: Gymnophiona). Electron Microscopy Society of Southern Africa 21: 247-248.

van der Horst, G., Wilson, B. A. and Channing, A. 1995. Amphibian sperm: phylogeny and fertilization environment. Pp. 333-342. In Jamieson, B. G. M., Ausio, J., and Justine, J.-L. (eds), *Advances in Spermatozoal Phylogeny and Taxonomy*. Mémoires du Muséum national d'Histoire naturelle, Paris, vol. 166.

Van Oordt, G. J., Beenakkers, A. M. Th., Van Oordt, P. G. W. J. and Stadhouders, A. M. 1954. On the mechanism of the initial states of spermiation in the grass frog, *Rana temporaria*. Acta Endocrinologica 17: 294-301

Vinogradov, A. E., Borkin, L. J., Gunther, R. and Rosanov, J. M. 1990. Genome elimination in diploid and triploid *Rana esculenta* males: cytological evidence from DNA flow cytometry. Genome 33: 619-627.

Vo, L. H. and Hedrick, J. L. 2000. Independent and hetero-oligomeric-dependent sperm binding to egg envelope glycoprotein ZPC in *Xenopus laevis*. Biology of Reproduction 62: 766-774.

Vinogradov, A. E., Razonov, Y. M., Tsaune, I. A. and Borkin, L. Y. 1988. Genome elimination of one of the parents in the hybridogenic species *Rana esculenta* prior to premeiotic DNA synthesis. Tsitologiya 30: 691-698.

Visser, J. and van der Horst, G. 1987. Description of *Heleophryne* sperm (Amphibia; Leptodactylidae). Proceedings of the Electron Microscopy Society of Southern Africa 17: 83-84.

Waggener W. L. and Carroll, E. J. Jr. 1998. Spermatozoon structure and motility in the anuran *Lepidobatrachus laevis*. Development, Growth and Differentiation 40: 27-34.

Wells, K. D. 1977. The courtship of frogs. Pp. 233-262. In D. H. Taylor and S. I. Guttman (eds), *The Reproductive Biology of Amphibians*. Plenum Press, New York.

Werner, G. 1970. On the development and structure of the neck in urodele sperm. Pp. 85-91. In B. Baccetti (ed.), *Comparative Spermatology*. Accademia Nazionale dei Lincei, Rome.

Wilkinson, R. F., Stanley, H. P. and Bowman, J. T. 1974. Genetic control of spermiogenesis in *Drosophila melanogaster*: the effects of abnormal cytoplasmic microtubule populations in mutant MS (3) 1 OR and it colcemid-induced phenocopy. Journal of Ultrastructure Research 48: 242-258 (Fide Rastogi *et al.* 1988).

Wilson, B. A. 1994. *The relationship between fertilization environment and structure and physiology of selected anuran spermatozoa.* Unpublished Ph. D. thesis, University of the Western Cape, Bellville, South Africa. 147 pp.

Wilson, B. A., van der Horst, G. and Channing, A. 1991. Scanning electron microscopy of the unique sperm of *Chiromantis xerampelina* (Amphibia: Anura). Electron Microscopy Society of Southern Africa 21: 255-256.

Wilson, B. A., van der Horst, G. and Channing, A. 1994. Scanning electron microscopy of South African anuran sperm. Electron Microscopy Society of Southern Africa 24: 94.

Wilson, B. A., van der Horst, G. and Channing, A. 1995. A comparison of ranid versus rhacophorid sperm: two extremes. Madoqua 19: 61-64.

Witschi, E. 1915. Studien uber die Geschlechtsbestimmung bei Froschen. Archiv für Mikroskopische Antomie 86: 1-50.

Wortham, J. W. E. Jr., Brandon, R. A. and Martan, J. 1977. Comparative morphology of some plethodontid salamander spermatozoa. Copeia 1977: 666-680.

Xavier, F. 1977. An exceptional reproductive strategy in Anura: *Nectophrynoides occidentalis* Angel (Bufonidae), an example of adaptation to terrestrial life by viviparity. Pp. 545-552. In M. K. Hecht, P. C. Goody and B. M. Hecht (eds), *Major Patterns in Vertebrate Evolution*, Plenum Press, New York.

Yoshizaki, N. 1987. Isolation of spermatozoa, their ultrastructure, and their fertilizing capacity in two frogs, *Rana japonica* and *Xenopus laevis*. Zoological Science 4: 193-196.

Yoshizaki, N. and Katagiri, C. 1982. Acrosome reaction in sperm of the toad, *Bufo bufo japonicus*. Gamete Research 6: 343-352.

Yoshizaki, N. and Katigiri, C. 1984. Necessity of oviducal pars recta secretions for the formation of the fertilization layer in *Xenopus laevis*. Zoological Science 1: 255-264.

Yurewicz, E. C., Oliphant, G. and Hedrick, J. L. 1975. The macromolecular composition of *Xenopus laevis* egg jelly coat. Biochemistry 14: 3101-3107.

Zheng, Z., Fei, L. and Ye, C. 2000. Study on morphology of spermatozoa of Megophrys (Amphibia:Pelobatidae) from China. Chinese Journal of Applied and Environmental Biology 6: 161-165.

Zirkin, B. R. 1971a. The fine structure of nuclei during spermiogenesis in the Leopard Frog, *Rana pipiens*. Journal of Ultrastructure Research 34: 159-174.

Zirkin, B. R. 1971b. The fine structure of nuclei in mature sperm. I. Application of the Langmuir Trough-critical point method to histone-containing sperm nuclei. Journal of Ultrastructure Research 36: 237-248.

Zuber-Vogeli, M., and Xavier, F. 1965. La spermatogenèse de *Nectophrynoides occidentalis* au cours du cycle annuel. Bulletin de la Société Zoologique de France 90: 261-267.

CHAPTER 6

An Overview of Breeding Glands

Rossana Brizzi, Giovanni Delfino and Silke Jantra

6.1 INTRODUCTION

The cutaneous apparatus of Amphibia is a complex and dynamic machinery that acts as an interface between the animal and the external world. Owing to the peculiar biology of the amphibians and their incomplete adaptation to the subaerial environment, the skin plays a crucial role in the homeostasis of these vertebrates. Primarily, it exchanges water, respiratory gas and salts but, in addition, a wide array of unicellular and multicellular skin components (including dermal glands) are engaged in the release of biochemical compounds relevant for survival strategies or communication activities, during larval or adult phases, aquatic or terrestrial life (Duellman and Trueb 1986; Fox 1994; Houck and Sever 1994). Although the main goal of this chapter is to consider cutaneous glands involved in anuran courtship and mating activities, more general considerations of the integumentary glands in the Amphibia are essential to an understanding of their specialized role in reproduction. In this connection, sections 6.2-6.4 will be devoted to a synthetic review of amphibian gland structure, development and different typologies, whereas sections 6.5-6.7 will consider in detail breeding glands in the Anura.

6.2 AMPHIBIAN SKIN GLANDS: A SYNTHESIS ON THEIR STRUCTURE AND DIFFERENT TYPES

The amphibians retained two basic secretory cell lines in the skin from their bony fish ancestors, producing mucus or proteins. Probable cytogenetic and phylogenetic relationships of the above secretory cells come into clearest focus when viewed in terms of evolutionary cell lines based upon microstructural and molecular characteristics. Morphogenetic evolutionary lines of secretory epidermal cells and glands form, thus, an essential and central point in the theme of glandular composition, function and evolution.

Dipartimento di Biologia Animale e Genetica, Università di Firenze, Via Romana 17, 50125 Firenze, Italia

The single mucocytes and protein secreting (serous) cells scattered in the epidermis of fish evolved into multicellular glands in amphibians, as reported in basic studies starting from the onset of last century (Mushe 1909; Dawson 1920; Noble and Noble 1944; Noble 1954; Quay 1972; Le Quang Trong 1973, 1976; Whitear 1977). In the living anurans and urodeles the cutaneous glands are simple alveolar and intradermal and according to recent morphological and biochemical criteria the four different types are distinguished: granular glands (also termed "serous" or "poison glands") (Fig. 6.1A), mucous glands (Fig. 6.1B), mixed glands (Fig. 6.1C) and lipid or wax (Fig. 6.1D) glands (Blaylock *et al.* 1976; Toledo and Jared 1995). Caecilians also have skin glands, but a few details are known about them (Taylor 1968). Although the various gland structural plans appear substantially homogeneous throughout the class, some differences exist between different taxa. In the anuran skin, the secretory units of lipid and serous glands are syncytial in structure (Figs. 6.1D-E, G), whereas discrete secretory cells form the mucous glands (Figs. 6.1B, H). Among Anura, mixed glands are rare. An epithelial organization characterizes, on the contrary, the three gland types of the urodeles (serous, mucous and mixed; Figs. 6.1F, H-I respectively). In these amphibians lipid glands have not been, to date, specifically described. Independently of their nature, amphibian skin glands exhibit the same structural pattern based upon integration of four parts (Figs. 6.1D, G). Proceeding according to the functional mode of secretory release, a gland consists of: 1) a myoepithelial sheath surrounding the gland acinus and involved in gland discharge, 2) a secretory unit; the most characteristic portion of a gland providing to its typical function, 3) an intercalary tract, or neck; a trunco-conical, subepidermal region between gland body and duct. It consists of imbricated adenoblasts and myoblasts that serve as a regenerative matrix after gland holocrine discharge, 4) a duct; an intradermal channel for the secretory product, bordered by epidermal cells. The myoepithelial sheath of serous and lipid glands possesses direct innervation, whereas, as a rule, this is lacking in mucous glands. In this case the release of the neurotransmitter from nerve terminals occurs, very likely, into the stromal environment and diffuses more slowly towards the gland. Nevertheless, occurrence of innervation in some specialized mucous glands (see 6.7.4) indicates that plasticity is present in patterns of gland innervation, as well as in most traits of these secretory structures.

As a rule, the granular glands, with syncytial secretory units (Fig. 6.1E) or with epithelial structures (Fig. 6.1F), contain a product in the form of discrete particles (granules). These appear strongly eosinophilic and mainly consisting of protein substance when observed using light microscope (LM) procedures. However, different structural and ultrastructural patterns, mainly related to

Fig. 6.1 The four different types of cutaneous ordinary glands in the Amphibia. **A.** Serous (also defined granular or poison) glands of *Pelobates cultripes* with transparent vesicles in the syncytial secretory units. **B.** Mucous gland of *Bufo bufo*; note the row of mucocytes around an obvious lumen. **C.** Mixed gland of *Hydromantes imperialis*;

Contd.

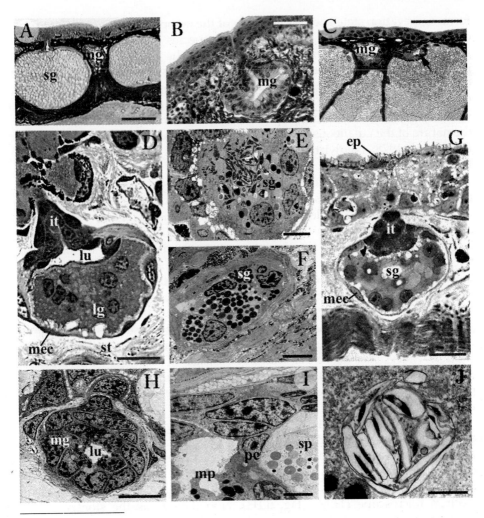

Fig. 6.1 *Contd.*

the arrow points to the mucous acinus. **D.** A wax gland of *Phyllomedusa hypocondrialis* characterized by syncytial arrangement and obvious lumen. The multinuclear secretory unit suggests close relationship with serous glands. **E.** Serous gland of *Xenopus laevis.* Notice the secretory syncytium and the central bulk of secretory product. **F.** Serous gland of *Salamandrina terdigitata.* The secretory granules are contained in discrete adenocytes. **G.** Syntcytial serous gland of tadpole *Physalaemus biligonigerous.* Note the secretory granules characterized by various densities. **H.** *Salamandrina terdigitata.* A mucous gland with typical epithelial structure and the central lumen. **I.** *Salamandrina terdigitata.* Detail at the boundary between serous and mucous portions of a mixed gland. **J.** Wax gland of *Phyllomedusa hypocondrialis.* Detail of an intracytoplasmic lipid aggregate consisting of heterogeneous material in form of elongated elements. Labels: ep, epidermis; it, intercalary tract; lg, lipid gland; lu, lumen; mec, myoepithelial cell; mg, mucous gland; mp, mucous portion; pc, partition cell; sg, serous gland; sp, serous portion; st, stroma. Bars: A = 200 μm, B = 50 μm, C = 100 μm; D,G = 25 μm; E-F = 10 μm; H = 20 μm; I = 5 μm; J = 500 μm. Original.

form, aggregation state and biochemistry of the cytoplasmic secretory product, characterize these glands according to the species. In relation to this same gland type, the term "serous" usually defines the fluid appearance of this product after release, whereas the term "poison" (or venom) glands indicates their possible antipredatory role. In this connection, Phisalix and Bertrand (1902) firstly demonstrated that the cutaneous glands of *Bufo bufo* contain venomous substances. Subsequently, many pharmacological studies revealed the nature of the cutaneous venom in many other species of different taxa (*e.g.* Daly *et al.* 1978; Neuwirth *et al.* 1979; Daly *et al.* 1987; Barthalmus 1994; Erspamer 1994). Actually, the syncytial cells lining the wall of the granular glands co-produce and co-secrete different compounds, such as amines, bioactive peptides, steroidal alkaloids and all their combinations. The result is the synthesis of substances with complex biochemical characteristics acting as repellent, alarm substance or "venom" with peculiar toxicity and pharmacological actions. Nonetheless, the ancestral function of granular glands is still problematic. The toxic nature of serous secretions in many salamanders and frogs led usually to the assumption that they are "poison glands" in all the amphibians, although this function seems absent in some lineages (Neuwirth *et al.* 1979). According to other authors (Bachmayer *et al.* 1967; Barberio *et al.* 1987; Zasloff 1988), the granular glands are involved in the production of substances acting as an antimicrobial film over the body surface. Also, serous glands in the tail of the urodele *Ambystoma* serve for nutrient storage of proteinaceous material (Williams and Larsen 1986). However, the ancestral function of granular glands, probably involved in some forms of defense, regulative functions, storage, or all, is an open question.

Observed under the LM, the product of mucous glands appears as a dense and poorly structured material, usually basophilic and mainly consisting of glycosaminoglycans, as indicated by the histochemical tests. The secretory cells (mucocytes) have an orderly arrangement, in a single row, around an obvious lumen, as detectable both under LM (Fig. 6.1B) and transmission electron microscope (TEM) (Fig. 6.1H). Nonetheless, as the storage of the secretory product is intracytoplasmic, the gland lumen usually appears empty, apart from the discharge phases. Changes detectable in the height of mucocytes and different granule types (Dominguez *et al.* 1981) have been mostly related to the different biosynthetic phases of the secretory cells. However, other investigators have emphasized heterogeneity in anuran mucous glands. In *Rana* spp., Mills and Prum (1984) have described two types of mucous glands; a "typical" mucous gland and a seromucous gland, which possesses secretory cells filled with serous granules as well as mucus. In addition, Thomas *et al.* (1990) detected the presence of sauvagine-like immunoreactivity only in some mucous glands of the phyllomedusine frogs. On the basis of this result, these authors suggested that biochemically discrete populations of mucous glands exist in the frog skin.

According to a general point of view, the mucous glands seem to be involved in saline and gas exchanges between tissues and environment, as well as in regulation of water loss during terrestrial phases. In addition, the mucous film on the body surface possesses friction-reducing properties, useful during swimming. These basic functions, common to all the components of the different amphibian orders, may largely account for the main homogeneity of the mucous glands and their consistent features through the class. Nevertheless, some morphological and functional specializations of these glands occur in both frogs and salamanders (see below).

Mixed glands are common secretory structures in most urodeles, where they represent a stable type (Delfino *et al.* 1986). In contrast, in the anurans they possibly correspond to a transitory stage in granular gland development (Duellman and Trueb 1986) or restoration (Delfino 1980). Mixed glands consist of a larger granular portion and a smaller mucigenous portion (Fig. 6.1C), as indicated by their stain with the routine methods and reaction to the histochemical tests (PAS negative the serous portion, PAS positive the mucous acinus), and their different structure (Fig. 6.1I). Mixed glands in urodeles possibly represent an adaptation for the synergic production of both mucous and serous products and, on this account, their occurrence has been selected by environmental and social constraints throughout the order.

Currently, lipid producing glands have been reported only in some hylids (Blaylock *et al.* 1976; Cei 1980, Delfino *et al.* 1998). Observed at low magnification, wax glands show a relatively large lumen (Fig. 6.1D). This is sharply bounded by a thin layer of flat cells and in some instances contains discrete secretory bodies exhibiting positive response in trials for lipids. Wax glands reveal their syncytial organization (Fig. 6.1D) and a content of secretory aggregates consisting of heterogeneous material, which under TEM includes minute particles and elongated elements, often cylindrical in shape (Fig. 6.1J). According to structural and ultrastructural criteria, the lipid producing glands may be regarded as a specialization of the syncytial type, possibly derived from the proteinaceous secretory line. Although their rich biosynthetic apparatus strongly indicates the capability to produce proteins, the ultimate secretion of these glands is not water-soluble owing to prevailing hydrophobic fractions (Blaylock *et al.*1976). In addition, lipid biosynthesis is suggested by the rich supply of smooth endoplasmic reticulum (SER). Secretions from these glands released onto the epidermis serve as a lubricant in the water and reduce dehydration during the terrestrial life phases. Functional optimization of this secretory product is achieved by its active distribution on the body skin, as suggested by behavioral evidence in *Phyllomedusa* (wiping behavior; see Blaylock *et al.* 1976).

6.3 MAIN PATTERNS OF AMPHIBIAN GLAND DEVELOPMENT

Development of ordinary and specialized cutaneous glands occurs from the midlarval period and consists of the segregation of cell pools (gland buds or Anlagen; Fig. 6.2A) from the epidermis (Bovbjerg 1963; Verma 1965). Possibly gland Anlagen arise from single stem cells (mother cells) through mitotic processes (Fig. 6.2B), regulated by thyroid hormones along the hypothalamus-hypophysis-thyroid axis. The development process can be analyzed according to microenvironmental influences, which account for the various differentiation

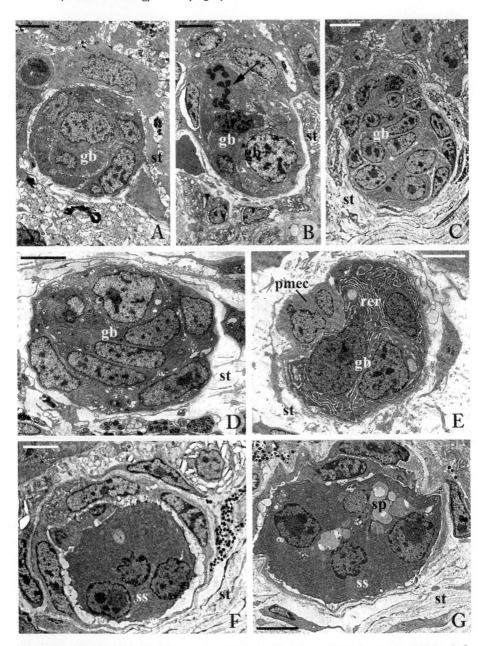

Fig. 6.2 Sequential steps of gland development observed under TEM. **A-C**. Undifferentiated gland buds during down-growth. **D-E**. Mucous gland buds **F-G**. Serous gland buds. **A**. *Xenopus laevis*. Spheroidal gland bud contained in the epidermis. **B**. *Physalaemus biligonigerus*. Mitotic processes (arrow) promove the *descensus* into the dermal primordium. **C**. *Physalaemus biligonigerus*. Although this gland bud is still undifferentiated, the discrete adenoblast arrangement suggests its mucous

Contd.

fates detectable among cell lines in the four gland regions (Delfino *et al.* 1985). Following the segregation process, a few cells of the Anlage remain at the interface between epidermis and dermis, whereas most of the pool occupies the dermal environment (Figs. 6.2C-D) Actually gland segregation from the epidermis is a down-growth, since the process is triggered and supported by cell multiplication. The growing cell bulk pushes down the collagen layers beneath the epidermis (the primordial dense dermis), and leads to development of a periglandular space (the primordial spongy dermis). This space will soon accommodate several dermal components: blood capillaries, fibroblasts, chromatophores, neurites and cells typical of the proper connective tissue. The microenvironment in the primordial spongy dermis is characterized by potential inductive structures, blood vessels and axons, with their regulative molecules (hormones and neurotransmitters, respectively). Neurite processes usually have been detected in close contiguity with serous glands, contacting both the secretory and contractile compartments (Delfino *et al.* 1988a). These factors contribute to the differentiation of the intradermal gland cells, according to the usual signal-receptor interactions. On the other hand, the primitive gland bud patterning during down-growth can also play a role, since external cells evolve as myoblasts, whereas the internal ones becomes adenoblasts. Such complex inductive activity is not effective on cells at the boundary between epidermis and dermis, which maintain their undifferentiated status and will form the stem compartment in the fully developed gland, involved in gland rehabilitation after massive secretory discharge, as described in anuran serous glands (Faraggiana 1938a, b, 1939; Delfino 1980). Contribution of epidermal cells to gland development includes formation of the duct, which is however relatively delayed in comparison with other gland regions. Actually, the duct lumen is a peculiar intraepidermal interstice, since its wall consists of keratinocytes with a differentiation timing that follows epidermis differentiation. Through these peculiar interstices the external environment penetrates into the epidermis, and therefore a continuous horny layer lines the duct as a result of a centripetal keratinisation gradient, conforming the base-surface horny evolution of epidermis.

Although the above basic sequence of development and morphogenesis steps is common to all cutaneous gland types in anurans and urodeles, some differences are detectable, which fit differences between the two orders and fundamental secretory classes (serous and mucous). Ordinary mucous glands in both Salientia and Caudata appear later than the serous ones (Figs. 6.2D-E)

Fig. 6.2 *Contd.*

developmental fate. **D.** *Xenopus laevis.* Presumptive mucous gland bud, as suggested by regular adenoblast arrangement. **E.** *Xenopus laevis.* Presumptive mucous adenocytes exhibit a widespread RER. **F.** *Physalaemus biligonigerus.* Notice the syncytial secretory unit in this gland Anlage, freshly segregated in the dermis primordium. **G.** *Physalaemus biligonigerus.* In a later developmental stage, nuclei become arranged at the periphery whereas serous product is central. Labels: gb, gland bud; pmec, presumptive myoepithelial cell; rer, rough endoplasmic reticulum; sp, secretory product; ss, secretory syncytium; st, stroma. Bars: A-G = 5 μm. Original.

but their development rate is higher and compensates for the delay. During mucous gland morphogenesis, no direct contact occurs between Anlagen and neurite endings, and in adult specimens myoepithelia encircling mucous secretory units lack close nerve supply. Adenoblasts assume a radial arrangement and transform into discrete pyramidal mucocytes encircling a real lumen. Segregation of urodele serous gland buds is a fast process achieved by means of transformation of the interstices around the Anlage into a primordial stromal space (Delfino et. al.1985), prepared by remodelling of the dermis-epidermis junction. Serous adenoblasts maintain their individuality, although adenocytes in mature secretory units exhibit a random arrangement. Serous buds grow toward the dermis at a slow rate, and Anlagen consisting of numerous cells are still inserted within the epidermis. Therefore, early serous Anlagen, which arise in dorsal skin regions, undergo secretory cytodifferentiation in a microenvironment influenced by the embryonic neurogenesis (Pearse 1977). These inductive conditions agree with the neurosecretory program expressed by serous glands of anurans (Delfino *et al.* 1985), which manufacture active biogenic amines and polypeptides. Beside this unusual biochemical trait, serous differentiation in adenoblasts of anuran cutaneous glands includes a peculiar step of cytoplasm merging, so that secretory unit early acquires the syncytial arrangement (Fig. 6.2F). Although patterns of serous biosynthesis have often been described in discrete adenocytes, secretory activity mostly involves the syncytium (Fig. 6.2G), which can better synchronize manufacturing and maturational processes (Delfino *et al.* 1988a).

6.4 BIOSYNTHESIS AND MATURATIVE STEPS OF THE SECRETORY PRODUCT

Regardless of the gland type, secretory activity is a basically continuous process consisting of a sequence of metabolic steps, involving different cytoplasmic organelles. It should, however, be emphasized that several active fractions of the serous products are small molecules, produced in the hyaloplasm and therefore their manufacture is hardly characterisable under the TEM. This is the case of 5-HT, which is synthesized in the cytosol environment, to be later accumulated into serous granules in an active process (Delfino 1980). Among various gland types, mucous secretory units display usual patterns of mucus biosynthesis, involving rough endoplasmic reticulum (RER) and Golgi stacks (dyctiosomes), arranged according to the usual base-apex gradient. Secretory products dispatched by dyctiosomes undergo consistent maturational changes, resulting in marked condensation. Lipid manufacturing in phyllomedusines occurs within the endoplasmic reticulum; both the rough and smooth types participate to the production of rod-shaped subunits, which accumulate in membrane-bounded compartments. Serous biosynthesis is characterized by consistent features in urodeles and leads to the production of secretory granules homogeneous in aspect throughout the order; in contrast, serous products are remarkably variable in features in anurans. However these differences appear to be related to the post-Golgian (maturational) phases, rather than to the proper biosynthesis processes. Actually this secretory activity involves, in both orders, the usual apparatus devoted to the biosynthesis of proteinaceous products to be segregated within membrane-bounded compartments, namely RER and Golgi stacks (Figs. 6.3A-B). In several species of Bufonidae (Figs. 6.3C-D) and Dendrobatidae, remarkable amount of SER contribute to the secretory processes, in agreement with occurrence of not polar component molecules in the secretory product, possibly lipophilic alkaloids.

Most distinctive characters between urodele and anuran skin poisons and poisons of the latter order depend on the long-term maturational processes. Consistent activities of maturation affect the product released from the Golgi stacks, and result in condensation of this moderately opaque material (Figs. 6.3C-D). In several species this process results in granules provided with repeating substructures (Delfino *et al.* 1998; Terreni *et al.* 2002). On the other hand, serous glands in a lesser number of anuran species produce relatively dense secretory granules destined to undergo fluidation and change into vesicles, holding translucent product (Delfino *et al.* 1992). The concept of maturation involves a centripetal polarization of the gland: consistent morphological differences are obvious in both organelles and secretory aggregates according to a peripheral-central gradient (Figs. 6.4A-B). In advanced tadpoles and juveniles, maturational processes account for the wide polymorphism detectable among serous aggregates (Figs. 6.4C-D).

6.5 DISTINCTION BETWEEN ORDINARY AND SPECIALIZED GLANDS IN AMPHIBIA

A first distinction of the integumentary glands of Amphibia as: "ordinary" or "specialized" glands, derives from considering their localization on the body skin. With the term "ordinary" glands we define multicellular secretory structures broadly and randomly distributed in the integument of the whole body surface and usually involved in functions typical of the various gland types (see 6.2). In this connection, wax glands appear as ordinary glands owing to their wide distribution on the body surface. With the term "specialized" glands we indicate, in contrast, secretory organs occurring only in limited skin regions and involved in unusual roles, related to the species biology. Several different gland specializations have been reported in both Anura (see Barthalmus 1994) and Urodela (see Houck and Sever 1994), all of which are referable to the following main functions: antipredatory defence or social communication and reproduction. When specialized cutaneous glands occur, the most prominent differences concern their secretory units; as a rule, these are large, elongated and closely packed, forming voluminous clusters and, in some cases, obvious outgrowths (such as plicae, swellings and patches).

On the basis of cytology and/or secretory products, most specialized glands of the amphibians pertain to the serous type, such the antipredatory units of both anurans and urodeles (Fig. 6.5). However, other secretory specializations of cutaneous glands seem to derive from the mucous type. This is typical of glands that, in some salamanders and frogs, are related to reproductive strategies

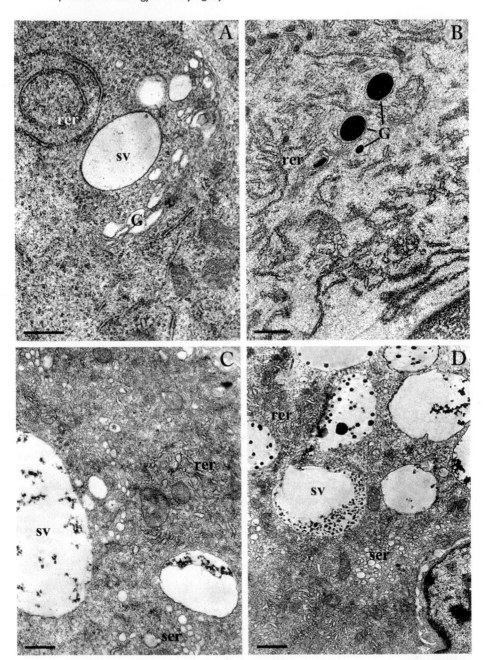

Fig. 6.3 Biosynthesis machinery in anuran serous glands. **A**. *Xenopus laevis*. Notice circularly arranged rer and active Golgi stacks producing secretory vesicles. **B**. *Alytes sp*. Rer arrangement is more regular in this gland, where the Golgi apparatus produces dense granules. **C-D**. *Bufo calamita*. Patterns of rough and smooth endoplasmic reticulum. Labels: G, Golgi stack; rer, rough endoplasmic reticulum; ser, smooth endoplasmic reticulum; sv, secretory vesicle. Bars: A-D = 1 μm. Original.

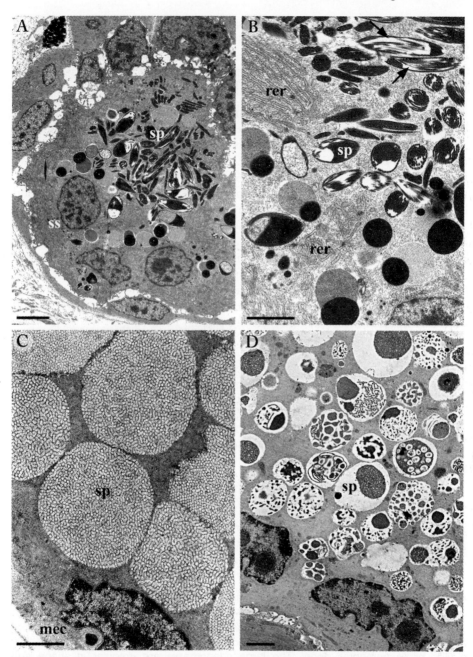

Fig. 6.4 Ultrastructural features of serous products. **A–B**. *Xenopus laevis*. Serous product consists of elongated granules (A) which at higher magnification (B) reveal elongated subunits (arrows). **C**. *Physalaemus biligonigerous*. In many anurans, secretory granules look like glomerules, due to close aggregation of short subunits, resembling bowed tubules. **D**. *Bufo bufo*. Notice remarkable polymorphism of the product in serous gland of the European common toad. Labels: mec, myoepithelial cell; rer, rough endoplasmic reticulum; sp, secretory product; ss, secretory syncytium. Bars: A = 10 µm; B = 5 µm; C-D = 2 µm. Original.

and produce pheromones (see 6.7). In other cases, clusters of both mucous and granular glands form sexually dimorphic structures, as the male tail tubercle of the salamandrid *Mertensiella caucasica* and *M. luschani* (Sever *et al.* 1997).

6.6 GLAND SPECIALIZED FOR ANTIPREDATORY DEFENCE

An efficacious antipredatory mechanism of amphibians involves the secretion of poisons and sticky products from the cutaneous glands. The defensive skin arsenal provided by the granular glands of the neotropical poison arrow frogs of the family Dendrobatidae (Daly *et al.* 1978; Neuwirth *et al.* 1979; Myers and Daly 1983; Daly *et al.* 1987; Erspamer 1994) is well known. The parotoid glands, which also produce defensive toxins, occur in some anurans (pelobatids and bufonids; Fig. 6.5A) as well as in several families of urodeles, including ambystomatids, plethodontids and salamandrids (Fig. 6.5B). Parotoid glands are serous in nature and look like other dermal granular glands except for their relatively larger size (Hostetler and Cannon 1974; Cannon and Hostetler 1976).

Nonetheless, many anurans and salamanders possess, in some skin regions, other patches of granular glands that seem to perform the same protective or defensive function as the parotoid ones (Brodie and Smatresk 1990). Among anurans, these are, for example, the leg glands occurring in the skin of the zeugopodium of the discoglossid *Bombina variegata* (Fig. 6.5C ; Delfino *et al.*

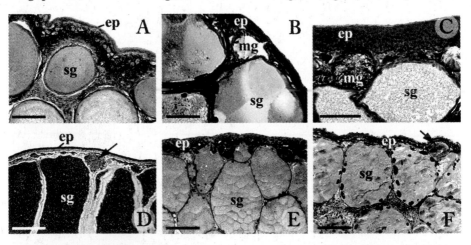

Fig. 6.5 Patterns of urodele serous glands possibly involved in antipredatory defence. **A–B.** Parotoid glands of *Bufo bufo* (A) and *Salamandra salamandra* (B). **C.** Leg glands in the zeugopodium of *Bombina variegata*. **D.** Inguinal glands of *Physalaemus biligonigerous*. The arrow points to a common serous gland. **E.** Tail glands of *Hydromantes genei*. The large adenomeres are irregular in shape and accommodate reciprocally their folded surfaces producing an apparent pattern of gland superposition. **F.** Tail glands of *Triturus marmoratus*. Observe their conspicuous size in comparison with a common serous unit (arrow). The large size of the adenocyte nuclei (compare with those of the epidermal cells) possibly results from polyploidy due to endomitosis. Labels: ep, epidermis; mg, mucous gland; sg, serous gland. Bars: A-B, D = 500 μm; B = 50 μm; C = 100 μm; E-F = 200 μm. Original.

1982) and the inguinal glands observed in the dorsal pelvic region of the leptodactylid *Physalaemus biligonigerous* (Fig. 6.5D; Delfino *et al.* 1999). Similar antipredatory glands occur also in the caudal region of some plethodontids (genus *Hydromantes*; Fig. 6.5E; Brizzi *et al.* 1994) and salamandrids (genus *Triturus*; Fig. 6.5F; Brizzi *et al.* 2000). The combination of toxin biosynthesis and effective neuromuscular release system makes these glands a remarkable example of a defensive device. From such perspective, one can imagine this mechanism being honed over evolutionary time at the expense of would-be predators. The full array of these specializations is yet to be known, and future studies on different species will permit an interpretation of their evolutionary history.

6.7 AMPHIBIAN GLANDS SPECIALIZED FOR SOCIAL COMMUNICATION AND REPRODUCTION; DEFINITION OF THE "BREEDING GLANDS"

In the lower vertebrates, smell cues corresponding to cutaneous chemosignals contribute to recognition and/or sex attraction among conspecifics, both in the aquatic and terrestrial environments. These substances, first termed "pheromones" by Karlson and Lüscher (1959), may be produced by a variety of tissue and organs. In fishes, the most commonly identified sources of pheromones are secretions of the urogenital tract and cutaneous exocrine glands (Colombo *et al.* 1980; Ali *et al.* 1987; Resink *et al.* 1987; Becker *et al.* 1992).

Gland specializations play a key role during social and reproductive activity of many amphibians too. A rich literature is available on amphibian social and reproductive patterns (reviewed by Jameson 1955; Lofts 1974; Salthe and Mechan 1974; McDiarmid 1978; Duellman and Trueb 1986) and the data point to the complexity of the situation in frogs as compared to that of salamanders and caecilians. Some of this complexity certainly is referable to anatomical specializations corresponding to cutaneous glands (Duellman and Trueb 1986; Houch and Sever 1994). A primary source of chemical cues in males and females is thought to be the product from diffuse mucous glands, but the occurrence of local gland clusters, frequently sexually dimorphic, has been reported in many urodeles and anurans. In the salamanders, peculiar cutaneous glands hypertrophy during the breeding season. Noble (1929) first reported courtship glands in the tail base of a plethodontid, *Eurycea bislineata*, and described their importance in eliciting courtship behavior from females. Presumably these specialized glands deliver sex attractants directly to the female during courtship (Sever 1989) and have a highly critical function in promoting individual mating success (Arnold and Houck 1982; Houck 1986; Houck and Reagan 1990; Houck and Verrell 1993; Houck and Sever 1994). Collectively, glands producing courtship pheromones previously were called "hedonic" glands (Gadow 1887; Noble 1927; Rogoff 1927; Noble 1929, 1931b). As the designation "hedonic" refers explicitly to pleasure-giving organs, Arnold (1977) suggested a less subjective term, "courtship" glands, actually the most used by the researchers (see Houch and Sever 1994). The courtship pheromones produced by the above specialized glands ("courtship glands") have been defined by Arnold and Houck (1982) as chemical signals that are transmitted only between potential sexual partners and only during courtship.

Among the specialized glands of ectodermal origin (although not strictly integumentary) involved in urodele reproduction also to be included are the cloacal glands occurring in most salamanders. As a rule, male cloacal glands contribute to spermatophore assemblage (Kingsbury's, pelvic and ventral glands; see: Sever 1981; Brizzi *et al.* 1990; Sever *et al.* 1990; Sever 1991; Brizzi and Calloni 1992; Brizzi *et al.* 1992a, b; Sever 1992a, b, c, d, Brizzi *et al.* 1994, Sever 1994; Brizzi *et al.* 1995a, 1996a, b, 1999), whereas females possess glands involved in sperm storage (see: Brizzi *et al.* 1995b; Sever and Brizzi 1998; Sever *et al.* 2000). In addition, both males and females may have cloacal glands that release courtship pheromones (male dorsal and vent glands, female ventral glands; see: Brizzi *et al.* 1996b; Sever 1992a, b, c, d).

Also in many anurans the males develop patches of special skin glands during the reproductive season, and on this account these glands are usually defined as breeding glands. The patches are often keratinized and in many frogs and toads lie on the thumbs and forearms (nuptial pads; see 6.7.1 and Fig. 6.6), but they can occur on virtually any other area of the body (Duellman and Trueb 1986), although in most cases they seem occur ventrally or

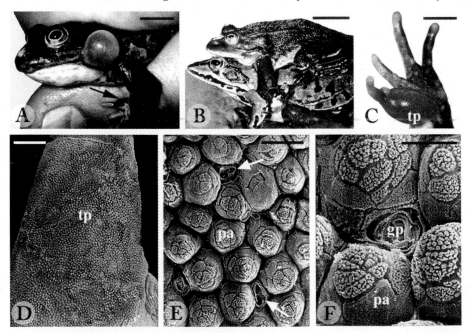

Fig. 6.6 Feature of nuptial thumb pads in *Rana*. **A–B**. Male (A) and clasping pair (B) of *Rana esculenta* (photos by G. Frangioni). Arrow in (A) points to the black epidermis covering the pad. **C**. Swollen thumb pad of male *Rana perezi*. **D–F**. Details of the pads of *R. perezi* observed under SEM. Note the highly papillate epidermis and the nuptial gland pores (arrows in E) randomly distributed among the adhesive hill-like protuberances. Labels: gp, gland pore; pa, papilla; tp, thumb pad. Bars: A = 6 mm; B = 10 mm; C = 2 mm; D = 500 μm; E = 50 μm; F = 20 μm. Original.

ventrolaterally (Fig. 6.7). Historically (*e.g.* Noble 1931a), distinctions have been proposed between the typical frog nuptial pads and other sexually dimorphic skin glands in various body regions. The differential distribution of the glands was believed to reflect functional divergence, and, consequently, no unifying function was hypothesized for these two groups of glands.

The term "breeding glands" was initially used exclusively to refer to the abdominal glands of microhylid frogs. In these anurans (*e.g. Gastrophryne carolinensis*), peculiar glands secrete an adhesive mucus that binds male and female together during amplexus (Conaway and Metter 1967). However, Thomas *et al.* (1993) redefined the preexisting term "breeding glands" to indicate, more broadly, the various sexually dimorphic skin glands (SDSG) occurring in the males of many frogs, that are structurally and histologically similar, independently of their function and their anatomical position, namely in the form of nuptial pads, glandular patches with other locations (see Table 6.1) or single glands dispersed in the dorsal skin.

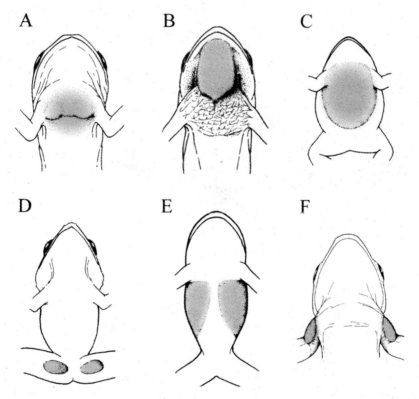

Fig. 6.7 Male glands possibly involved in adhesive functions with the female during amplexus. **A**. Pectoral glands of *Leptopelis karissimbensis*. **B**. Mental glands of *Kassina senegalensis*. **C**. abdominal glands of *Kaloula verrucosa*. **D**. Femoral glands of *Mantidactylus pseudoasper*. **E**. Ventrolateral glands of *Ptychohyla schmidtorum*. **F**. Humeral glands on arm of *Hylarana albolabris*. After Duellman, W. E. and Trueb, L. 1986. Biology of Amphibians. McGraw-Hill, New York. 670 pp, Figs 3-12, 3-13, 3-14.

Table 6.1 Original term (when existing) and anatomical localization of breeding glands in anuran males

abdominal glands

Kaloula verrucosa (Duellman and Trueb 1986)

in many microhylids, including *Breviceps*, *Gastrophryne* and *Kaloula* (Duellman and Trueb 1986). Most likely these glands correspond to those —not specifically named— reported by other authors in the belly or pectoral/abdominal skin of some microhylids (see last section of Table 1)

femoral glands

in some African ranids, such as *Petropedetes* and *Phrynodon*, in some species of *Phrynobatrachus* and Madagascaran ranids such as *Laurentomantis* and in some species of *Mantidactylus* (Duellman and Trueb 1986). Most likely these glands correspond to those —not specifically termed— reported by Thomas *et al.* (1993) in the femoral skin of *Mantidactylus betsileanus* (see last section of Table 1)

humeral glands

in some species of *Rana* and *Hylarana* (Duellman and Trueb 1986). Most likely these glands correspond to those —not specifically named— reported by Thomas *et al.* (1993) in the in the humeral skin of *Hylarana albolabris* (see last section of Table 1)

mental (gular) glands

Kassina senegalensis (Duellman and Trueb 1986)

in most hyperoliids and in members of the Neotropical *Hyla bogotensis* and Australian *Litoria citropa* groups (Duellman and Trueb 1986)

nuptial/thumb pads (or nuptial glands)

Acris gryllus (Greenberg 1942)

Bufo fowlery (Blayr 1946)

Bufo funereus (Inger and Greenberg 1956)

Bufo regularis (Inger and Greenberg 1956)

Discoglossus pictus (Bolognani Fantin and Fraschini 1965)

Leptodactylus bolivianus (Thomas *et al.* 1993)

Rana aurora (Thomas *et al.* 1993)

R. *cyanophlyctis* (Saidapur and Nadkarni 1975)

R. *esculenta* (Aron 1926; Horie 1939; Vialli 1946a, b, c; Loft 1964; Botte and Delrio 1967; Rastogi and Chieffi 1971; Botte *et al.* 1972; d'Istria *et al.* 1972; Rastogi *et al.* 1972; Delrio and d'Istria 1973; d'Istria *et al.* 1974; Chieffi *et al.* 1975; d'Istria *et al.* 1977, 1982; Brizzi *et al.* present observations)

R. *iberica* (Brizzi *et al.* present observations)

R. *perezi* (Brizzi *et al.* present observations)

R. *pipiens* (Cristensen 1931; Kermit 1931; Glass and Rugh 1944; Burgos and Ladman 1957; Parakkal and Ellis 1963; Thomas *et al.* 1993)

R. *tigrina* (Saidapur and Nadkarni 1975)

R. *temporaria* (Aron 1926; Horie 1939; Gallien 1940; Savage 1961)

Xenopus laevis (Berk 1939; Kelly and Pfaff 1976; Thomas *et al.* 1993)

pectoral glands

Leptopelis karissimbensis (Duellman and Trueb 1986). Most likely these glands correspond to those —not specifically named— reported by Thomas *et al.* (1993) in the pectoral skin of *Leptopelis flavomaculatus* and *L. viridis* (see last section of Table 1)

Contd.

Table 6.1   Contd.

Microhyla carolinensis (Conaway and Metter 1967) and *Microhyla olivacea* (Fitch 1956; Metter and Conaway 1969). These glands are also present, but not functional, in adult females and juvenile males of both species

Scaphiopus (in which both males and females possess glands) and *Megophrys* (in which glands apparently are limited to males) (Jacob *et al.* 1985). According to these authors, the same glands have been reported as axillary breast glands or glands of the thorax by Wright and Wright (1949)

postaxial (postaxillary) glands

Rana adenopleura (Duellman and Trueb 1986; Thomas *et al.* 1993)

Hymenochirus and *Pseudhymenochirus* (Duellman and Trueb 1986)

Hymenochirus boettgeri (Rabb and Rabb 1963a; Thomas *et al.* 1993).

Hymenochirus curtipes (Sokol 1959, termed "Fettpolster"; Burns and Thomas 1997; Vahamaki and Thomas 1997; Chan *et al.* 1999; Pytlik *et al.* 1999; Pearl *et al.* 2000a,b; Thomas (personal communication) in some dwarf African clawed frogs (genus *Hymenochirus*).

ventrolateral glands

Ptychohyla schmidtorum (Duellman and Trueb 1986). Most likely these glands correspond to those —not specifically termed— reported by Thomas *et al.* (1993) in the ventrolateral skin of the same species (see last section of Table 1)

Breeding glands (not specifically defined) reported in some male skin regions

in the belly skin of *Breviceps* (Wager 1965; Rabb 1973; Visser *et al.* 1982)

in the belly skin (belly glands) of male *Kaioula picta*, *K. conjuncta* and *K. rigida* (Inger 1954; Rabb 1973)

in the brachial region ("ghiandola brachiale") of *Pelobates* (Bolognani Fantin and Fraschini 1965)

in the dorsal skin of *Rana aurora* and *R. pipiens* (Thomas *et al.* 1993)

in the dorsal skin of *Rana dalmatina*, *R. iberica* and *R. italica* (Brizzi *et al.* 2002)

in the dorsal skin of *Rana pipiens* (as "cellular granular glands"; Dapson *et al.* 1973)

in the dorsal skin of *Rana pipiens* and *Xenopus laevis* (Thomas and Licht 1993)

in the dorsal skin of the hand in *Dimorphognathus* and *Hemisus* (Duellman and Trueb 1986)

in the femoral skin of *Mantidactylus betsileanus* (Thomas *et al.* 1993)

in the humeral skin of *Hylarana albolabris* (Thomas *et al.* 1993)

in the inner surface of the forearms of many hyperoliids (Duellman and Trueb 1986)

in the mental skin of *Kassina senegalensis* (Thomas *et al.* 1993)

in the pectoral/abdominal skin of *Gastrophryne* species and, in general, of microhylid frogs (Fitch 1956; Metter and Conaway 1969; Holloway and Dawson 1971; Rabb 1973; Thomas *et al.* 1993)

in the pectoral skin of several Asian pelobatids (Taylor 1962)

in the pectoral skin of *Leptopelis flavomaculatus* and *L. viridis* (Thomas *et al.* 1993)

in the snout skin of *Rana macrodactyla* and *Polypedates dennysi* (Duellman and Trueb 1986)

in the ventrolateral skin of *Ptychohyla schmidtorum* (Thomas *et al.* 1993)

On the basis of the above considerations and the wide literature pertaining to the reproductive modes in the Anura (for review see Duellman and Trueb 1986), the term "breeding glands" should include, in our opinion, all the types of cutaneous secretory structures involved in different reproductive activities such as courtship and mating. In this connection, we use the term "courtship" sensu Wells (1977a), to refer to interactions between males and females leading to pair formation and effective mating. Among the above gland specializations (breeding glands) we include also those cutaneous glands involved in later reproductive steps, namely eggs and embryos transport or incubation on the dorsum of adult frogs, as typical in the family Hylidae and Dendrobatidae (see 6.7.5). In this case, however, the definition SDSG seems partially inappropriate. At least in some species, in fact, the nurse adult frog could be either a female or a male (Wells 1977b, 1978; Weygoldt 1980; Wells 1981; Myers and Daly 1983; De Perez *et al.* 1992a). The next sections of this chapter will be devoted to a review of the different cutaneous anuran glands involved in reproductive strategies, with the aim of summarizing their main morphological characteristics and of drawing an exhaustive pattern of their possible functional significances.

6.7.1 Structure and Ultrastructure of the Breeding Glands

In all probability the nuptial pad glands occurring in the males of many anurans are the best known breeding glands, both as regards morphology and histochemical characteristics. These SDSG have been object of several study in *Rana* and *Bufo*, also in relation to seasonal changes, animal age and experimental castration (see 6.7.2-6.7.4). Historical and basic reports on these structures in the Anura are referable to: Aron (1926), Kermitt (1931), Berk (1939), Horie (1939), Gallien (1940), Glass and Rugh (1944), Vialli (1946a, b, c), Inger and Greenberg (1956), Burgos and Ladman (1957), Parakkal and Ellis (1963) Lofts (1964). Nuptial pads may occur also in the urodeles, and their development, mature stage and regression during the years have been reported in the red-spotted newt, *Notophthalmus viridescens* (Forbes *et al.* 1975). In this salamander, however, the nuptial pads consist in cell specializations of dermis and epidermis that do not involve secretory structures.

On the whole, the anuran nuptial pads also consist of dermo-epidermal specialized elements. The outer layer of the dermis has small, conical protuberances over which the *stratum germinativum* of the epidermis forms still higher ones capped with dense melanin. Usually, the cells of the *stratum corneum* are not modified but cuboidal, like those of other areas of the fingers. The cuticle, however, is distinctly thicker at the peaks of spinules than in the valleys and melanin is absent from the valleys. The cutaneous glands embedded in the pad dermis are large and close together. A nuptial pad contains 10-15 (or less) multicellular and alveolar glands, ovoid in profile and arranged in single or double rows (Figs. 6.8A-E). In the nuptial pad glands of sexually active *Rana esculenta*, Vialli (1946a, b, c) observed different secretory cell types: cells with small apical granules, cells with larger and more rounded granules and cells devoid of secretory product. A similar cell dimorphism was also noticed by Parakkal and Ellis (1963) in *Rana pipiens* through TEM observations. According to these authors, most secretory cells of the thumb pads contained apical granules and their cytoplasm showed a rich endoplasmic reticulum but scarce mitochondria and Golgi stacks. In contrast, other, not secretive cells, showed an active Golgi apparatus and rare endoplasmic reticulum strands; patterns possibly related to the annual regenerative function of these cells.

In a study on SDSG in 14 frog species, Thomas *et al.* (1993) described ordinary skin glands and nuptial pads of *Xenopus laevis* and used these data as a reference to describe distinct variations among mucous, serous and SDSG in various species and to demonstrate changes due to the androgen status of the frogs. The results indicate that, on the whole, the nuptial glands of *X. laevis* and the SDSG of the other frogs examined possess similar structural traits. In most ranids examined and in the pipid *Hymenochirus*, the glands appeared hypertrophied and forming external skin swellings. In these and most of the other species, the glands are condensed together in the skin, to the virtual exclusion of mucous and serous glands, resulting in the formation of patch-like macroglandular structures.

Further morphological characteristics of the male nuptial pads in the European frogs *Rana esculenta R. iberica* and *R. perezi* are shown in Figures 6.6, 6.8-6.12. The animals we used for this study were collected during the breeding season and, on dissection, showed large testes and Wolffian ducts with bundles of sperm. The pads on the inner fingers of their forelimbs appeared well developed and covered by a black, strongly keratinized epidermis (Figs. 6.8A-E). Observations under the scanning electron microscope (SEM) provided further details of the external feature of the pad (Figs. 6.6D-F), particularly of its highly papillate epidermis and nuptial gland pores. Sections of thumb pads observed under LM showed conspicuous mucous glands filling the dermis (Figs. 6.8B-E). The same region of *Rana iberica* (Fig. 6.8A), showed, on the contrary, a few and relatively small mucous glands dispersed in the dermis, although the pad epidermis was keratinized. This finding could be related to physiological condition of the animal (we examined only a male) or to scarcely developed pads in this species. Further observations are needed to clarify this point.

The thumb pad epidermis of *Rana esculenta* and *R. perezi* consists of several (5 to 8) layers of cells, with elongated nuclei, arranged orthogonal to the skin surface (Fig. 6.8A-E), but the in outer one. Cells in the external layer are flattened in shape (Fig. 6.8D), as a possible effect of the horny evolution of this epithelium. However, several differences are detectable, consistent with the species considered. In *R. iberica*, the outer skin surface exhibits a regular profile (Fig. 6.8A); in *R. esculenta* the surface shows hill-like protuberances (Fig. 6.8B); in *R. perezi* the epidermis forms a complex apparatus of stumped papillae (Figs. 6.8C-D). On the whole, the epidermal characteristics of *R. esculenta* and *R. perezi* agree with those reported in *Rana pipiens* (Parakkal and Ellis 1963) and *Discoglossus pictus* (Bolognani Fantin and Fraschini 1965). As a shared trait in all the *Rana* specimens we observed, the loose dermal layer underlying the epidermis contains cutaneous glands, the features consistent with epidermal characteristics. These are mucous glands, although their morphological traits are different according to the species observed. In *R. iberica* they are ordinary mucous units, both in size and cell arrangement (Fig. 6.8A). Furthermore, mucocytes appear to be in various functional phases, as shown by different cell height and secretory content. Mucous glands in the thumb pads of *R. perezi* are larger and ellipsoidal, with the major axis orthogonal to the epidermal layers (Fig. 6.8C). Sparse serous glands, recognizable by morphology and stain

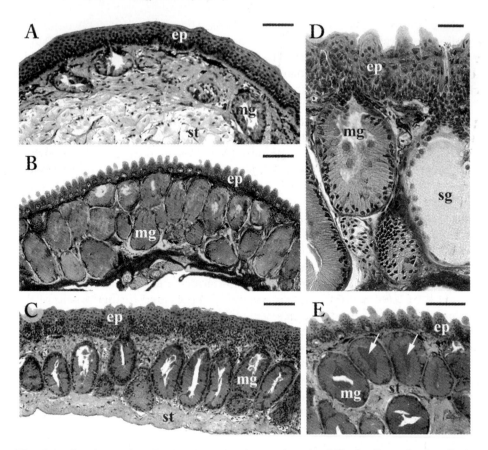

Fig. 6.8 Sections of male thumb pads observed under LM. **A**. *Rana iberica*. Both epidermis and underlying mucous glands exhibit usual features: the former has a flat external surface, the latter are scanty and consist of secretory cells in various activity stages. **B**. *Rana esculenta*. The epidermal surface is remarkably papillate and the mucous glands are numerous. The glands adhere to each other with irregular surfaces, thus resembling to be arranged in two-three layers. **C–D**. *Rana perezi*. The thumb pads consist of strongly keratinized epidermis and large mucous glands, with the major axis orthogonal to the body surface. In some sections, sparse serous glands are visible among the specialized mucous ones (D). Notice elongated mucocytes with basal nuclei. **E**. *R. esculenta*. Intermediate magnification, showing secretory product inside the gland lumina (arrows). Labels: ep, epidermis; mg, mucous glands; sg, serous glands; st, stroma. Bars: A = 50 µm; B-C = 100 µm; D = 25 µm; E = 100 µm. Original.

characteristics, are visible among the mucous units, similar to these latter only in their large size (Fig. 6.8D). In *R. esculenta* the mucous glands exhibit even larger sizes and posses an irregular shape; these units are closely packed together and, as a sectioning effect, it seems that two-three gland layers are stratified beneath the epidermis (Figs. 6.8B, E). In both *R. esculenta* and *R. perezi*, mucous secretory units display peculiar features. Mucocytes are tall, prismatic cells, with their major axis parallel to the gland lumen. Nuclei lay basally, according to an obvious functional polarization: the supranuclear cytoplasm contains larger amounts of mucous material, that is homogeneous in staining properties, when single or several cells are observed, suggesting a remarkable co-ordination of biosynthesis processes. In some instances, secretory product is visible inside the gland lumina (Fig. 6.8E).

TEM observation confirms previous LM analysis in both epidermis and gland features. Epidermal cells (keratinocytes) of the thumb pad of *Rana esculenta* and *R. perezi* exhibit tall polyhedral shapes and ellipsoidal nuclei (Figs. 6.9A,B). There is a somewhat variable cytoplasm density among contiguous keratinocytes, whereas rounded mitochondria with a light, inner compartment are common features (Fig. 6.9C). Relationships among epidermis and underlying dermis are rather complex and characterized in section by wave profiles that suggest adequate trophic flows between the two tissues. When thumb pad glands occur, the basal epidermal layer appears to be continuous with flattened cells of the gland neck (or intercalary tract; Figs. 6.9 C, D). On account of the basal-apical polarization of their secretory cells, mucous glands will be described from periphery to central lumen. However, no ultrastructural pattern will be provided on *R. iberica*, which possesses thumb pad glands with usual features. A contractile sheath, the myoepithelium, encircles the mucous secretory units (Fig. 6.10A), with smooth muscle cells characterized by a variable content of intracytoplasmic filaments . Furthermore, there is also an obvious variability in the thickness of the myoepithelial cells, which form a discontinuous layer (Fig. 6.10B). Where gaps occur, slender cytoplasmic processes emanate from the basal portions of mucocytes, and remarkably amplify their exchange surfaces. Usually the system of interstices between secretory and contractile compartments is exiguous; it is electron-transparent (Fig. 6.10B) and in repeated observations on several specimens it appears to lack any axonal profile. The basal cytoplasm of secretory cells contains the biosynthesis machinery typical of mucocytes, consisting of RER cisternae and stacked saccules of the Golgi apparatus. The former display variable shapes, ranging from slender (Fig. 6.11A), to moderately dilated (Figs. 6.10A, 6.11B), to remarkably enlarged (Fig. 6.11C). Sacculi in the Golgi apparatus are slightly dilated, and show patterns of scanty activity (Fig. 6.11D).

Mature secretory material can be observed at the nuclear level (included the basal areas), and displays typical features of mucous product (Fig. 6.12A). It consists of rounded discrete structures (mucous granules), characterized by various densities and close, reciprocal adhesion, which leads them to acquire a peculiar polyhedral shape. Patterns of secretory release can be detected around the gland lumen, which was seen to hold a structureless, rather opaque material (Fig.12B). At this apical level, the still intracytoplasmic mucous granules undergo fluidation, before to be released by means of merocrine processes (Fig. 6.12C).

According to Conaway and Metter (1967), the breeding glands of male *Microhyla carolinensis* are multicellular and restricted to the sternal region, where they are mixed with common mucous and granular skin glands. In this species the breeding glands are flask-shaped and average 70-100 µm in diameter. The secretory cells are apocrine and columnar. The secretion is released by the

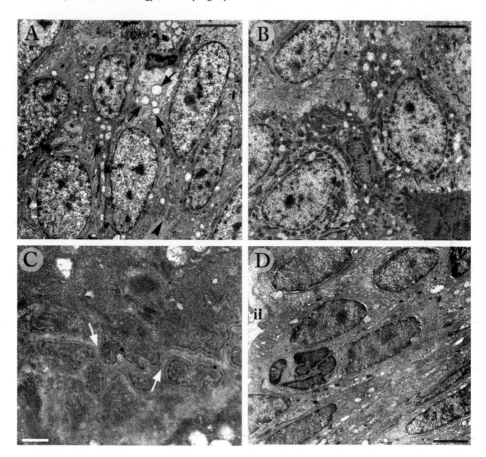

Fig. 6.9 Ultrastructural patterns of thumb pad epidermis. **A**. *Rana esculenta*. Elongated nuclei of epidermal cells. Arrows point to mitochondria with light inner chambers. Arrow-head points to the dermal-epidermal boundary. **B**. *R. esculenta*. Various electron-opacity characterizes the cytoplasm of epidermal cells. **C**. *R. esculenta*. Subepidermal area; this labyrinthine pattern (arrows) results from complex dermal- epidermal relationships. **D**. *R. perezi*. Flat, embricated cells of a cutaneous gland neck are continuous with the basal epidermal layer. Label: il, intercalary neck lumen. Bars: A-B = 5 μm; C = 1 μm; D = 3 μm. Original.

top half of the cells breaking off. As a result, the secretory cells of a spent breeding gland are cuboidal.

In the egg-brooding hylid *Cryptobatrachus*, De Pérez and Ruiz Carranza (1985) and De Perez *et al.* (1992b) attributed adhesion and transport of egg and tadpoles on the female dorsum especially to peculiar mucous glands (named G2) occurring in the female dorsal skin (see 6.7.4-6.7.5). LM and TEM observations by the above authors revealed significant differences of these glands in comparison with the common, mucous glands (named G1) also occurring in the skin together with the serous ones. G2 glands showed a lower secretory epithelium but a larger lumen, this latter containing a metachromatic

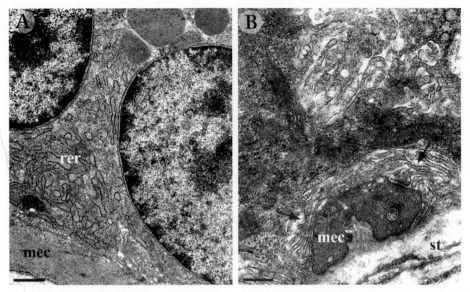

Fig. 6.10 Thumb pad glands observed under TEM. **A**. *Rana perezi*. Basal gland portion, ensheathed by myoepithelial cells; notice rer profiles with variable shapes. **B**. *R. esculenta*. Complex network of slender plasma membranes processes (arrows) characterizes the gap-areas of the discontinuous myoepithelium. Labels: rer, rough endoplasmic reticulum; mec, myoepithelial cell; st, stroma. Bars: A = 1 μm; B = 500 nm. Original.

material, foamy in appearance. The most obvious differences between the two gland types were the following: 1) secretory product (neutral glycoproteins in form of rounded and electron-dense granules in G1, acid mucopolysaccharides more irregular in shape and density in G2), and 2) mitochondrial equipment and biosynthetic organelles (more abundant in G2).

Two different types of mucous glands, possibly involved in transport of tadpoles, have been also reported in males of *Minyobates virolinensis* (De Perez *et al.* 1992a). These glands differ in the height of their epithelium, amount of secretory product and some ultrastructural details. Nonetheless, according to the above authors both gland types lack a myoepithelial sheath, which suggests a continuous secretion release.

In *Kaioula picta*, Inger (1954) found aggregations of single-celled epidermal glands distributed on the bellies. The margin of the area containing these glands is usually easily seen at naked eye. The belly gland occupies between one-half and five-sixths of the area of the abdomen and seems to produce a gelatinous secretion causing adhesion of breeding pair (see 6.7.5).

In some anurans, SDSG do not occur in form of localized patches, but are widely diffused on the body. Thomas *et al.* (1993) reported specialized glands distributed through the male dorsal skin of *Rana aurora* and *R. pipiens*. Examination of the dorsal skin of sexually mature female of the same species failed to demonstrate comparable glands, suggesting that these dorsal skin glands may be sexually dimorphic.

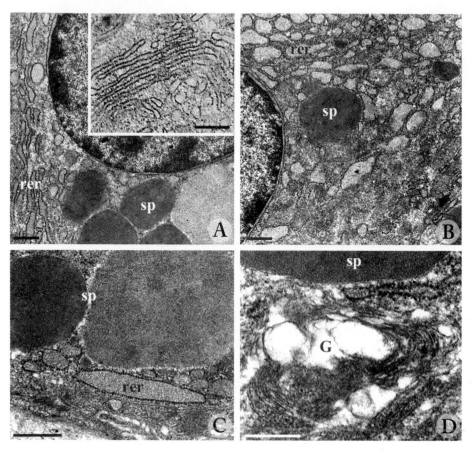

Fig. 6.11 Biosynthesis machinery in thumb pad glands of *Rana perezi*. **A** and insert: Flat profiles of RER cisterns. **B**. Moderately dilated rer profiles. **C**. Dilated RER profiles. **D**. Sacculi of a Golgi stack with light compartments. Labels: G, Golgi stack; RER, rough endoplasmic reticulum; sp, secretory product. Bars: A (and insert) = 1 µm; B-C = 1 µm; D = 500nm. Original.

Recently, Brizzi *et al.* (2002) described "specialized" mucous glands dispersed in the male dorsal skin of the trunk of three European frogs (*Rana dalmatina, R. iberica* and *R. italica*; Figs. 6.13A-C) and distinguishable morphologically and histologically from the "ordinary" mucous units occurring over the body surface in both males and females (Figs. 6.14A-B). Specialized mucous glands

Fig. 6.12 Ultrastructural features of mucous product in thumb pad glands. **A**. *Rana perezi*. Paranuclear mucous granules, characterized by variable electron- density and polyhedral shape. **B**. *R. esculenta*. The gland lumen contains structureless material with a moderate electron-opacity. **C**. *R. esculenta*. This product derives by thinning of mucous granules, followed by merocrine release (arrows in sequence). Labels: lu, gland lumen; mv, microvilli; sp, secretory product. Bars: A = 3 µm; B = 2 µm; C = 1 µm. Original.

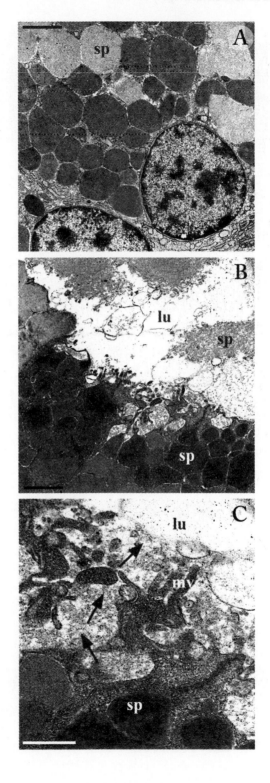

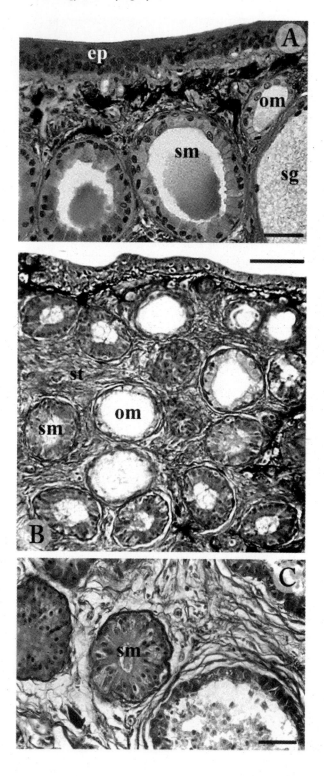

were absent in both sexes of *R." esculenta"* and *R. perezi* (Figs 6.14C,D). In all the above species, the ordinary mucous units consist of secretory cells (mucocytes) orderly arranged in a low prismatic, simple epithelium around an obvious lumen (Figs. 6.13A,B, 6.14C,D). Ordinary mucous glands observed under TEM in *R. dalmatina* (and fully corresponding to those typical of the anurans) show low to moderately tall epithelial cells encircling wide lumina (Figs. 6.15A-B). Their nuclei are arranged in a single row parallel to the stromal surface (Fig. 6.15B) and the supranuclear cytoplasm is packed with large secretory granules, irregular in shape and varying in density. In addition, secretory vesicles holding a finely granular product can be detected in the periluminal cytoplasm (Fig. 6.15B). Thin cytoplasmic screens separate the vesicles from the lumen, until they become discontinuous and secretory release is achieved. The mature vesicles contain a relatively transparent product which give the "foamy" appearance to the mucous units. A discontinuous and thin contractile sheath encircles the mucous secretory unit (Fig. 6.15B).

The specialized mucous glands occurring in the male dorsal skin of *Rana dalmatina* and *R. iberica* are very larger than the ordinary ones (Figs. 6.13A, 6.14 A,B) whereas in *R. italica* the two mucous gland types are more similar in size (Fig 6.13 B). However, all the specialized mucous glands consist of high columnar mucocytes, usually with basal nuclei (Figs. 6.13C, 6.15A,C). A dense, eosinophilic material fills their cytoplasm and in some cases their lumina (Fig. 6.13 A-C). We found only serous and ordinary mucous glands in sections from lateral, ventral and leg skin of both males and females, in accordance with the typical patterns usually reported in the anurans (see Fox 1986, for review).

In specialized mucous units of *Rana dalmatina, R. iberica* and *R. italica* observed under TEM, the mucocytes are high prismatic (Fig. 6.15C) and separated reciprocally by wide interstices containing slender and elongate plasmalemmal projections (Figs. 6.15C, 6.16A). The biosynthetic apparatus is localized in the perinuclear cytoplasm and consists of RER cisterns as well as Golgi stacks (Fig. 6.16B). The former are slightly dilated in profile and contain a moderately dense material. The latter consists of enlarged sacculi that dispatch wide vesicles. These hold a finely granular product that gradually undergoes remarkable condensation, leading to the typical ellipsoidal electron-opaque granules filling the supranuclear cytoplasm (Figs. 6.15C, 6.16C). Basal portions of specialized mucocytes contain vesicular structures, moderately opaque to transparent; resembling lipid droplets (Fig. 6.16D).

The mucous granules tend to aggregate at the luminal border, where they maintain remarkable electron-opacity (Fig. 6.16E). However, the final dilution processes render the secretory product transparent, as it is obvious in the

Fig 6.13 Male dorsal glands of *Rana* observed under LM. **A**. *Rana iberica*. Specialized mucous glands are very larger than the ordinary ones. **B**. *Rana italica*. Ordinary and specialized mucous glands can be recognized by their different cytology and staining characteristics of the secretory product. **C**. *R. dalmatina*. Specialized mucous glands show a tall epithelium. Labels: ep, epidermis; om, ordinary mucous gland; sg, serous gland; sm, specialized mucous gland; st, stroma. Bars: A, C = 30 μm; B = 100 μm. Original.

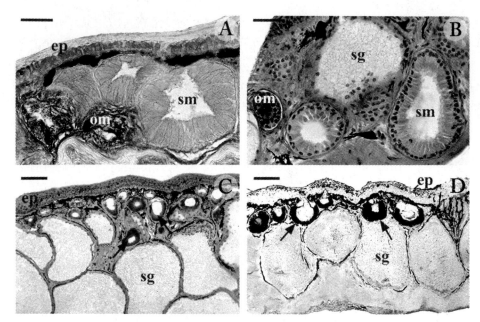

Fig. 6.14 Dorsal skin glands of *Rana* observed under LM after histochemical trials (**A-C**. Transverse sections after PAS reaction, contrasted with haemalum-orange, **D**. Uncontrasted sections after alcian blue 8GX at pH 2,5). **A-B**. Ordinary and specialized mucous glands of male *Rana dalmatina* (A) and *R. iberica* (B) are recognizable on account of their different size and histochemical response. Note that only the ordinary mucous glands are PAS positive. **C**. Serous and ordinary mucous glands of *R. perezi* (male) treated with the PAS reaction. Notice the positive response of the ordinary mucous glands (arrows). **D**. Features of the skin glands in female *R. esculenta* after the alcian blue test. The ordinary mucous glands react positively (arrows). Labels: ep, epidermis; om, ordinary mucous gland; sg, serous gland; sm, specialized mucous gland. Bars: A-B = 50 μm; C-D = 100 μm. Original.

center of the lumen. As is typical of amphibian mucous glands, the myoepithelial sheath is discontinuous and lacks any direct nerve supply. Basal interstices between specialized mucocytes and the myoepithelium exhibit short plasmalemmal projections derived from secretory cells (Fig. 6.16F), according to the usual patterns observed in ordinary and specialized mucous cells. The large gaps occurring between contiguous myocytes allow the mucocytes to reach the stromal environment through their basal projections.

Fig. 6.15 Ordinary and specialized mucous glands of male *Rana dalmatina*. TEM observations. **A**. Peripheral portions of ordinary and specialized mucous units separated by a thin stromal septum. Notice the different height of the mucocytes. **B**. Ordinary mucous glands showing low prismatic mucocytes with basal nuclei and apical secretory product. This latter varies from granules of different density to electron-transparent vesicles. **C**. In the specialized mucous glands, the mucocytes are high prismatic and separated reciprocally by wide interstices. Labels: in, interstice; lu, lumen; mec, myoepithelial cell; om, ordinary mucous gland; sm, specialized mucous gland; st, stroma, sv, secretory vesicle. Bars: A,C = 5 μm; B = 2 μm. Original.

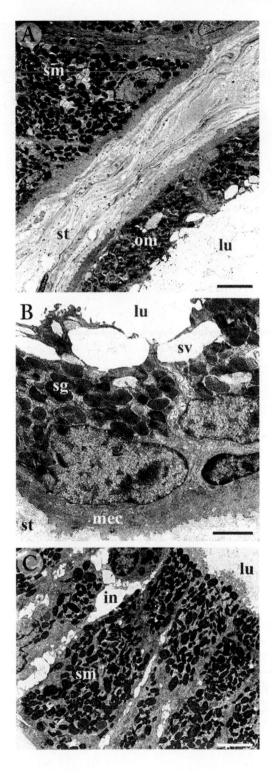

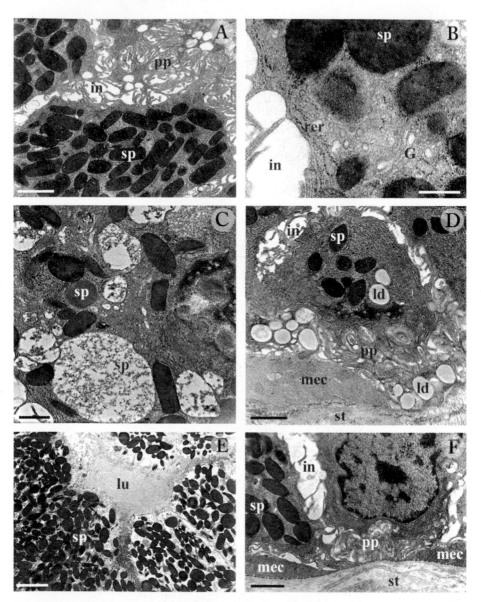

Fig. 6.16 TEM features of specialized mucous glands in male *Rana dalmatina*. **A.** The wide interstices between contiguous mucocytes contain intertwined plasmalemma projections. **B.** The perinuclear cytoplasm accommodates the protein synthetic machinery, consisting of slender rer profiles and Golgi stacks. **C.** Granules of varying electron-density fill the supranuclear cytoplasm. **D.** Lipid-like droplets are common structures in the basal portions of these glands. **E.** The mucous, secretory granules tend to aggregate at the luminal border, where they maintain remarkable electron-opacity. However, the final dilution processes render the secretory product transparent, as it is obvious in the center of the lumen. **F.** Basal interstices between mucocytes

Contd.

6.7.2 Histochemistry of the breeding glands

From the wide literature on frog skin glands (*e.g.* Dapson 1970; Dapson *et al.* 1973; Hostetler and Cannon 1974; Cannon and Hostetler 1976), it is well known that the ordinary mucous glands contain neutral and acidic mucosubstances and lack proteinaceous and lipid secretions. In contrast, serous glands possess bound lipids, proteins and sometimes biogenic amines. In comparison with the extensive histochemical data available on ordinary skin glands in the frogs, few studies, concerning the histochemistry of the nuptial pads and other anuran SDSG, have been performed until 1993. In that year, Thomas *et al.* reported observations on specialized glands in 14 frog species, representing six families (see 6.7.1).

Nonetheless, some histochemical data on SDSG (mostly nuptial pad glands) were also provided by previous authors: Vialli (1946a, b, c) in *Rana esculenta*, Parakkal and Ellis (1963) in *Rana pipiens*, Bolognani Fantin and Fraschini (1965) in *Discoglossus pictus*. On the whole, the result of disparate histochemical tests suggested the mucous nature of these glands (see 6.7.4).

In male of *Microhyla carolinensis*, Conaway and Metter (1967) identified three distinct types of integumentary glands (mucous, granular and breeding glands) and their secretions stained very differently. The material in the breeding glands did not react with mucoid stains (Mayer's mucicarmine and metachromatic toluidine blue), whereas the typical mucous glands were positive to both stains. The staining reaction of the breeding glands was also quite different from that of the poison glands. The breeding gland secretion stained very intensely with Biebrich scarlet and orange II in a modified Shorr's stain, whereas the poison glands were mainly non-reactive. In summary, the results of Conaway and Metter (1967) suggest that breeding gland secretion is not mucus. However, according to Holloway and Dapson (1971), if the pH of the toluidine blue and mucicarmine tests used by Conaway and Metter (1967) was below 3.5, electrostatic binding of dye to carboxylic acids was probably prevented. Sulfated mucus would continue to stain below this pH. This same mechanism might explain staining of the basophilic breeding glands by Shorr's acid dyes. The low pH (2.6-2.7) would prevent dissociation of carboxylic groups in the breeding gland secretion. Conversely, any basic groups in the secretion would ionize at low pH, and react with the acid dye molecules. On these grounds, Holloway and Dapson (1971) concluded that, except for an absence of sulfate groups, breeding gland secretions are histochemically more similar to mucus than to granular product.

On the basis of a wide histochemical study, Thomas *et al.* (1993) stated that nuptial pads of *Xenopus laevis* are not consistently like either mucous or serous glands. In this anuran, the nuptial glands possess neutral mucosubstances and

Fig. 6.16 *Contd.*

and myoepithelium exhibit short cytoplasm projections from secretory cells. Labels: in, interstice; G, Golgi stack; ld, lipid droplet; lu, lumen; mec, myoepithelial cell; pp, plasmalemma projections; rer, rough endoplasmic reticulum; sp, secretory product; st, stroma. Bars: A,D,F = 2 μm; B-C = 1 μm; E = 5 μm. Original.

sulfated proteins, indicative of neutral mucoproteins (Pearse 1985), whereas ordinary mucous glands possess acidic and neutral mucosubstances, and serous glands possess proteins and bound lipids. According to Thomas *et al.* (1993), the differential staining of mucous glands with ninhydrin-Schiff (Bancroft 1975) and PAS, clearly demonstrates that the positive response obtained for proteins in nuptial pads is not simply due to periodic acid-sensitive moieties being present in nuptial glands. Likewise, the positive result observed for nuptial gland secretions using PAAB (proteins—ninhydrin-Schiff, performic acid-alcian blue; Bancroft, 1975) independently supports the presence of proteins in the nuptial glands. Neither the ninhydrin-Schiff nor the PAAB test alone can firmly demonstrate proteins. However, because these tests stain for different chemical compounds within protein molecules (free amino groups and sulfhydryl groups, respectively), together these tests give strong evidences for the proteinaceous nature of nuptial pad secretions. In the same study, Thomas *et al.* (1993) reported that the SDSG of the other 13 species reveal histochemical profiles similar to that of the nuptial pads of *X. laevis*, apart from some species-specific histochemical differences observed for mucous, serous and SDSG. It is also possible that these variations may depend to the use of museum specimens, some of which with uncertain fixation history. However, in all specimens with well-preserved skin glands, the SDSG could be distinguished histologically from mucous and serous glands using criteria based on the histochemistry of *X. laevis* nuptial pads. Notably, the variations observed by Thomas *et al.* (1993) in SDSG staining were independent of both the anatomical location of the glands and phylogeny of the species and this evidence fits with other minor histochemical differences observed among serous and mucous glands in several anurans (compare Dapson *et al.* 1973 and Cannon and Hostetler 1976).

Histochemical trials performed on the dorsal specialized glands of male *Rana dalmatina*, *R. iberica* and *R. italica* (Brizzi *et al.* 2002) revealed a mainly proteinaceous content bounded with different amount of neutral carbohydrates and carboxylated acidic glycosaminoglycans. Remarkable differences between ordinary and SDSG consisted in their responses to PAS and alcian blue, showing that the specialized ones lack secretory components positive to these stains (see Figs.14A-D). In contrast, we noticed a wider histochemical similarity between thumb glands of *Rana esculenta R. iberica* and *R. perezi* and their ordinary mucous glands. This finding agrees with the impression that, at least in these species, nuptial pads developed a low degree of specialization (mostly dimensional) in comparison with the common mucous units (see 6.7.1).

6.7.3 Androgen Dependence Changes in the Breeding Glands

It is well known that reproductive behavior depends on both a set of internal motivating factors and a set of external inducers or stimuli. In anurans the relationships among these factors are often close, and particular temporal sequences of interactions are part of normal successful reproduction (Rabb 1973). In this connection, considerable importance must be assigned to certain sexual traits not directly involved in gametogenesis or fertilization and therefore defined as secondary sexual characters (SSC). The SSC of the anurans fall into two broad categories: 1) those that tend to separate individuals of the same sex or of different species, and 2) those that bring the male and female together and keep them associated until oviposition has been completed, thus facilitating courtship and mating success. Among the various secondary sexual characters known in the animal kingdom, the skin, its pigmentation and outward appearance, play a predominant role. In amphibians, some skin characteristics may be considered SSC, since they change enormously during the breeding season and between male and female (dimorphic characters). Obviously, all these secondary sexual characters underlie hormonal and neurological control (Houch and Woodley 1995). The main internal factors have to do with the priming of receptor and effector systems by gonadotropic, gonadal and probably other hormones whose influence is reflected in sexual dimorphism traits. Within this hormonal control, cutaneous glands are certainly very important targets. Actually, testicular androgens produce physiological changes in the cutaneous glands of many vertebrates (Mykytowycz 1970) and frequently the androgen-sensitive glands are clearly distinguishable from the other skin glands (Müller-Schwarze 1967). The temporal correlation between increased plasma testosterone levels (shortly before and during the mating season) and androgen-induced glandular changes seem to indicate that the function of these glands is associated in some way with reproduction, although in many cases the specific function of the glands is obscure. In the urodeles, the development of sexually dimorphic glands, for example caudal and mental courtship glands (see Houck and Sever 1994), is well known to be dependent upon sex hormones associated with maturation and sexual activity (Noble 1931a, b; Sever 1976b; Norris 1987). Courtship gland clusters are unrecognizable prior to sexual maturity, and their size and secretory activity is much reduced outside of the breeding season (Noble 1927; Weichert 1945; Lanza 1959; Sever 1975a, b, 1989).

In anuran amphibians, there is much evidence that some exocrine glands dispersed throughout the integument are SSC. The most dramatic changes observed in the sexually dimorphic skin glands are hypertrophy and morphological differentiation. In addition, these structures undergo pronounced epidermal thickening and keratinisation in many species (see 6.7.1). The result may be the formation of localized, distinct patches of skin containing almost exclusively androgen-sensitive glands. These skin patches are consequently characterized as secondary sexual characters. In anurans that lack such obvious cutaneous modifications, it is more difficult to determine whether the skin possesses sexually dimorphic structures.

The most common examples of sexually dimorphic skin glands are the glandular nuptial pads occurring in many frogs and toads (see 6.7.1), namely modified patches of skin covering fingers, hands and/or forearms (review in Duellman and Trueb 1986). These pads develop in response to annual variation of testicular hormones in natural condition (Gallien 1940; Glass and Rugh 1944, Sluiter *et al.* 1950; Dodd 1960; Forbes, 1961; Lofts 1964; d'Istria *et al.* 1972; Delrio and d'Istria 1973; Forbes *et al.* 1975) and can be induced in both adult males and females and even tadpoles by treatment with exogenous androgens (Aron 1926; Cristensen 1931; Kermit 1931; Horie 1939; Greenberg 1942; Glass and Rugh 1944; Blair 1946; Sluiter *et al.* 1950; Burgos and Ladman 1957; Penhos and

Cardeza 1957; Botte *et al.* 1972; d'Istria *et al.* 1971, 1972; Delrio and d'Istria 1973; Chieffi *et al.* 1975; Forbes *et al.* 1975; Kelly and Pfaff 1976; d'Istria *et al.* 1977). However, it has been proved that these glands posses androgen receptor sites (Delrio and d'Istria, 1973; d'Istria *et al.* 1971, 1972, 1977, Delrio *et al.*1979; D'Istria *et al.* 1979). Although the above glandular changes may occur in the females, estrogens have never been found to stimulate the development of nuptial pad or other breeding glands in anurans (Botte and Delrio 1967; Saidapur and Nadkarni 1975; Kelly and Pfaff 1976; d'Istria *et al.* 1977). In contrast, estrogen treatment in *Rana esculenta* (Botte and Delrio 1967), *R. cyanophlyctis* and *R. tigrina* (Saidapur and Nadkarni 1975), suggested a reduction in androgen synthesis, which is reflected in the regression of the androgen-dependent secondary sexual characters, the thumb pads. In addition, cyproterone acetate (antiandrogen) and methallibure (antigonadotropic) have also been shown to cause spermatogenetic inhibition and regression of thumb pads in *R. esculenta* (Rastogi and Chieffi 1971; Rastogi *et al.* 1972).

Whereas the glandular variations in the nuptial pads are usually correlated with an externally visible epithelial keratinization, no such external changes are evident in the dorsal skin of most of these species. Consequently, there is no equally obvious indication of the occurrence of androgen dependent changes in all the anuran skin glands. Nonetheless, Some morphological parameters of the skin change after gonadectomy, and androgen treatment restores their normal aspect in both males and females. That skin variations also occur in females of *Rana esculenta*, is attributed to the high plasma levels of testosterone in females during the mating season. (d'Istria *et al.* 1974, 1982). Moreover, receptors for sex hormones have been reported in many sites of the amphibian skin (Galgano 1942; Lofts and Bern 1972; Delrio and d'Istria 1973; d'Istria *et al.* 1974; Chieffi *et al.* 1975; d'Istria *et al.* 1972, 1975, 1977, 1982), suggesting that probably the entire body skin may be a secondary sexual character (d'Istria *et al.* 1977; Delrio *et al.* 1979). In the adults of both sexes of *R. esculenta*, gonadectomy provokes a reduction in the height of epidermis and the atrophy of the mucous glands, whereas the epidermis of this species thickens and the mucous glands enlarge in response to increasing testosterone levels (d'Istria *et al.* 1977).

In the high Andean marsupial frog *Gastrotheca riobambe*, the incubatory pouch where young develop (see 6.7.5) originates from the dorsal skin under hormonal control. Jones *et al.* (1973) stimulated, through estradiol, formation of the pouch. As a result, the skin lining of the pouch developed histological characteristic also observed in not-treated mature females, including reduction of the poison glands and increase of the mucous ones.

In the frog *Gastrophryne carolinensis*, the abdominal breeding glands are dependent upon testosterone for normal development (Metter and Conaway 1969). These authors observed an increase in serous gland size in juveniles in response to testosterone treatment. Thomas and Licht (1993) demonstrated that the dorsal skin of *Rana pipiens* and *Xenopus laevis* responded differently to testosterone stimulation. None of the glands in the dorsum of *X. laevis* was affected by castration or testosterone treatment; gland size, epithelium height and mucous gland cell properties were unaltered, as previously reported in this frog by Fujikura *et al.* (1988). Those investigators found the mucous, serous, and small serous glands of this species to be sexually dimorphic, being larger in females than in males. However, whereas hypophysectomy in *X. laevis* decreased nuptial pad gland size in males, possibly by ending gonadotropin-stimulated androgen secretion, no variations were noticed in the other skin glands, suggesting that gonadal hormones do not affect the sexual dimorphism of mucous, serous and small serous gland size in this species. This dimorphism is more likely allometric, since females are much larger than males. In contrast to the skin of *Xenopus laevis*, the dorsal skin of *Rana pipiens* showed distinct differences between castrated and testosterone-treated animals. The most obvious changes were an increase in the number of breeding glands and a proportional decrease in the abundance of seromucous glands following testosterone treatment.

Adult males of the genus *Hymenochirus* show a sexual dimorphic trait consisting in the lateral post-axillary subdermal gland, visible externally as a round whitish spot or raised area where it is attached to the skin. In *H. boettgeri*, this gland is actually teardrop-shaped, about 0.5 mm thick, 1.5 mm wide, and up to 2.5 mm long (Rabb and Rabb 1963a). In periods of sexual activity the gland enlarges, the skin about it becomes more vascularized, and the center of the gland appears as a reddish spot. The gland's products are presumably shed to the outside at such times. Similar glands also occur in *Pseudhymenochirus merlini* and *H. curtipes*, although erroneously termed "Fettpolster" by Sokol (1959).

Vahamaki and Thomas (1997) observed that adult males of *Hymenochirus curtipes* experienced a reddening and swelling of the sexually dimorphic postaxial skin glands when treated with GnRHa (GnRH-ethyl amide). In addition, Pytlik *et al.* (1999) observed increase of the postaxillary sexually-dimorphic skin glands in male *Hymenochirus curtipes* injected with testosterone and suggested that changes in the breeding glands (hypertrophy and hyperemia) during mating may be testosterone-induced. Recently, the same above changes have been obtained by Thomas (personal communication) in males of *Hymenochirus curtipes* injected with 25 microliters of a synthetic GnRH agonist and compared with other males treated only with 25 microliters of amphibian Ringer saline (Figs. 6.17A-F).

Conaway and Metter (1967) and Metter and Conaway (1969) reported functional "breeding glands" on the pectoral region of male *Microhyla carolinensis* and *Microhyla olivacea*, respectively. These glands were also observed, although not functional, in adult females and juvenile males of both species. On the basis of hormonal treatments, Metter and Conaway (1969) concluded that the secretory activity of *Microhyla* "breeding glands" is apparently entirely dependent on male gonadal hormones. They observed glandular regression in castrated males and development to a functional state in juvenile males and all females treated with testosterone. The fact that a secondary sex characteristic is present in both sexes and is controlled by gonadal hormones is not surprising in view of the well know cases in other vertebrates. However, also the functional

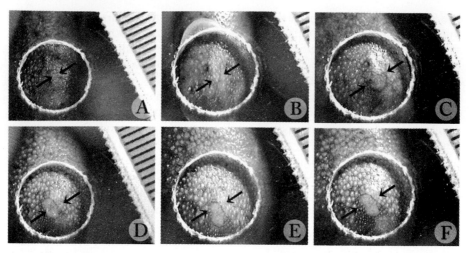

Fig. 6.17 Effects of hormonal treatment on the postaxillary glands in male *Hymenochirus curtipes*. **A–C**. One male was injected s.c. with 25 microliters of a synthetic GnRH agonist (12.5 micrograms/ml in amphibian Ringers saline) at "time zero". **D-F**. Another male was injected s.c. with 25 microliters of amphibians Ringers saline at "time zero". The frogs were photographed at time zero (A and D), after 90 minutes (B and E) and after 210 minutes (C and F) from injection. The frog treated only with Ringers solution showed no change in the size or appearance of its breeding gland over the 210 minute period (D-F). In contrast, the frog treated with the GnRH solution showed an increase in size and blood perfusion of the breeding gland even after 90 minutes (Fig. B), and the effect was still evident after 210 minutes (Fig. C). The ruler scale to the right of each picture is 1 mm. Photos and data kindly provided by Eric O. Thomas.

significance of many androgen responsive SDSG in male frogs remains to be determined. Clearly, not all these glands can function in a mechanical context as might be the case for the thumb glands, so that the roles of these latter glands have not been fully elucidated.

6.7.4 Nature of the Breeding Glands

The literature on the nature of breeding glands is rather sparse and some contradictory indications do not help to draw an exhaustive pattern. In 1946, Vialli (a, b, c) described the thumb pad secretion of *Rana esculenta* as "mucoid" in nature. A similar opinion was also shared by Bolognani Fantin and Fraschini (1965) for the same glands in *Discoglossus pictus*, whereas other authors simply referred the mucous appearance of the above glands: Parakkal and Ellis (1963) in *Rana pipiens*, Inger and Greenberg (1956) in *Bufo regularis* and *Bufo funereus*.

More recently, Thomas and Licht (1993) described in *Rana pipiens* and *Xenopus laevis* four different types of skin glands. In addition to the conventionally defined serous and mucous glands, seromucous glands (sensu Mills and Prum 1984) and breeding glands (sensu Thomas *et al.* 1993) were evident in the dorsal skin. In the skin of castrated and testosterone treated frogs, Thomas and Licht (1993) observed a reciprocal relationship between the abundance of seromucous and breeding glands: in castrated, seromucous glands were abundant and breeding glands almost absent, whereas in testosterone-treated animals, breeding glands were numerous and seromucous glands less common. According to the two authors, this suggested that the two gland types are correlated, possibly representing extreme morphs of a single gland type that varies morphologically with to the androgen status of the animal. If so, seromucous glands may be the regressed forms of breeding glands. Further indication for this suggestion was the presence of glands that appeared to be intermediate in form between the two types. Two factors contributed to this ambiguity: the smallest breeding glands of castrated frogs were similar in size to the seromucous glands, and some small glands were indistinguishable histologically as either gland type, although most seromucous and well developed breeding glands could be recognized using PAS and Alcian Blue (AB) tests. Typically, seromucous glands are stained by both PAS and AB, and well-developed breeding gland cells are uniformly stained with PAS but not with AB. Such a pattern of PAS+/AB- is also reported in developed nuptial pad breeding glands (Thomas *et al.* 1993). The ambiguous glands possessed some AB+ material in the gland lumen, but lacked clearly defined AB staining in the cells. A similar condition has been also described in regressed nuptial pad breeding glands by Thomas *et al.* (1993). The changes in the frequencies of breeding and seromucous glands may also explain why the status of these glands has been so vague in previous reports. Breeding glands in the dorsal skin of *R. pipiens* were also reported by Dapson *et al.* (1973) but at the time the glands were named "cellular granular glands" because they were considered more closely related to serous than to mucous glands. Unfortunately, the sexes of the animals used in this study was not reported. On the other hand, in a survey of ranid mucous glands (Mills and Prum 1984), no mention was made of breeding glands as a type of skin gland in *R. pipiens*. These authors described "seromucous glands" as a second type of mucous glands, with morphological and secretory characteristics unique from other mucous glands. Nothing equivalent to seromucous glands was reported in studies of the mucous glands (Dapson 1970) and serous glands (Dapson *et al.* 1973) in *R. pipiens* skin.

No doubts exist as to the mucous nature of the nuptial pad glands and other SDSG observed by Brizzi *et al.* in some European *Rana* species (see 6.7.1), as revealed by LM and TEM observations. The biosynthetic machinery of the secretory cells, the features in the maturation steps, and the morphological traits of the secretory granules, all suggest that these glands belong to the same mucous gland line. Some common features are also obvious in the contractile apparatus. Their peripheral myocytes, lacking any direct innervation, strongly resemble the myoepithelial cells of skin mucous glands, both in their scanty thickness and poorly developed sarcoplasmic reticulum (Bani 1976).

However, a comparison between thumb glands and others mucous units of Ranidae, both the ordinary and the diffuse specialized ones, reveals some differences. In particular, the specialization degree of the thumb pad glands appears moderate. Actually, the most impressive morphological characteristic of these glands merely consists of their remarkable large sizes, although the synchronous biosynthesis processes represent a more informative trait, relevant to their secretory activity. Other features resemble morphological traits usual in the mucous glands, namely low electron opacity and weak consistence of granules (which become polyhedral in shape when crowded together), lack of direct nerve supply and discontinuous, thin myoepithelium. In contrast to specialized mucous glands scattered throughout the skin of several *Rana* species (Brizzi *et al.* 2002), mucocytes in thumb pad glands do not exhibit any lipid droplets; therefore it seems that these secretory organs are not engaged in the production of steroid compounds, as possible molecules involved in chemical communication.

On the whole, patterns of mucous glands are also typical of sexually dimorphic (hedonic) glands of some urodeles (Brizzi *et al.* 2001), such as the dorsal glands of the Salamandridae (Brizzi *et al.* 1986 ; Delfino *et al.* 1988b; Sever *et al.* 1990; Sever 1992b; Brizzi *et al.* 1996a, b , 1999), the vent glands (Sever 1988, 1994) and the mental glands (Sever 1975b, 1976a; Borgioli 1977; Williams 1978; Testa Riva *et al.* 1993; Brizzi *et al.* 1994; Houck and Sever 1994) of the Plethodontidae. Despite their different localization —along the cloacal border (dorsal and vent glands) or on the chin (mental glands)— physiological and ethological studies indicate that in the salamanders these male glands are sources of substances for female attraction and/or persuasion during courtship (Salthe and Mecham 1974; Arnold 1977; Arnold and Houck 1982; Duellman and Trueb 1986; Houck 1986).

Scanty literature data are available to allow a comparison between the specialized glands of the urodeles and glands pertaining to the ordinary mucous type. Truffelli (1954) was the first author to point out a close relationship between the mucous and mental glands in some Mexican and South American urodeles, and Sever (1976a, b) experimentally demonstrated that mental glands are modified mucous glands. In addition, some histological evidence collected from dorsal, mental and vent glands strongly suggest their mucous nature in other urodeles. Nevertheless, it should be noted that dorsal and mental glands of the urodeles are equipped with a direct nerve supply, quite unusual for mucous units. Since the peculiar mucous glands described in *Rana* lack direct innervation, it seems that these secretory organs have acquired a relatively low degree of functional specialization, at least in comparison with those observed in the urodeles.

Some uncertainty still exists as to the nature (and function) of cutaneous specialized glands in other anurans. The pectoral glands of the pelobatids *Scaphiopus* and *Megophrys* (see Table 6.1) have been described by Jacob *et al.* (1985) as granular glands occurring in paired masses on the breast, below the insertion of the arm, almost suggesting mammae. The main structure of these glandular clusters appeared similar to the parotoids of pelobatids and bufonids, containing granular glands primarily and a few smaller mucous glands.

Thomas *et al.* (1993) reported that the SDSG femoral glands of the ranid *Mantidactylus betsileanus* are histochemically and structurally similar to the serous glands in this species. This suggests that the femoral glands of *M. betsileanus* could be macroglandular clusters of serous glands, not SDSG. These glands require more study to determine whether they are, in fact, a secondary sexual character as in other *Mantidactylus,* as described by Duellman and Trueb (1986), or, alternatively, specialized serous glands possibly involved in peculiar functions. However, in the leptodactylid *Physalaemus biligonigerus* Delfino *et al.* (1999) reported, in addition to the ordinary granular glands, peculiar clusters of large serous units defined "inguinal glands". These secretory structures are located in the dorsolateral areas of the pelvic girdle of both males and females. An ultrastructural analysis of the above glands suggested that they belong to the serous type and represent a gland population specialized in the storage of remarkable amounts of product. The large size and condensing activity of the secretory units optimize their storage capabilities. Altogether, the specialization for secretory storage, the localization in the dorsal pelvic region (affected by predation during the flight reaction) and the neural control on secretory release (typical of the serous gland type) suggest that the inguinal glands play a role in chemical defence.

According to Thomas *et al.* (1993), the ventrolateral glands of the hylid *Ptychohyla schmidtorum* are also histochemically more similar to serous glands than SDSG, but do not possess the same structure as the serous glands in *Mantidactylus betsileanus.* Other hylids must be examined to determine whether the ventrolateral glands of *Ptychohyla schmidtorum* are typical of this family and whether these glands are serous or SDSG (as in *M. betsileanus*). Moreover, In *Microhyla carolinensis* (Conaway and Metter 1967) the secretion of pectoral breeding glands stain differently from either mucus or the secretion of the granular glands, indicating it is a distinct substance. Nonetheless, Metter and Conaway (1969) stated that the "breeding glands" of *Microhyla* are a derivation of mucous rather than granular glands. This assumption is based on the similarity of regressed and undeveloped "breeding glands" to immature mucous glands. In 1971, Holloway and Dapson reported in *Gastrophryne carolinensis* demonstrates that breeding gland secretion is histochemically more similar to mucous than to granular secretion (see 6.7.2).

More congruent histological reports exist about the nature of the cutaneous glands involved in brood incubation, a reproductive specialization typical of hylids and dendrobatids (see 6.7.5 and Figs. 6.18A-F). In these frogs, the secretion of well characterized mucous glands in the adult back skin suggests adhesion and transport of eggs and tadpoles (Del Pino 1980; Myers and Daly 1983; De Pérez and Ruiz-Carranza 1985; De Pérez *et al.* 1992a, b). In addition, De Pérez *et al.* (1992b) observed an obvious increase of the mucous glands size (named G1 and G2), a major height of their mucocytes and a more intense secretory activity in the skin of incubatory females of *Cryptobatrachus boulengeri,* which supports the hypothesis that just these glands (one or both types) play a role in this reproductive strategy. In *Gastrotheca riobambae,* Jones *et al.* (1973) and Del Pino *et al.* (1975) noticed a reduction of the serous glands correlated to an increase of the mucous ones in the maternal pouch of incubatory frogs. Similar characteristic were reported by Del Pino (1980) in different species of the subfamily Hemiphractinae. In one museum specimen of *Stefania scalae,* the same author observed that the PAS positive material forming the egg matrix on

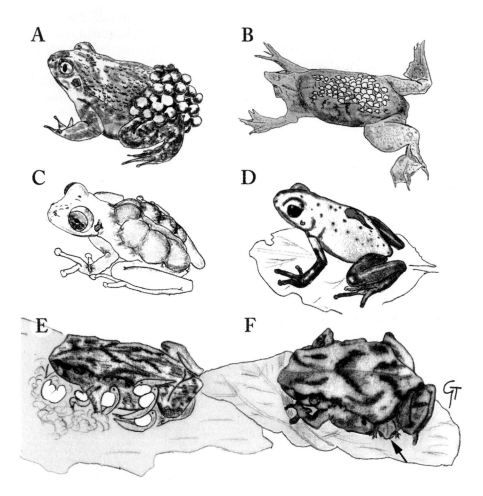

Fig. 6.18 Examples of mobile parental care involving transportation of eggs and/or larvae by male or female frogs. **A**. Male *Alytes obstetricans* with eggs adherent to hind limbs. **B**. Female *Pipa pipa* with eggs imbedded in dorsum. **C**. Female *Flectonotus pygmaeus* with eggs in the dorsal pouch. **D**. Female *Dendrobates pumilio* carrying a tadpole adhering to the dorsal skin by a sticky mucus. **E, F**. Male *Assa darlingtoni* with embryos entering in its lateral pouches (E). About 50-70 days after embryo incorporation, the first froglet emerges from the male pouch (arrow in F). Pencil drawings by G. Tanteri.

mother integument was continuous with the secretion found in the ducts and lumina of cutaneous mucous glands.

That, at least in some instances, SDSG can be related to serous glands (according to Dapson *et al.* 1973) or mucous glands (*e.g.* Metter and Conaway 1969, Thomas *et al.* 1993; Brizzi *et al.* 2002) demonstrates the prevailing ambiguity regarding the exact nature of these glands, as well as the difficulty in establishing an evolutionary relationship between SDSG and ordinary mucous or serous cutaneous glands (Quay 1972). On the basis of more general criteria, Thomas *et al.* (1993) suggest that the nuptial pads and the SDSG observed in many frogs represent a common manifestation of cutaneous exocrine glands associated with reproduction in frogs and should be classified as a unique gland type, apart from mucous and serous glands. However, a wide comparison of SDSG in anurans and urodeles (Brizzi *et al.* 2002) suggests that in both these amphibian orders the production of chemosignals for social and reproductive interactions is mainly a prerogative of the mucous gland line.

6.7.5   Possible Functions of the SDSG in the Anura

Many studies on courtship and mating behaviors of anurans and urodeles indicate that the most obvious difference between these two amphibian orders is that they rely upon different sensory modes for intraspecific communication. While both groups use visual and tactile cues, most toads and frogs depend primarily on auditory signals during the mating season, whereas urodeles use pheromonal signals (Jaeger and Gergits 1979; Jaeger 1986; Halliday and Tejedo 1995; Kikuyama *et al.* 1995; Sullivan *et al.* 1995, Dawley 1998; Rollmann *et al.* 1999). This difference is so pronounced that Houck and Sever (1994) suggested that chemical cues may be the salamander counterpart of auditory vocalization in anurans. According to these authors, pheromonal communication of the urodeles during courtship can strongly influence male mating success. The diversity and phylogenetic persistence of courtship pheromones attests to the evolutionary significance of chemical persuasion. A rich bibliography on SDSG in frogs and toads suggests that, also for these amphibians, a variety of integumentary specializations play a critical role in the reproductive activities. Pheromone-producing courtship glands, however, are not the only dermal specializations influencing mating success in Anura. In species that engage in amplexus, for example, some specializations increase the male's chances of physically retaining a potential mate until she has been inseminated or until oviposition and fertilization have occurred. These specializations include glandular "adhesives" and dermal keratinizations. On the whole, functions that might be referable to the anuran breeding glands are the release of individual or species recognition factors, mate attractants, or pheromones that enhance receptiveness or induce oviposition. In addition, the product of some specialized cutaneous glands seems to facilitate the male in clasping the female, in holding the eggs and in developing embryos on the skin of the dorsum in those species which incubate their young on the back. More details on the possible skin gland role in anuran reproduction are reported below.

Cutaneous glands as possible source of mate attractants. In 1999, Wabnitz *et al.* demonstrated that male *Litoria splendida* produces a polypeptide with mate-attracting properties. This substance, named "splendipherin" is the first sequenced anuran pheromone and its discovery supports earlier works (see below) suggesting that frog and toads also produce mate attractants. However, splendipherin was identified from secretion of *Litoria* parotoid and rostral glands, that occur in both males and females. In contrast, other studies have related production of mating pheromones (Duellman and Trueb 1986; Thomas *et al.* 1993) to sexually dimorphic exocrine skin glands of male anurans ("breeding glands" *sensu* Thomas *et al.* 1993). These data indicate that the relation between production of mate-attractant chemosignals and breeding gland function is complex and, in many cases, still an open question. In some species, the breeding glands undergo such rapid and dramatic changes in appearance and size during reproductive behavior (Rabb and Rabb 1963a), and in response to sexual hormones, that these authors suggested that the post-axillary glands of *Hymenochirus boettgeri* may produce mate-attractant or male repellent pheromones. In the meantime, because the glands are distinctly colored, the same authors do not exclude the possibility that these glands may function as visual cues during conspecific recognition. Burns and Thomas (1997) demonstrated experimentally that male *Hymenochirus curtipes* are able to attract conspecific females, but these authors did not determine whether the attraction was gender-specific or test the breeding glands as the possible source of the mate attractant. Similar results were reported by Chan *et al.* (1999), confirming that reproductive females "purposefully" track and follow the "odors" of males. In their work, Burns and Thomas (1997) suggested that female *Hymenochirus* were either attracted to a mate attractant from males and an aggregation chemosignal from females or attracted to a single gender-neutral social aggregation pheromone. Using a two-choice aquatic Y-maze, Pearl *et al.* (2000a, b) demonstrated that pheromonal substances released by the male's postaxillary breeding glands of *Hymenochirus curtipes* attract sexually active females. To establish that the positive chemotaxis toward males was gender-specific, Pearl *et al.* (2000b) studied the response of females to conspecific females, both in the absence and presence of males in the contralateral arm of the maze, as well as the response of males in the maze. The results demonstrated that females *Hymenochirus* are specifically attracted to males and that the chemosignal is gender-specific. In addition, breeding gland ablation was sufficient to abolish the male's attractive potential, and isolated breeding (postaxillary) glands were more attractive to females than glandless males. It is likely that the breeding gland is the primary, and possibly the sole, source of the mate-attractants.

A possible interpretation as mate attractants for the specialized mucous glands reported by Brizzi *et al.* (2002) in *Rana dalmatina, R. iberica,* and *R. italica* (see 6.7.1) comes from some biological data on these frogs. As reported by Lanza (1983) and Barbadillo Escriva (1987), males of the above three species lack vocal sacs, sexually dimorphic structures occurring in most anurans and usually involved in conspecific acoustic signals (Littlejohn 1977). Typically, anuran males use "advertisement calls" (Wells 1977a,b; Halliday and Tejedo 1995) to attract females as well as to announce occupied territory to other males. Since vocalization plays an important role in social interactions related to reproductive success, lack of vocal sacs in some species, possibly due to secondary loss, may be replaced by the evolution of alternative strategies, such as production of chemical signals released by sexually dimorphic mucous units. Further data point to the above specialized mucous glands of *Rana* as pheromonal sources. A consistent trait of these glands, just reported in SDSG of some urodeles (Brizzi *et al.* 1986; Delfino *et al.* 1988b), is the occurrence of moderately electron-opaque granules in the basal subnuclear cytoplasm. According to ultrastructural and cytochemical criteria, in salamandrid dorsal glands these granules may hold lipids, which are possibly residual molecules from the steroid metabolism of pheromones produced by these glands. The same functional correlation may be hypothesized for the dorsal breeding glands of *Rana* (Brizzi *et al.* 2002). Although lipids (steroids) may be component molecules of their product, we never observed SER complements in *Rana* specialized mucous glands. However, this organelle, directly involved in lipid biosynthesis, is hardly detectable in mucocytes filled with secretory granules.

Cutaneous gland secretions as "glue" for clasping pairs. In some frogs, males possess dermal glands that cause them to adhere to the female during amplexus. These glands may be located in various ventral regions (see Fig. 6.7) and secrete a glue-like substance that positions the male in a manner facilitating fertilization (Inger 1954; Fitch 1956; Wager 1965; Conaway and Metter 1967; Holloway and Dapson 1971; Duellman and Trueb 1986). Such glands occur in many species in which the relatively smaller size of the male hinders effective axillary or inguinal embrace ("amplexus by gluing"; see Jurgens 1978 ; Figs. 5-8 in Rabb 1973). Males of the African microhylids of the genus *Breviceps* are glued to the posterior part of the dorsum of the female (Wager 1965; Fig. 3.25E in Duellman and Trueb 1986) and in this position they both dig with their hind feet and may remain attached for 3 days. On the other hand, secretions of these glands may also prevent sexual interference to the extent that rival males are thwarted in attempts to remove the glued male from the female. In addition, females of some species (*e.g. Breviceps gibbosus*) possess patches of glands on the dorsal or ventral surface of the body (Visser *et al.* 1982; Duellman and Trueb 1986). Whether female gland patches are homologous with those of males is unknown and functional studies on this topic are needed. In a histological study, Jacob *et al.* (1985) found that, in the pelobatids *Scaphiopus* (in which both males and females possess specialized glands) and in *Megophrys* (in which glands apparently are limited to males), the pectoral glands are granular glands occurring in paired masses. The selective advantage for the occurrence of the above gland clusters on the venter of these pelobatids remain to be determined.

In *Microhyla carolinensis* and *M. olivacea* (Conaway and Metter 1967; Metter and Conaway 1969), male specialized glands occurring in the sternal area exhibit slight morphological modifications of mucocytes but significant changes in their secretory products. During amplexus, this specialized secretion possibly interacts with secretory materials of ordinary glands localized in the male ventral skin and in the dorsal one of females. As a result of such blending, a peculiar multi-component glue originates and solidly fastens male and female together. Due to their role, these glands have been labeled as "breeding glands". Conaway and Metter (1967) observed clasping pairs of *Microhyla carolinensis* still adhering to each other two hours after collection, despite the fact that the short-armed males were not actually gripping the females. Skin secretions from male pectoral glands appeared to be holding the pairs together, as also described for *M. olivacea* by Fitch (1956). Preservation in 6% formaldehyde failed to separate the pairs in amplexus. The same gland function was confirmed by Metter and Conaway (1969) in *Microhyla carolinensis* and *M. olivacea*.

In the males of *Kaioula picta*, Inger (1954) found belly glands that seemed to produce a gelatinous secretion causing adhesion of breeding pair. According to Fitch (1956), the adhesion through gland secretion of breeding pairs of *Microhyla* has survival value. Alternatively, Conaway and Metter (1967) suggested that the secretion allows the continuation of breeding when the pair is disturbed. In addition, Conaway and Metter underlined that the male of *Microhyla carolinensis* has such a rotund body and such short arms that it would be very difficult for him to maintain his position on the female until mating was completed without some means of adhesion.

Another kind of integumentary specialization that functions in promoting a male's ability to retain the female during amplexus involves keratinized excrescences on the limbs (Fig. 6.6). These structures are termed nuptial pads and are well evident during the breeding season on the male thumbs, or other portions of hand and/or arms (see Fig. 3.9 in Duellman and Trueb 1986), thus increasing the gripping surface (Inger and Greenberg 1956; see also figures in Forbes *et al.* 1975 and Cei 1980). In some species (for example *Bufo regularis* and *B. funereus*; see Inger and Greenberg 1956) the nuptial pads occur on more fingers and are located upon warts, each of which bears one or more pigmented spinules. In addition, patches of spinous skin may occur at the knee, outer border of the leg and plantar surface of the foot (see Fig. 3.10 in Duellman and Trueb 1986). It is possible that these spines assist the male in holding and stimulating the female, although a role as an isolating mechanism of sympatric species has also been proposed (Noble 1931; Inger and Greenberg 1956).

As previously reported (see 6.7.1), specialized, mucous glands (nuptial or thumb glands) release their product only onto the keratinized surface of the pads, which indicates an adhesive function of the secretion. Our study on thumb pads in *Rana* also emphasizes the mechanical role of these structures, related to the epidermal specializations (namely hillocks and ridges). The adhesive properties of the mucous-like secretory product act like a glue on the thumb surface. Such performances seem to be synergic with the complex ridge apparatus, capable of working like a multi-lamellar sucker.

In this connection, Bolognani Fantin and Fraschini (1965), suggested a possible relation between the lumbar embrace in *Discoglossus pictus* and occurrence of hook-like shaped nuptial pads on first and second fingers, possibly to strengthen this precarious breeding mode (in comparison to the axillary one). However, ecological evidences point to the following functional interpretation. Nuptial pads are secondary sexual characters typical of "explosive breeders", namely species that form very dense aggregations of reproductive individuals, frequently in temporary and/or small pools (Wells 1977b). As a rule, once a male had seized a female, he would have to guard against attempted takeovers by unpaired males. This may favored the evolution of enlarged nuptial pads, as also suggested by Savage (1961) in his study of *Rana temporaria*. According to Wells (1977b), it seems unlikely that such pads would be needed simply for holding the female, because females are usually passive and do not attempt to escape from clasping males. Selection of structures for guarding the female probably also explains the tenacity of males holding onto dead females, bits of floating debris, and other inappropriate objects. Houck and Sever (1994) describe the function of nuptial pad as "facilitating" reproduction, which implies that the evolution of these structures has been guided by natural selection. Alternatively, sexually dimorphic characters such as nuptial pads may have evolved by sexual selection; these characters influence male-male competition for mates and promote the mating success of an individual male by reducing the opportunity for sexual interference by rivals.

Cutaneous gland secretions as parental care adaptations. Integumentary specializations influencing an individual's ability to locate a potential mate and enhance female receptiveness (*e.g.* sexual attractants), as well as mechanisms affecting success during courtship and mating (*e.g.* epidermal modifications or gluing secretions), are not unique to the amphibians, as reported by researches carried out in both invertebrate and vertebrate organisms (for review see: Karlson and Lüscher 1959; Forbes 1961; Müller-Schwarze 1967; Arnold and Houch 1982; Houck 1986; Halliday 1990; Halliday and Tejedo 1995).

Table 6.2 Specialized cutaneous glands possibly involved in eggs and embryos incubation or transport on the back of the nurse frog

in the female dorsal skin of *Stephania* (Rivero 1970; Duellman and Maness 1980)
in the middorsal region of the trunk of female *Cryptobatrachus boulengeri* (in form of specialized mucous glands named G2; De Pérez and Ruiz-Carranza 1985; De Pérez *et al.* 1992a)
in the dorsal skin of adult males of *Minyobates virolinensis*, in form of two different types of mucous glands (De Pérez *et al.* 1992a)
in the maternal pouch of the marsupial frog *Gastrotheca riobambae* (Jones *et al.* 1973; Del Pino *et al.* 1975; Del Pino 1980)
in the maternal pouch of the marsupial frogs *Amphignathodon*, *Flectonotus*, *Fritziana*, *Gastrotheca* and in the incubatory integument of *Cryptobatrachus*, *Hemiphractus* and *Stefania* (Del Pino 1980; Del Pino and Escobar 1981)
in the nurse dorsal integument of some dendrobatids as "glue" for tadpole carriage (Myers and Daly 1980, 1983)
in the dorsal, maternal integument of *Pipa pipa* (Rabb and Snedigar 1960; Rabb and Rabb 1961; Weygoldt 1976)
in the nurse back skin of *Colostethus subpunctatus* (Stebbins and Hendrickson 1959; Wells 1981).

In the Amphibia, beyond gland specializations critical in the initial stages of the reproductive process, some integumentary traits may promote the survivorship of the developing eggs and larvae (see Table 6.2). As pointed out by Trivers (1972), Duellman and Trueb (1986) and Crump (1995), parental care may be defined as any behavior exhibited by a parental toward its offspring that increases the offspring's chances of survival. Recent studies have provided evidence for an astonishing array of parental care in anurans (see review by Crump 1995 and Lehtinen and Nussbaum, chapter **8** of this volume), that usually involves protecting the developing clutch from predators or from desiccation, or both (Kluge 1981; Townsend *et al.* 1984). In addition, Tyler (1976) hypothesized a further role of the parent frog attentions, namely production of skin secretions aging as antibacterial and/or antifungal substances towards eggs and tadpoles.

Certainly, only in anurans does parental care involve transport of eggs and/or larvae on or within their bodies (specifically termed "brooding" by Crump 1995). Although the presence of at least some type of parental investment is rather common among amphibians (*e.g.*, Salthe and Mecham 1974; McDiarmid 1978; Wells 1981), most parental attentions do not require specific modifications of the integument. McDiarmid (1978) recognized twelve types of anuran parental care, based on oviposition habitat (aquatic or terrestrial), sites of parental care (nest, burrow, or on the parent body), nature of the larval stage (free-swimming or direct development), and sex of the care-giving parent. More recently, Wells (1981) proposed four major categories: (1) egg attendance, (2) tadpole attendance, (3) egg transport, (4) tadpole transport (for a summary of the different modes see Crump 1995, tables 3-4, and Lehtinen and Nussbaum, chapter 8 of this volume). Discussion in this section is limited to some typical cases of (3) and (4) parental care modes, in which derived integumentary features are critical.

In several anuran species from temperate and tropical areas (McDiarmid 1978; Duellman and Trueb 1986) the parental cares include incubation and/or transport of eggs and embryos on the back skin of a nurse adult ("mobile parental care"; Figs. 6.18A-F); female or male according to the species or environmental conditions (Duellman and Maness 1980; Wells 1981; Myers and Daly 1983; De Pérez *et al.* 1992a; Duellman and Trueb 1986). As reported by Crump (1995), egg transport is found in four families of anurans (Discoglossidae, Hylidae, Myobatrachidae and Pipidae), whereas tadpole transport occurs in seven families (Dendrobatidae, Hylidae, Leptodactylidae, Myobatrachidae, Ranidae, Rhinodermatidae, Sooglossidae). For example, in the male of some dendrobatids, particularly in *Colostethus* spp., when eggs hatch, the tadpoles wriggle onto the male's back and become attached by mucus "adhesives" (Wells 1981). In *Dendrobates tricolor*, the nurse frog occasionally wriggles its hindquarters into the egg mass, an action that ultimately aids the hatching process and helps the tadpoles to squirm up onto their parent's back. There they adhere to a patch of mucus secreted by glands in the frog's skin. The nurse frog eventually travels to a suitable body of water, where a period of soaking loosens the mucous bond and the tadpoles swim free (Myers and Daly 1983). For reviews of the reproductive strategies and trends in these Anura see: Lutz (1947), Rabb (1973), Salthe and Duellman (1973), Crump (1974, 1995), Trueb (1974), Lamotte and Lescure (1977), Duellman (1978), McDiarmid (1978), Wake (1982), Duellman (1985), Duellman and Trueb (1986); and Lehtinen and Nussbaum, chapter 8 of this volume.

Several observations on the brooding frogs have minutely reported the different modes of eggs and embryos incubation. According to De Pérez *et al.*

(1992a), species of various genera in the subfamily Hemiphractinae carry out "movable" incubation involving female dorsal skin. In three of the above genera (*Stefania, Cryptobatrachus* and *Hemiphractus*) the fertilized eggs simply adhere to the mother's dorsal tegument by gelatinous substances of unknown origin and are carried exposed until development is achieved (Del Pino 1980). Very little is known of the reproductive adaptations of these frogs (Trueb 1974; Rivero 1970). In other genera (*Amphignathodon, Flectonotus, Fritziana* and *Gastrotheca*), incubatory pouches (more or less closed and complete) occur on the female dorsum and the release of tadpoles may occur at different development stages (Grenard 1958; Hoogmoed 1967; Del Pino *et al.* 1975; Del Pino 1980; Duellman and Maness 1980; Del Pino and Escobar 1981; Duellman and Gray 1983). In addition, males of the leptodactylid *Assa darlingtoni* posses brood pouches in the inguinal regions. Here the embryos enter actively (Fig. 6.18E) and remain from hatching to metamorphosis (Fig. 6.18F) (Ingram *et al.* 1975; Ehmann and Swan 1985). According to McDiarmid (1978) *Assa*, pouches are "the expected evolutionary results of initial efforts at transport of the larvae back to the water".

Based on the site of embryonic development, Duellman and Maness (1980) grouped the egg-brooding modes as follows: 1) eggs adhere to back of the nurse frog, 2) Eggs are carried in shallow basin on back of female, 3) eggs develop within dorsal pouch of female (or male). The frogs having incubatory pouches are usually termed marsupial frogs (Duellman and Maness 1980; Del Pino 1989). Among the European frogs, only *Alytes obstetricans* practises parental cares in form of egg-brooding. The male carries the strings of fertilized eggs entwined around his groin and hind legs (Fig. 6.18A) and periodically enters in water; when the eggs are near hatching, he moves to a suitable aquatic site (Boulenger 1912; Crespo 1979).

Apart from the different modes of mobile parental care, Duellman and Trueb (1986), underline that the term "brooding" might best be reserved for those frogs that carry eggs or tadpoles and provide gaseous exchange between parental and embryonic tissues. This applies to the egg-brooding hylids (del Pino *et al.* 1975) and pipids (Weygoldt 1976) and presumably to *Assa*, *Rheobatrachus* and *Rhinoderma*. However, as a necessary morpho-functional requisite for the performance of the egg-brooding, the cutaneous apparatus of all the above frogs shows a strong morpho-functional plasticity of its components, including vascularization and gland secretions. Observations on the reproduction of hylids of genus *Stephania* (Rivero 1970) suggested that egg adhesion to the female dorsal skin may involve a chemical interaction between egg membranes and glandular secretions of maternal tegument. Further studies on the skin histology of *Cryptobatrachus, Stephania* and *Hemiphractus* species during reproduction (Del Pino 1980) indicated some modifications of the cutaneous components, in particular: number reduction of the poison glands and number and size increase of the mucous ones. According to the same author, the secretion of these latter glands allows the embryos adhesion to the mother's tegument. Further data of Del Pino and Escobar (1981) on *Cryptobatrachus* confirm that maternal and embryonic tissues are separated by an additional layer of mucous secretion that cements the jelly capsule of the embryos to the incubatory integument of the mother.

Furthermore, De Pérez and Ruiz-Carranza (1985) and De Pérez *et al.* (1992a) observed histological and cytological modifications of the skin related to the incubation of the eggs and developing embryos on the female dorsum of the hylid *Cryptobatrachus boulengeri*. According to these reports, the epidermis and dermis are significantly thicker and have a spongy appearance. In addition, the two types of mucous glands described in these anurans and defined G1 and G2 (De Pérez and Ruiz-Carranza 1985) appear active and secreting acid glycoproteins. The product of the G2 mucous glands is holocrine in females with dorsal eggs in the first developmental stages. Two mechanisms seem contribute to the firm sticking of the egg mass to the frog's back: mechanical, related to the anchorage of the envelope of eggs into the microscopic epithelial infoldings scattered throughout the maternal tegument, and chemical, by the sticky secretions of mucous glands, especially of the G2 glands.

A direct relation between brood adhesion to the adult tegument and its mucous secretions was also suggested by Myers and Daly (1980, 1983) in some dendrobatids. The tadpoles are attached not by their mouths, but by the probably universal dendrobatid method of sticking to a small patch of mucus on the nurse frog's back. In some species, this attachment is accomplished solely by mere surface adhesion between the mucus and the tadpoles' flattened or slightly concave bellies and the larvae are easily moved about and dislodged; but even so, a certain amount of soaking in water is required for the bond to loosen and for the tadpoles to be released normally. In other dendrobatids, including *Dendrobates bombetes* and *D. opisthomelas* (Myers and Daly 1980), the mucus attachment seems almost glue-like and the tadpoles are very resistant to being dislodged (even when the frog is roughly handled or subjected to formalin fixation), and presumably the nurse frog must immerse itself for a longer time before the larvae swim free. When dorsal larvae of *D. bombetes* are gently pried from the live frog, the sticky mucus pulls out into definite strands, and a mucous patch (white in preservative solution) remains on the adult's back, as described by Stebbins and Hendrickson (1959) for *Colostethus subpunctatus*. The degree of firmness of the attachment may be correlated with the length of time that the tadpoles are carried about by the nurse frog, which appears to be a variable trait among species of dendrobatids.

Later, obvious integumentary modifications associated with transport of tadpoles by adult males were described in the dendrobatid *Minyobates virolinensis* (De Pérez *et al.* 1992b). Specializations of tadpole ventral skin affect two cell types in the outer epidermal layer; long microvilli inserted in a dense, integumentary matrix allow anchorage between tadpole and male frog's back. The epidermis of the nurse frog is similar to that of adult males and females without tadpoles on the back. However, several corny layers are accumulated on the skin surface of nurse frogs and two different types of mucous glands secrete acid glycoproteins to the matrix. On the whole, the integumentary matrix may originate from the skin of both tadpole and adult.

In the marsupial frog *Gastrotheca riobambae*, the pouch is a permanent cutaneous structure that undergoes profound changes with reproduction; these include vascularization and the formation of embryonic chambers (Del Pino *et al.* 1975; see also Lehtinen and Nussbaum, chapter **8** of this volume). In this frog, but also in species of *Amphignathodon*, *Flectonotus* and *Fritziana*, maintenance of early incubation and associated changes of the pouch seem to be mediated by the postovulatory follicles of the ovary (del Pino and Sánchez 1977; del Pino 1980). The pouch is open when ovaries are small; it becomes closed prior to ovulation, when ovaries contain fully grown oocytes, and remains closed during incubation. The pouch is entirely lined with integument and in some species it extends from the skin into the dorsal and sometimes the lateral lymph spaces. During incubation, the pouch becomes distended with embryos and it occupies the entire back and in most cases also the sides of the body. The pouch is connected to the body in the midline by a dorsal septum that contains blood vessels, connective tissue and possibly nerves. Histological observations on the pouch of *Gastrotheca riobambae* (Del Pino *et al.* 1975) indicated the resemblance of this structure to the common integument. The pouch is lined with stratified squamous epithelium, which is closely associated with numerous simple alveolar mucous glands, usually with large lumina containing a PAS positive substance. Granules in the secretory cells of these glands are also PAS positive. Some mucous glands have small lumina, and their cells are swollen with secretory material. A few scattered serous glands are present. The initial changes of the pouch associated with incubation of the eggs are the development of activity in the mucous glands and a thinning of the pouch. On the basis of pouch morphology and the characteristics of the dorsal septum, Del Pino (1980) distinguished six types of pouches which, however, show rather similar histological structure comparable to that reported in *Gastrotheca riobambae* (del Pino *et al.* 1975). Also in the brooding frogs without pouches (*Cryptobatrachus*, *Hemiphractus* and *Stefania*), Del Pino (1980) noticed an incubatory integument with abundant, large and active mucous glands, whereas serous glands were rare. In addition, the secretion of mucous glands as well as the granules in the secretory cells were positive to PAS. During incubation, the mucous gland secretion forms an egg matrix of about 50 µm in thickness, which separates the embryos from the dorsal integument of the mother. According to Del Pino *et al.* (1975) and Del Pino (1980), the complex pouches of some *Gastrotheca* species may have originated as a fold of dorsal skin to protect the embryos in a terrestrial environment against predation and/or desiccation. However, shallow folds of skin border the egg mass carried on the back of other Hylidae (Del Pino 1980; Duellman and Maness 1980). Development of a closed pouch, which involves invagination and specialization of the integument (including glands), is in agreement with that idea. Nonetheless, analyses of embryos and tadpoles of *Gastrotheca riobambae*, performed by Del Pino and Escobar (1981), suggest that, at least in this species, the mother does not contribute nutrient, but gases and other factors are probably exchanged through the pouch. A similar exchange was hypothesized also by Ingram *et al.* (1975) between male pouches and embryos of *Assa darlingtoni*.

Also in the genus *Pipa* skin gland specializations possibly are used in transporting and protecting eggs and larvae. Pouch formation (a single small cell for each egg; Fig. 6.18B) results from gradual sinking of an egg into the mother's proliferating skin. The cover of a pouch consists solely of exposed egg membrane. During the first day of incubation, the eggs on the back sink further into the skin, which appears to swell up around the eggs, particularly the peripheral ones (see Fig. 1b in Rabb and Snedigar 1960). During this process, mucus gland secretions initially seem to attach the eggs to the female's back. Later, the dorsal integument thickens around each egg, providing a more secure attachment (Rabb and Snedigar 1960; Rabb and Rabb 1961; Weigoldt 1976). Perhaps, this additional dermal development is correlated with the aquatic habitat in which *Pipa* provides parental care. In terrestrial brooding species of some hylids (*e.g.*, species of *Hemiphractus*) mucus gland secretions of the dorsal integument are sufficient for attaching the eggs to the female back.

Interesting data and speculations on evolution of pipid dorsal brooding (primitive or derived character?), come from considering some differences (including cutaneous glands) between male and female of the African pipid *Xenopus* (Barthalmus 1994). The two currently isolated geographic groups of pipids; the African one symbolized by *Xenopus* and the South American one by *Pipa*, probably did not evolve in isolation but rather arose prior to the break-up of Gondwana from a common ancestor that already brooded young on the back of the female. Although at present only the *Pipa*-group members exhibit egg brooding, according to Fujikura *et al.* (1988) and Barthalmus (1994), sexually dimorphic differences in the skin of *X. laevis* provide evidence that this toad retains integumental vestiges from a dorsal brooding ancestor. Cutaneous glands are larger in the female and 10% more numerous on her dorsal surface, features that would be related to dorsal brooding. In addition, males have 13% fewer granular glands and 3% fewer mucous glands on the ventral surface than on the dorsum, traits perhaps reflecting previous selection for a mechanism preventing the sticking of the eggs to the ventral surface during transfer of eggs to the female's back. Male and female *Pipa* perform turnover manoeuvres during amplexus that result in the eggs first becoming lightly attached to the male's abdomen after which they are pressed against the back of the female where they are fertilized and partially sink into the highly glandular and vascularized skin (Rabb and Snedigar 1960; Rabb and Rabb 1961, 1963b). Although none of the African pipids exhibit egg brooding, some dimorphic differences (larger and more numerous skin glands in female *X. laevis*), together with turnover dance performed in some species (Barthalmus 1994), suggest that dorsal brooding may be a primitive trait in the family Pipidae rather than a highly derived one.

Further reproductive modes may include gland specializations of ectodermal origin; for example tending of eggs and embryos and/or transport of tadpoles to water performed by some dendrobatids (Wells 1978, 1981) and Ascaphidae (Bell 1985), and foam nest construction typical of some African, Australian and

Neotropical frogs (Heyer 1977; Heyer and Rand 1977; Tyler 1985; Duellman and Trueb 1986). In all the above cases, however, a critical role of cutaneous glands is only conjectural.

The foam nests may be constructed in water or on land vegetation (Figs. 6.19A-B) and in very different ways (Tyler and Davies 1979): paddling of hands by female, kicking of feet by male, paddling of hands and feet by male and female (Pisano and Del Rio 1968; Coe 1974; Heyer and Rand 1977; Hoogmoed and Gorzula 1979; Tyler and Davies 1979; Duellman 1985; Duellman and Trueb 1986). For example, the male of *Dendrobates auratus* moves around the clutch of eggs and works his legs in and out of the jelly. The egg mass increases in size with this treatment, suggesting that the male provides moisture from his skin (Wells 1978). Multiple males amplexing a female have been observed in *Chiromantis rufescens*, with all individuals engaging in movements to whip up the foam nest (Coe 1967,1974) and in *Polypedates dennysi* (Pope 1931). In at least some species of *Polypedates* and *Rhacophorus*, a large amount of seminal fluid with suspended sperm is discharged during amplexus and may contribute to the foam nest (Bhaduri 1932). On the other hand, females of some species construct a nest in the absence of males (Pope 1931; Coe 1974; Duellman and Trueb 1986), whereas other observations seem indicate that the jelly material is discharged by the female cloaca, and possibly produced by cloacal glands, during the amplexus.

As a result, the viscosity of the nest and the hardening of the outer layer of the foam differs according to the manner of construction. In addition, experimental evidences suggest that some biological property of the foam regulates the growth of tadpoles (Pisano and Delrio 1968), according to fluctuating water or seasonally wet environments. Although foam nests seem to be adapted both to static and moving waters, their functional significance is mostly uncertain.

However, the habit of constructing foam nests certainly evolved independently, on several occasions, in the different group of frogs in Africa, South America and Australia, (Duellman and Trueb 1986), providing a striking example of convergence (Martin 1970).

Fig. 6.19 Semischematic drawings of **A**. aquatic and **B**. terrestrial foam nests. In some cases, male or female provide attendance at the nest against intruders. Pencil drawings by G. Tanteri.

As a concluding remark in this review of ordinary and specialized glands in the amphibians, based upon a comparison of the wide literature and our investigations on some SDSG, it appears evident that selective pressures due to environmental and social constraints produced the rich supply of the secretory structures observed in the skin of the various body regions. Thus, the typical gland specializations of the different species strongly reflect their survival and reproductive strategies. More particularly, we share the opinion of Thomas *et al.* (1993) that the structural and histochemical similarities of anuran breeding glands and urodele courtship glands may reflect a common evolutionary process. Nevertheless, microevolutionary patterns in the integumentary glands are correlated with different evolving interactions between environment and organism. In this connection morphological and physiological characters within the integumentary apparatus are often essential for understanding either the evolutionary etiology or the functional significance of the glands.

6.8 ACKNOWLEDGMENTS

We wish to thank D.M. Sever and the editor, B.G.M. Jamieson, for inviting us to contribute this review for the volume. We express gratitude to Prof. W.E. Duellman for commenting on the manuscript and granting permission for publication of some of his illustrations. Prof. B.G. M. Jamieson has been a constant source of stimulation, useful suggestions and support for all the authors of this volume. He provided a kindly revision of the earlier and final drafts of our paper and edited with special care the manuscript. For all this we thank him very much. We would also like to thank the following people for their help in collecting data: E.O. Thomas for providing unpublished information and illustrations, G.Tanteri for his histological assistance and production of the pencil drawings, P. Nuti for the technical support in SEM observations, P. Malenotti for providing amphibian specimens.

6.9 LITERATURE CITED

Ali, S. A., Schoonen, W. G. E. J., Van der Hurk, R. and van Oordt, P. G. W. J. 1987. The skin of the male African catfish, *Clarias gariepinus*: A source of steroid glucuronides. General Comparative Endocrinology 66: 415-424.

Arnold, S. J. 1977. The evolution of courtship behaviour in New World salamanders with some comments on Old World salamanders. Pp 185-222. In D. H. Taylor and S. I. Guttman (eds), *The Reproductive Biology of Amphibians*. Plenum Press, New York, London.

Arnold, S. J. and Houck, L. D. 1982. Courtship pheromones: evolution by natural and sexual selection. Pp 173-211. In M. Nitecki (ed.), *Biochemical Aspects of Evolutionary Biology*, University of Chicago Press, Chicago.

Aron, M. 1926. Récherches morphologiques et expérimentales sur le Determinisme des caracterès sexuels secondaires mâles chez les Anoures (*Rana esculenta* L. et *Rana temporaria* L.). Archives de Biologie 36: 1-97.

Bachmayer, H., Michl, H. and Roos, B. 1967. Chemistry of cytotoxin substances in amphibian toxins. Pp. 395-399. In F.E. Russel and P.R. Saunders (eds), *Animal Toxins*. Pergamon Press, Edinburgh.

Bancroft, J. D. 1975. *Histochemical techniques.* 2nd ed. Butterworths, London, England. 348 pp.

Bani, G. 1976. Cellule mioepiteliali accluse alle ghiandole cutanee di alcuni Anfibi. Archivio Italiano di Anatomia e di Embriologia 81: 133-164.

Barbadillo Escriva, L. J. 1987. *La guia de Incafo de los Anfibios y Reptilos de la penisula Iberica, islas Baleares y Canarios.* Incafo, Madrid. 694 pp.

Barberio, C., Delfino, G. and Mastromei, G. 1987. A low molecular weight protein with antimicrobical activity in the cutaneous "venom" of the yellow-bellied toad (Bombina variegata pachipus). Toxicon 25: 899-909.

Barthalmus, G. 1994. Biological roles of amphibian skin secretions. Pp.382-410. In H. Heatwole and G. T. Barthalmus (eds), *Amphibian Biology*, Volume I., Surrey Beatty and Sons, New South Wales, Australia.

Becker, D. N., Galili, N. and Degani, G. 1992. GCMS-identified steroid glucuronides in gonads and holding water of *Trichogaster trichopterus.* Comparative Biochemical Physiology B 103: 15-19.

Bell, B. D. 1985. Development and parental-care in the endemic New Zeland frogs. Pp. 269-278. In G. Grigg, R. Shine and H. Ehmann (eds), *Biology of Australasian Frogs and Reptiles*, Royal Zoological Society of New South Wales.

Berk, L. 1939. Studies in the reproduction of *Xenopus laevis.* II. The secondary sex characters of the male *Xenopus*: the pads. South African Journal of Medical Science 4: 47-59.

Bhaduri, J. L. 1932. Observations on the urogenital system of the tree frogs of the genus *Rhacophorus* Kuhl, with remarks on their breeding habits. Anatomischer Anzeiger 74: 336-343.

Blair, A. P. 1946. The effects of various hormones on primary and secondary sex characters in juvenile *Bufo fowleri.* Journal of Experimental Zoology 103: 365-400.

Blaylock, L. A., Ruibal, R. and Platt-Aloia, K. 1976. Skin structure and wiping behaviour of phyllomedusine frogs. Copeia 1976: 283-295.

Bolognani-Fantin, A. M. and Fraschini, A. 1965. Contributo alla conoscenza istochimica della ghiandola del pollice di *Discoglossus pictus.* Bollettino di Zoologia 32: 765-776.

Borgioli, G. 1977. First data on the ultrastructure of the mental hedonic gland cluster of *Hydromantes italicus* Dunn. Bollettino di Zoologia 44: 119-122.

Botte, V. and Delrio, G. 1967. Effect of estradiol-17b on the distribution of 3 b - hydroxysteroid dehydrogenase in the testis of *Rana esculenta* and *Lacerta sicula.* General and Comparative Endocrinology 9: 110-115.

Botte, V., d'Istria, M., Delrio, G. and Chieffi, G. 1972. Hormonal regulation of thumb pads in males of *Rana esculenta.* General and Comparative Endocrinology 18: 577.

Boulenger, G. A. 1912. Observations sur l'accouplement et la ponte de l'Alyte accoucheur, *Alytes obstetricans.* Bulletin de la classe de sciences Accademie royale de Belgique 1912: 570-579.

Bovbjerg, A. M. 1963. Development of the glands of the dermal plicae in *Rana pipiens.* Journal of Morphology 113: 231-243.

Brizzi, R. and Calloni, C. 1992. Male cloacal region of the spotted salamander, *Salamandra salamandra giglioli*, Eiselt and Lanza, 1956 (Amphibia, Salamandridae). Bollettino di Zoologia 59: 377-385.

Brizzi, R., Calloni, C. and Delfino, G. 1986. Accessory structures in the genital apparatus of *Salamandrina terdigitata* (Amphibia: Salamandridae). I. Ultrastructural patterns of the male abdominal gland. Zeitschrift für mikroskopisch-anatomische Forschung 100: 397-409.

Brizzi, R., Calloni, C. and Delfino, G. 1990. Accessory structures in the genital apparatus of *Salamandrina terdigitata* (Amphibia: Salamandridae). IV. Male cloacal glands. A study under light and scanning electron microscopes. Zeitschrift für mikroskopisch-anatomisce Forschung 104: 871-897.

Brizzi, R., Calloni, C. and Delfino, G. 1992a. Male cloacal region of *Triturus* italicus (Peracca) with reference to the cloacal anatomy and reproductive patterns of the Salamandridae. Archivio Italiano di Anatomia e di Embriologia 97: 121-138.

Brizzi, R., Delfino, G. and Calloni, C. 1994. Structural and ultrastructural comparison between tail base- and mental glands in *Hydromantes*. Animal Biology 3: 31-40.

Brizzi, R., Delfino, G. and Jantra, S. 1996a. Comparative anatomy and evolution of the male dorsal glands in the Salamandridae (Amphibia, Urodela). Acta Biologica Benrodis 8: 61-77.

Brizzi, R., Delfino, G. and Pellegrini, R. 2002. Specialized mucous glands and their possible adaptive role in the males of some species of *Rana*. Journal of Morphology (in press).

Brizzi, R., Calloni, C., Delfino, G. and Jantra, S. 2000. Structural and ultrastructural observations on antipredator cutaneous glands in *Triturus marmoratus*. Museo Regionale Scienze Naturali di Torino 2000: 199-205.

Brizzi, R., Calloni, C., Delfino, G. and Lotti, S. 1996b. The male cloaca of *Salamandra lanzai* with notes on the dorsal glands in the genera *Salamandra* and *Mertensiella* in relation to their courthship habits. Herpetologica 52: 505-515.

Brizzi, R., Calloni, C., Delfino, G. and Menna, M. 1992b. Occurrence of dorsal glands in the female cloacal region of Salamandrina terdigitata and their phylogenetic significance in the urodeles. Zoologisches Jahrbuch der Anatomie 122: 23-33.

Brizzi, R., Calloni, C., Delfino, G. and Tanteri, G. 1995a. Notes on the male cloacal anatomy and reproductive biology of *Euproctus montanus* (Amphibia: Salamandridae). Herpetologica 51: 8-18.

Brizzi, R., Delfino, G., Rebelo, R. and Sever, D. M. 1999. Absence of dorsal glands in the cloaca of male *Chioglossa lusitanica*. Journal of Herpetology 33: 220-228.

Brizzi, R., Delfino, G. and Calloni, C. 1994. Structural and ultrastructural comparison between tail base- and mental glands in *Hydromantes*. Animal Biology 3: 31-40.

Brizzi, R., Delfino, G. and Jantra, S. 1996a. Comparative anatomy and evolution of the male dorsal glands in the Salamandridae (Amphibia, Urodela). Acta Biologica Benrodis 8: 61-77.

Brizzi, R., Delfino, G. and Pellegrini, R. 2002. Specialized mucous glands and their possible adaptive role in the males of some species of *Rana*. Journal of Morphology, 254: 328-341.

Brizzi, R., Delfino, G., Selmi, M. G. and Sever, D. M. 1995b. Spermathecae of *Salamandrina terdigitata* (Amphibia: Salamandridae): patterns of sperm storage and degradation. Journal of Morphology 223: 21-33.

Brizzi, R., Delfino, G., Jantra, S., Alvarez, B. B. and Sever, D. M. 2001. The amphibian cutaneous glands: some aspects of their structure and adaptive role. Pp. 43-49. In P. Lymberakis, E. Valakos, P. Pafilis and M. Mylonas (eds), *Herpetologica Candiana*. Typokreta, Irakleio, Greece.

Brodie, E. D. jr. and Smatresk, N. J. 1990. The antipredator arsenal of fire salamanders: spraying of secretions from highly pressurized dorsal skin glands. Herpetologica 46: 1-7.

Burgos, M. H. and Ladman, A. J. 1957. The effect of purified gonadotropins on the morphology of the testes and thumb pads of the normal and hypophysectomized autumn frog (*Rana pipiens*). Endocrinology 61: 20-34.

Burns, A. E. and Thomas, E. O. 1997. Behavioral evidence for mate-attracting pheromones in the dwarf African clawed frog, *Hymenochirus curtipes*. Abstract presented at Combined Western/Southwestern Regional Conference on Comparative Endocrinology, Denver, CO.

Cannon, M. S. and Hostetler, J. R. 1976. The anatomy of the parotoid gland in Bufonidae with some histochemical findings. Journal of Morphology 148: 137-160.

Cei, J. M. 1980. *Amphibians of Argentina*. Monitore Zoologico Italiano (N.S.), Monografia 2, Italy. 609 pp.

Chan, M., Ho, U., Shoji, R. and Thomas, E. O. 1999. Attraction of female dwarf African clawed frogs (*Hymenochirus curtipes*) to the reproductive odors of males. Abstract presented at 24th Annual West Coast Biological Sciences Undergraduete Research Conference, Irvine, CA.

Chieffi, G., Delrio, G., d'Istria, M. and Valentino, M. 1975. Appearance of sex hormone receptors in frog (*Rana esculenta*) tadpole skin during metamorphosis. Experientia 31: 989-990.

Coe M.J. 1967. Co-opertion of three males in nest construction by *Chiromantes rufescens* Günther (Amphibia: Rhacophoridae), Nature 214: 112-113.

Coe, M. J. 1974. Observations on the ecology and breeding biology of the genus Chiromantis (Amphibia: Rhacophoridae). Journal of Zoology 172: 13-34.

Colombo, L. A., Marconato, A., Colombo Belvedere, P. and Frisco, C. 1980. Endocrinology of teleost reproduction: A testicular steroid pheromone in black goby, *Gobius jozo*. Bollettino di Zoologia 47: 355-364.

Conaway, C. H. and Metter, D. E. 1967. Skin glands associated with breeding in *Microhyla carolinensis*. Copeia 1967: 672-673.

Crespo, E.G. 1979. Contribução para o Conhecimento da Biologia dos *Alytes ibéricos*, *Alytes obstetricans boscai* Lataste, 1879 e *Alytes cisternasii* Boscá, 1879 (Amphibia-Salientia): A Problemática da Especiação de *Alytes cisternasii*. Thesis, University of Lisboa, Spain.

Cristensen, K. 1931. Effect of castration on the secondary sex characteristics of males and females of *Rana pipiens*. Anatomical Record 48: 241-250.

Crump, M. L. 1974. Reproductive strategies in tropical anuran Community. Miscellaneous Publications, Museum of Natural History University of Kansas 61: 1-68.

Crump, M. L. 1995. Parental care. Pp. 518-567. In H. Heatwole and B. K. Sullivan (eds), *Amphibian Biology*, Volume II, Surrey Beatty and Sons, Chipping Norton, Australia.

Daly, J. W, Brown, G. B. and Mensah-Dwumah, M. 1978. Classification of skin alkaloids from neotropical poison-dart frogs (Dendrobatidae). Toxicon 16: 163-188.

Daly, J. W., Myners, C. W. and Whittaker, N. 1987. Further classification of skin alkaloids from neotropical poison frogs (Dendrobatidae), with a general survey of toxic/noxius substances in the Amphibia. Toxicon 25: 1023-1095.

Dapson, R. W. 1970. Histochemistry of mucus in the skin of the frog *Rana pipiens*. Anatomical Record 166: 615-626.

Dapson, R. W., Feldman, A. T. and Wright, O. L. 1973. Histochemistry of the granular (poison) secretion in the skin of the frog, *Rana pipiens*. Anatomical Record 177: 549-560.

Dawley, E. M. 1998. Olfaction. Pp. 711-742. In H. Heatwole and E. M. Dawley (eds), *Amphibian biology*, Volume III, Surrey Beatty and Sons, New South Wales, Australia.

Dawson, A. B. 1920. Integument of *Necturus maculosus*. Journal of Morphology 34: 487-580.

Delfino, G. 1980. L'attività rigeneratrice del tratto intercalare nelle ghiandole granulose cutanee dell'ululone *Bombina variegata pachipus* (Bonaparte). (Anfibio, Anuro, Discoglosside): studio sperimentale al microscopio elettronico. Archivio Italiano di Anatomia e di Embriologia 85: 283-310.

Delfino, G., Amerini, S. and Mugelli, A.1982. In vitro studies on the "venom" emission from the skin of *Bombina variegata* pachypus (Bonaparte) (Amphibia Anura Discoglossidae). Cell Biology International Reports 6: 843-850.

Delfino G, Brizzi R, Calloni C. 1985. Dermo-epithelial interactions during the development of cutaneous gland anlagen in Amphibia: A light and electron microscope study on several species with some cytochemical findings. Zeitschrift für Mikroskopisch-Anatomische Forschung 99: 225-253.

Delfino, G., Brizzi, R. and Borrelli, G. 1988a. Cutaneous glands in anurans: differentiation of the secretory syncytium in serous Anlagen. Zoologisches Jahrbuch für Anatomie 117: 255-275.

Delfino, G., Brizzi, R. and Borrelli, G. 1988b. Accessory structures in the genital apparatus of *Salamandrina terdigitata* (Amphibia: Salamandridae). III Cytochemical study of the male abdominal gland using ruthenium red staining. Zoologischer Anzeiger 117: 255-275.

Delfino, G., Brizzi, R. and Calloni, C. 1986. Mixed cutaneous glands in Amphibia: an ultrastructural study on urodele larvae. Zoologisches Jahrbuch für Anatomie 114: 325-344.

Delfino, G., Alvarez, B. B., Brizzi, R. and Cespedez, J. A. 1998. Serous cutaneous glands of Argentine *Phyllomedusa* Wagler 1830 (Anura Hylidae): secretory polymorphism and adaptive plasticity. Tropical Zoology 11: 333-351.

Delfino, G., Brizzi, R., Alvarez, B. B. and Gentili, M. 1999. Granular cutaneous glands in the frog *Physalaemus biligonigerous* (Anura,Leptodactylidae): comparison between ordinary serous and "inguinal" glands. Tissue and Cell 31: 576-586.

Delfino, G., Brizzi, R, De Santis R. and Melosi, M. 1992. Serous cutaneous glands of the western spade-foot toad *Pelobates cultripes* (Amphibia, Anura): an ultrastructural study on adults and juveniles. Archivio Italiano di Anatomia e di Embriologia 97: 109-120.

Del Pino, E. M. 1980. Morphology of the pouch and incubatory integument in marsupial frogs. Copeia 1980: 10-17.

Del Pino, E. M. 1989. Marsupial frogs. Spektrum der Wissenschaft 1989: 19-28.

Del Pino, E. M. and Escobar, B. 1981. Embryonic stages of *Gastrotheca riobambae* (Fowler) during maternal incubation and comparison of development with that of other egg-brooding hylid frogs. Journal of Morphology 167: 277-295.

Del Pino, E. M., Galarza, M .L., Albuja de, C. M. and Humphries, A. A. jr. 1975. The maternal pouch and development in the marsupial frog *Gastrotheca riobambae* (Fowler). Biological Bulletin 149: 480-491.

Del Pino, E. M. and Sánchez, G. 1977. Ovarian structure of the marsupial frog *Gastrotheca riobambae* (Fowler) Anura, Hylidae. Journal of Morphology 153: 153-162.

Delrio, G. and d'Istria, M. 1973. Androgen receptor in the thumb pads of *Rana esculenta*. Experientia 29: 1412-1413.

Delrio, G., d'Istria, M., Iela, M. and Chieffi, G. 1979. The possible significance of testosterone in the female green frog *Rana esculenta*. Bollettino di Zoologia 46: 1-9.

De Pérez, G. R and Ruiz-Carranza, P. M. 1985. Ultraestructura e histoquimica de dos tipos de glandulas mucosas de la piel de *Cryptobatrachus* (Amphibia, Anura). Caldasia 14: 251-264.

De Pérez, G. R, Ruiz-Carranza, P. M. and Ramírez-Pinilla, M. P. 1992a. Modificaciones tegumentarias de larvas y adultos durante el cuidado parental en *Minyobates virolinensis* (Amphibia: Anura: Dendrobatidae). Caldasia 17 : 75-86.

De Pérez, G. R, Ruiz-Carranza, P. M. and Ramírez-Pinilla, M. P. 1992b. Especializaciones del tegumento de incubacion de la hembra de *Cryptobatrachus boulengeri* (Amphibia: Anura: Hylidae). Caldasia 17: 87-94.

d'Istria, M., Botte, V. and Chieffi, G. 1971. La regolazione ormonale dei caratteri sessuali secondari degli anfibi anuri. Azione del propionato di testosterone sulla callosità del pollice di maschi adulti castrati di *Rana esculenta*. Atti della Accademica nazionale dei Lincei, classe di scienze fisiche, matematiche e naturali 50: 205-212.

d'Istria, M., Botte, V., Delrio, G. and Chieffi, G. 1972. Implication of testosterone and its metabolites in the hormonal regulation of thumb pads of *Rana esculenta*. Steroids Lipids Researches 3: 321-327.

d'Istria, M., Delrio, G., Botte, V. and Chieffi, G. 1974. Radioimmunoassay of testosterone, 17 b-estradiol and estrone in the male and female plasma of *Rana esculenta* during sexual cycle. Steroids Lipids Researches 5: 42-48

d'Istria, M., Delrio, G. and Chieffi, G. 1975. Receptors for sex hormones in the skin of the amphibia. General and Comparative Endocrinology 26: 281-28.

d'Istria, M., Delrio, G. and Chieffi, G. 1977. La pelle: Un nuovo carattere sessuale secondario di *Rana esculenta*. Atti della Accademia Nazionale dei Lincei, Classe di Scienze Fisiche, Matematiche e Naturali 63: 126-129.

d'Istria, M., Citarella, F., Iela, L. and Delrio, G. 1979. Characterization of a cytoplasmic androgen receptor in the male secondary sexual character of green frog (*Rana esculenta*). Journal of Steroid Biochemistry 10: 53-59.

d'Istria, M., Picilli, A., Basile, C., Delrio, G. and Chieffi, G. 1982. Morphological and biochemical variations in the skin of *Rana esculenta* during the annual cycle. General and Comparative Endocrinology 48: 20-24.

Dodd, J. M. 1960. Gonadal and gonadotrophic hormones in lower vertebrates. Pp. 417-582. In A. S. Parkes (ed.), *Marshall's Physiology of Reproduction*. Little Brown and Company, Boston.

Domínguez, E., Navas, P., Hidalgo, J., Aijon, J. and López-Campos, J. L. 1981. Mucous glands of the skin of *Rana ridibunda*. A histochemical and ultrastructural study. Basic and Applied Histochemistry 25: 15-22.

Duellman, W. E. 1978. The biology of an equatorial herpetofauna in amazonian Ecuador. Miscellaneous Publications, Museum of Natural History University of Kansas 65: 1-352.

Duellman, W. E. 1985. Reproductive modes in anuran amphibians: phylogenetic significance of adaptive strategies. South African Journal of Science 81: 174-178.

Duellman, W. E. and Gray, P. 1983. Developmental biology and systematics of the egg-brooding hylid frogs, genera *Flectonotus* and *Fritziana*. Herpetologica 39: 333-359.

Duellman, W. E. and Maness, S. J. 1980. The reproductive behavior of some hylid marsupial frogs. Journal of Herpetology 14: 213-222.

Duellman, W. E. and Trueb, L. 1986. *Biology of Amphibians*. McGraw-Hill, New York. 670 pp. (reprint, Johns Hopkins University Press, 1994).

Ehmann, H. and Swan, G. 1985. Reproduction and development in the marsupial frog, *Assa darlingtoni* (Leptodactylidae, Anura). Pp. 279-285. In G. Grigg, R. Shine and H. Ehmann (eds), *Biology of Australasian Frogs and Reptiles*. Royal Zoological Society, New South Wales.

Erspamer, V. 1994. Bioactive secretions of the integument. Pp. 178-350. In: Heatwole, H. and Barthalmus, G. T. (eds), *Amphibian Biology*, Volume I. Surrey Beatty and Sons, New South Wales, Australia.

Faraggiana, R. 1938a. Ricerche istologiche sulle ghiandole cutanee granulose degli

Anfibi anuri. I. *Bufo vulgaris* e *Bufo viridis*. Archivio Italiano di Anatomia e di Embriologia 39: 327-376.

Faraggiana, R. 1938b. La struttura sinciziale e il meccanismo di secrezione delle ghiandole cutanee granulose di Anfibi anuri. Monitore Zoologico Italiano 49: 105-108.

Faraggiana, R. 1939. Ricerche istologiche sulle ghiandole cutanee granulose degli Anfibi anuri. II. *Rana esculenta, Rana agilis* e *Bombinator pachypus*. Archivio Italiano di Anatomia e di Embriologia 41: 390-410.

Fitch, H. S. 1956. A field study of the Kansas ant-eating frog, *Gastrophryne olivacea*. University of Kansas publications, Museum of Natural History 8: 275-306.

Forbes, T. R. 1961. Endocrinology of reproduction in cold blooded vertebrates. Pp. 1035-1087. In W.C. Young (ed.), *Sex and Internal Secretions*, Volume II , Williams and Wilkins, Baltimore.

Forbes, M. S., Dent, J. N. and Singhas, C. A. 1975. The developmental cytology of the nuptial pad in the red-spotted newt. Developmental Biology 46: 56-78.

Fox, H. 1986. Dermal glands. Pp. 116-135. In J. Bereiter-Hahn, A. G. Matoltsy and K. S. Richards (eds), *Biology of the Integument*, Volume II. Springer Verlag, Berlin.

Fox, H. 1994. Structure of the integument. Pp. 1-32. In H. Heatwole and G. T. Barthalmus (eds), *Amphibian biology*, Volume I. Surrey Beatty and Sons, New South Wales, Australia.

Fujikura K., Kurabuchi, S., Tabuchi, M. and Inoue, S. 1988. Morphology and distribution of the skin glands in *Xenopus laevis* and their response to experimental stimulations. Zoological Science 5: 415-430.

Gadow, H. 1887. Remarks on the cloaca and copulatory organs of Amniota. Philosophical Transactions of the Royal Society of London 178 (Ser. B): 5-37.

Galgano, M. 1942. Effetti misti, deboli, nulli ottenuti con la somministrazione di ormoni sessuali negli adulti di *Triton cristatus* Laur. Bollettino della Società Italiana di Biologia Sperimentale 17: 1-2.

Gallien, L. 1940. Récherches sur la physiologie hypophysaire dans ses relations avec les gonades et le cycle sexuel chez la grenouille rouge, *Rana temporaria* L. Bulletin biologique de la France et de la Belgique 74: 1-42.

Glass, F. M. and Rugh, R. 1944. Seasonal study of the normal and pituitary stimulated frog (*Rana pipiens*). I. Testis and thumb-pad. Journal of Morphology 74: 409-427.

Greeberg, B. 1942. Some effects of testosterone on sexual pigmentation and other sex characters o the cricket frog (*Acris gryllus*). Journal of Experimental Zoology 91: 435-446.

Grenard, S. 1958. Life history of *Gastrotheca marsupiata* spp. Herpetologica 14: 151-152.

Halliday, T. R. 1990. Morphology and sexual selection. Pp. 9-21. In Museo Regionale di Scienze Naturali Torino (ed.), *Atti VI Convegno Nazionale Ass. "Alessandro Ghigi"* (22-24 giugno 1989). Museo regionale Scienze Naturali Torino, Italy.

Halliday, T. R. and Tejedo, M. 1995. Intrasexual selection and alternative mating behavior. Pp. 419-468. In H. Heatwole and B. K. Sullivan (eds), *Amphibian Biology*, Volume II. Surrey Beatty and Sons, Chipping Norton, Australia.

Heyer, W. R. 1977. Foam nest construction in the leptodactylid frogs *Leptodactylus pentadactylus* and *Physalaemus pustulosus* (Amphibia, Anura, Leptodactylidae). Journal of Herpetology 11: 225-228.

Holloway, W. R. and Dapson, R. W. 1971. Histochemistry of integumentary secretions of the narrow-mouth toad, *Gastrophryne carolinensis*. Copeia 1971: 351-353.

Hoogmoed, M. S. 1967. Mating and early development of *Gastrotheca marsupiata* (Duméril and Bibron) in captivity (Hylidae, Anura, Amphibia). British Journal of Herpetology 4: 1-7.

Hoogmoed, M. S. and Gorzula, S. J. 1979. Checklist of the savanna inhabiting frogs of the El Manteco region with notes on their ecology and the description of a new species of tree-frog (Hylidae, Anura), Zoologische Mededelingen 54: 183-216.

Horie, H. 1939. Appearance of the thumb pads in the metamorphosing *Rana temporaria* by injecting the male hormone. Proceedings of the Imperial Academy of Tokio 15: 362-370.

Hostetler, J. R. and Cannon, M. S. 1974. The anatomy of the parotoid gland in Bufonidae with some histochemical findings. I. *Bufo marinus*. Journal of Morphology 142: 225-240.

Houck, L. D. 1986. The evolution of salamander courtship pheromones. Pp. 173-190. In D. Duvall, D. Müller-Schwarze and R. M. Silverstein (eds), *Chemical Signals in Vertebrates*, Volume IV. Plenum Press, New York.

Houck, L. D. and Reagan, N. L. 1990. Male courtship pheromones increase female receptivity in a plethodontid salamander. Animal Behavior 39: 729-734.

Houck, L. D. and Sever, D. M. 1994. Role of the skin in reproduction and behaviour. Pp.351-381. In H. Heatwole and G. T. Barthalmus (eds), *Amphibian Biology*, Volume I. Surrey Beatty and Sons, New South Wales, Australia,.

Houck, L. D. and Verrell, P. A. 1993. Studies of courtship behaviour in plethodontid salamanders: a review. Herpetologica 49: 175-184.

Houck, L. D. and Woodley, S. K. 1995. Field studies of steroid hormones and male reproductive behaviour in amphibians. Pp. 677-703. In H. Heatwole and B. K. Sullivan (eds), *Amphibian Biology*, Volume II. Surrey Beatty and Sons, Chipping Norton, Australia.

Inger, R .I. 1954. Systematics and zoogeography of Philippine Amphibia. Fieldiana Zoology 33: 181-531.

Inger, R. F. and Greenberg, G. B. 1956. Morphology and seasonal development of sex characters in two sympatric African toads. Journal of Morphology 99: 549-574.

Ingram, G. J., Anstis, M. and Corben, C. J. 1975. Observations on the Australian leptodactylid frog, *Assa darlingtoni*. Herpetologica 31: 425-429.

Jacob, J. S., Greenshaw, J. J., Plummer, M. V. and Goy, J. M. 1985. Pectoral glands of *Scaphiopus* and *Megophrys*. Journal of Herpetology 19: 419-420.

Jaeger, R. G. 1986. Pheromonal markers as territorial advertisement by terrestrial salamanders. Pp. 191-203. In D. Duvall, D. Müller-Schwarze and D. Silverstein (eds), *Chemical Signals in Vertebrates*, Volume IV, Plenum Press, New York.

Jaeger, R. G. and Gergits, W. F. 1979. Intra- and interspecific communication in salamanders through chemical signals on the substrate. Animal Behavior 27: 50-156.

Jameson, D. L. 1955. Courtship and mating in Salientia. Systematic Zoology 4: 105-119.

Jones, R. E., Gerrard, A. M. and Roth, J. J. 1973. Estrogen and brood pouch formation in the marsupial frog *Gastrotheca riobambae*. Journal of Experimental Zoology 184: 177-184.

Jurgens, J.D. 1978. Amplexus in *Breviceps adspersus*–who's the sticky partner. Journal of the Herpetological Association of Africa, 18: 6.

Karlson, P. and Lüscher, M. 1959. "Pheromones": a new term for a class of biologically active substances. Nature 183: 55-56.

Kelly, D. B. and Pfaff, D. W. 1976. Hormone effects on male sex behavior in adult South African Clawed Frogs, *Xenopus laevis*. Hormones and Behavior 7: 159-182.

Kermit, C. 1931. Effect of castration on the secondary sex characters of males and females of *Rana pipiens*. Anatomical Record 48: 24-27.

Kikuyama, S., Toyoda F., Ohmiva Y., Matsuda K., Tanaka S. and Hayashi, H. 1995. Sodefrin: A female-attracting peptide pheromone in newt cloacal glands. Science 267: 1643-1645.

Kluge, A. G. 1981. The life history, social organization, and parental behavior of *Hyla rosenbergi* Boulenger, a nest-building gladiator frog. Miscellaneous publications, University of Michigan, Museum of Zoology 60: 1-170.

Lamotte, M. and Lescure, J. 1977. Tendances adaptatives a l'affranchissement du milieu aquatique chez les amphibiens anoures. Terre et la Vie 2: 225-312.

Lanza, B. 1959. Il corpo ghiandolare mentoniero dei Plethodontidae (amphibia, Caudata). Monitore Zoologico Italiano 67: 15-53.

Lanza, B. 1983. *Guide per il riconoscimento delle specie animali delle acque interne italiane. 27. Anfibi, Rettili.* Consiglio Nazionale delle Ricerche, Roma, Italy. 196 pp.

Le Quang Trong, Y. 1973. Structure et dévelopment de la peau et des glandes cutanées de *Bufo regularis* Reuss. Bulletin de la Societé zoologique de France. 98: 449-485.

Le Quang Trong, Y. 1976. Etude de la peau et des glandes cutanees de quelques Amphibiens de la famille Rhacophoridae. Bulletin de l'Institut francais d'Afrique noire 38A: 166-187.

Littlejohn, M. J. 1977. Long-range acoustic communication in anurans: an integrated and evolutionary approach. Pp. 263-294. In D. H. Taylor and S. I. Guttman (eds), *The Reproductive Biology of Amphibians.* Plenum Press, New York, NY.

Lofts, B. 1964. Seasonal changes in the functional activity of the interstitial and spermatogenetic tissue of the green frog, Rana esculenta. General and Comparative Endocrinology 4: 550-562.

Lofts, B. 1974. Reproduction. Pp. 107-218. In B. Lofts (ed.), *Physiology of the Amphibia*, Volume II, Academic Press, New York.

Lofts, B. and Bern, H. A. 1972. The functional morphology of steroidogenic tissue. Pp. 37-126. In D. R. Idler (ed.), *Steroids in Nonmammalian Vertebrates.* Academic Press, New York, NY.

Lutz, B. 1947. Trends towards non-aquatic and direct development in frogs. Copeia 1947: 242-252.

Martin, A. A. 1970. Parallel evolution in the adaptive ecology of leptodactylid frogs of South America and Australia. Evolution 24: 643-644.

McDiarmid, R. W. 1978. Evolution and parental care in frog. Pp. 127-147. In G. M. Burghardt and M. Bekoff (eds), *The Development of Behavior: Comparative and Evolutionary Aspects.* Garland STPM Press, New York.

Metter, D. E. and Conaway, C. H. 1969. The influence of hormones on the development of breeding glands in *Microhyla*. Copeia 1969: 621-622.

Mills, J. W. and Prum, B. E. 1984. Morphology of the exocrine glands of the frog skin. American Journal of Anatomy 171: 91-106.

Müller-Schwarze, D. 1967. Social odors in young mule deer. Presented at the AAAS Meeting, New York.

Mushe, E. F. 1909. The cutaneous glands of the common toads. American Journal of Anatomy 9: 321-360.

Myers, C. W. and Daly, J. W. 1980. Taxonomy and ecology of *Dendrobates bombetes*, a new Andean poison frog with new skin toxins. American Museum Novitates 2692: 1-23.

Myers, C. W. and Daly, J. W. 1983. Dart poison frogs. Scientific American 1983: 97-105.

Mykytowycz, R. 1970. The role of skin glands in mammalian communication. Pp. 327-360. In J. W. Johnson jr., D. G. Moulton and A. Turk (eds), *Advances in Chemoreception*, Volume I. Appleton-Century-Crofts, New York, N.Y.

Neuwirth, M., Daly, J. W., Myers, C. W. and Tice, L. W. 1979. Morphology of the granular secretory glands in the skin of poison-dart frogs (Dendrobatidae). Tissue and Cell 11: 755-771.

Noble, G. K. 1927. The plethodontid salamanders, some aspects of their evolution. American Museum Novitates 249: 1-26.

Noble, G. K. 1929. The relation of courtship to the secondary sexual characters of the two-lined salamander, *Eurycea bislineata* (Green). American Museum Novitates 362: 1-5.

Noble, G. K. 1931a. *The Biology of theAmphibia*. McGraw-Hill Book Co., New York. NY 577 pp.

Noble, G. K. 1931b. The hedonic glands of the plethodontid salamanders and their relation to the sex hormones. Anatomical Record 48: 57-58.

Noble, G. K. 1954. *The Biology of the Amphibia*. Third ed. Dover Publ. Inc. New York, NY 577 pp.

Noble, G. K. and Noble E. R. 1944. On the histology of frog skin glands. Transactions of the American microscopical Society 63: 254-263.

Norris, D. O. 1987. Regulation of male gonads, ducts and sex accessory structures. Pp. 327-354. In D. O. Norris and R. E. Jones (eds), *Hormones and Reproduction in Fishes, Amphibians, and Reptile*, Plenum Press, New York, N.Y.

Parakkal, P. F. and Ellis, R. A. 1963. A cytochemical and electron microscopic study of the thumb pad in *Rana pipiens*. Experimental Cell Research 32: 280-288.

Pearl, C. A., Cervantes, M. and Thomas, E. O. 2000a. Evidence for a mate-attracting chemosignal in the dwarf African clawed frog *Hymenochirus*. Abstract presented at the Western Regional Conference on Comparative Endocrinology, Corvallis, OR.

Pearl, C. A., Cervantes, M., Chan, M., Ho, U., Shoji, R. and Thomas, E. O. 2000b. Evidence for a mate attracting chemosignal in the dwarf African clawed frog *Hymenochirus*. Hormones and Behavior 38: 67-74.

Pearse, A. G. E. 1977. The diffuse neuroendocrine system and the APUD concept: related "endocrine" peptides in brain, intestine, pituitary, placenta and anuran cutaneous glands. Medical Biology 55: 115-125.

Pearse, A. G. E. 1985. *Histochemistry, theoretical and applied*. Volume II, 4th ed. Churchill Livingstone, New York, NY, 614 pp.

Penhos, J. C. and Cardeza, A. F. 1957. Characteres sexuales secundarios del sapo macho castrado tratado con hormonas sexuales y acido folico o aminopterina. Revista de la Sociedad Argentina de Biología 33: 121-128.

Phisalix, C. and Bertrand, G. 1902. Sur les principes actifs du venom de crapaud comun (*Bufo vulgaris* L.). Comptes Rendus des Seances de la Societe de Biologie et des ses Filiales 54: 932-934.

Pisano, A. and Delrio, A. G. 1968. New biological properties in the foam jelly of Amphibians. Archivio Zoologico Italiano 53: 189-201.

Pope, C. H. 1931. Notes on amphibians from Fukien, Haihan, and other parts of China. Bulleti of American Museum of Natural History 61: 397-611.

Pytlik, L., Nguyen, M. and Thomas, E. O. 1999. The effects of testosterone on cutaneous breeding glands in the dwarf African frog, *Hymenochirus curtipes*. Abstract presented at the 24th Annual West Coast Biological Sciences Undergraduate Research Conference, Irvine, CA.

Quay, W. B. 1972. Integument and the environment: glandular composition, function, and evolution. American Zoologist 12: 95-108.

Rabb, G. B. 1973. Evolutionary aspects of the reproductive behavior of frogs. Pp. 213-227. In J. L. Vial (ed.), *Evolutionary Biology of the Anurans*. University of Missouri Press, Columbia.

Rabb, G. B. and Rabb, M.S. 1961. On the mating and egg-laying behavior of the Suriman toad, *Pipa pipa*. Copeia, 1960: 271-276.

Rabb, G. B. and Rabb, M. S. 1963a. On the behavior and breeding biology of the African pipid frog *Hymenochirus boettgeri*. Zeitschrift für Tierpsychologie 20: 215-241.

Rabb, G. B. and Rabb, M. S. 1963b. Additional observations on breeding behavior of the Surinam toad *Pipa*. Copeia 1963: 636-642.

Rabb, G.B. and Snedigar, R. 1960. Observations on breeding and development of the Suriman toad, *Pipa pipa*. Copeia 1960: 40-44.

Rastogi, R. K. and Chieffi, G. 1971. Effect of an antiandrogen, cyproterone acetate, on the pars distalis of pituitary and thumb pad of the male green frog, *Rana esculenta* L. Steroidologia 2: 276-282.

Rastogi, R. K., Chieffi, G. and Marmarino, C. 1972. Effects of methallibure (ICI 33,828) on the pars distalis of pituitary, testis and thumb pad of the male green frog, *Rana esculenta* L. Zeitschrift für Zellforschung 123: 430-440.

Resink, J. W., Van den Hurk, R., Groeninx van Zoelen, P. F. and Husiman, A. 1987. The seminal vesicle as source of sex attracting substances in the African catfish, *Clarias gariepinus*. Aquaculture 63: 97-114.

Rivero, J. A. 1970. On the origin, endemism and distribution of the genus *Stephania* Rivero (Amphibia, Salientia) with a description of a new species from southeastern Venezuela. Boletín de la Sociedad Venezolana de Ciencias Naturales 28: 456-481.

Rogoff, J. L. 1927. The hedonic glands of *Triturus viridescens*: a structural and morphological study. Anatomical Record 34: 132-133.

Rollmann, S. M., Houck, L. D. and Feldhoff, R. C. 1999. Proteinaceous pheromone affecting female receptivity in a terrestrial salamander. Science 285: 1907-1909.

Saidapur, S. K. and Nadkarni, V.B. 1975. The effect of 17b-estradiol acetate on the testis and thumb pad of *Rana cyanophylyctis* (Schn.) and *Rana tigrina* (Daud.) General and Comparative Endocrinology 27: 350-357.

Salthe, S. N. and Duellman, W. 1973. Quantitative constraints associated with reproductive mode in anurans. Pp. 229-249. In J. L. Vial (ed.), *Evolutionary Biology of the Anurans: Contemporary Research on Major Problems*, University of Missouri Press, Columbia.

Salthe, S. N. and Mecham, J. S. 1974. Reproductive and courtship patterns. Pp. 309-521. In B. Lofts (ed.), *The Physiology of Amphibia*, Volume II, Academic Press, New York.

Savage, R. M. 1961. *The Ecology and Life History of the Common Frog (Rana temporaria temporaria)*. Pitman and Sons, London. 221 pp.

Sever, D. M. 1975a. Morphology and seasonal variation of the nasolabial glands of *Eurycea quadridigitata* (Holbrook). Journal of Herpetology 9: 337-348.

Sever, D. M. 1975b. Morphology and seasonal variation of the dwarf salamander, *Eurycea quadridigitata* (Holbrook). Herpetologica 31: 241-251.

Sever, D. M. 1976a. Morphology of the mental hedonic gland clusters of plethodontid salamanders (Amphibia: Caudata). Journal of Herpetology 10: 227-239.

Sever, D. M. 1976b. Induction of secondary sexual characters in *Eurycea quadridigitata*. Copeia 1976: 830-833.

Sever, D. M. 1981. Cloacal anatomy of male salamanders in the families Ambystomatidae, Salamandridae and Plethodontidae. Herpetologica 37: 142-155.

Sever, D. M. 1988. Male *Rhyacotriton olympicus* (Dicamptodontidae: Urodela) has a unique cloacal vent gland. Herpetologica 44: 274-280.

Sever, D. M. 1989. Caudal hedonic glands in salamanders of the *Eurycea bislineata* complex (Amphibia: Plethodontidae). Herpetologica 45: 322-329.

Sever, D. M. 1991. Comparative anatomy and phylogeny of the cloacae of salamanders (Amphibia: Caudata). I. Evolution at the family level. Herpetologica 47: 165-193.

Sever, D. M. 1992a. Comparative anatomy and phylogeny of the cloacae of salamanders (Amphibia: Caudata). III. Amphiumidae. Journal of Morphology 211: 63-72.

Sever, D. M. 1992b. Comparative anatomy and phylogeny of the cloacae of salamanders (Amphibia: Caudata). IV. Salamandridae. Anatomical Record 232: 229-244.

Sever, D. M. 1992c. Comparative anatomy and phylogeny of the cloacae of salamanders (Amphibia: Caudata). V. Proteidae. Herpetologica 48: 318-329.

Sever, D. M. 1992d. Comparative anatomy and phylogeny of the cloacae of salamanders (Amphibia: Caudata). VI. Ambystomatidae and Dicamptodontidae. Journal of Morphology 212: 305-322.

Sever, D. M. 1994. Comparative anatomy and phylogeny of the cloacae of salamanders (Amphibia: Caudata) VII. Plethodontidae. Herpetological Monographs 8: 276-337.

Sever, D. M. and Brizzi, R. 1998. Comparative Biology of sperm storage in female salamanders. Journal of Experimental Zoology 282: 460-476.

Sever, D. M., Rania, L. C. and Brizzi, R. 2000. P. 95. Sperm storage in the class Amphibia. XVIIIth International Congress of Zoology. The New Panorama of Animal Evolution. Athens, Greece.

Sever, D. M., Sparreboom M., and Schultschik G. 1997. The dorsal tail tubercle of *Mertensiella caucasica* and *M. luschani* (Amphibia: Salamandridae). Journal of Morphology 232: 93-105.

Sever, D. M., Verrell, P. A., Halliday, T. R., Griffiths, M. and Waights, V. 1990. The cloaca and cloacal glands of the male smooth newt, *Triturus vulgaris vulgaris* (Linnaeus), with special emphasis on the dorsal gland. Herpetologica 46:160-162.

Sluiter, J. W., Von Oordt, G. J. and Mighorst, C. A. 1950. A study of the testis tubules, interstitial tissue and sex characters (thumb-pads and wolffian ducts) of normal and hypophysectomized frogs. Quarterly Journal of microscopical Science 91: 131-144.

Sokol, O. M. 1959. Studien an pipiden Fröschen I. Die Kaulquappe von *Hymenochirus curtipes* Noble. Zoologischer Anzeiger 162: 272-284.

Stebbins, R. C. and Hendrickson, J. R. 1959. Field studies of amphibians in Columbia, South America. University of California publications in Zoology 56: 497-540.

Sullivan, B. K., Ryan, M. J. and Verrell, P. A. 1995. Female choice and mating system structure. Pp. 469-517. In H. Heatwole and B. K. Sullivan (eds), *Amphibian Biology*, Volume II, Surrey Beatty and Sons, Chipping Norton, Australia.

Taylor, E. H. 1962. The amphibian fauna of Thailand. University of Kansas Science Bulletin 43:265-599.

Taylor, E. H. 1968. *The caecilians of the world: a taxonomic review*. University of Kansas Press, Laurence, Kansas, U.S.A. 848 pp.

Terreni, A., Nosi, D., Brizzi, R., and Delfino, G. 2002. Cutaneous serous glands in South-American anurans: an ultrastructural comparison between hylid and pseudid species. Italian Journal of Zoology 69: 115-123.

Testa Riva, F., Serra, G. P., Loffredo, F. and Riva, A. 1993. Ultrastructural study of the mental body of *Hydromantes genei* (Amphibia: Plethodontidae). Journal of Morphology 217: 75-86.

Thomas, E .O., Carroll, E. J. jr. and Ruibal, R. 1990. Immunohistochemical localization of the peptide sauvagine in the skins of phyllomedusine frogs. General and Comparative Endocrinology 77: 298-308.

Thomas, E. O. and Licht, P. 1993. Testicular and androgen dependence of skin gland morphology in the anurans, *Xenopus laevis* and *Rana pipiens*. Journal of Morphology 215: 195-200.

Thomas, E. O., Tsang, L. and Licht, P. 1993. Comparative histochemistry of the sexually dimorphic skin glands of anuran amphibians. Copeia 1993: 133-143.

Toledo, R. C. and Jared, C. 1995. Cutaneous granular glands and amphibian venoms. Comparative Biochemistry and Physiology 1. IIIA: 1-29.

Townsend, D. S, Stewart, M. M. and Pough, F. H. 1984. Male parental care and its adaptive significance in a neotropical frog. Animal Behavior 32: 421-431.

Trivers, R. L. 1972. Parental investment and sexual selection. Pp. 136-179. In B. G. Campbell (ed.), *Sexual Selection and the Descent of Man*, Aldine Press, Chicago.

Trueb, L. 1974. Systematic relationships of Neotropical horned frogs, genus *Hemiphractus* (Anura: Hylidae). Occasional Papers of the Museum of Natural History, University of Kansas 29: 1-60.

Truffelli, T. G. 1954. A macroscopic and microscopic study of the mental hedonic gland-clusters of some Plethodontid salamanders. University of Kansas Science Bulletin 36: 3-39.

Tyler, M. J. 1976. *Frogs*. Collins, Ltd., Sydney. 256 pp.

Tyler, M . J. 1985. Reproductive modes in Australian Amphibia. Pp. 265-267. In G. Grigg, R. Shine and H. Ehmann (eds), *Biology of Australasian Frogs and Reptiles*. Royal Zoological Society of New South Wales.

Tyler, M. J. and Davies, M. 1979. Foam nest construction by Australian leptodactylid frogs (Amphibia, Anura, Peptodactylidae). Journal of Herpetology 13: 509-510.

Vahamaki, V. and Thomas, E. O. 1997. Effects of a gonadotropin-releasing hormone agonist on cutaneous breeding glands in the dwarf African frog, *Hymenochirus curtipes*. Abstract presented at the Combined Western/Southwestern Regional Conference on Comparative Endocrinology, Denver, CO.

Verma, K. 1965. Regional differences in skin gland differentiation in *Rana pipiens*. Journal of Morphology 117: 73-86.

Vialli, M. 1946a. Contributo alla conoscenza istologica delle ghiandole del pollice di *Rana esculenta*. Bollettino-Società Italiana Biologia Sperimentale 21: 309.

Vialli, M. 1946b. Caratteristiche istochimiche delle ghiandole del pollice di *Rana esculenta* durante il periodo della copula. Bollettino-Società Italiana Biologia Sperimentale 21: 310.

Vialli, M. 1946c. Significato morfologico della ghiandola del pollice di *Rana esculenta*. Bollettino-Società Italiana Biologia Sperimentale 21: 311.

Visser, J., Cei, J. M., and Gutierrez L. S. 1982. The histology of dermal glands of mating *Breviceps* with comments on their possible functional value in microhylids (Amphibia: Anura). South African Journal of Science 17: 24-27

Wabnitz, P. A., Bowie, J. H., Tyler, M. J., Wallace, J. C. and Smith, B. P. 1999. Aquatic sex pheromone from a male tree frog. Nature 401: 444-445.

Wager V. A. 1965. *Th frogs of South Africa*. Purnell and Sons, Capetown, Johannesburg. 242 pp.

Wake, M. H. 1982. Diversity within a framework of constraints. Amphibian reproductive modes. Pp. 87-106. In D. Mossakowski and G. Roth (eds), *Environmental Adaptation and Evolution*, Gustav Fischer Verlag, Stuttgart.

Weichert, C .F. 1945. Seasonal variation in the mental gland and reproductive organs of the male *Eurycea bislineata*. Copeia 1945: 78-84.

Wells, K. D. 1977a. The courtship of frogs. Pp. 233-262. In D. H. Taylor and S. I. Guttman (eds), *The reproductive biology of amphibians*. Plenum Press, New York.

Wells, K. D. 1977b. The social behaviour of anuran amphibians. Animal Behaviour 25: 666-693.

Wells, K. D. 1978. Courtship and parental behavior in a Panamian poison-arrow frog (*Dendrobates auratus*). Herpetologica 34: 148-155.

Wells, K. D. 1981. Parental behavior in male and female frogs. Pp. 184-197. In D. R. Alexander and D. W. Tinkle (eds), *Natural selection and social behavior: recent research and new theory.* Chiron Press, Newton, Massachusetts.

Weygoldt, P. 1976. Beobachtung zur Biologie und Ethologie von Pipa (Hemipipa) carval hoi Mir. Rib. 1937 (Anura, Pipidae). Zeitschrift für Tierpsychologie 40: 80-99.

Weygoldt, P. 1980. Complex brood care and reproductive behavior in captive poison-arrow frogs, *Dendrobates pumilio*. Behavioral Ecology and Sociobiology 7: 329-332.

Whitear, M. 1977. A functional comparison between the epidermis of fish and of amphibians. Pp. 291-313. In R. I. C. Spearman (ed.), *Comparative Biology of the skin.* Academic Press, London.

Williams, A. A. 1978. Morphology and histochemistry of mental hedonic glands in *Eurycea, Desmognathus,* and *Plethodon* (Amphibia: Plethodontidae). American Zoologist 18: 601.

Williams, T. A. and Larsen, J. H. 1986. New function for the granular skin glands of the eastern long-toed salamander, *Ambystoma macrodactylum colombianum*. Journal of Experimental Zoology 239: 329-333.

Wright, A. H. and Wright, A. A. 1949. *Handbook of frogs and toads of the United States and Canada.* Comstock Publishing Co., Ithaca, New York, NY 640 pp.

Zasloff, M., Martin, B. and Chen, H.-C. 1988. Antimicrobical activity of synthetic magainin peptides and several analogues. Proceedings of the National Academy of Sciences of the United States of America 85: 910-913.

CHAPTER 7

Internal Fertilization in the Anura with Special Reference to Mating and Female Sperm Storage in *Ascaphus*

David M. Sever[1], William C. Hamlett[2], Rachel Slabach [2], Barry Stephenson[3] and Paul A. Verrell[3]

7.1 INTERNAL FERTILIZATION WITHIN THE ANURA AND OTHER ANAMNIOTES

Of the 5000+ species of anurans currently recognized, only a few species are known or suspected to practice internal fertilization. These species include *Ascaphus truei* and *A. montanus* in the Ascaphidae from the northwestern contiguous United States and SW British Columbia; four genera of Bufonidae from Africa (Wake 1980; Grandison and Ashe, 1983), and *Eleutherodactylus jasperi* (Wake, 1978) and *E. coqui* (Townsend *et al.* 1981) within the Leptodactylidae from Puerto Rica (Table 7.1). All of these bufonids and leptodactylids probably accomplish internal fertilization by cloacal apposition.

Ascaphus truei and *A. montanus* are the only species known to engage in copulation (the latter species recently was recognized on the basis of phylogeographic analysis of molecular markers; Nielson *et al.* 2001). The male of each possesses a copulatory organ, usually referred to as a "tail" but which we herein formally designate as a penis. When engorged, the penis forms a sulcus for passage of sperm and is inserted in the cloaca of the female (Noble 1925; Noble and Putnam 1931; Slater 1931). The original description of *A. truei* by Stejneger (1899) was based upon a single specimen that

[1]Department of Biology, Saint Mary's College, Notre Dame, Indiana 46556 USA
[2]Indiana University School of Medicine, Indiana Medical Center, Notre Dame, Indiana 46556 USA
[3]School of Biological Sciences and Center for Reproductive Biology, Washington State University, Pullman, Washington 99164 USA

Table 7.1 Internal fertilization within the Anura

Family	Species	Distribution	Reference
Ascaphidae	Ascaphus truei	Coastal NW North America	Sever et al. 2001
	A. montanus	Inland NW North America	Stephenson and Verrell 2003
Bufonidae	Altiphrynoides malcomi	Balé Province, Ethiopia	Wake 1980
	Nectophrynoides tornieri	Uluguru Mts, Tanzania	
	Nectophrynoides vivipara	Central to SW Tanzania	
	Nectophrynoides liberiensis	Liberia	
	Nimbaphrynoides occidentalis	Guinea, Ivory Coast	
	Mertensophryne micranotis	Kenya, Tanzania	Grandison and Ashe 1983
Leptodactylidae	Eleutherodactylus jasperi	Puerto Rico	Wake 1978
	Eleutherodactylus coqui	Puerto Rico	Townsend et al. 1981

"evidently was a female" (Gaige 1920:259), and the first observations of the male "tail" were not reported until Van Denburgh (1912:261), who suggested the tail "may be a sexual organ." Noble (1925:17) suggested that the "tail is pressed against the cloaca of the female in copulation," and was the first to report sperm in the oviducts of females. Intromission of the tail in copulation was first described by Noble and Putnam (1931). Van Dijk (1955:65) reported that the "turgid 'tail' can only be applied to the cloacal orifice of the female and not inserted into it." Sever et al. (2001), however, reported that the entire turgid member is completely inserted into the female, and introduced the term copulexus to describe the unique form of mating that results in internal fertilization in the genus Ascaphus. A pair of A. truei in copulexus is illustrated in Fig. 7.1. Note that the male clasps the female inguinally rather than axillary as in most frogs. Stephenson and Verrell (in press) provide a detailed study of copulexus in A. montanus, and their findings are summarized below. Copulation has been assumed to be an adaptation that ensures fertilization in fast-moving stream water that forms the breeding habitat of Ascaphus (Stebbins and Cohen 1995).

Ascaphus truei is the only anuran in which oviducal sperm storage has been reported (although the same almost certainly is true for A. montanus). Indeed, the only other anamniotes in which oviducal sperm storage is known are elasmobranchs (Pratt 1993; Hamlett et al. 1998; Hamlett and Koob 1999) in the class Chondrichthyes, which is not considered the sister taxon of Amphibia. Females of some teleosts in the Osteichthyes store sperm (Howarth 1974), but they lack homologues to the oviduct (Kardong 1995). Instead sperm are stored in either the ovary or in a gonaduct (ovarian duct) formed from ovarian tissue (Howarth 1974; Constanz, 1989).

The extant representatives of Actinistia and Dipnoi, descendant taxa of sarcopterygiian sister groups of amphibians (Schultze 1994), possess oviducts

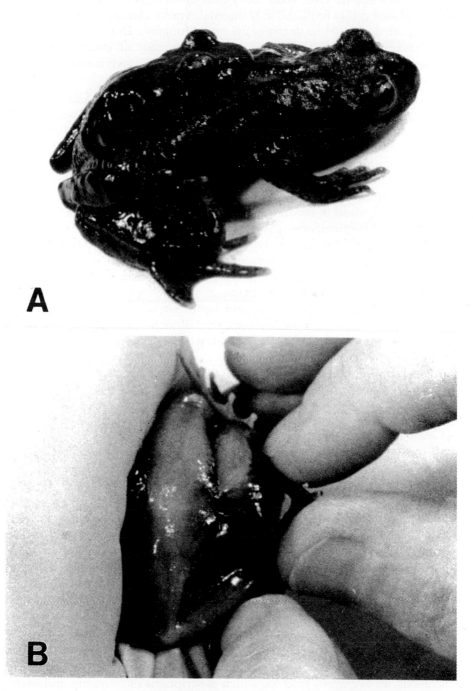

Fig. 7.1 *Ascaphus truei* in copulexus. **A**. The male is dorsal and grasps the female inguinally. **B**. Insertion of the penis into the cloaca of the female. Original.

(Millot and Anthony 1960; Wake 1987), and *Latimeria* is viviparous (Smith *et al.* 1975) indicating that fertilization is internal. Sperm storage, however, has not been reported in *Latimeria* or any of the extant lungfish. Thus, although internal fertilization could have characterized the ancestor of extant amphibians (Jamieson *et al.* 1993), oviducal sperm storage is unknown neontologically in the descendants of these presumed ancestral groups.

The Lissamphibia is generally considered monophyletic and consists of three groups, the Anura (frogs), Urodela (salamanders), and Gymnophiona (caecilians). Most evidence supports a frog + salamander clade (Pough *et al.* 1998). Sperm storage is unknown in female caecilians, even though internal fertilization apparently occurs in all taxa, and many species are viviparous (Wilkinson and Nussbaum 1998). Sperm storage occurs in all females found in the seven families of salamanders that comprise the suborder Salamandroidea (Sever 2002). Instead of oviducal sperm storage, however, sperm are stored in cloacal glands (spermathecae) that consist of a single compound tubulo-alveolar gland (Plethodontidae) or numerous simple tubular glands (other families). The ancestral condition for salamanders is lack of sperm storage glands, a condition found in three families (Sever 2002). Obviously, research needs to be conducted to determine whether oviducal sperm storage occurs in caecilians and internally-fertilizing bufonids and leptodactylids. *Ascaphus*, however, is not the sister taxon of any caecilian or of the other internal fertilizing anurans, and therefore oviducal sperm storage must be considered independently derived in this genus.

7.2 COURTSHIP AND MATING

7.2.1 Bufonidae

At least six species of small-bodied, African toads apparently exhibit internal fertilization, and sperm transfer probably is achieved by cloacal apposition in all (Wake 1980).

Five of these species formerly were assigned to the genus *Nectophrynoides*; however, recent systematic work suggests that they may not be especially closely related, and perhaps not even a monophyletic assemblage (Dubois 1986; Graybeal and Cannatella 1995). Thus, internal fertilization may have evolved independently more than once in these toads. Reproductive modes among the five species are diverse, and include oviparity, ovoviviparity and, perhaps, "true" viviparity (we acknowledge the controversy surrounding use of these terms: see Lombardi 1998, Greven 2002). Only in *Altiphrynoides malcolmi* have mating behavior patterns been observed in any detail. Males appear to amplex females in a belly-to-belly posture in this species (Grandison 1978; Wake 1980).

The sixth African bufonid that practices internal fertilization is *Mertensophryne micranotis*, an oviparous toad for which descriptions of courtship and mating are quite extensive (Grandison and Ashe 1983). Amplexus is axillary, with the male typically clasping the female from above. However, belly-to-belly amplexus also occurs in this species and may facilitate sperm transfer by cloacal apposition. Grandison (1980) reported that breeding males develop spiny protuberances on their ventral surfaces, and she suggested that these may aid in securing a firm purchase on the female during amplexus.

7.2.2 Leptodactylidae

The neotropical genus *Eleutherodactylus* is perhaps the most speciose of all vertebrate genera, with more than 500 species currently described. All are believed to exhibit direct development. One species is known to, and one is assumed to, practice internal fertilization. Both are from Puerto Rico: *E. coqui* (oviparous) and *E. jasperi* (ovoviviparous). They probably achieve sperm transfer by cloacal apposition.

Mating in *E. coqui* was described by Townsend and Stewart (1986; see also Townsend *et al.* 1981). The male perches on the female's back, his cloaca close to hers, but does not clasp his partner with his limbs; thus, while there is close physical contact, there is no true amplexus. Nothing is known about the mating habits of *E. jasperi*, but Wake (1978) hypothesized that this species must practice internal fertilization given that embryos are retained within female oviducts (the same logic applies to some of the African bufonids discussed above).

7.2.3 Ascaphidae

The tailed frog *Ascaphus truei* and its sibling species *A. montanus* (Nielson *et al.* 2001) are in many ways "odd men out" among internally-fertilizing taxa of anurans. The genus *Ascaphus* is basal phylogenetically, restricted to the temperate zone, and its two species mate and lay eggs in the cold, clear water of mountain streams.

Van Denburgh (1912) described the first male specimen of this genus and suggested that the male's "tail" might serve as an intromittent organ. Such use later was confirmed by Slater (1931) and Noble and Putnam (1931). Jameson (1955) suggested that males searching for females along the rocky substrates of streams accomplish mate location visually. Subsequent observations by Metter (1964a,b) and Wernz (1969) showed that males are capable of multiple matings over relatively short periods of time, and that *Ascaphus* can mate in the spring, instead of exclusively in the fall, as had been assumed previously. Stephenson and Verrell (2003) collected breeding *A. montanus* from a stream in northwestern Idaho and obtained detailed descriptions of courtship and mating in the laboratory. Their major findings are summarized below.

In agreement with Jameson (1955), Stephenson and Verrell observed that males seem to respond to females (and other males) visually. Males apparently do not call to attract females but both sexes may produce soft clicking sounds when handled. Most mating attempts begin with the male making a rapid lunge towards the female and then clasping her with one or both of his forelimbs to secure inguinal amplexus in a 'male dorsal' position. It is while engaged in amplexus that the male inserts his "tail" into the female's cloaca to initiate what Sever *et al.* (2001) have termed copulexus. While so engaged the male kicks with his hind limbs (although these kicks seldom result in movement of the pair), makes thrusting movements of his pelvic area and squeezes the female with his clasping forelimbs. These three male activities may provide tactile stimulation to the female. Episodes of copulexus can be of long duration (24-30 hr; Metter 1964b), and several bouts of intromission may occur during a single sexual encounter.

Stephenson and Verrell also observed several instances in which inguinal amplexus was achieved in a belly-to-belly posture that they termed ventral amplexus. This posture was seen occasionally in encounters between single males and females; ventral amplexus also was adopted by one male in a trio, while the other male amplexed the same female from above. Stephenson and Verrell were unable to determine whether intromission occurs during ventral amplexus. However, as noted earlier, belly-to-belly clasping with apparent sperm transfer has been observed in some internally-fertilizing African bufonids.

7.2.4 Why do Some Anurans Fertilize their Eggs Internally?

Many aspects of the natural history of the 10 species of anurans known or believed to practice internal fertilization appear to be highly variable. Most species are of relatively small body size; Salthe and Duellman (1973) suggested that "evolutionary experimentation" in aspects of reproductive biology might be accomplished most readily in small species of anurans. Reproductive modes among these 10 species certainly are diverse, and include both direct and indirect development as well as oviparity (probably the ancestral mode for anurans), retention of eggs until hatching (ovoviviparity) and, at least in the bufonid *Nimbaphrynoides occidentalis* (Wake 1993), nourishment of young hatched internally ("true" viviparity). In addition, these species breed in a variety of both terrestrial and aquatic habitats.

No clear correlates of internal fertilization in these anurans are apparent to us. However, we suggest that internal fertilization may have been favored historically (if not maintained contemporarily) by at least two conditions that are not mutually exclusive:

1. The threat of severe sperm competition from rival males if sperm are liberated into the external environment (a male benefit in the context of sexual selection). Internal sperm competition then becomes likely if females mate with multiple males before fertilizing their eggs (Halliday and Verrell 1984).

2. Reduced success of external fertilization via gamete desiccation (e.g., *Eleutherodactylus*), stream drift (e.g., *Ascaphus*) or predation (natural selection of benefit to both sexes).

Regardless of the selection pressure(s) involved, the evolution of internal fertilization in anurans surely facilitated the acquisition of new modes of reproduction. With the subsequent ability to retain developing embryos, some internally-fertilizing species may have been able to exploit ecological niches that would have been closed to them otherwise.

7.3. STRUCTURE OF THE PENIS OF *ASCAPHUS TRUEI*

Males of *Ascaphus truei* are unique among anurans in the possession of a copulatory organ containing cavernous tissue, cartilaginous support rods, and specialized cloacal ("proctodeal") glands (male *A. montanus* probably will prove to be similar when studied morphologically). Because a male copulatory organ containing cavernous tissue is typically referred to as a penis, we apply that term for this species, although the structure has traditionally been called simply the "tail." Anatomical descriptions of the organ are provided by Noble (1922, 1925), DeVilliers (1934, 1935), and Van Dijk (1955, 1959), although it is unclear whether any of these authors actually histologically examined a fully erect penis. Flaccid and erect penes of *A. truei* are shown in Fig. 7.2. Note that

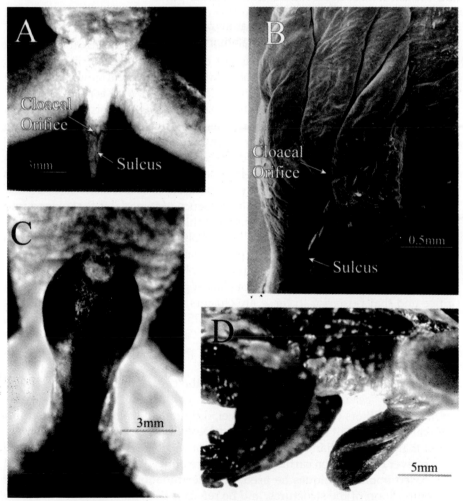

Fig. 7.2 The penis of male *Ascaphus truei*. **A**. Flaccid penis, ventral view. **B**. Scanning electron micrograph of a flaccid penis. **C**. Erect penis, ventral view. **D**. Erect penis, lateral view. Original.

the penis is pulled ventrally and anteriorly during erection, and that the length and the diameter of the distal end enlarge 2.5-3X. A correlation exists in vertebrates between possession of a relatively large penis and possession of a rigid trunk, reduced tail and a posterior mating position (Frey 1994).

Here we supplement Van Dijk's classic drawings of histological sections through flaccid penes by providing light micrographs of sections through an erect penis (Figs. 7.3-7.7). In order to show transverse sections through the entire penis, the micrographs were taken at very low magnification (8X), so they do not clearly illustrate the histological details of all the tissues. However, the overall relationships are shown sufficiently to allow comparison of our observations to past work. We plan to describe histology in more detail and add ultrastructural observations in a separate paper.

Van Denburgh (1912: 261) was the first one to describe a male *A. truei* and declared, "By far the most remarkable external feature of the toad is the *tail!*" Noble (1922), however, noted that the "tail" of metamorphosed *A. truei* is not retention of a larval structure and is not otherwise a true tail in any sense, because the structure is not associated with the vertebral column and the cloaca opens along its length. Thus, the "tail" of male *A. truei* is purely a cloacal organ. Noble (1922: 34) goes on to state that the organ "consists chiefly of an outer and an inner layer of muscular tissue. The outer layer is free both dorsally and laterally from the inner layer. The vascularity is greatest distally and ventrally. The inner wall of the deep layer, in other words, the surface of the cloaca, exhibits numerous longitudinal furrows. The rectus abdominis seems to form the anterior part of this structure."

Noble (1922) also commented on the presence of two plates that extend into the tail from a tendinous connection to the ischium, and described as "calcified cartilage" by Van Denburgh (1912: 264). Noble considered these structures cartilaginous and proposed that they provide support for the copulatory organ. DeVilliers (1934:35) stated that the rods "probably act as *ossa penis.*" Noble (1925) went on to report that a ring of cloacal glands occurs inside the cloacal orifice and that posterior to these are clusters of horny spines (illustrated by Noble 1931: Fig. 154). Noble and Putnam (1925) stated that the cavernous rods in the copulatory organ of male *A. truei* are "apparently" homologous to the corpora cavernosa of higher vertebrates, a position not held by Van Djyk (1955).

Fejérváry (1923: 179) proposed a name for the cartilaginous structures found by Noble, the "*cartilago abdominalis Nobleiana* (comprising a *cornu dexter* and *c. sinister*)". DeVilliers (1933, 1934), however, did not believe that these structures were cartilaginous but rather bone. Indeed, DeVilliers (1934: 26) quite adamantly stated that "the Nobelian 'cartilages' were not cartilages: in fact, they show no trace of ever having been cartilaginous, but are bony." Van Dijk (1955) reported that, based upon techniques he used, no determination could be made about the composition of the structures, and he referred to them as the rods of Noble or *rudes Nobleianae*. Our examination of the rods indicates that they are not similar in histology to typical bone elsewhere in *Ascaphus*, and that they appear to be a modified type of cartilage. Certainly, the "enormous marrow cavities" mentioned by DeVilliers (1933) are lacking in our specimens. The rods possess an outer dense connective tissue capsule (called the "cortex" by Van Dijk and others) and an inner "medulla" composed of tightly packed lacunae each apparently containing a single cell. Very little inter-lacunal substance is present, and canals are lacking (as also noted by DeVilliers 1934).

Other aspects of the histology of the penis are not as controversial. DeVilliers (1933: 693) provides a concise summary: "The cloaca is continued into the tail, at the end of which it has a groove-like opening. The cloacal opening has large dermal proctodeal glands. The "tail" is a mass of spongiose fibres [sic], the interstices of the network filled with blood. This erectile tissue is distributed into a pair of strands ventral to the cloaca and lateral to the Nobelian bones, and a strand is pierced by the cloaca. The skin is separated from the erectile tissue by large subdermal lymph spaces, but is adherent ventrally and middorsally."

Van Djyk (1955) provided much additional detail and should be consulted for a comprehensive treatment of cloacal and penile anatomy in *A. truei*. In this review, we will discuss and compare our observations with those primarily of Van Djyk concerning muscular layers, the epithelial lining, and the vascular supply.

7.3.1 Muscle Layers

Van Dijk indicates that inner circular and outer longitudinal layers of smooth muscle occur, and this tunica muscularis is bordered most superficially by the striated *M. compressores cloacae*. The compressores cloacae may be the same as the *M. sphincter ani cloacalis* of other frogs (Noble 1922), but Van Dijk preferred compressores cloacae (a name apparently coined by DeVilliers 1934) because the muscle is a cloacal depressor and not an anal sphincter. The compressores cloacae has attachments to the Nobelian rods anteriorly. Van Djyk (1955) stated that during erection of the penis, contraction of this muscle apparently is responsible for rotation of the rods ventrally, because this is the only muscle inserting on them. However, we find that another muscle inserts on the bases of the rods (Fig. 7.3), and the thick, parallel nature of the fibers makes us believe that this muscle could be the rectus abdominis (as reported by Noble 1922) or the pectineus (as suggested by DeVillers 1934), although we have not traced the muscle further anteriorly.

We find that the compressores cloacae varies along the length of the penis. At the base, the fibers are circular dorsally and longitudinal ventrally where they attach to the Nobelian rods (Fig. 7.3). Distally, blood vessels invade the muscle layers just superior to the rods and form large blood sinuses deep to the compressores cloacae along the remainder of the penis (Figs. 7.4-7.6). As the blood sinuses appear, the fibers become circular around the entire cloacal cavity (Fig. 7.4B) and remain as such to the cloacal orifice, when once again the fibers are more longitudinal in orientation (Fig. 7.6A,B). The fibers continue almost to the distal tip of the penis (Fig. 7.6C).

Van Djyk (1955) indicates that beginning in the anterior part of the cloacal tube, the longitudinal and circular muscle layers of the tunica muscularis are

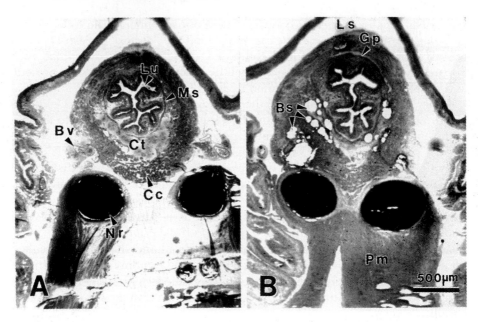

Fig. 7.3 Sections through the base of the penis of *Ascaphus truei*. **A**. The most proximal. **B**. Blood vessels in A form sinuses within the compressors cloacae. Abbreviations: Bs, blood sinuses; Bv, blood vessels; Cc, compressores cloacae; Ct, connective tissue; Gp, gliding plane; Ls, lymph sac; Lu, cloacal lumen; Ms, muscularis; Nr, Nobelian rods; Pm, parallel skeletal muscle fibers. Original.

not in contact but separated by "gliding planes." These planes are recesses that extend from the cloacal orifice to form cavities around the dorsal and lateral cloaca. DeVilliers (1934: 40) suggested that the gliding planes are "enlargements" of more lateral and anterior lymph spaces, but we found no connection between the gliding planes and the lymph sacs. We easily recognize the circular layer of the tunica muscularis, which occurs deep to the lamina propria from the base of the penis (Fig. 7.3B) to the region of the cloacal glands (Fig. 7.4D). However, we do not find a longitudinal layer superficial to the gliding plane. The gliding plane is present dorsally from the base of the penis (Fig. 7.3A) to the tip of the penis (Fig. 7.6) and reaches its greatest development laterally at the proximal appearance of the cloacal glands (Fig. 7.4D). The striated compressores cloacae is the only muscle superficial to the gliding plane.

7.3.2. Epithelium

The epithelial lining of the anterior portion of the cloaca consists of goblet cells interspersed among ciliated columnar cells that appear stratified into two layers. The lamina propria contains numerous collagen fibers, blood vessels, and nerve bundles. The mucosa (epithelium + lamina propria) is folded into longitudinal rugae. Staining of the goblet cells is apparent in Fig. 7.4C. With the appearance of the cloacal glands, which occur just inside the cloacal orifice, the epithelium gradually changes into a stratified cuboidal form that lacks cilia

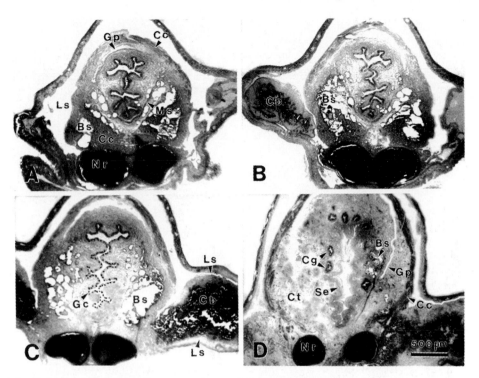

Fig. 7.4 Sections through successive regions of the penis of *Ascaphus truei* distal to those in Fig. 7.3. **A**. The more proximal. **B**. Note especially the proximal appearance of the cavernous body. **C**. The bodies are laterally bordered by the lymph sacs. **D**. The lining of the cloaca changes concomitant with the appearance of cloacal glands. Abbreviations: Bs, blood sinuses; Cb, cavernous body; Cc, compressores cloacae; Cg, cloacal gland; Ct, connective tissue; Gc, goblet cells; Gp, gliding plane; Ls, lymph sac; Ms, muscularis; Nr, Nobelian rods; Se, stratified epithelium. Original.

and goblet cells, and receives the ducts of the cloacal glands (Figs. 7.4D, 7.5B). Distal to the cloacal orifice, the lining is highly stratified and is similar to the surrounding epidermis but even thicker (Figs. 7.5-7.7). "Horny spines" were noted by both Noble (1925) and Van Dijk (1955), but not DeVilliers (1934), just posterior to the cloacal glands. DeVillers (1934:36) noted "…the spines described by Noble are bulges made in the cloacal wall by the large dermal glands. At any rate, it is quite impossible to miss spinose cuticular structures in sections." Projections in these areas do not appear spine-like to us. Rather, we recognize a series of thickenings in the stratum corneum inside the cloacal orifice (Fig. 7.6A, Ep).

When sperm occur in the cloaca, they are in irregular bundles and often associated with a secretory product (Fig. 7.5A), perhaps from the cloacal glands. Cloacal glands "arise as involutions of the epithelium" (Van Dijk 1955: 32) and are simple tubuloalveolar or branched tubuloalveolar glands. Van Dijk (1955) noted that as many as 30 lobules may exist. The acini of the glands consist of columnar cells with basal nuclei (Fig. 7.5B), and the secretory product

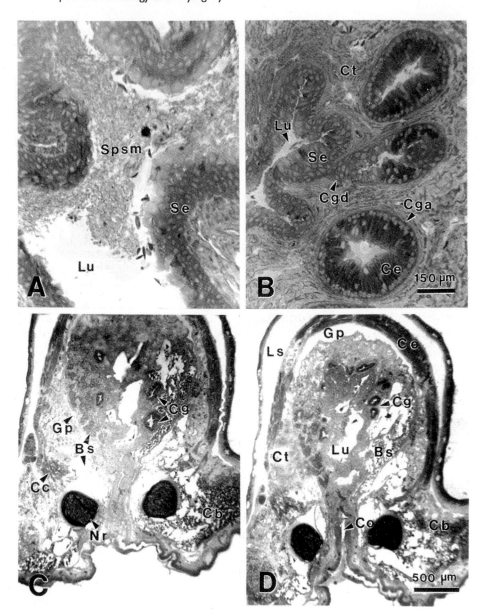

Fig. 7.5 Sections through the penis of *Ascaphus truei*. **A**. Sperm within secretory material in the cloaca. **B**. Detail of cloacal glands and a duct. **C**. The distal portion of the closed cloacal tube. **D**. The most proximal end of the cloacal orifice. Bs, blood sinuses; Cc, compressores cloacae; Cg, cloacal glands; Cga, cloacal gland ascinus; Cgd, cloacal gland duct; Co, cloacal orifice; Ct, connective tissue; Gp, gliding plane; Ls, lymph sac; Lu, cloacal lumen; Nr, Nobelian rods; Se, stratified epithelium; Spsm, sperm within secretory material. Original.

consists of granules that stain positively with bromphenol blue, an indicator of proteins. The granules do not stain positively with the carbohydrate stains that we utilized (periodic acid and Schiff's reagent or with alcian blue at pH 2.5), but staining affinity could have been affected by the nature of our samples (glycol methacrylate sections).

7.3.3   Vascular Supply and Cavernous Tissue

Van Djyk (1955: 33) stated, "There are fundamentally two lymph spaces in the "tail", a very large dorsal sac and a smaller ventral one, and these are subdivided posteriorly by bridges between the cloaca and the skin. These lymph spaces correspond to the *sacci interfemorales* of *Rana* ..." Van Djyk indicated that the dorsal space ends in the region of the cloacal glands, and that the ventral one continues to the cloacal orifice. Contrary to Van Djyk, we observe a large lymphatic sac deep to the dermis from the base of the penis (Fig. 7.3) to near the very tip (Fig. 7.6B). Further comparative work is necessary before considering this space to be the same as the *sacci interfemorales* of *Rana*, as proposed by Van Djyk, and here we simply refer to the cavity as a "lymph sac." The lymph sac is largest dorsally, and narrows both laterally and ventrally, where it passes around the cavernous bodies on each side (Figs. 7.4, 7.5). Right and left cavities do not join midventrally prior to the cloacal orifice (Fig. 7.5). Filling of lymphatic chambers is involved in the erection of the penis in some birds (Kardong 1998).

The notion that the cardiovascular supply to the cloacal region is related to the copulatory function of the penis originated from gross examinations. Noble (1922: 34) stated, "It is obviously very vascular and turgid with blood", and Slater (1931: 63) remarked, "The color at the base of the tail and the tail itself is reddish brown, indicating a rich blood supply." DeVillers (1934:35) stated, "...the vascularized tissue is very obviously erectile in nature." Van Djyk (1955) presented an intricately detailed description of the blood supply of the cloaca, but concedes (p. 44), "The direction in which blood flows in the sinuses in the 'tail' is difficult to conceive."

Of importance here is description of the cavernous tissue, which was first described by Noble and Putnam (1931). They reported (p. 99), "The organ is strengthened by two pairs of vascular pads extending the long axis of the structure under the skin on its ventral surface." We find these structures between the compressores cloacae and Nobelian rods medially and the lymph sacs laterally from the region where blood sinuses develop deep to the compressores cloacae (Fig. 7.4) to just distal to the cloacal orifice (Fig. 7.5D). Some smaller blood spaces in this area, however, are associated with the rods until the tip of the penis (Fig. 7.6). The cavernous bodies consist of large blood sinuses surrounded on the lymph sac border by a dense connective tissue capsule that is similar to the tunica albuginea of the mammalian penis. Fibrous trabeculae extend inward from the outer capsule, branching to form a complex and irregular framework around the blood sinuses. Because of the similarities in structure, the cavernous tissue of the penis of *A. truei* may function similar to that of mammals during erection (Noble and Putnam 1931; DeVilliers 1933; Van Djyk 1955). The blood sinuses in the connective tissue deep to the

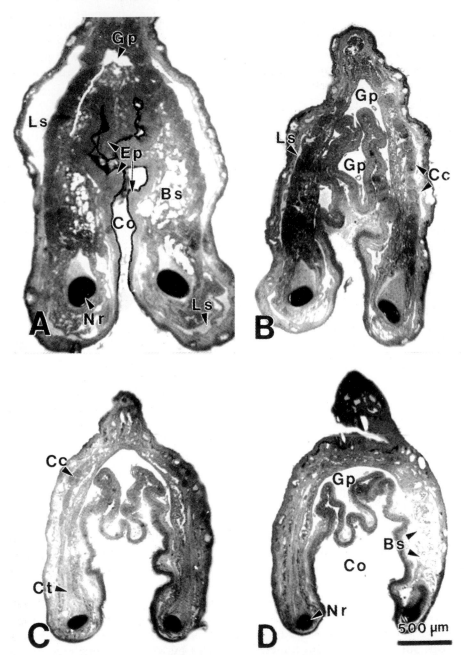

Fig. 7.6 A,B. Sections through successive regions of the penis of *Ascaphus truei* distal to those in Fig. 7.5. Note the thick projections of the epidermal lining of the cloacal cavity in A and the loss of the cavernous bodies. **C,D**. Tip of the penis, with D showing the more distal section. Some longitudinal fibers of the compressores cloacae are still present in A, but are absent in B. The Nobelian rods and blood sinuses remain to the tip. Bs, blood sinuses; Cc, compressores cloacae; Co, cloacal orifice; Ct, connective tissue; Ep, epidermal projections; Gp, gliding plane; Ls, lymph sac; Nr, Nobelian rods. Original.

compressores cloacae may also be involved in engorgement of the penis (see also DeVillers 1934).

7.4 FEMALE SPERM STORAGE IN *ASCAPHUS*

Noble (1925:17) reported, "My sections of the urinogenital organs of the breeding female reveal great masses of spermatozoa in the lumen of the oviducts and particularly in the glands along the posterior part of the oviducts. Sections of the oviducts of females taken after their eggs had been laid show many of the glands of the posterior oviduct still filled with spermatozoa." Noble, however, did not provide any illustrations. Van Dijk (1959, Fig. 32) illustrated a paraffin section through the oviduct of an *Ascaphus truei*, indicating the "spermia", but made no other comment. Metter (1964b) rinsed sperm from the oviducts of *A. montanus*, and he reported that sperm are stored in the lower, straight portion of the oviduct with none in the upper coiled portion. Sever *et al.* (2001) found sperm storage tubules limited to the anterior portion of the ovisac, the "straight portion" of the oviduct. The coiled portion, the ampulla, is the area where the gelatinous coats are applied to the eggs after they enter the oviduct. Fertilization probably occurs as the jelly-coated eggs pass through the ovisac. The jelly-coats are necessary for successful fertilization of anuran eggs (Barbieri and DelPino 1975; Wake and Dickie 1998).

7.4.1 Histology and Ultrastructure of Sperm Storage Tubules

Sperm storage tubules (Ssts) in *Ascaphus truei* (and presumably, *A. montanus*) are simple tubular or simple branched tubular glands that occur in the ovisac of the oviduct, a region distal to the portion (ampulla) where jelly-coats are applied to eggs (Fig. 7.8; Sever *et al.*, 2001).

The linings of the Ssts and the luminal epithelium of the oviduct are simple columnar epithelia and consist of secretory cells with microvilli interspersed among ciliated cells (Fig. 7.9). Intercellular canaliculi are wide, especially among groups of ciliated cells, and tortuous basally. Circular vacuoles of various sizes and electron densities occur in secretory cells and are especially numerous and variable in size in the distal portions of the Ssts (Fig. 7.9B,C). These vacuoles are intensely positive with periodic acid and Schiff's reagent, indicating the presence of neutral carbohydrates. They do not stain with alcian blue 8GX at pH 2.5, which suggests absence of carboxylated glycosaminoglycans, or with bromophenol blue, which suggests an absence of proteins (Sever *et al.* 2001). Both ciliated and secretory cells often possess vacuoles containing a flocculent material, especially in the epithelium of the oviducal lining.

Sperm are present in the lumen of both the oviduct and the Ssts (Fig. 7.9A,B). Sperm in the lumen usually are loosely aggregated and small clusters of sperm show similar orientations (Fig. 7.9D). Occasional groups of sperm in the oviducal lumen are found embedded in an acellular matrix composed of a uniformly electron-dense substance and small vesicles. Sperm are also frequently found embedded in the secretory epithelial cells, especially in the Ssts (Fig. 7.9C). Sperm nuclei are sometimes found deep in these cells, in the infranuclear

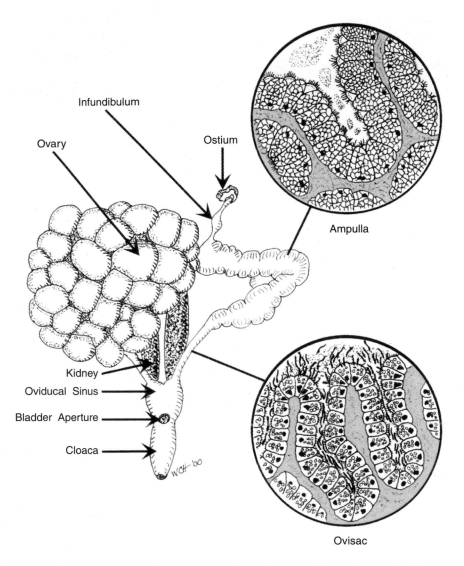

Fig. 7.7 Female reproductive system of *Ascaphus truei*. From Sever, D. M. *et al.* 2001. Journal of Morphology 248: 1-21, Fig. 2.

cytoplasm bordering the basal lamina. The nuclei are not vacuolated, appear normal in cytology, and are often in close proximity to secretory vacuoles.

The basic structure of the oviduct of *A. truei* is not notably modified from that of other anurans (Wake and Dickie, 1998), and the sperm storage tubules (Ssts) do not seem highly modified for sperm maintenance. Rather, the Ssts appear merely to be sites of temporary sperm residence. Sperm storage in relatively unmodified oviducal glands in *A. truei* indicates that this adaptation could convergently evolve in the similar oviducal "bauplan" of internal fertilizing frogs in the Bufonidae and Leptodactylidae.

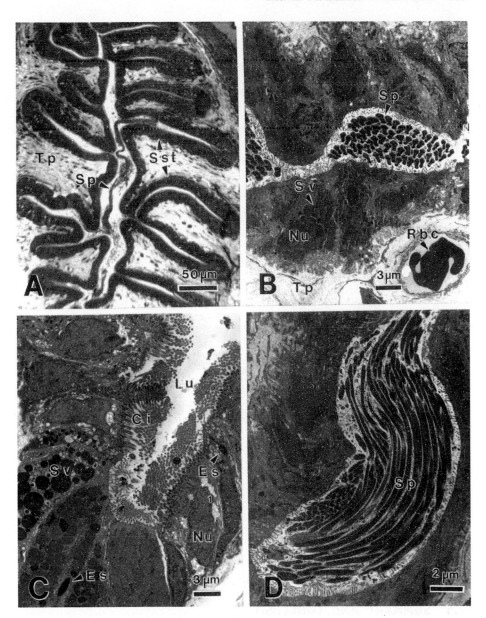

Fig. 7.8 Sperm storage glands in the ovisac of female *Ascaphus truei*. **A**. Light micrograph showing a transverse section through the sperm storage region in the ovisac. **B**. Electron-micrograph of the ovisac showing sperm in the lumen of a Sst duct. **C**. Sst containing embedded sperm. **D**. Sst filled distally with sperm, many of which exhibit similar alignment along their long axes. Ci, cilia; Es, embedded sperm; Lu, lumen; Nu, nucleus; Rbs, red blood cell; Sp, sperm; Sst, sperm storage tubules; Sv, secretory vacuoles; Tp, tunica propria. From Sever, D. M. 2002. Journal of Experimental Zoology 292: 165-179, Fig. 1.

7.4.2 Variation in Time and Space

Metter (1964b) noted that sperm taken from the oviducts of *Ascaphus* were highly motile when placed in saline and that secretions from the oviduct may provide "nutrients" for the sperm. This supposition also has been made for sperm storage in spermathecae of salamanders, but no evidence exists that nourishment of stored sperm occurs (Sever and Kloepfer 1993). In contrast, Hardy and Dent (1986) found that sperm stored in salamander spermathecae are quiescent during storage, and that spermathecal secretions may therefore provide the chemical/osmotic environment for sperm quiescence (Sever and Kloepfer 1993). In our sample, sperm in both vitellogenic and non-vitellogenic females of *A. truei* appear normal and are similar in abundance and distribution. Occasionally, sperm are embedded in the Sst epithelium, but we found no evidence of involvement of the epithelium in either nourishment of sperm or spermiophagy

The populations studied by Metter (1964b) came from Idaho and southwest Washington (now recognized as *Ascaphus montanus*), and mating occurred in the fall. Of the 19 females examined by Metter, 15 contained sperm and some of these (the number is unclear) were non-vitellogenic. Vitellogenic females must wait nearly a year until the subsequent summer to oviposit, while non-vitellogenic females mating in fall were hypothesized to wait almost two years. He also mentioned that one female with large vitellogenic eggs contained no sperm despite repeated copulations with males over two weeks. Sever *et al.* (2001) studied a coastal California population of *A. truei* and reported that the height of the mating season is June and July in northern California. Both vitellogenic and non-vitellogenic females from this period contained sperm in their Ssts, even though the oviducts of the non-vitellogenic females were not as hypertrophied and actively secretory as those of vitellogenic females. They suggested that oviposition generally follows summer mating in vitellogenic females. Non-vitellogenic females mating at the same time would presumably lay down yolk in their follicles and oviposit in the following year, resulting in a maximum of one year of sperm storage. The California coastal populations also differ from inland populations in having a 1-2 year larval period (Wallace and Diller 1998) rather than 2-4 years as reported by Metter (1964a) in Idaho and southwest Washington. Thus, considerable geographic variation occurs in the reproductive biology of the two species in the genus *Ascaphus*.

7.5 SUMMARY AND CONCLUSIONS

Ascaphus truei and *A. montanus* are unique to our knowledge among anurans in having males that possess a penis and females that undergo oviducal sperm storage after copulation and internal fertilization. These adaptations appear to be independently derived in *Ascaphus*, and thus structural and functional similarities with other vertebrates in these regards are classic examples of homoplasy through convergence (Sanderson and Hufford 1996). Structural and functional similarities in penile cavernous tissue or in sperm storage glands between *Ascaphus* and other vertebrates with similar mechanisms therefore are not based upon direct descent but related either to similar functional adaptations and/or to internal design restraints (Wake 1991).

DeVillers (1934) notes that the penis of *Ascaphus* is not similar structurally to the copulatory organs that occur in some bony fish or to the phallodeum of caecilians, because erectile tissue is lacking in these anamniotes. The phallodeum is derived from an eversible portion of the cloacal wall. Although blood sinuses may be involved in eversion of the phallodeum in some caecilians, the structure is basically composed of muscular tissue and fibrous connective tissue (Wake 1972).

Aside from *Ascaphus*, cavernous tissue involved in erection of a male copulatory organ is an amniote character that is known in reptiles, some groups of birds, and mammals. In reptiles and birds, however, the cavernous tissue is everted from the cloacal orifice whereas in *Ascaphus*, the vascular pads serve to lengthen and strengthen a cloacal protrusion that exists otherwise externally in a flaccid state. In mammalian monotremes, the penis is extruded from a cloacal sheath during erection (Weidersheim and Parker 1897). Penes of marsupial and placental mammals show a wide range of variation, but all include a third sinusoidal tissue, the corpus spongiosum, which encloses the cavernous urethra, a tube that is involved in both urine and seminal fluid transport (Kardong 1998). Thus, the structure of the penis of *Ascaphus* is truly unique among tetrapods.

In the case of sperm storage, structural and physiological constraints on the basic vertebrate oviduct and sperm morphologies (the "bauplan") may limit the options for expression of oviducal sperm storage (Sever 2002). The group of anamniotes phyletically that is closest to frogs and with which frogs share the most developmental similarities (the closest generative system, Wake 1996) is the Urodela. The ultrastructure of sperm storage in salamanders has been studied extensively and was reviewed by Sever and Brizzi (1998).

Numerous differences occur between the spermathecae of salamanders and the Ssts of *Ascaphus*. The distal portions of the spermathecae of salamanders are typically alveolar, lack cilia, and possess basal myoepithelium (Sever and Brizzi 1998). Secretory activity in salamander spermathecae is sometimes regionalized and seasonal, depending upon the taxon (Sever 1994). A great deal of variation also occurs in reaction to carbohydrate stains; however, most species exhibit AB+ reactions for carboxylated glycosaminoglycans (Sever 1994).

In some forms the sperm are in orderly arrays in the spermathecae (Sever and Hamlett 1998) whereas in others, sperm are in tangled masses (Sever *et al.* 1999). Alignment of sperm may depend to some degree upon the anatomy of the spermatheca (more orderly in compound glands than simple tubular; Sever 2002). Spermiophagy by the spermathecal epithelium has been described in various taxa of salamanders (Sever and Brizzi 1998).

The Ssts of *Ascaphus* more closely resemble those of squamate reptiles (Fox 1956; Girling *et al.* 1997; Sever and Ryan 1999; Sever and Hamlett 2002). Oviducal sperm storage glands are known from all groups in the reptile-bird clade except Amphisbaenia (in which they probably occur) and Rhyncocephalia

(Gist and Jones 1987; Sever and Hamlett 2002). Like reptiles, the Ssts of *Ascaphus* are simply continuations of the oviducal lining, and contain ciliated non-secretory cells and unciliated secretory cells. Myoepithelium is absent, but the oviduct possesses layers of smooth muscle (tunica muscularis) superficial to the mucosa. The linings and glands of reptilian oviducts are generally described as PAS+, like those of *Ascaphus*, with little reaction to acidic mucosubstances. Sperm in the Ssts of *Ascaphus* are generally in close alignment, although alignment is variable in squamates (Fox 1956; Sever and Ryan 1999; Sever and Hamlett 2002). Although sperm are sometimes found embedded in Ssts of reptiles (Sever and Ryan 1999) and of *Ascaphus*, no evidence exists for spermiophagy in these taxa.

Thus, oviducal Ssts in distantly related taxa show more similarities than Ssts in *Ascaphus* and spermathecae in salamanders, members of sister taxa. This is here considered to indicate that the basic structure of the vertebrate oviduct may limit the range of features associated with oviducal sperm storage (Sever *et al.* 2001). Alternatively, the similarity of Ssts in *Ascaphus* and reptiles could reflect the close relationship of *Ascaphus* to basal amniotes deduced from sperm ultrastructure (Jamieson *et al.* 1993; Jamieson, 1999; see Chapter 5).

7.6 ACKNOWLEDGEMENTS

We thank Lowell Diller for specimens of *Ascaphus truei* from northern California that were used in the descriptions of the penis and of female sperm storage.

7.7 LITERATURE CITED

Barbieri, F. D. and DelPino, E. J. 1975. Jelly coats and a diffusible factor in anuran fertilization. Archives de Biologie Bruxelles 86: 311-321.

Constanz, G. D. 1989. Reproductive biology of poeciliid fishes. Pp. 33-50. In G. K. Meffe and F. F. Snelson Jr. (eds), *Ecology and Evolution of Livebearing Fishes (Poeciliidae)*. Prentice Hall, New Jersey.

DeVilliers, C. G. S. 1933. The "tail" of the male American toad, *Ascaphus*. Nature (London) 131: 692-693.

DeVilliers, C. G. S. 1934. On the morphology of the epipubis, the Nobelian bones, and the phallic organ of *Ascaphus truei* Stejneger. Anatomischer Anzeiger 78: 23-47.

Fejérváry, G. J. de. 1923. Ascaphidae, a new family of tailless batrachians. Annales Historico-Naturales. Musei Nationalis Hungarici 20: 178-181.

Fox , W. 1956. Seminal receptacles of snakes. Anatomical Record 124: 519-539.

Frey R. 1994. Der Zusammenhang zwischen Begattungsstellung, Lokomotionsweise und Kopulationsorgan bei Vertebrata, mit Ausnahme der Mammalia. Eine vergleichende Betrachtung. Journal of Zoological Systematics and Evolutionary Research 33: 17-31.

Gaige, H. T. 1920. Observations upon the habits of *Ascaphus truei* Stejneger. Occasional Papers of the Museum of Zoology, University of Michigan (84): 1-9.

Girling, J. E., Cree, A. and Guillette, L. J. Jr. 1997. Oviductal structure in a viviparous New Zealand gecko, *Hoplodactylus maculatus*. Journal of Morphology 234: 51-68.

Gist, D. H. and Jones, J. M. 1987. Storage of sperm in the reptilian oviduct. Scanning Microscopy 1: 1839-1849.

Grandison, A.G.C. 1978. The occurrence of *Nectophrynoides* (Anura: Bufonidae) in Ethiopia. A new concept of the genus with a description of a new species. Monitore Zoologica Italia New Series Supplement XI: 119-172.

Grandison, A. G. C. 1980. Aspects of breeding morphology in *Mertensophryne micranotis* (Anura: Bufonidae): secondary sexual characters, eggs and tadpole. Bulletin of the British Museum of Natural History (Zoology) 39: 299-304.

Grandison, A. G. C. and Ashe, S. 1983. The distribution, behavioural ecology and breeding strategy of the pygmy toad, *Mertenosphryne micranotis* (Lov.). Bulletin of the British Museum of Natural History (Zoology) 45: 85-93.

Graybeal, A. and Cannatella, D. C. 1995. A new taxon of Bufonidae from Peru, with descriptions of two new species and a review of the phylogenetic status of supraspecific bufonid taxa. Herpetologica 51: 105-131.

Greven, H. 2002. The urodele oviduct and its secretions in and after G. von Wahlert's doctoral thesis "Eileiter, Laich und Kloake der Salamandriden." Bonner Zoologische Monographien 50: 25-61.

Halliday, T. R. and Verrell, P. A. 1984. Sperm competition in amphibians. Pp. 487-508. In R. L. Smith (ed.), *Sperm Competition and the Evolution of Animal Mating Systems*. Academic Press, New York, NY.

Hamlett, W. C., Knight, D. P., Koob, T. J., Jezior, M., Luoug, T., Rozycki, T., Brunette, N. and Hysell M. K.. 1998. Survey of oviducal gland structure and function in elasmobranchs. Journal of Experimental Zoology 282:399-420.

Hamlett WC, Koob TJ. 1999. Chapter 15. Female reproductive cycle. Pp. 315-345. In W. D. Hamlett (ed.), *Sharks, Skates and Rays: The Biology of Elasmobranch Fishes*. Johns Hopkins University Press, Baltimore, Maryland.

Hardy, M. P. and Dent, J. N. 1986. Transport of sperm within the cloaca of the female red-spotted newt. Journal of Morphology 190: 259-270.

Howarth, B. Jr. 1974. Sperm storage: as a function of the female reproductive tract. Pp. 237-270. In A. D. Johnson and C. W. Foley (eds). *The Oviduct and Its Functions*. Academic Press, New York, NY.

Jameson, D. L. 1955. Evolutionary trends in the courtship and mating behavior of Salientia. Systematic Zoology 4: 105-119.

Jamieson, B. G. M., Lee, M. S. Y. and Long, K. 1993. Ultrastructure of the spermatozoan of the internal fertilizing frog *Ascaphus truei* (Ascaphidae: Anura: Amphibia) with phylogenetic considerations. Herpetologica 49: 52-65.

Jamieson, B. G. M. 1999. Spermatozoal phylogeny of the Vertebrata. Pp. 303-331. In C. Gagnon (ed.), *The Male Gamete: From Basic Science to Clinical Applications*. Cache River Press, St. Louis, Missouri.

Kardong, K. V. 1998. *Vertebrates Comparative Anatomy, Function, Evolution*. 2nd ed. W. C. Brown, Dubuque, Iowa.

Lombardi, J. 1998. *Comparative Vertebrate Reproduction*. Kluwer Academic, Boston, Massachusetts. 488 pp.

Metter, D. E. 1964a. A morphological and ecological comparison of two populations of the tailed *Ascaphus truei* frog, Stejneger. Copeia 1964: 181-195.

Metter, D. E. 1964b. On breeding and sperm retention in *Ascaphus*. Copeia 1964: 710-711.

Millot, J. and Anthony, J. 1960. Appareil genital et reproduction des coelacanthes. Comptes Rendus Hebdomadaires des Seances. Academie des Sciences (Paris) 251: 442-443.

Nielson, M., K. Lohman and J. Sullivan. 2001. Phylogeography of the tailed frog (*Ascaphus truei*): implications for the biogeography of the Pacific Northwest. Evolution 55: 147-160.

Noble, G. K. 1922. The phylogeny of the Salientia I. The osteology and the thigh musculature; their bearing on classification and phylogeny. Bulletin of the American Museum of Natural History 13: 1-87.

Noble, G. K. 1925. An outline of the relation of ontogeny to phylogeny within the Amphibia I. American Museum Novitates (165): 1-17.

Noble, G. K. 1931. *The Biology of the Amphibia*. McGraw Hill Book Company, New York. 577 pp.

Noble, G. K. and Putnam, P. G. 1931. Observations on the life history of *Ascaphus truei* Stejneger. Copeia 1931: 97-101

Pough, F. H., Andrews, R. M., Cadle, J. E., Crump, M. L., Savitzky, A. H. and Wells, K. D. 1998. *Herpetology*. Prentice Hall, Upper Saddle River, New Jersey. 577 pp.

Pratt, H. L. 1993. The storage of spermatozoa in the OGs of western North Atlantic sharks. Enivronmental Biology of Fishes 38: 139-149.

Salthe, S. N. and Duellman., W. E. 1973. Quantitative constraints associated with reproductive mode in anurans. Pp. 229-250. In J. L. Vial (ed.), *Evolutionary Biology of the Anurans. Contemporary Research on Major Problems*. University of Missouri Press, Columbia, Missouri.

Sanderson, M. J. and Hufford, L (eds). 1996. *Homoplasy: the Recurrence of Similarity in Evolution*. Academic Press, San Diego, California. 339 pp.

Schultze, H-P. 1994. Comparison of hypotheses on the relationships of sarcopterygians. Systematic Biology 43: 155-173.

Sever, D. M. 1994. Observations on regionalization of secretory activity in the spermathecae of salamanders and comments on phylogeny of sperm storage in female salamanders. Herpetologica 50: 383-397.

Sever, D. M. 2002. Female sperm storage in amphibians. Journal of Experimental Zoology 292: 165-179.

Sever, D. M. and Brizzi, R. 1998. Comparative biology of sperm storage in female salamanders. Journal of Experimental Zoology 282: 460-476.

Sever, D. M., Moriarty, E. C., Rania, L. C. and Hamlett, W. C. 2001. Sperm storage in the oviduct of the internal fertilizing frog, *Ascaphus truei*. Journal of Morphology 248: 1-21.

Sever, D. M., Halliday, T., Waights, V., Brown, J., Davies, H. A. and Moriarty, E. C. 1999. Sperm storage in females of the smooth newt (*Triturus v. vulgaris* L.): I. Ultrastructure of the spermathecae during the breeding season. Journal of Experimental Zoology 283: 51-70.

Sever, D. M. and Hamlett, W. C. 1998. Sperm aggregations in the spermatheca of female desmognathine salamanders (Amphibia: Urodela: Plethodontidae). Journal of Morphology 238: 143-155.

Sever, D. M. and Hamlett, W. C. 2002. Female sperm storage in reptiles. Journal of Experimental Zoology 292: 187-199.

Sever, D. M. and Kloepfer, N. M. 1993. Spermathecal cytology of *Ambystoma opacum* (Amphibia: Ambystomatidae) and the phylogeny of sperm storage organs in female salamanders. Journal of Morphology 217: 115-127.

Sever, D. M. and Ryan, T. J. 1999. Ultrastructure of the reproductive system of the black swamp snake (*Seminatrix pygaea*): Part I. Evidence for oviducal sperm storage. Journal of Morphology 241: 1-18.

Slater, J. R. 1931. The mating of *Ascaphus truei* Stejneger. Copeia 1931: 62-63.

Smith, C. L., Rand, C. S., Schaeffer, B. and Atz, J. 1975. *Latimeria*, the living coelacanth, is ovoviviparous. Science 190: 1105-1106.

Stebbins, R. C. and Cohen, N. W. 1995. *A Natural History of Amphibians*. Princeton University Press, New Jersey. 332 pp.

Stejneger, L. 1899. Description of a new genus and species of discoglossid toad from North America. Proceedings of the United States National Museum 21: 899-902.

Stephenson, B. and Verrell, P. 2003. Courtship and mating of the tailed frog, *Ascaphus truei*. Journal of Zoology, London, 259: 15-22.

Townsend, D. S., Stewart, M. M., Pough, F. H., and Brussard, P. F. 1981. Internal fertilization in an oviparous frog. Science 212: 469-471.

Townsend, D. S. and Stewart, M. M. 1986. Courtship and mating behavior of a Puerto Rican frog, *Eleutherodactylus coqui*. Herpetologica 42: 165-170.

Van Denburgh J. 1912. Notes on *Ascaphus*, the discoglossid toad of North America. Proceedings of the California Academy of Science 4th Series 3: 259-264.

Van Dijk D. E. 1955. The "tail" of *Ascaphus*: A historical resume and new histological-anatomical details. Annals of the University of Stellenbosch 31: 1-71.

Van Dijk, D. E. 1959. On the cloacal region of Anura in particular of larval *Ascaphus*. Annals of the University of Stellenbosch 35: 169-249.

Wake, D. B. 1991. Homoplasy: The result of natural selection, or evidence of design limitations? American Naturalist 138: 543-567.

Wake, D. B. 1996. Introduction. Pp. xvii-xxv. In M. J. Sanderson and L. Hufford (eds), *Homoplasy: The Recurrence of Similarity in Evolution*. Academic Press, San Diego, C.A.

Wake, M. H. 1972. Evolutionary morphology of the caecilian urogenital system. IV. The cloaca. Journal of Morphology 136: 353-366.

Wake, M. H. 1978. The reproductive biology of *Eleutherodactylus jasperi* (Amphibia, Anura, Leptodactylidae), with comments on the evolution of live-bearing systems. Journal of Herpetology 12: 121-133.

Wake, M. H. 1980. The reproductive biology of *Nectophrynoides malcolmi* (Amphibia: Bufonidae), with comments on the evolution of reproductive modes in the genus *Nectophrynoides*. Copeia 1980: 193-209.

Wake, M. H. 1987. Urogenital morphology of dipnoans, with comparisons to other fishes and to amphibians. Pp. 199-216. In W. E.Bemis, W. W. Burggren and N. E. Kemp (eds), *The Biology and Evolution of Lungfishes*, A. R. Liss, New York, NY.

Wake, M. H. 1993. Evolution of oviductal gestation in amphibians. Journal of Experimental Zoology 266: 394-413.

Wake, M. H. and Dickie, R. 1998. Oviduct structure and function and reproductive modes in amphibians. Journal of Experimental Zoology 282: 477-506.

Wallace, R. L. and Diller. L. V. 1998. Length of the larval cycle of *Ascaphus truei* in coastal streams of the redwood region, northern California. Journal of Herpetology 32: 404-409.

Weidersheim, R. and Parker, W. N. 1897. *Elements of the Comparative Anatomy of Vertebrates*. MacMillan and Company, London, U.K. 488 pp.

Wilkinson, M. and Nussbaum, R. A. 1998. Caecilian viviparity and amniote origins. Journal of Natural History 32: 1403-1409.

CHAPTER 8

Parental Care: A Phylogenetic Perspective

Richard M. Lehtinen and Ronald A. Nussbaum

8.1 INTRODUCTION

"The one merit that is claimed [for comparative studies] is that it suggests new ways of looking at facts and new sorts of fact to look for." Simpson 1944:xviii.

Naturalists have long observed the various parental care behaviors of animals but were generally unsuccessful in constructing a predictive theoretical framework for explaining them. A framework for studying parental care behavior based on Darwinian theory has been provided through the work of Lack (1947, 1954, 1968), Williams (1966a,b, 1975), Tinkle (1969), and Trivers (1972). These authors developed the concepts of reproductive effort and parental investment, both of which subsume parental care according to most definitions. Most studies of parental care emphasize its function, particularly in relation to the costs of parental care to the parents and the benefits to the offspring. On this basis, ethologists have attempted to explain why some species have parental care and others do not, in what circumstances parental care is likely to evolve and how various parental care strategies may have evolved from one another. These analyses have been primarily based on logical or mathematical arguments, usually from a cost-benefit or game theoretic perspective (Clutton-Brock 1991).

However, many questions of interest concerning the evolution of parental care are fundamentally historical questions. How many times has parental care evolved or been lost in anurans? What was the ancestral condition of parental care in a given lineage? In what sequence did parental care behaviors evolve from one another? Is the evolution of derived parental care correlated with other traits? These and other important questions concerning the evolution of parental care are not directly testable in the logical-mathematical framework of cost-benefit analyses and require an explicitly historical approach to answer. Such an approach is provided by adopting a phylogenetic perspective.

University of Michigan Museum of Zoology, Division of Reptiles and Amphibians, Ann Arbor, Michigan 48109-1079 USA

8.2 THE PHYLOGENETIC PERSPECTIVE

Species are historical entities. This concept of evolution as a genealogical process is one of two primary themes argued by Charles Darwin in the *Origin of Species* (1859). It follows from this postulate that species are not independent units for analysis, but rather bear the legacy of their evolutionary pathways. Thus, because species give rise to other species, their characteristics are not only determined by their contemporary environment, but by their ancestry as well. Therefore, macroevolutionary patterns in parental care can only be fully understood with reference to the evolutionary relationships among species (i.e., their phylogeny – see Brooks *et al.* 1995, Brooks and McLennan 2002, Ridley 1983 for a full justification of this approach).

The fundamental advantage of using a phylogeny to infer the origin or evolution of a trait is that it provides an independent frame of reference. Without this historical approach, adaptive hypotheses are vulnerable to error in closely related species. That is, adaptive explanations about the origins of characters can be tested only with convergent cases, and not with origins resulting from common ancestry (Felsenstein 1985). For example, whether species A and B exhibit egg attendance because they independently evolved this trait (Fig. 8.1A) or because they inherited it from a common ancestor (Fig. 8.1B) can only be assessed with explicit reference to phylogenetic relationships.

Phylogenies are constructed using cladistic principles based on morphological, behavioral, biochemical, or other types of characters (Kitching *et al.* 1998, Wiley 1981). These hypotheses of genealogical relationships can then be used to address such questions as the origins of particular traits, the pattern and sequence of trait evolution and hypotheses of adaptation (Cracraft 1981). This approach is also useful to determine taxa on which to focus in order to answer a particular question. However, an important assumption in using this approach is that the phylogenetic relationships on which the conclusions are based are robust and well supported. If the genealogical relationships among the species under study are unclear, conclusions based on them are tentative at best and misleading at worst. Nonetheless, this approach has been used successfully to explore the evolution of parental care in many other taxa (e.g., birds (McKitrick 1992, Székely and Reynolds 1995, Tullberg *et al.* 2002), marine invertebrates (Ó Foighil and Taylor 2000 and references therein), fish (Gittleman 1981, Klett and Meyer 2002, McLennan 1994)). In combination with field observations, experiments and cost-benefit analyses, the phylogenetic approach is a powerful method of testing alternative hypotheses of the evolution of parental care.

While a historical perspective has proved useful in many other contexts (e.g., Ryan and Rand 1995, Emerson 1996) few herpetologists have yet used this historical approach in their studies of parental care of frogs. In this chapter, we will review the diversity of parental care in anurans and analyze and discuss the evolution of parental care, using a phylogenetic perspective.

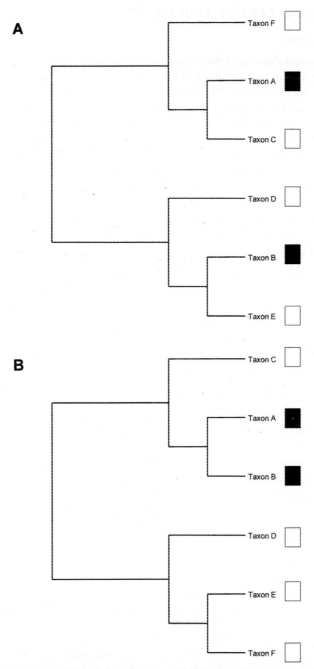

Fig. 8.1 Hypothetical cladogram of six taxa (A-F). Black boxes indicate where trait X is present, white boxes indicate where trait X is absent. Fig. 8.1A shows phylogenetic relationships that imply trait X evolved twice independently in taxa A and B. Fig. 8.1B shows phylogenetic relationships that imply that trait X evolved only once and is homologous in taxa A and B. Original.

8.3   PARENTAL CARE IN ANURANS

8.3.1   General

Anuran parental care has been reviewed numerous times (Beck 1998, Crump 1995, 1996, Duellman and Trueb 1986, Lamotte and Lescure 1977, McDiarmid 1978, Salthe and Mecham 1974, Wells 1977, 1981; see also Brizzi *et al.*, Chapter 6 of this volume, emphasizing glandular structure and function) and, therefore, this chapter is not meant to be a comprehensive review of this topic. Rather, our primary purpose is to examine the available data from a phylogenetic viewpoint. Nevertheless, many new observations on parental care in anurans were reported since the last major review (Crump 1996) and our examples of the modes of anuran parental care will focus on those reported since 1996 or those unknown to Crump (1995, 1996). For convenience, we have adopted the family designations of Duellman (chapter 1 of this volume).

We define parental care as any investments in offspring after fertilization. This definition implies a cost to the parent but also an increase in parental fitness through increased survivorship of offspring, as explained by Smith and Fretwell (1974). Parting from previous authors (Crump 1995, Duellman and Trueb 1986), we have included nest construction and viviparity in our discussion of parental care. Nest construction sometimes occurs prior to fertilization, however these nests clearly function in protecting offspring at some cost to the parent(s) that construct them. Similarly, live birth of offspring (viviparity) surely increases the fitness of offspring and reduces the ability of the parent to invest in other offspring. While strictly speaking this is a different form of parental investment, it should be considered parental care under the broad definition adopted above.

Parental care in anurans is quite diverse, yet exhibited only by about one tenth of extant species (McDiarmid 1978, but see below). Many species with parental care have a terrestrial or semi-terrestrial mode of reproduction. This reflects the fundamental differences in selective pressures in aquatic and terrestrial environments, where terrestrial environments are gamete-unfriendly, relative to aquatic ones. Also, most species that exhibit parental care behaviors are found in the tropics, although this may only be a consequence of the fact that the vast majority of frogs are found there.

The classification of anuran parental care into "modes" is a difficult and somewhat arbitrary process. Most of these behaviors have convergently evolved numerous times in unrelated groups. The behaviors may not necessarily be obligate, and many species exhibit more than one mode of parental care. Also, in some cases, mode of parental care and reproductive mode are inextricably linked, further complicating attempts at classification.

Nonetheless, we recognize ten types of parental care in anurans: nest construction, egg attendance, egg transport, egg brooding, tadpole attendance, tadpole transport, tadpole brooding, tadpole feeding, froglet transport and viviparity. We restrict use of the term "egg transport" to those situations where physical relocation of eggs from one habitat to another is the sole or primary function, and the larvae become free-living. We use the term "egg brooding" to refer to situations where eggs of direct developing taxa complete all of their development in or on the body of the parent and are not free-living. We similarly distinguish between "tadpole transport" and "tadpole brooding". Tadpole transport involves the physical relocation of tadpoles from one habitat to another, where they become free-living. Tadpole brooding is defined as situations in which tadpoles complete most or all of their development in or on the body of the parent and are not free-living. If larvae are not free-living and develop in or on the parent, the function is clearly developmental and any transport while with the parent is purely coincidental. In this sense, we believe the terms "egg transport" and "tadpole transport" (*sensu* Crump 1995, 1996) do not accurately reflect the function of these behaviors, and we prefer "egg brooding" or "tadpole brooding" to describe these cases. We make these distinctions in order to clarify related patterns with different functions. Below we discuss and describe each type of parental care.

8.3.2 Nest Construction

A wide variety of different types of nests are produced by anurans. Here we restrict our discussion to nests that are actually constructed or manipulated by the parents, rather than just the opportunistic use of an appropriate depression. Nests are thought to function in protecting eggs and/or larvae from predators or desiccation and are often maintained by attending parents (Heyer 1969). However, other functions have been suggested for nests such as aeration of embryos, manipulation of temperature regimes (and therefore developmental rates of tadpoles) and the use of foam (in the case of foam nests) as a food source for tadpoles (Dobkin and Gettinger 1985, Downie 1988, Tanaka and Nishihara 1987). Usually only the embryonic and early larval stages remain in the nest, and later stages of development are completed in aquatic habitats. However, non-feeding, terrestrial tadpoles that complete development in foam nests are known from the myobatrachids *Kyarranus* (Moore 1961) and *Philoria* (Littlejohn 1963) as well as the leptodactylids *Adenomera* (Heyer and Silverstone 1969) and some *Leptodactylus* (Heyer 1969).

Perhaps the most common type of anuran nest is the foam nest (Fig. 8.2). Foam nests have independently evolved at least three times and are found in Hyperoliidae, Rhacophoridae, Leptodactylidae, and Myobatrachidae. Foam nests may be placed on the ground, in tree holes or other phytotelms, on vegetation, under stones in streams, or on the surface of ponds (Haddad and Pombal 1998, Heyer 1969, Martin 1970). The way in which the foam nest is constructed differs, however, among lineages. For example, foam nests in *Limnodynastes* are produced by the female "paddling" cloacal secretions into foam with the front limbs (Tyler and Davies 1979) while in leptodactylids it is the male using the hind limbs (Heyer 1969). By contrast, in *Chiromantis* (Rhacophoridae) foam nests are produced by both sexes using their hind limbs to beat secretions into foam (Coe 1964). A different type of nest is produced by the neotropical microhylid *Chiasmocleis leucosticta*. In this species, the male and female produce a "bubble nest" by swimming underneath the eggs and exhaling through their nostrils (Haddad and Hödl 1997).

as fungi. This may involve the parent either consuming infected eggs or physically moving the eggs around to destroy fungal hyphae (Simon 1983). It has also been suggested that skin secretions of the attending parent may have antimicrobial properties that, when applied to the embryos, prevent infections. This has been tested for salamanders, but apparently not for frogs. Presumably, these activities prevent further infection of the clutch and avoidance of fungal infection may have been much more influential in the evolution of parental care and reproductive modes than is generally recognized (Green 1999).

Egg attendance in *Nectrophryne afra* is associated with aeration of the aquatic eggs by swimming in place (Scheel 1970). Terrestrial eggs, on the other hand, are vulnerable to desiccation and numerous species are known to hydrate the attended eggs (e.g. several *Dendrobates* (Wells 1978, Weygoldt 1987), *Colostethus beebei* (Bourne *et al.* 2001), several *Hyperolius* (Stevens 1971, C. Richards, personal communication), several *Eleutherodactylus* (Myers 1969, Taigen *et al.* 1984), *Hylactophryne augusti* (Jameson 1950)). Lastly, egg attendance may prevent developmental abnormalities in some taxa (*Cophixalus parkeri* Simon 1983, *Breviceps sylvestris* Wager 1986). Burrowes (2000) recently suggested that while there are clear benefits to egg attendance in *Eleutherodactylus cooki*, there are no obvious costs.

8.3.4 Egg Transport

Egg transport (as defined above) is known from only three families (Discoglossidae, Hylidae and Pipidae). In two species of *Alytes* (Discoglossidae), the male carries the eggs entwined on his hind legs until they are ready to hatch. He then enters a pond where the hatched larvae swim away to complete their development. Similarly, some hemiphractine hylids (e.g., *Flectonotus*, some *Gastrotheca*) deposit transported eggs into bromeliads or other water-filled plants (Weygoldt and Plotsch 1991). These free-living, aquatic tadpoles contain large yolk reserves and do not feed before metamorphosis (Duellman and Gray 1983). In some pipid frogs (*Pipa carvalhoi, P. parva*), eggs are transported until the larvae hatch from the females dorsum, at which time they assume a free-living, feeding tadpole lifestyle (Trueb and Cannatella 1986).

Egg transport may function to increase survivorship of vulnerable embryonic stages. Eggs carried by adults would have a decreased likelihood of being eaten by predators. In addition, desiccation of eggs is less probable in or on the body of an adult. The ability of parents who transport eggs to assess quality of aquatic sites may add further to offspring fitness. For example, parents transporting eggs may be able to track warmer sites for more rapid embryonic development and preferentially transport their eggs to these water bodies. This idea may apply to egg brooding, tadpole transport and tadpole brooding as well. In cases where eggs are rapidly transported to an aquatic site, transport may be related to the evolution of separate sites for courtship/mating and nesting. This might occur if good sites for attracting mates, especially in territorial species, are not good nest sites.

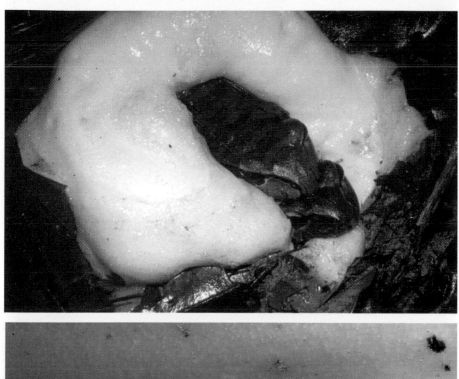

Fig. 8.2 (Top). Male and female *Leptodactylus knudseni* (Leptodactylidae) constructing a foam nest. Photo by Luis C. Schiesari.

Fig. 8.3 (Bottom). Egg attendance in a water-filled *Pandanus* leaf axil by *Mantidactylus bicalcaratus* (Mantellidae). Photo by Richard M. Lehtinen.

Fig. 8.4 (Top). Egg attendance by *Mantidactylus majori* on vegetation overhanging a small stream. Photo by Richard M. Lehtinen.

Fig. 8.5 (Bottom). *Stefania scalae* (Hylidae) brooding eggs on dorsum. Photo by William E. Duellman.

Fig. 8.6 *Plethodontohyla notosticta* (Microhylidae) attending tadpoles in a water-filled bamboo stump. Note recently metamorphosed froglets. Photo by Bret Weinstein.

8.3.5 Egg Brooding

Brooding of direct developing eggs is found in only two families (Hylidae and Pipidae) where the eggs develop in or on the body of the female until the froglets hatch (Fig. 8.5). This is characteristic of numerous hemiphractine hylids, some of which brood the offspring in a dorsal pouch (e.g., many *Gastrotheca*) and others which carry the developing eggs exposed on the dorsum of the female (*Cryptobatrachus, Stefania,* and *Hemiphractus*; Duellman and Hoogmoed 1984, Duellman and Maness 1980). Direct development of eggs embedded in the dorsum of the female is found in the pipids *Pipa pipa, P. snethlageae, P. aspera* and *P. arrabali* eggs (Rabb and Snedigar 1960, Trueb and Cannatella 1986, Trueb and Massemin 2001). Similarly to egg transport, the function of egg brooding probably relates to decreasing mortality of eggs from predators or pathogens.

8.3.6 Tadpole Attendance

Attendance of tadpoles involves a parent remaining with tadpoles at a relatively fixed location through part or all of the larval period (Fig. 8.6). This form of parental care has been documented in six families (Bufonidae, Dendrobatidae, Leiopelmatidae, Leptodactylidae, Microhylidae, Ranidae) but in relatively few species. Some of these species attend aquatic tadpoles. For example, males of the African ranid *Pyxicephalus adspersus* actively defend groups of tadpoles from potential predators until metamorphosis (Balinsky and Balinsky 1954). Males also dig channels to other water bodies when the pool with their tadpoles is drying out (Kok *et al.* 1989). Cook *et al.* (2001) clearly showed the costs (injury or death from bird predators) and benefits (increased survivorship of offspring) of paternal care in this species. A similar pattern is

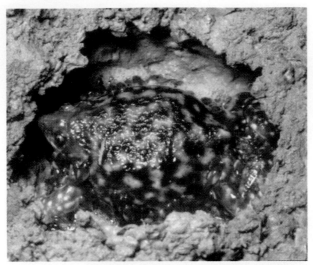

Fig. 8.7 (Top). Tadpoles of *Hemisus marmoratus* (Hemisotidae) clinging to a female in an underground nest. Photo by Mark-Oliver Roedel.

Fig. 8.8 (Bottom). A clutch of tadpoles being transported by *Epipedobates* sp. (Dendrobatidae). Photo by Janalee P. Caldwell.

seen in *Hemisus* where the female (after attending the subterranean eggs) digs a tunnel to the adjacent water body. The tadpoles wriggle along the path (or are transported on the back of the female – see van Dijk 1997; Fig. 8.7) until they reach the pond, where they continue their development (Wager 1986).

Attendance of aquatic tadpoles has also been reported for several species of *Leptodactylus*. In *L. ocellatus* and *L. validus*, the female remains with the foam nest containing the developing eggs. After the eggs hatch, the female remains with her tadpoles until metamorphosis, protecting them from predators (Downie 1996, Vaz-Ferreira and Gehrau 1974). The pattern is similar in *L. bolivianus*, except that the female also seems to direct the tadpoles to different microhabitats

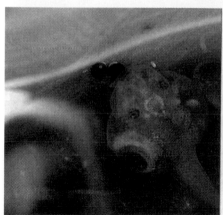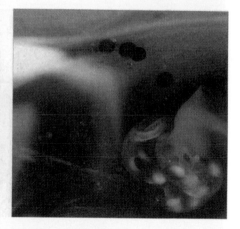

Fig. 8.9 (Top). Female *Sooglossus sechellensis* (Sooglossidae) brooding non-feeding tadpoles on dorsum. Photo by Ronald A. Nussbaum.

Fig. 8.10 (Bottom). Tadpoles of *Osteocephalus oophagous* (Hylidae) feeding on eggs deposited by female. Photo by Walter Hödl.

Fig. 8.11   (Top). Recently metamorphosed *Sooglossus sechellensis* (Sooglossidae) froglets transported by female. Photo by Ronald A. Nussbaum.

Fig. 8.12   (Bottom). Recently metamorphosed *Liophryne schlaginhaufeni* (Microhylidae) froglets transported by male. Photo by David Bickford.

in the water (Wells and Bard 1988). A recent study of *L. insularum* indicates that aggressive behavior by adult females and schooling by larvae have anti-predator functions (Ponssa 2001). In the bufonid *Nectophryne afra*, males (in addition to attending the eggs, see above) remain with the tadpoles for several weeks after hatching (Scheel 1970). During this time, the male periodically swims in place. Two phytotelm-breeding microhylids from Madagascar (*Anodonthyla boulengeri* and *Plethodontohyla notosticta*) attend eggs and aquatic tadpoles until metamorphosis (Blommers-Schlösser 1975a, Heying 2001). In *A. boulengeri*, the male also "somersaults" around the eggs and/or tadpoles in the tree hole (H. Heying, personal communication). Presumably this functions to aerate the water in the small cavities where eggs are deposited. In *P. notosticta*, attending males in bamboo stumps bark at, and sometimes lunge toward, intruders (H. Heying, personal communication).

Attendance of terrestrial tadpoles is known for several other species. In all cases, this involves non-feeding tadpoles. For example, in the neotropical microhylid *Synapturanus salseri*, the male remains with the eggs and larvae in an underground burrow until metamorphosis (Pyburn 1975). A similar pattern occurs in the three species of *Leiopelma*, again with the males providing the care (Bell 1985). Female *Leptodactylus fallax* remain with their terrestrial foam nest, in which the tadpoles develop (Brooks 1968, Lescure 1979). Males of the African ranid *Phrynobatrachus alticola* also attend terrestrial, non-feeding tadpoles, based on observations by Rödel and Ernst (2002).

8.3.7 Tadpole Transport

This mode of parental care is found in at least four anuran families (Dendrobatidae, Hemisotidae, Ranidae, Rhinodermatidae – Table 8.1). Typically, eggs are laid terrestrially and then, after hatching, larvae are transported on the body of a parent to an aquatic habitat to continue development as free-living tadpoles (Fig. 8.7, 8.8). Note that tadpoles of several species are known to be capable of terrestrial movement without parental assistance (e.g., *Hyla bromeliacia*, Duellman 2001, *Mantidactylus bicalcaratus*, *M. punctatus*, Lehtinen unpublished data). Here, however, we restrict our discussion to parental transport.

Male *Colostethus beebei* attend clutches laid on bromeliad leaves and later transport the tadpoles to water-filled parts of the plant (Bourne *et al.* 2001). In fact, most dendrobatids exhibit tadpole transport (Caldwell 1996, Weygoldt 1987 - Table 8.1). The details of transport differ among species, however, and tadpoles may be transported singly (*Dendrobates granuliferus*; van Wijngaarden and Bolaños 1992), or en masse (*Colostethus caeruleodactylus*; Lima *et al.* 2002). Males of two South East Asian ranids (*Limnonectes finchi* and *L. palavanensis*) transport tadpoles from terrestrial oviposition sites to streams, pools and other aquatic environments (Inger 1966, Inger and Voris 1988, see below). Male *Rhinoderma rufum* also transport tadpoles from the oviposition site to an aquatic site but in the vocal sacs rather than on the dorsum (Formas *et al.* 1975).

Selection should favor any behavior that reduces mortality during vulnerable life history stages. In some areas, predators of anuran eggs may be more abundant in aquatic environments than terrestrial ones (Magnusson and Hero 1991). If the aquatic environment is predator–rich, depositing tadpoles instead of eggs would be selected for, because the tadpoles would survive better than embryos. This may be because they are better able to avoid predation, or it may simply be that the length of time the offspring are exposed to predators in the aquatic environment is reduced, because a portion of development was completed terrestrially. A similar argument could be made with avoidance of competition or desiccation of water bodies being the selective agent.

8.3.8 Tadpole Brooding

Brooding of non-feeding tadpoles is known for six families (Dendrobatidae, Leiopelmatidae, Leptodactylidae, Myobatrachidae, Rhinodermatidae and Sooglossidae). Parents of some species keep non-feeding tadpoles on their back throughout development until metamorphosis (e.g., *Cyclorhamphus stejnegeri* (Heyer and Crombie 1979); *Colostethus degranvillei* (Lescure 1984); *Leiopelma archeyi* and *L. hamiltoni* (Bell 1985)). For example, *Sooglossus sechellensis* females attend their embryos in a hidden, terrestrial nest until hatching, at which time the tadpoles move on to their mother's back where they remain through metamorphosis (Nussbaum unpublished data; Fig. 8.9). Tadpole brooding does not always take place on the parent's dorsum, however. In *Rhinoderma darwinii*, the males brood non-feeding tadpoles in their vocal sacs until metamorphosis (Formas *et al.* 1975). In the myobatrachid *Assa darlingtoni*, larvae wriggle into inguinal brood pouches, where they develop from yolk reserves. In two *Rheobatrachus* species, the female swallows her hatched larvae and the non-feeding tadpoles develop in the stomach for approximately eight weeks (Corben *et al.* 1974, McDonald and Tyler 1984). The female does not feed during this time, and gastric acid is not secreted in the stomach until the offspring complete metamorphosis and hop from the mother's mouth (Tyler and Carter 1981).

8.3.9 Tadpole Feeding

Tadpole feeding occurs in a least seven families (Dendrobatidae, Hylidae, Mantellidae, Microhylidae, Ranidae, Rhacophoridae and Rhinodermatidae). In all known cases, tadpole feeding is associated with egg deposition in phytotelmata (plant-held water bodies, such as leaf axils, bromeliads, tree holes, etc.) and clearly functions to provide offspring with the resources necessary to complete metamorphosis in a small, food-poor habitat (Fig. 8.10). For example, the African microhylid *Hoplophryne rogersi* breeds between the leaves of banana plants, and also probably feeds eggs to its tadpoles (Noble 1929). At least eight species in the genus *Dendrobates* are known to feed their tadpoles eggs and it is likely that others do as well (K. Summers, personal communication). This behavior in *Dendrobates* is preceded by egg attendance and tadpole transport (see above). Eggs provided to tadpoles in dendrobatids are generally unfertilized and the female returns to the water body regularly to deposit eggs until metamorphosis occurs. Tadpole feeding also occurs in *Colostethus beebei*, although it may be facultative in this species (Bourne *et al.* 2001). In yet another striking case of convergence between the neotropical dendrobatids and *Mantella* of Madagascar, a similar tadpole feeding behavior has recently been reported in the bamboo and treehole-breeding *Mantella laevigata* (Heying 2001).

Tadpole feeding is also relatively widespread in hylids. In Central America, the bromeliad breeding *Hyla zeteki* feeds its tadpoles with eggs (Dunn 1937) and the closely related *H. picadoi* may as well (Duellman 2001). All known Jamaican hylids (*Calyptahyla crucialis, Osteopilus brunneus, Hyla wilderae,* and *H. marianae*) are bromeliad breeders, and their tadpoles are known, or strongly suspected, to be obligate egg eaters (Dunn 1926, Lannoo *et al.* 1987, Thompson 1996 - see below). Three South American species, *Osteocephalus oophagus, O. deridens* and *Anotheca spinosa* are bromeliad or treehole breeders with oophagous tadpoles (Jungfer and Schiesari 1995, Jungfer 1996, Jungfer and Weygoldt 1999, Jungfer *et al.* 2000). *Phrynohyas resinifictrix* is another Amazonian treehole-breeder that frequently consumes conspecific eggs, but larvae of this species can also feed on other food sources (Schiesari *et al.* 1996).

Several Asian rhacophorids also feed their tadpoles. Wassersug *et al.* (1981) reported eggs in the guts of tadpoles from Thailand attributed to *Philautus* cf. *carinensis*. The extreme morphological modifications of these larvae led the authors to conclude that it was obligatorily oophagous. Perhaps the best-studied frog that feeds its tadpoles is *Chirixalus effingieri* from Taiwan. This species breeds in water-filled bamboo stumps, and females return periodically to provide eggs (Kam *et al.* 1996, Ueda 1986). In a series of experiments conducted on this species by Yeong-Choy Kam and colleagues, they demonstrated that cohorts of tadpoles not fed by females never survived to metamorphosis (Kam *et al.* 1997), competition for eggs was density dependent (Kam *et al.* 1998) and tadpoles from older cohorts are more able to compete for and consume the limited eggs available (Chen *et al.* 2001). The petropedetine ranid *Phrynobatrachus guineensis* has recently been reported from tree holes in the Ivory Coast by Rödel (1998). Like many oophagous tadpoles, *P. guineensis* has a reduced number of tooth rows, and individual larvae have been observed consuming conspecific eggs (Rödel 1998). In a somewhat different case, there is some evidence to suggest that tadpoles of vocal sac-brooding *Rhinoderma darwinii* (Rhinodermatidae) indirectly receive nutritional input from secretions of the male (patrotrophy - Goicoechia *et al.* 1986). Obligatory tadpole feeding seems to only evolve when the habitats in which tadpoles are placed (e.g., phytotelms, empty snail shells, the vocal sacs of male *Rhinoderma*, etc.) are extremely food limited. Numerous other species from the Old and New world tropics are known only from food-limited micro-aquatic habitats such as tree holes, and we suspect that tadpole feeding will be reported from other species in the future.

8.3.10 Froglet Transport

Transport of newly hatched froglets (Fig. 8.11, 8.12) has been reported in the literature only relatively recently and is known from four families (Hylidae, Leptodactylidae, Microhylidae, and Sooglossidae). Diesel *et al.* (1995) reported froglet transport in *Eleutherodactylus cundalli* from Jamaica. These *Eleutherodactylus* oviposit in caves and the female attends the clutch until hatching. Egg attendance in this species may reduce predation by cave crabs and/or reduce the probability of fungal infection (Diesel *et al.* 1995). After hatching, young froglets climb onto the head, flanks or dorsum of the female and are transported out of the cave. Up to 72 froglets were counted on a single female and may be carried for a substantial distance (> 100 m). Froglets apparently remain on the female for some time, as recently hatched froglets still had yolk in their guts as they climbed onto the female, while those outside the cave were larger and no yolk was visible (Diesel *et al.* 1995). Diesel *et al.* (1995) suggested that froglet transport evolved in this species as a means of delivering offspring to a more productive terrestrial habitat that is more conducive to froglet growth. MacCulloch and Lathrop (2002) described female transport of eggs and froglets in *Stefania ayangannae, S. coxi and S. evansi* (Hylidae) from Guyana. The froglets remain with the mother until the yolk is completely consumed. Jungfer and Boehme (1991) suggested that dorsal secretions played a role in keeping froglets "on board" in *S. evansi*.

Female *Sooglossus sechellensis* transport tadpoles through metamorphosis, and the froglets remain on her dorsum for an unknown period (Nussbaum unpublished data; Fig. 8.11). The function of froglet transport is unknown in this species and difficult to understand, because the adult and froglet microhabitats are seemingly identical (forest litter). Rainer *et al.* (2001) described male transport of froglets in the microhylid *Sphenophryne cornuta* from New Guinea. Bickford (2002) described froglet transport by males in *Liophryne schlaginhaufeni* (Fig. 8.12) and provided additional observations on *S. cornuta*. Male *S. cornuta* attend egg masses and after hatching, the froglets climb onto their father's back. Males carried froglets for up to nine days to distances up to 55 m from the oviposition site. Froglets departed from the transporting male at regular intervals, and Bickford (2002) speculated that this may function to decrease competition for food, lower predation risk or reduce chances for inbreeding. While only relatively few species are known to exhibit froglet transport, this behavior may be more widespread than previously recognized.

One could add "froglet attendance" to the list of anuran modes of parental care and define it as a parent staying at a fixed location with froglet offspring. However, it is not yet clear that froglet attendance is distinct from egg attendance. It may be that froglet attendance is simply prolonged egg attendance, such that, the parent is present when the froglets emerge. For example, male *Bryobatrachus nimbus* has been observed attending fresh egg masses, developing embryos and tailed froglets (Mitchell 2002). If it could be shown that, after emergence, some parental care was given to froglets by parents, this could be unambiguously called froglet attendance. However, froglets from an entire clutch of eggs can take some time to emerge and thus the presence of the parent near the froglets does not necessarily reflect any parental care of froglets, only completion of egg attendance.

8.3.11 Viviparity

We define viviparity as any reproductive mode in which eggs are retained in the oviduct and offspring have live birth. Viviparity has evolved at least three times in anurans (Bufonidae, Leptodactylidae, Ranidae) and two types can be recognized: that in which the offspring are nourished entirely on yolk reserves (lecithotrophy – "ovoviviparity" of some authors) and cases in which offspring are nourished at least partly by maternal secretions (matrotrophy). At least seven lecithotrophic viviparous species are known, five *Nectophrynoides* (see Graybeal and Cannatella 1995 for a discussion of the systematics of this genus), *Eleutherodactylus jasperi* (Drewry and Jones 1976, Wake 1978) and an undescribed *Limnonectes* (Emerson 2001). Two matrotrophic viviparous species are known, *Nectophrynoides occidentalis* and *N. liberiensis* (Wake 1980). All viviparous species, by definition, have internal fertilization (Sever *et al.*, Chapter 7). Additional species in *Eleutherodactylus* and *Nectophrynoides* may be viviparous, however, data on the reproductive biology of many species are unavailable.

8.4 SEX OF THE CARE GIVER

Sociobiologists have wondered what factors determine whether it is the female, the male, or both sexes that provide care for their offspring. Factors that select for parental care are not necessarily the same factors that determine whether care will be maternal, paternal, or biparental. Many authors have addressed the theoretical aspects of this issue including: Baylis (1978, 1981), Beck (1998), Blumer (1979), Chase (1980), Clutton-Brock (1991), Dawkins and Carlisle (1976), Grafen and Sibly (1978), Gross and Shine (1981), Gross and Sargent (1985), Loiselle (1978), Maynard Smith (1977, 1978), Perrone and Zaret (1979), Ridley (1978), Trivers (1972), Wells (1977, 1981), Werren *et al.* (1980), and Williams (1975).

Maynard Smith (1977) and Chase (1980) argued that biparental care should be rarer than uniparental care, because if one parent is already providing care, then the selective advantage of providing additional care is reduced for the other parent. Whether or not this is the correct explanation, biparental care is generally much rarer among invertebrates and poikilothermic vertebrates (Clutton-Brock 1991).

Trivers (1972) presented two lines of reasoning regarding factors that might determine which parent cares for the offspring. These are the "anisogamy" and "paternal certainty" arguments. From this foundation, considerable additional speculation about what causes male or female parental care has been published. Much of this work was based on fishes, a group in which diverse parental care behavior has evolved repeatedly.

The anisogamy hypothesis stems from the different costs of producing male and female gametes. Females are destined to become the care givers, because they have more invested in each of their ova than males have invested in individual spermatozoa. Female reproductive success may depend on the survival of her limited propagules, but male reproductive success may be more a function of multiple-matings than the success of any one brood. Dawkins and Carlisle (1976), citing the "Concorde Fallacy", and Maynard Smith (1977) pointed out a minor flaw in the anisogamy argument. They argued that females who are deserted at the nest by males stay on to care *not* because of previous investments in ova, but because continued investment is likely to result in greater returns than starting over again with a new set of fertilized ova. Deserted females are more likely to stay at the nest than deserted males, because the cost of additional investment is lower for females than for males.

The anisogamy argument has other problems that have not been adequately addressed. For example, the cost of producing offspring is not measured solely by the caloric cost of individual gametes (Spencer 1852, 1867). The cost also include behaviors that lead to successful mating, including fighting for territories or a position of dominance, which are costly for the male both in calories and risk of injury or death. Also, the cost of spermatozoa should not be underrated. Although males produce more gametes than females, most male gametes will not contribute genetically to future generations. Furthermore, spermatozoa are not unlimited. In some mating systems, a single set of fertilized eggs may be just as valuable to the male as they are to the female that produced them.

Trivers' (1972) paternal certainty argument rests on the assumption that males gain more returns on post-fertilization investment (parental care) as their paternal confidence increases. Trivers assumed that mating systems in which males are territorial and fertilization is external provide maximum paternal confidence, and in these instances, males are expected to be the care givers. Where fertilization is internal, paternal confidence is low, and females are expected to be the care givers. This hypothesis was evaluated by Maynard Smith (1977) and Werren *et al.* (1980) among many others. One problem is that paternal certainty is not necessarily higher among species with external fertilization compared to those with internal fertilization.

Dawkins and Carlisle (1976) argued that in species with external fertilization, the sex that spawns first will be selected to desert first, leaving the other mate to care for the embryos. Often in such species it is the female that spawns first, so the male is expected to be the care giver. Loiselle (1978) studied fishes in which males and females spawn simultaneously and found that males are the most frequent care-giving sex. He argued, therefore, that males are the most frequent care giver because they are the territorial sex and not because they spawn second. Among fishes with limited, high quality nest sites, the male is almost always the territorial sex, because of multiple mating opportunities for males that do not exist for females. By guarding his territory, the male is predisposed to becoming the care giver. Baylis (1978, 1981), Blumer (1979), Perrone and Zaret (1979), and Ridley (1978) presented arguments similar to, but differing in details, from that of Loiselle (1978).

Gross and Sargent (1985) suggested that the relative cost of parental care for males and females may be important in determining which parent cares. The sex with the lowest cost of caring is expected to be the care giver. Gross and Sargent (1985) presented evidence from fish behavior to support this hypothesis. Wells (1981) discussed aspects of the cost of caring for each sex for the various kinds of parental care among anurans. To date, these considerations remain largely speculative.

According to the "association hypothesis" of Williams (1975), an association of one parent or the other with its offspring, for whatever reason, preadapts that parent for parental care. This hypothesis was favored by Gross and Shine (1981) over the other hypotheses outlined above.

The various desertion hypotheses outlined above are not mutually exclusive, and it is difficult, if not impossible, to sort out which factor, or factors, may be responsible for determining the sex of the care giver in particular species. The issue is also complicated by historical questions. Whatever costs and benefits that originally accrued to male or female parental care may have changed through time as selective pressures changed.

The majority of care-giving poikilotherms, including frogs, with internal fertilization have maternal care (Clutton-Brock 1991, Nussbaum 1985). These observations could support both the mode of fertilization argument and the association hypothesis, because internal fertilization facilitates separation of the courtship and nest sites in space and time. In these circumstances, females may be the only sex associated with the embryos at the time of oviposition

Beck (1998:440) claims to have "tested the hypothesis that mode of fertilization is related to [the] sex providing uniparental care in anurans." He did so using Ridley's comparative method (Ridley 1983, Ridley and Grafen 1996), which involves analyzing traits such as parental care in a phylogenetic framework. He found (p. 443) "no support for the assumptions of a significant relationship between the sex caring for offspring and mode of fertilization." Therefore, he concluded that alternative hypotheses are needed to account for which sex is the care giver in anurans. There are alternative explanations (see above), but Beck addressed only one in detail, namely the differential cost of care by the two sexes as proposed by Gross and Sargent (1985). Although Beck's conclusions may be correct, there are two problems with his analysis that may have influenced his results. First, although he reported data for a large number of species (334), most of his sources were secondary compilations, and there is no evidence that the original sources were critically examined. Secondly, he reported data for only nine species with internal fertilization, and five of these are viviparous. Because Beck did not define parental care, we have no way of knowing whether he considers viviparity to be a form of parental care. In any case, inclusion of the viviparous species biases his analysis, because internal fertilization is a necessary correlate of viviparity.

At present, there are no convincing generalizations about factors that determine whether parental care in anurans will be maternal, paternal, biparental, or amphisexual. As pointed out by Wells (1981), parental care among anurans is highly variable, and it is difficult to account for all forms of parental care in this group by a single hypothesis. It seems likely that each mating system may have different evolutionary forces at play, and that different explanations will be needed for each one.

8.5 THE PHYLOGENETIC DISTRIBUTION OF PARENTAL CARE IN ANURANS

While parental care is not particularly common in anurans (Crump 1995), it is phylogenetically widespread. Many families of anurans have some sort of parental care, although apparently ten families do not (Allophrynidae, Ascaphidae, Bombinatoridae, Brachycephalidae, Heleophrynidae, Megophryidae, Pelobatidae, Pelodytidae, Pseudidae and Rhinophrynidae). Many of these families that lack parental care are ancient, species-poor lineages (e.g., Rhinophrynidae). Others have substantial diversity, but are poorly known. We suspect that parental care will eventually be discovered in at least some of these lineages (e.g., Brachycephalidae).

Egg attendance is the most common form of parental care in anurans and has evolved independently numerous times. It is known in 14 families (Table 8.1). Nest construction is known for eight families. Egg transport and egg brooding are known for three and two families, respectively. Tadpole attendance is found in six families, while tadpole transport and tadpole brooding are found in four and six, respectively. Seven families exhibit tadpole feeding. Froglet transport, while only recently described, is known for four families. Viviparity has been reported from three families. For a full presentation of these data see Table 8.1.

8.6 THE PREDOMINANCE OF NO PARENTAL CARE IN ANURANS – TRIUMPH OF THE STATUS QUO?

The above descriptions of the types of parental care behaviors known in anurans indicate a diverse and phylogenetically widespread suite of behaviors. However, it has been estimated that up to 90% of extant species perform no parental care whatsoever (McDiarmid 1978). The addition of recently published data and the inclusion of nest construction and viviparity as forms of parental care, may decrease this estimate to perhaps 80%. Nevertheless, it remains clear that the vast majority of anurans lack parental care. What accounts for this pattern?

The stated function of parental care is to increase the survivorship of offspring. As discussed above, this can be attained by the parent providing some protection, food or transport for the offspring during vulnerable life history stages. However, evolution is often a conservative process and if the status quo is adequate for survival and reproduction in the contemporary environment, then no potentially costly parental care behaviors need evolve. Alternatively, parents can increase survivorship of offspring via oviposition site choice. Many species of frogs lay eggs singly or in small clumps, possibly to avoid catastrophic predation or fungal infection (e.g., *Hyla regilla*; Kiesecker and Blaustein 1997) or to maximize oxygen absorption (Strathmann 2001). Parents may also be able to respond to environmental cues to avoid oviposition in sites where predators or competitors are abundant (Petranka *et al.* 1994). These strategies are not normally considered parental care, but they are probably effective in reducing mortality of offspring.

However, these parent-centered views implicitly presume that selection is only acting on adult behaviors. Selection, of course, may also act on the embryos, larvae, or juveniles themselves. For example, the unpalatability of *Bufo* embryos and tadpoles to predators is a characteristic that surely provides a selective advantage (Wassersug 1971). If this trait provides an alternative way in which mortality rates are decreased during early life history stages, then selection would not favor the evolution of potentially costly parental care behaviors to accomplish the same task (Orton 1955, Salthe and Mecham 1974). It is perhaps not a coincidence that no *Bufo* are known to exhibit parental care. Similar arguments could be made for other traits in tadpoles (alteration of activity patterns in presence of predators, phenotypic plasticity, accelerated growth and developmental rates; Relyea 2001).

While relatively few taxa with aquatic larvae exhibit parental care, a larger proportion of those species that have fully, or semi-terrestrial development do (Crump 1995). In particular, many species that lack a free-living larval period have evolved a diverse suite of parental care behaviors (Table 8.1). This pattern may result from an absence of opportunity for selection to act on tadpoles and instead selection has lowered mortality rates by acting on adult behavior. Other vertebrates in which parental care is widespread (birds, mammals) similarly lack a free-living larval stage. These observations may help to explain the oft-cited observation that relatively few anurans exhibit parental care.

8.7 CASE STUDIES

8.7.1 General

In the following section, we discuss the use of phylogenetic information in explaining patterns and processes in the evolution of parental care in anurans. We present several case studies from the literature as well as several novel analyses. In some cases, our purpose is to test hypotheses regarding the evolution of parental care. In other cases, our goal is to point out where a phylogenetic perspective would be helpful. Where the data are inadequate, we point to where gaps in our knowledge lie.

We assume that no parental care is the ancestral state in anurans. This assumption is based on the facts that the absence of parental care is: (1) the most common state among anurans, (2) the most widely distributed state phylogenetically, and (3) it is associated with the presumed ancestral, biphasic life cycle in which small eggs are abandoned in still waters and hatch into tadpoles. However, this may not be the case in a given lineage. For example, there is no reason why parental care cannot be secondarily lost, and if this happens, parental care (not the lack of it) is the ancestral state. We also assume that the stated phylogenetic relationships represent the actual genealogy of the species under consideration.

8.7.2 The Evolution of Parental Care in Poison Dart Frogs (Dendrobatidae)

The poison dart frogs of Central and South America (Dendrobatidae) have long fascinated ethologists because of their diverse and unusual breeding behaviors. These include egg attendance, tadpole transport, tadpole brooding and tadpole feeding. These various parental care duties are carried out by the male in some species, the female in others or some combination of males and females. At least one species has biparental care (*Dendrobates vanzolinii* – Caldwell 1997, Caldwell and de Oliveira 1999). Previous authors have speculated on how these various parental care behaviors may have evolved in this group. For example, Weygoldt (1987) and Zimmerman and Zimmerman (1984, 1988) suggested that male parental care was ancestral in dendrobatids and that biparental care evolved from it. These authors also suggested that female parental care evolved from biparental care and that female care was the most derived form of parental care in dendrobatids.

Recently, phylogenetic information has become available on the relationships within the family (Clough and Summers 2000, Summers *et al.* 1999). These data allow tests of the various hypotheses regarding the evolution of parental care in this group. Clough and Summers (2000), constructed a phylogenetic hypothesis using three genes of mitochondrial DNA (~1200 bp) from 27 species of dendrobatids including: 9 *Epipedobates*, 2 *Colostethus*, 3 *Phyllobates*, 12 *Dendrobates* and *Aromobates femoralis* (the non-toxic, presumed most basal member of the clade). The most parsimonious hypothesis of the relationships of these taxa based on these data is presented in Figure 8.13.

Assuming that this gene tree reflects the actual genealogical relationships among these species, there are several clear patterns. First, larval feeding with eggs appears to have evolved twice independently, once in the Amazonian *Dendrobates fantasticus* + *D. ventrimaculatus* + *D. vanzolinii* lineage and once in the *D. histrionicus*-group from Central and northern South America (Fig. 8.13). In both instances, tadpole feeding evolved from ancestors in which males attended eggs and transported tadpoles (Clough and Summers 2001, Summers *et al.* 1999). If correctly placed in the genus *Colostethus*, tadpole feeding in *C. beebei* (Bourne *et al.* 2001) would be a third independent evolution of this characteristic in dendrobatids.

Secondly, male-only care is ancestral in these frogs, as predicted by Weygoldt (1987) and Zimmerman and Zimmerman (1984, 1988). However, female (or predominantly female) parental care did not evolve from biparental care, but rather from male only care. There seems to be an intermediate stage where males cared for the eggs and females cared for the tadpoles. Eventually, females cared for both eggs (egg attendance) and tadpoles (tadpole transport and feeding) and the most derived members of this lineage have female only parental care (Summers *et al.* 1999). Biparental care evolved once (in *D. vanzolinii*), also from male only parental care. This may represent a possible transitional stage toward female only care (Reynolds *et al.* 2002).

Recently, *Colostethus stepheni* was shown to have non-feeding tadpoles that are not transported (Juncá 1996, 1998). Depending on the position of *C. stepheni*, this may represent the ancestral condition in *Colostethus* (which is itself nearly basal among dendrobatids, Caldwell 1996) or it may be derived. New species, further phylogenetic analysis, and additional field observations will further clarify the evolution of parental care in dendrobatids.

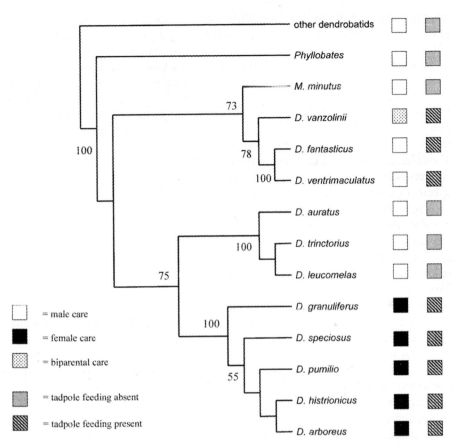

Fig. 8.13 Phylogenetic tree based on Clough and Summers (2000) from mitochondrial DNA data. Numbers below each node indicate the percentage of times this topology was recovered in bootstrap replications (only values over 50% are shown). See text and Clough and Summers (2000) for details.

8.7.3 The Origin of Amphisexual Egg Attendance Behavior in Frogs from Madagascar

Recent field work by one of us revealed that two species of mantelline frogs (*Mantidactylus bicalcaratus* and *M. punctatus*) exhibit amphisexual egg attendance (Lehtinen unpublished data – Fig. 8.3). In these species, egg masses are laid on the leaves of screw pine plants (*Pandanus*) above small pools of rainwater in the leaf axils. The embryos develop on the leaf for 3-12 d (depending on rainfall) before the jelly dissolves and the hatchlings wriggle or are washed down into the water-filled leaf axil. During this terrestrial phase, the egg mass is attended by either the male or the female, but not both. Other field observations indicate that both ants and day geckos (*Phelsuma*) consume unattended egg clutches implying some benefit of egg attendance behavior by adults. Predation by frog-eating spiders (Heteropodidae) that also live on *Pandanus* plants may be a cost of this behavior (Lehtinen unpublished data). Amphisexual egg attendance (as opposed to a single sex providing care) is known from only three other anurans, the New Guinean microhylid *Cophixalus parkeri* (Simon 1983), *Eleutherodactylus johnstonei* and *E. alticola* (Townsend 1996, Bourne 1998).

To study the origin and evolution of this parental care behavior, we used molecular data to construct a phylogenetic hypothesis for the mantelline frogs. In collaboration with C. Richards, we sequenced the 12S, 16S and tRNA mitochondrial genes (~ 1800 bp) of 35 mantellid frogs (29 species of *Mantidactylus* and 6 species of *Mantella* - the putative sister taxon to *Mantidactylus*). This taxon sample includes multiple members of all the putative subgenera of *Mantidactylus* (Glaw and Vences 1994) and all of the known members of the subgenus *Pandanusicola* (the subgenus which includes the *Pandanus* breeding species) except *M. punctatus*.

A strict consensus of the four equally parsimonious trees indicates that all of the *Pandanus* breeding *Mantidactylus* (with the inclusion of the non-*Pandanus*-breeding *M. liber*) form a well-supported clade (Fig. 8.14 full details of this analysis will be presented elsewhere). The *Pandanusicola* + *M. liber* clade is sister to species of the subgenus *Guibemantis* (the *M. depressiceps* species-group; Blommers-Schlösser 1979) which lay egg masses on vegetation overhanging ponds and are not known to provide parental care (Blommers-Schlosser 1979, Glaw and Vences 1994, Lehtinen unpublished data). *M. liber* is currently allocated to *Guibemantis* and similarly lays its eggs on vegetation above ponds and provides no parental care (Blommers-Schlösser 1975b). However, these data clearly indicate that *M. liber* is nested in the *Pandanus*-breeding clade. If egg attendance is ancestral in *Pandanusicola* + *M. liber*, this indicates a reversal from a condition in which some egg attendance behavior was provided (Fig. 8.14). This reversal is correlated with a switch from a phytotelm-breeding habit to a pond-breeding habit and suggests that evolutionary transitions may often accompany ecological changes. These data also suggest that egg attendance is a relatively labile trait that can be readily lost or re-evolved based on contemporary selective pressures. However, egg attendance may have arisen independently in *M. bicalcaratus* and *M. punctatus*. Inclusion of sequences from *M. punctatus*, should sufficiently resolve these relationships to permit a test of these alternative hypotheses. Unfortunately, no additional data on the parental care behaviors (if any) of the other species in the *Pandanusicola* + *M. liber* clade are available.

In this case, a lack of resolution of phylogenetic relationships, by and large, does not prevent us (as in the Jamaican hylids example, below) from drawing conclusions, rather we are limited mostly by a lack of field observations. Other interesting questions concerning the evolution of parental care in this lineage, such as whether amphisexual egg attendance evolved from no parental care, male, female or biparental egg attendance remain unanswered. Further progress on reconstructing the evolution of egg attendance in "*Mantidactylus*" awaits additional field data on parental care in these frogs. However, a phylogenetic tree clearly indicates where critical data gaps lie and which species should be investigated to clarify the evolution of egg attendance in this group (namely *M. albolineatus*, *M. flavobrunneus* *M. pulcher* and *M.* sp. nov.; Fig. 8.14).

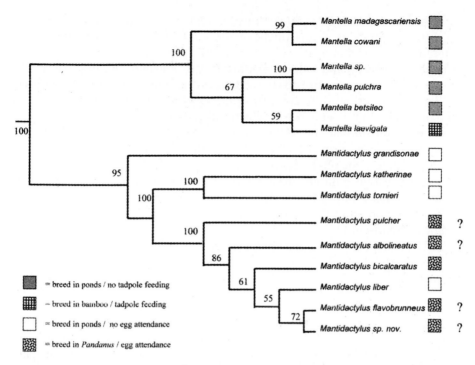

Fig. 8.14 Phylogenetic tree of mantelline frogs based on unpublished DNA data (R. Lehtinen, C. Richards and R. Nussbaum) from three mitochondrial genes (only taxa of interest depicted here). Numbers below each node indicate the percentage of times this topology was recovered in bootstrap replication. ? = These species breed in *Pandanus* but it is unknown if they exhibit egg attendance. See text for details.

8.7.4 *The Evolution of Tadpole Feeding in Caribbean Hylids and* Mantella

All four hylids endemic to the island of Jamaica (*Calyptahyla crucialis, Osteopilus brunneus, Hyla wilderae,* and *H. marianae*) breed in water-filled bromeliads. An additional undescribed species of Jamaican *Osteopilus* may also be a bromeliad breeder (Crombie 1999). The available information also strongly suggests that feeding of tadpoles with eggs by the mother is obligatory in these species (Dunn 1926, Noble 1931, Lannoo *et al.* 1987).

While only rudimentary ecological information is available from the other Jamaican hylids, *Osteopilus brunneus* is comparatively well-studied (Lannoo *et al.* 1987, Thompson 1996). The larvae of *O. brunneus* are morphologically modified to ingest eggs and have no labial tooth rows, (Lannoo *et al.* 1987). The mother periodically returns to the bromeliad in which she oviposited and lays additional eggs, which the tadpoles already in the bromeliad quickly consume. These eggs are fertilized when the pre-existing cohort is young, but as the tadpoles approach metamorphosis, unfertilized eggs are laid by the female (Thompson 1996).This life history pattern is not characteristic of other hylids in the Caribbean region. For example, *Hyla vasta* and *H. heilprini* from Hispaniola breed in swift-flowing streams and provide no parental care (Schwartz and

Henderson 1991). *Hyla pulchrilineata, Osteopilus dominicensis,* and *O. septentrionalis* breed in ponds, ditches or flooded fields and also do not provide parental care (Schwartz and Henderson 1991). So, what was the origin of bromeliad breeding and tadpole feeding in this lineage? Did this trait evolve from a single common ancestor that colonized Jamaica, or did it convergently evolve from multiple different ancestors? Do the phylogenetic relationships shed any light on the sequence in which this parental care behavior evolved or reveal any transitional states? By reconstructing the evolutionary history of this group we can address these, and other, questions.

Unfortunately, the phylogenetic relationships of Caribbean hylids are poorly known. Only two datasets are available that present a formal phylogenetic analysis of this group: Maxson (1992), based on albumin data, and da Silva (1998), based on adult and larval morphology. Maxson's (1992) data place the Jamaican hylids in a single clade and are each other's closest relatives (Fig. 8.15). The conclusion from this phylogenetic hypothesis is that tadpole feeding only evolved once in this lineage (Fig. 8.15). The closest common ancestor of the Jamaican clade is a Hispaniolan clade containing *Hyla pulchrilineata* and

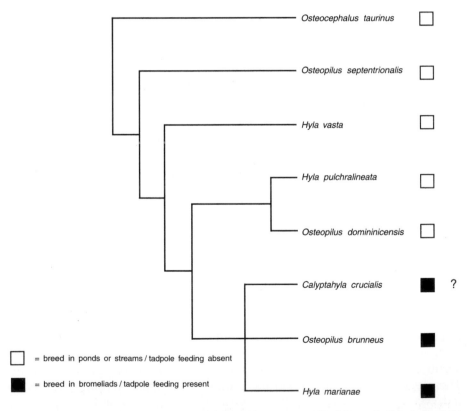

Fig. 8.15 Phylogenetic tree of the relationships among Caribbean hylids based on serum albumin data from Maxson (1992). ? = *Calyptahyla* is strongly suspected to exhibit tadpole feeding but critical observations are unavailable. See text for details.

Osteocephalus dominicensis. Karyological data from Anderson (1996) also indicate that *Osteopilus brunneus* is not as closely related to *O. dominicensis* and *O. septentionalis* as they are to each other. Both *O. dominicensis* and *O. septentionalis* are pond breeders with no parental care, indicating that this was the ancestral condition of the more specialized Jamaican species. Based on these data, all the putative tadpole-feeding species in Jamaica form a single, monophyletic clade.

The phylogenetic hypothesis by da Silva (1998), however, offers a much different scenario. In his tree, *Osteopilus* (including species from both Hispaniola and Jamaica) is monophyletic, and *Calyptahyla crucialis* is only distantly related to *Osteopilis* (Fig. 8.16). *Osteopilus* is sister to a clade including a monophyletic *Pternohyla*, *Argentiohyla siemersi* and *Anotheca spinosa*. The sister taxon to *Calyptahyla* is *Nyctimantis rugiceps* (Fig. 8.16). Contrary to Maxson's (1992),

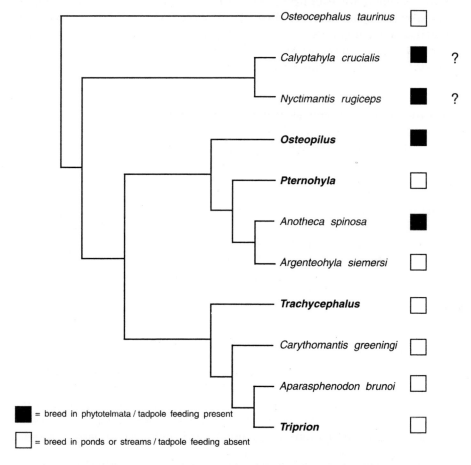

Fig. 8.16 Phylogenetic tree of the relationships among South American, Central American and Caribbean hylids based on adult and larval morphology from da Silva (1998). **Bold** taxa are monophyletic. ? = *Calyptahyla* and *Nyctimantis* both probably breed in phytotelmata (bromeliads and tree holes, respectively), but critical observations on tadpole feeding are unavailable. See text for details.

phylogeny, these results suggest that at least two separate hylid lineages colonized Jamaica, and that bromeliad breeding / tadpole feeding is convergently evolved (similar conclusions were reported by Treub and Tyler (1974)). *Anotheca spinosa* is a bromeliad breeder with oophagous tadpoles (Jungfer 1996) and *Nyctimantis rugiceps* is probably a treehole breeder (Crump 1974). *Pternohyla* spp. and *Argentiohyla siemersi*, however, are both pond breeders (Cei 1980, Duellman 2001). This suggests that phytotelm breeding with oophagous tadpoles may have been the ancestral condition in the *Calyptahyla* + *Nyctimantis* clade. In the clade including *Osteopilus*, however, the situation is less clear. Tadpole feeding may have evolved independently in *Osteopilus* and *Anotheca* or it may have been the ancestral condition and was independently lost in *Pternohyla* and *Argentiohyla* (Fig. 8.16).

However, there are several problems with the phylogenetic analyses presented by Maxson (1992) and da Silva (1998). First, some taxa (e.g., *Hyla wilderae*) were not included in either analysis. This (and other) missing species could provide important information concerning the relationships within this clade. Secondly, while da Silva (1998) included a large number of non-Caribbean hylids, Maxson (1992) only used a single outgroup taxon (*Osteocephalus taurinus* from South America). Because the affinities of the Caribbean species among the other new world hylids is controversial, it would be appropriate to include numerous representatives of other hylids from Central and South America in a full analysis. Thirdly, data presented by da Silva (1998) indicate that *Osteopilus* is monophyletic, but he does not present the relationships within this genus. Without this information, it is impossible to know if tadpole feeding evolved once or more than once, if there were any reversals, or what the ancestral state was in this clade. Lastly, there are no decay indices, bootstrap, or jackknife values presented with either tree (Fig. 8.15, 8.16). These values provide a quantitative measure of the reliability of the topology at each node. The absence of these indices does not allow us to assess the level of support for these two particular phylogenetic hypotheses and therefore these results remain equivocal. Further knowledge of the evolution of tadpole feeding in Jamaican hylids awaits a full phylogenetic analysis of this group as well as additional field observations.

Mantella laevigata from Madagascar was recently shown to feed eggs to its tadpoles (Heying 2001). After egg deposition, females periodically return to the water-filled bamboo stumps to lay unfertilized eggs, which the tadpoles consume. A phylogeny of the mantellid frogs (Fig. 8.14; see above for details) indicates that tadpole feeding is a uniquely derived behavior in *M. laevigata*. The closest ancestors of *M. laevigata* are pond breeders and are not known to provide any parental care (Glaw and Vences 1994). These examples again show that evolutionary transitions in parental care often accompany ecological changes in habitat use.

8.7.5 Evolution of Nest Construction and Tadpole Transport in "Fanged Frogs" (*Limnonectes*) from South East Asia

The South East Asian ranid frogs in the genus *Limnonectes* (formerly *Rana*) have remarkable reproductive diversity including aquatic tadpoles, nest construction, direct development and viviparity (Emerson 2001). In two Bornean species of *Limnonectes*, (*L. finchi* and *L. palavanensis*) males attend eggs and transport tadpoles. Males attend egg masses underneath leaf litter near streams (Inger and Voris 1988). After the eggs hatch, males transport the tadpoles on their back to nearby water bodies such as rain pools, animal wallows or streams (Inger 1966, Inger *et al.* 1986, Inger and Voris 1988) where they are released and continue development in the aquatic environment. In several other species in this clade, males construct basins together in which the eggs are laid (Emerson 2001, Orlov 1997).

Recently, Emerson (2001 – see also Emerson *et al.* 2000, Emerson 1996) presented a phylogenetic hypothesis of the relationships among the fanged frogs and their relatives based on mitochondrial DNA sequences. Based on this phylogeny (Fig. 8.17), we hypothesize that tadpole transport evolved only

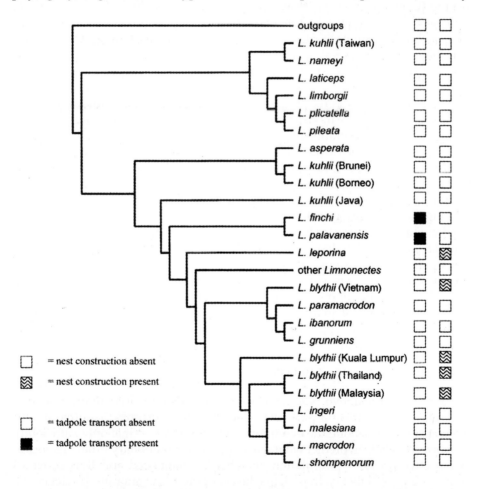

Fig. 8.17 Phylogenetic tree of the relationships of the "fanged frogs" (*Limnonectes*) and their relatives based on mitochondrial DNA (Emerson 2001). See text for details.

once in this group, as *L. finchi* and *L. palavanensis* are each other's closest relatives. This indicates that tadpole transport is a homologous behavior in these two species rather than convergently evolved. Information is unavailable on the parental care behaviors (if any) of the species immediately basal to *L. finchi* and *L. palavanensis*, so we are unable to infer the ancestral condition. However, the more derived sister group that includes *L. leporina* is known to have male nest construction. These nests are built in shallow areas in the stream and eggs are laid in aquatic sites (Inger and Stuebing 1997). This suggests a reversal from terrestrial oviposition, egg attendance and tadpole transport back to aquatic oviposition in streams.

Nest construction may have evolved at least three times in *Limnonectes* (Fig. 8.17). However, it is equally parsimonious to postulate that nest construction was present in the ancestor of *L. leporina* and was lost three times independently (in the *L. paramacrodon* + *L. ibanorum* + *L. grunniens* clade, in the *L. ingeri* + *L. malesiana* + *L. macrodon* + *L. shompenorum* clade, and in the clade labeled "other *Limnonectes*"). In any case, the lack of data on reproductive biology and parental care behaviors in several species further clouds this issue and further field observations on these species are necessary before any conclusions are warranted.

Emerson (2001) explored whether the evolution of sexual dimorphism and terrestrial reproductive modes was contingent upon prior evolutionary changes in *Limnonectes*. The evolution of parental care may also be significantly correlated with historical changes in other characteristics, such as body size or reproductive mode. This approach is potentially very useful to discover if the evolution of certain types of parental care are contingent on the prior appearance (or loss) of other characteristics. For example, the evolution of viviparity is probably contingent on the prior evolution of internal fertilization. From this we would predict that there are sister taxa to viviparous *Nectophrynoides*, *Eleutherodactylus* and *Limnonectes* that have internal fertilization but are oviparous. Other questions include: is egg attendance contingent on the evolution of small clutch size, large eggs with prolonged embryonic development, terrestrial reproductive modes or philopatry? Does tadpole feeding only evolve in species with certain types of mating systems? These, and other, questions can be addressed using this approach and offer the prospect of understanding the sequence in which these evolutionary changes occur.

8.8 ADVANCES AND CAVEATS

In the examples above, we have seen that detailed information on the evolutionary history of a particular clade can help identify patterns in, and answer questions about, the evolution of parental care. While only sufficiently advanced theory can predict what will happen in evolutionary time, a historical perspective provides information on what *has* happened and thus provides important tests of theory. In perhaps the best example in anurans, phylogenetic data from Summers *et al.* (1999) and Clough and Summers (2000) provided independent tests of hypotheses regarding the sequence of evolutionary change in the parental care behaviors of dendrobatids. In cases without sufficient data to test a hypothesis, a phylogeny clearly indicated where additional observations are needed. In addition to directly addressing questions in a phylogenetic framework, it is also clear that this perspective can inform and complement other approaches to studying parental care (cost-benefit analyses, experiments, field observations, etc.) and that this is an iterative process of repeated discovery.

However, not all questions can be usefully addressed with a phylogenetic approach. For example, when evolutionary change occurs within a species without accompanying speciation events (anagenesis), we are left without sister groups for comparison. This situation may not be diagnosable using a phylogenetic perspective. Because *Leiopelma* is a basal lineage and is species poor (three extant species), tadpole brooding and egg attendance may have developed through anagenetic evolutionary change rather than accompanied by speciation events (cladogenesis). If indeed *Leiopelma* evolved through transformation of an unbranched lineage, we have limited opportunity to understand the evolution of parental care in this group.

Similarly, if evolutionary change did proceed by cladogenesis but all sister taxa are extinct, our inferences are limited because there are no extant taxa to use in a comparative analysis. This is not necessarily a problem for the study of morphological characters, if appropriate fossils are available. Unfortunately, data from fossils usually provide no data on parental care behaviors. In cases where the transitional states are not recoverable, the comparative approach is not likely to be helpful. A phylogenetic perspective also cannot determine why a particular parental care behavior is maintained in a population. This is a function of the organism in its contemporary environment and requires detailed field data to assess.

It is also clear that the use of phylogenetic information is not a panacea and that this type of data can be misused. Phylogenies, like any hypothesis, must be evaluated individually based on the size, nature and quality of the data and the methodologies used to construct them. When dealing with a poorly supported phylogeny, it is more appropriate first to attempt to further resolve the genealogical relationships, rather than drawing conclusions from weak data. Even with well-supported phylogenetic information, however, it is not always possible to distinguish between possibilities.

Nevertheless, the use of a phylogenetic approach in the study of parental care evolution produces both new facts and new insights into what is already known. It should be clear from the information presented here that a historical approach is a powerful tool in identifying patterns and testing evolutionary hypotheses.

8.9 SUGGESTIONS FOR FUTURE RESEARCH

As has been lamented for many years (e.g., Noble 1931), the evolutionary relationships of many anurans remain poorly known. Even at the family level and higher there is little agreement among systematists (Duellman, Chapter 1). At the level of species and genera, there are similarly few examples of rigorously constructed phylogenetic hypotheses available. Fortunately, however, this is starting to change and several recent studies have provided resolution on the affinities of poorly known frogs (e.g., Pramuk *et al.* 2001, Read *et al.* 2001, Sheil *et al.* 2001, Wieczorek *et al.* 2000). This has been facilitated particularly by the availability of molecular data, although morphological and other types of data are also playing a role. As more data become available, the phylogenetic relationships of anurans will come more into focus and this will facilitate additional analyses of the types presented and reviewed here. We therefore both applaud and encourage further systematic studies of anurans, both for their use in answering specific evolutionary questions, and for their own sake.

Secondly, we need more quantitative field observations on poorly known tropical species. While there is nothing inherently wrong with studying parental care in temperate frogs, the fact remains that the vast majority of parental care in anurans occurs in species found only in tropical areas. Undoubtedly, many more examples of parental care in tropical anurans remain to be uncovered and even new modes of parental care are likely to be discovered. Crump (1996) indicated that the frequency and diversity of anuran parental care was most pronounced in the American tropics. However, there may exist a geographic bias here, given that Crump has conducted nearly all of her own work in that part of the world and is undoubtedly most familiar with that literature. There is a sampling issue here as well, since the American tropics have been more frequently visited in recent times by herpetologists (particularly those who publish in journals in English) than other tropical areas. The large number of cases of parental care in the American tropics, therefore, may stem from the many excellent field studies that have been conducted there, rather than any paucity of parental care in the old world tropics. Indeed, recent fieldwork in Taiwan, India, Papua New Guinea, Ivory Coast, Indonesia, Madagascar and other locations has revealed a large number of new examples of parental care in frogs (e.g., Bickford 2002, Heying 2001, Kam *et al.* 1996, Kanamadi *et al.* 1996, Rödel 1998, Rödel and Ernst 2002). Further studies of anuran reproductive biology in these particularly poorly studied areas are needed.

Thirdly, more experiments examining parental care in anurans are needed. As pointed out by Crump (1995), we have some knowledge of the benefits of parental care but know almost nothing about the costs. Well-designed, manipulative experiments (e.g., Burrowes 2000, Cook *et al.* 2001) are necessary to tease out these costs and benefits and will be crucial to further progress in the study of parental care evolution. Basic questions like what types of environmental conditions select for what kinds of parental care will also require much additional experimental work.

Lastly, and most importantly, we need a synthetic approach that uses both theory and data from observation, experiment and evolutionary history to explain the evolution of parental care. Macroevolutionary patterns can be identified with a historical approach, but this is only the beginning. Predictions generated from a phylogenetic analysis need to be tested using experimental and observational approaches (McLennan 1991). While our purpose here has been to highlight the phylogenetic approach to studying the evolution of parental care, a pluralistic approach surely has the most prospects for advancement in this area.

8.10 SUMMARY

Parental care is taxonomically widespread in anurans, but occurs in relatively few species. We recognize ten modes of parental care including: nest construction, egg attendance, egg transport, egg brooding, tadpole attendance, tadpole transport, tadpole brooding, tadpole feeding, froglet transport and viviparity. Numerous ideas have been put forward explaining the evolution of this diverse suite of parental care behaviors. However, a historical perspective based on detailed and robust phylogenetic information allows us to test alternative hypotheses. For example, phylogenetic information from mantellines and hylids suggests that evolutionary transitions in egg attendance and tadpole feeding often accompany ecological changes. Also, a phylogeny from dendrobatids suggests that female-only care was derived from male-only care, not from biparental care. These kinds of information allow us to predict the directionality of evolutionary change. We believe a historical approach both provides independent tests of theory and is an integral part of a pluralistic approach to understanding the evolution of parental care.

8.11 ACKNOWLEDGMENTS

We thank K. Summers, S. Emerson, and H. Heying for unpublished observations. We thank D. Bickford, J. Caldwell, W. Duellman, W. Hödl, M.-O. Roedel, L. Schiesari, and B. Weinstein for permission to use their photographs. We thank C. Richards for allowing us to use unpublished data. This manuscript was improved by comments from D. Ó Foighil.

8.12 LITERATURE CITED

Anderson, K. 1996. A karyological perspective on the monophyly of the hylid genus *Osteopilus*. Pp. 157-168. In: *Contributions to West Indian herpetology: a tribute to Albert Schwartz*. R. Powell and R.W. Henderson (eds), Society for the Study of Amphibians and Reptiles, Ithaca, New York. Contributions to Herpetology, volume 12. 457 pp.

Balinsky, B.I. and J.B. Balinsky. 1954. On the breeding habits of the South African bullfrog, *Pyxicephalus adspersus*. South African Journal of Science 51: 55-58.

Baylis, J.R. 1978. Paternal behavior in fishes: a question of investment, timing or rate? Nature 276: 738.

Baylis, J.R. 1981. The evolution of parental care in fishes, with reference to Darwin's rule of male sexual selection. Environmental Biology of Fishes 6: 223-251.

Beck, C.W. 1998. Mode of fertilization and parental care in anurans. Animal Behaviour 55: 439-449.

Bell, B.D. 1985. Development and parental care in the endemic New Zealand frogs. Pp. 269-278. In: *The Biology of Australasian Frogs and Reptiles*. G. Grigg, R. Shine and H. Ehmann (eds). Surrey Beatty and Sons, Chipping Norton, NSW, Australia. 527 pp.

Bickford, D. 2002. Male parenting of New Guinea froglets. Nature 418: 601-602.
Blanchard, F.N. 1934. The relation of the four-toed salamander to her nest. Copeia 1934: 137-138.
Blommers-Schlösser, R.M.A. 1975a. Observations on the larval development of some Malagasy frogs, with notes on their ecology and biology (Anura: Dyscophinae, Scaphiophrynidae and Cophylinae). Beaufortia 24: 7-26.
Blommers-Schlösser, R.M.A. 1975b. A unique case of mating behavior in a Malagasy tree frog, *Gephyromantis liber* (Peracca, 1893), with observations on the larval development. Beaufortia 23: 15-25.
Blommers-Schlösser, R.M.A. 1979. Biosystematics of the Malagasy frogs. I. Mantellinae (Ranidae). Beaufortia 29: 1-77.
Blumer, L.S. 1979. Male parental care in the bony fishes. The Quarterly Review of Biology 54: 149-161.
Bourne, G.R. 1998. Amphisexual parental behavior of a terrestrial breeding frog *Eleutherodactylus johnstonei* in Guyana. Behavioral Ecology 9: 1-7.
Bourne, G.R., A.C. Collins, A.M. Holder and C.L. McCarthy. 2001. Vocal communication and reproductive behavior of the frog *Colostethus beebei* in Guyana. Journal of Herpetology 35: 272-281.
Brooks, G.R. 1968. Natural history of a West Indian frog, *Leptodactylus fallax*. Virginia Journal of Science 19: 176.
Brooks, D.R., D.A. McLennan, J.M. Carpenter, S.G. Weller and J.A. Coddington. 1995. Systematics, ecology, and behavior. Bioscience 45: 687-695.
Brooks, D.R. and D.A. McLennan. 2002. *The Nature of Diversity: An Evolutionary Voyage of Discovery.* University of Chicago Press, Chicago, 668 pp.
Burrowes, P.A. 2000. Parental care and sexual selection in the Puerto Rican cave-dwelling frog, *Eleutherodactylus cooki*. Herpetológica 56: 375-386.
Caldwell, J.P. 1996. Diversity of Amazonian anurans: the role of systematics and phylogeny in identifying macroecological and evolutionary patterns. Pp. 73-88 In: Neotropical biodiversity and conservation. A.C. Gibson (ed.), University of California, Los Angeles. Los Angeles California. 202 pp.
Caldwell, J.P. 1997. Pair bonding in spotted poison frogs. Nature 385: 211.
Caldwell, J.P. and V.R.L. de Oliveira. 1999. Determinants of biparental care in the spotted poison frog, *Dendrobates vanzolinii* (Anura: Dendrobatidae). Copeia 1999: 565-575.
Cei, J.M. 1980. Amphibians of Argentina. Monitore Zoologica Italiano (N.S.) Monografia 2. 609 pp.
Chase, I.D. 1980. Cooperative and non-cooperative behavior in animals. The American Naturalist 115: 827-857.
Chen, Y.-H., Y.-J. Su, Y.-S. Lin and Y.-C. Kam. 2001. Inter- and intraclutch competition among oophagous tadpoles of the Taiwanese tree frog, *Chirixalus eiffingeri* (Anura: Rhacophoridae). Herpetologica 57: 438-448.
Clough, M. and K. Summers. 2000. Phylogenetic systematics and biogeography of the poison frogs: evidence from mitochondrial DNA sequences. Biological Journal of the Linnean Society 70: 515-540.
Clutton-Brock, T.H. 1991. *The Evolution of Parental Care.* Princeton University Press, Princeton, NJ. 352 pp.
Coe, M.J. 1964. Observations of the ecology and breeding biology of the genus *Chiromantis*. Journal of Zoology 172: 13-34.
Cook, C.L., J.W.H. Ferguson and S.R. Telford. 2001. Adaptive male parental care in the giant bullfrog *Pyxicephalus adspersus*. Journal of Herpetology 35: 310-315.

Corben, C.J., G.J. Ingram and M.J. Tyler. 1974. Gastric brooding: a unique form of parental care in an Australian frog. Science 186: 946-947.

Cracraft, J. 1981. The use of functional and adaptive criteria in phylogenetic systematics. American Zoologist 21: 21-36.

Crombie, R.I. 1999. Jamaica. Pp. 63-92 In: *Caribbean amphibians and reptiles.* B.I. Crother (ed.), Academic Press. San Diego, California. 495 pp.

Crump, ML. 1974. Reproductive strategies in a tropical anuran community. University of Kansas Museum of Natural History Miscellaneous Publication No. 61. 68 pp.

Crump, M.L. 1995. Parental care. Pp. 518-567. In H. Heatwole and B.K. Sullivan (eds). *Amphibian Biology* Volume 2: Social Behaviour, Surrey Beatty and Sons, Chipping Norton, NSW, Australia. 291 pp.

Crump, M.L. 1996. Parental care among the Amphibia. Advances in the Study of Behavior 25: 109-144.

da Silva, H.R. 1998. Phylogenetic relationships of the family Hylidae with emphasis on the relationships within the subfamily Hylinae (Amphibia: Anura). Ph.D. Dissertation, University of Kansas, Lawrence, Kansas. 175 pp.

Darwin, C. 1859. The origin of species. The New American Library, New York. 479 pp.

Dawkins, R. and Carlisle, T.R. 1976. Parental investment, mate desertion and a fallacy. Nature 272: 131-133.

Diesel, R., G. Bäurle and P. Vogel. 1995. Cave breeding and froglet transport: a novel pattern of anuran brood care in the Jamaican frog, *Eleutherodactylus cundalli.* Copeia 1995: 354-360.

Dobkin, D.S. and R.D. Gettinger. 1985. Thermal aspects of anuran foam nests. Journal of Herpetology 19: 271-275.

Downie, J.R. 1988. Functions of the foam in foam-nesting leptodactylid *Physalaemus pustulosus.* Herpetological Journal 1: 302-307.

Downie, J.R. 1996. A new example of female parental behaviour in *Leptodactylus validus,* a frog of the Leptodactylid '*melanotus*' species group. Herpetological Journal 6: 32-34.

Drewry, G.E. and K.L. Jones. 1976. A new ovoviviparous frog from Puerto Rico. Journal of Herpetology 10: 161-165.

Duellman, W.E. and S.J. Maness. 1980. The reproductive behavior of some hylid marsupial frogs. Journal of Herpetology 14: 213-222.

Duellman, W.E. and P. Gray. 1983. Developmental biology and systematics of the egg-brooding hylid frogs, genera *Flectonotus* and *Fritzania.* Herpetologica 39: 333-359.

Duellman, W.E. and M.S. Hoogmoed. 1984. The taxonomy and phylogenetic relationships of the hylid frog genus *Stefania.* Miscellaneous Publications of the University of Kansas Museum of Natural History 75: 1-39.

Duellman, W.E. and L. Trueb. 1986. *Biology of the Amphibians.* McGraw-Hill, New York, NY 670 pp.

Duellman, W.E. 2001. Hylid frogs of middle America. Vol. 1. Contributions to Herpetology, Vol. 18. Society of the Study of Amphibians and Reptiles, Ithaca, New York. 694 pp.

Duellman, W.E. 2003. 1. An Overview of Anuran Phylogeny and Classification. This volume.

Dunn, E.R. 1926. The frogs of Jamaica. Proceedings of the Boston Society of Natural History 38: 11-130.

Dunn, E.R. 1937. The amphibian and reptilian fauna of bromeliads in Costa Rica and Panama. Copeia 1937: 163-167.

Emerson, S.B. 1996. Phylogenies and physiological processes – the evolution of sexual dimorphism in Southeast Asian frogs. Systematic Biology 45: 278-289.

Emerson, S.B., R.F. Inger and D. Iskandar. 2000. Molecular systematics and biogeography of the fanged frogs of Southeast Asia. Molecular Phylogenetics and Evolution 16: 131-142.

Emerson, S.B. 2001. A macroevolutionary study of historical contingency in the fanged frogs of Southeast Asia. Biological Journal of the Linnean Society 73: 136-151.

Felsenstein, J. 1985. Phylogenies and the comparative method. American Naturalist 125: 1-15.

Formas, R., E. Pugin and B. Jorquera. 1975. La identidad del batracio chileno *Heminectes rufus* Philippi, 1902. Physis (Sección C) 34: 147-157.

Gittleman, J.L. 1981. The phylogeny of parental care in fishes. Animal Behaviour 29: 936-941.

Glaw, F. and M. Vences. 1994. *A Fieldguide to the Amphibians and Reptiles of Madagascar*, 2nd ed. Moos Druck, Germany. 480 pp.

Goicoechea, O., O. Garrdo, and B. Jorquera. 1986. Evidence for a trophic paternal-larval relationship in the frog *Rhinoderma darwinii*. Journal of Herpetology 20: 168-178.

Grafen, A. and Sibly, R. 1978. A model of mate desertion. Animal Behavior 26: 645-652.

Graybeal, A. and D.C. Cannatella. 1995. A new taxon of Bufonidae from Peru, with descriptions of two new species and a review of the phylogenetic status of supraspecific bufonid taxa. Herpetologica 51: 105-131.

Green, A.J. 1999. Implications of pathogenic fungi for life-history evolution in amphibians. Functional Ecology 13: 573-575.

Gross, M.R. and Sargent, R.C. 1985. Evolution of male and female parental care in fishes. American Zoologist 25: 807-822.

Gross, M.R. and Shine, R. 1981. Parental care and mode of fertilization in ectothermic vertebrates. Evolution, 35: 775-793.

Haddad, C.F.B. and W. Hödl. 1997. New reproductive mode in anurans: bubble nest in *Chiasmocleis leucosticta* (Microhylidae). Copeia 1997: 585-588.

Haddad, C.F.B. and J.P. Pombal, Jr. 1998. Redescription of *Physalaemus spiniger* (Anura: Leptodactylidae) and description of two new reproductive modes. Journal of Herpetology 32: 557-565.

Heyer, W.R. 1969. The adaptive ecology of the species groups of the genus *Leptodactylus* (Amphibia, Leptodactylidae). Evolution 23: 421-428.

Heyer, W.R. and P.A. Silverstone. 1969. The larva of the frog *Leptodactylus hylaedactylus* (Leptodactylidae). Fieldiana Zoology 51: 141-145.

Heyer, W.R. and R.I. Crombie. 1979. Natural history notes on *Craspedoglossa stejnegeri* and *Thoropa petropolitana* (Amphibia: Salientia, Leptodactylidae). Journal of the Washington Academy of Science 69: 17-20.

Heying, H. 2001. Social and reproductive behavior in the Madagascan poison frog, *Mantella laevigata*, with comparisons to the dendrobatids. Animal Behaviour 61: 567-577.

Inger, R.F. 1966. The systematics and zoogeography of the amphibia of Borneo. Fieldiana Zoology 52: 1-402.

Inger, R.F. H.K. Voris and P. Walker. 1986. Larval transport in a Bornean ranid frog. Copeia 1986: 523-525.

Inger, R.F. and H.K. Voris. 1988. Taxonomic status and reproductive biology of Bornean tadpole-carrying frogs. Copeia 1988: 1060-1061.

Inger, R.F. and R.B. Steubing. 1997. *A Field Guide to the Frogs of Borneo*. Natural History Publications, Kota Kinabalu, Malaysia. 205 pp.

Jameson, D.L. 1950. The development of *Eleutherodactylus latrans*. Copeia 1950: 44-46.

Juncá, F.A. 1996. Parental care and egg mortality in *Colostethus stepheni*. Journal of Herpetology 30: 292-294.

Juncá, F.A. 1998. Reproductive biology of *Colostethus stepheni* and *Colostethus marchesianus* (Dendrobatidae), with the description of a new anuran mating behavior. Herpetologica 54: 377-387.

Jungfer, K.-H. and W. Boehme. 1991. The backpack strategy of parental care in frogs, with notes on froglet-carrying in *Stefania evansi* (Boulenger 1904) (Anura: Hylidae, Hemiphractinae). Revue Française d'Aquariologie Herpetologie 18: 91-95.

Jungfer, K.-H. and L.C. Schiesari. 1995. Description of a central Amazonian and Guianan tree frog, genus *Osteocephalus* (Anura, Hylidae), with oophagous tadpoles. Alytes 13: 1-13.

Jungfer, K.-H. 1996. Reproduction and parental care of the coronated treefrog, *Anotheca spinosa* (Steindachner, 1864) (Anura: Hylidae). Herpetologica 52: 25-32.

Jungfer, K.-H. and P. Weygoldt. 1999. Biparental care in the tadpole feeding Amazonian treefrog *Osteocephalus oophagus*. Amphibia-Reptilia 20: 235-249.

Jungfer, K.-H., S. Ron, R. Siepp and A. Almendáriz. 2000. Two new species of hylid frogs, genus *Osteocephalus*, from Amazonian Ecuador. Amphibia-Reptila 21: 327-340.

Kam, Y.-C., Z.-S. Chuang and C.-F. Yen. 1996. Reproduction, oviposition site selection and larval oophagy of an arboreal nester, *Chirixalus eiffingeri* (Rhacophoridae) from Taiwan. Journal of Herpetology 30: 52-59.

Kam, Y.-C., Y.-H. Chen, Z.-S. Chuang and T.-S. Huang. 1997. Growth and development of oophagous tadpoles in relation to brood care of an arboreal breeder, *Chirixalus eiffingeri* (Rhacophoridae). Zoological Studies 36: 186-193.

Kam, Y.-C., C.-F. Lin, Y.-S. Lin and Y.-F. Tsal. 1998. Density effects of oophagous tadpoles of *Chirixalus eiffingeri* (Anura: Rhacophoridae): importance of maternal brood care. Herpetologica 54: 425-433.

Kanamadi, R.D., H.N. Nandihal, S.K. Saidapur and N.S. Patil. 1996. Parental care in the frog *Philautus variablis* (Gunther). Journal of Advanced Zoology 17: 68-70.

Kiesecker, J.M. and A.R. Blaustein. 1997. Influences of egg laying behavior on pathogenic infection of amphibian eggs. Conservation Biology 11: 214-220.

Kitching, I.J., P.L. Forey, C.J. Humphries and D.M. Williams. 1998. Cladistics, 2nd ed. Oxford University Press, Oxford, UK. 228 pp.

Klett, V. and A. Meyer. 2002. What, if anything, is a *Tilapia*? Mitochondrial ND2 phylogeny of tilapiines and the evolution of parental care systems in the African cichlid fishes. Molecular Biology and Evolution 19: 865-883.

Kluge, A.G. 1981. The life history, social organization, and parental behavior of *Hyla rosenbergi* Boulenger, a nest-building gladiator frog. Miscellaneous Publications Museum of Zoology, University of Michigan No. 160. 1-170.

Kok, D., L.H. Du Preez and A. Channing. 1989. Channel construction by the African bullfrog: another anuran parental care strategy. Journal of Herpetology 23: 435-437.

Lack, D. 1947. The significance of clutch size. I. and II. Ibis 89: 302-352.

Lack, D. 1954. *The Natural Regulation of Animal Numbers*. Clarendon Press, Oxford, i-viii + 343 pp.

Lack. D. 1968. *Ecological Adaptations for Breeding in Birds*. Methuen and Co. Ltd., Longdon , i-xii + 409 pp.

Lamotte, M. and J. Lescure. 1977. Tendances adaptatives a l'affranchissement du milieu aquatique chez les amphibiens anoures. Tierre et Vie 30: 225-312.

Lannoo, M.J., D.S. Townsend and R.J. Wassersug. 1987. Larval life in the leaves: arboreal tadpole types, with special attention to the morphology, ecology and behavior of the oophagous *Osteopilus brunneus* (Hylidae) larva. Fieldiana Zoology (N.S.) 38: 1-31.

Lescure, J. 1979. Étude taxonomique et éco-éthologique d'un amphibien des petites Antilles *Leptodactylus fallax* Muller, 1926 (Leptodactylidae). Bulletin du Muséum National d'Histoire Naturelle, Paris 1: 757-774.

Lescure, J. 1984. Las larvas de Dendrobatidae. II. Reunión Iberoamer. Cons. Zool. Vert. 37-45.

Lima, A.P., J.P. Caldwell and G.M. Biavati. 2002. Territorial and reproductive behavior of an Amazonian dendrobatid frog, *Colostethus caeruleodactylus*. Copeia 2002: 44-51.

Littlejohn, M.J. 1963. The breeding biology of the Baw Baw frog *Philoria frosti* Spencer. Proceedings of the Linnean Society of New South Wales 88: 273-276.

Liu, C.C. 1950. Amphibians of western China. Fieldiana Zoology Memoirs, Vol. 2. 400 pp.

Loiselle, P.V. 1978. Prevalence of male brood care in teleosts. Nature 276: 98.

MacCulloch, R.D. and A. Lathrop. 2002. Exceptional diversity of *Stefania* (Anura: Hylidae) on Mount Ayanganna, Guyana: three new species and new distribution records. 58: 327-346.

Magnusson, W.E. and J.-M. Hero. 1991. Predation and the evolution of complex oviposition behavior in Amazon rainforest frogs. Oecologia 86: 310-318.

Martin, A.A. 1970. Parallel evolution in the adaptive ecology of leptodactylid frogs of South America and Australia. Evolution 24: 643-644.

Martins, M., J.P. Pombal, Jr. and C.F.B. Haddad. 1998. Escalated aggressive behavior and facultative parental care in the nest building gladiator frog, *Hyla faber*. Amphibia-Reptilia 19: 65-73.

Maxson, L.R. 1992. Tempo and pattern in anuran speciation and phylogeny: an albumin perspective. Pp 41-57. In: Herpetology: Current research on the biology of amphibians and reptiles. Proceedings of the First World Congress of Herpetology. K. Adler (ed.), Society for the Study of Amphibians and Reptiles, Ithaca, New York. 245 pp.

Maynard Smith, J. 1977. Parental investment: a prospective analysis. Animal Behavior 25: 1-9.

Maynard Smith, J. 1978. *The Evolution of Sex*. Cambridge University Press, Cambridge, i-x + 222 pp.

McDiarmid, R.W. 1978. Evolution of parental care in frogs. Pp. 127-147. In G.M. Burghardt and M. Bekoff (eds), The development of behavior: comparative and evolutionary aspects. Garland STPM Press, New York. 429 pp.

McDonald, K.R. and M.J. Tyler. 1984. Evidence of gastric brooding in the Australian leptodactylid frog *Rheobatrachus vitellinus*. Transactions of the Royal Society of South Australia 108: 226.

McKitrick, M.C. 1992. Phylogenetic analysis of avian parental care. The Auk 109: 828-846.

McLennan, D.A. 1991. Integrating phylogeny and experimental ethology: from pattern to process. Evolution 45: 1773-1789.

McLennan, D.A. 1994. A phylogenetic approach to the evolution of fish behavior. Reviews in Fish Biology and Fisheries 4: 430-460.

Mitchell, N.J. 2002. Low tolerance of embryonic desiccation in the terrestrial nesting frog *Bryobatrachus nimbus* (Anura: Myobatrachinae). Copeia 2002: 364-373.

Moore, J.A. 1961. The frogs of eastern New South Wales. Bulletin of the American Museum of Natural History 121: 149-386.

Myers, C.W. 1969. The ecological geography of cloud forest in Panama. American Museum Novitates No. 2396: 1-52.

Noble, G.K. 1929. The adaptive modifications of the arboreal tadpoles of *Hoplophryne* and the torrent tadpoles of *Staurois*. Bulletin of the American Museum of Natural History 58: 291-334.

Noble, G.K. 1931. *The Biology of the Amphibia*. McGraw-Hill, Boston. 577 pp.

Nussbaum, R.A. 1985. The evolution of parental care in salamanders. Miscellaneous Publications of the Museum of Zoology, University of Michigan No. 169: 1-49.

Ó Foighil, D. and D.J. Taylor. 2000. Evolution of parental care and ovulation behavior in oysters. Molecular Phylogenetics and Evolution 15: 301-313.

Orlov, N. 1997. Breeding behavior and nest construction in a Vietnam frog related to *Rana blythi*. Copeia 1997: 464-465.

Orton, G.L. 1955. Some aspects of ecology and ontogeny in the fishes and amphibians. American Naturalist 89: 193-198.

Perrone, M., Jr. and Zaret, T.M. 1979. Paternal care patterns of fishes. The American Naturalist 113: 351-361.

Petranka, J.W., M.E. Hopey, B.T. Jennings, S.D. Baird and S.J. Boone. 1994. Breeding habitat segregation of wood frogs and American toads: the role of interspecific tadpole predation and adult choice. Copeia 1994: 691-697.

Ponssa, M.L. 2001. Parental care and behavior of tadpole schools in *Leptodactylus insularum* (Anura, Leptodactylidae). Alytes 19: 183-195.

Pramuk, J.B., C.A. Hass and S.B. Hedges. 2001. Molecular phylogeny and biogeography of West Indian toads (Anura: Bufonidae). Molecular Phylogenetics and Evolution 20: 294-301.

Pyburn, W.F. 1975. A new species of microhylid frog of the genus *Synapturanus* from southeastern Colombia. Herpetologica 31: 439-443.

Rabb, G.B. and R. Snedigar. 1960. Observations on breeding and development of the Surinam toad, *Pipa pipa*. Copeia 1960: 40-44.

Rainer, G., M. Kapisa and I. Tetzlaff. 2001. A case of strange parental behaviour in frogs: males of *Sphenophryne cornuta* transports young on its back (Anura, Microhylidae). Herpetofauna 23: 14-24.

Read, K., J.S. Keogh, I.A.W. Scott, J.D. Roberts and P. Doughty. 2001. Molecular phylogeny of the Australian frog genera *Crinia*, *Geocrinia* and allied taxa (Anura: Myobatrachidae). Molecular Phylogenetics and Evolution 21: 294-308.

Relyea, R.A. 2001. Morphological and behavioral plasticity of larval anurans in response to different predators. Ecology 82: 523-540

Reynolds, J.D., N.B. Goodwin and R.P. Freckleton. 2002. Evolutionary transitions in parental care and live bearing in vertebrates. Philosophical Transactions of the Royal Society of London, Series B 357: 269-281.

Richards, S.J. and R.A. Alford. 1992. Nest construction by an Australian rainforest frog of the *Litoria lesueuri* complex. Copeia 1992: 1120-1123.

Ridley, M. 1978. Paternal care. Animal Behavior 26: 904-932.

Ridley, M. 1983. *The Explanation of Organic Diversity. The Comparative Method and Adaptations of Mating*. Oxford, Clarendon. i-viii + 272 pp.

Ridley, M. and Grafen, A. 1996. How to study discrete comparative methods. Pp. 76-103. In E. P. Martins (ed.), *Phylogenies and the Comparative Method in Animal Behavior*. Oxford University Press, New York. 415 pp.

Rödel, M.O. 1998. A reproductive mode so far unknown in African ranids: *Phrynobatrachus guineensis* Guibé and Lamotte, 1961 breeds in tree holes. Herpetozoa 11: 19-26.

Rödel, M.O. and R. Ernst. 2002. A new reproductive mode for the genus *Phrynobatrachus*: *Phrynobatrachus alticola* has non-feeding, non-hatching tadpoles. Journal of Herpetology 36: 121-125.

Ryan, M.J. and A.S. Rand. 1995. Female responses to ancestral advertisement calls in Túngara frogs. Science 269: 390-392.

Salthe, S.N. and J.S. Mecham. 1974. Reproductive and courtship patterns. Pp. 309-521 In B. Lofts (ed.), *Physiology of the Amphibia*, Vol. II. Academic Press, New York. 592 pp.

Scheel, J.J. 1970. Notes on the biology of the African tree-toad, *Nectophryne afra* Buchholz and Peters, 1875 (Bufonidae, Anura) from Fernando Póo. Rev. Zool. Bot. Afr. 81: 225-236.

Schiesari, L.C., B. Grillitsch and C. Vogl. 1996. Comparative morphology of phytotelmonous and pond-dwelling larvae of four neotropical treefrog species (Anura, Hylidae, *Osteocephalus oophagous, Osteocephalus taurinus, Phrynohyas resinifictrix, Phrynohyas venulosa*). Alytes 13: 109-139.

Schiøtz, A. 1999. Treefrogs of Africa. Frankfurt am Main, Meckenheim, Germany. 350 pp.

Schwartz, A. and R.W. Henderson. 1991. *Amphibians and Reptiles of the West Indies*. University of Florida Press, Gainesville, Florida. 720 pp.

Sever, D.M.., Hamlett, W.C., Slabach, S., Stephenson, B. and Verrell. P.A. 2003. 7. Internal Fertilization in the Anura with Special Reference to Mating and Female Sperm Storage in *Ascaphus*. This volume.

Sheil, C.A., J.R. Mendelson III and H.R. da Silva. 2001. Phylogenetic relationships of the species of neotropical horned frogs, genus *Hemiphractus* (Anura: Hylidae: Hemiphractinae), based on evidence from morphology. Herpetologica 57: 203-214.

Simon, M.P. 1983. The ecology of parental care in a terrestrial breeding frog from New Guinea. Behavioral Ecology and Sociobiology 14: 61-67.

Simpson, G.G. 1944. *Tempo and Mode in Evolution*. Columbia University Press, New York, NY. 237 pp.

Smith, C.C. and Fretwell, S.D. 1974. The optimal balance between size and the number of offspring. The American Naturalist 108: 499-506.

Spencer, H. 1852. A theory of population, deduced from the general law of animal fertility. The Westminster Review. LVII (January and April): 250-268.

Spencer, H. 1867. *The Principles of Biology*. Vol. II. D. Appleton and Company, New York, NY i-viii + 569 pp.

Stevens, R.A. 1971. A new treefrog from Malawi. Zoologica Africana 6: 313-320.

Strathmann, R. 2001. A method for the masses: oxygen delivery for stay at home embryos. Natural History 110: 62.

Summers, K., L.A. Weigt, P. Boag, and E. Bermingham. 1999. The evolution of female parental care in poison frogs on the genus *Dendrobates*: evidence from mitochondrial DNA sequences. Herpetologica 55: 254-270.

Székely, T. and J.D. Reynolds. 1995. Evolutionary transitions in parental care in shorebirds. Proceedings of the Royal Society of London Series B 262: 57-64.

Taigen, T.L., F.H. Pough and M.M. Stewart. 1984. Water balance of terrestrial anuran (*Eleutherodactylus coqui*) eggs: importance of parental care. Ecology 65: 248-255.

Tanaka, S. and M. Nishihara. 1987. Foam nests as a potential food source for anuran larvae: a preliminary experiment. Journal of Ethology 5: 86-88.

Thompson, R.L. 1996. Larval habitat, ecology and parental investment of *Osteopilus brunneus* (Hylidae). Pp. 259-269 In: Contributions to West Indian herpetology: a tribute to Albert Schwartz. R. Powell and R.W Henderson (eds), Society for the

Study of Amphibians and Reptiles, Ithaca, New York. Contributions to Herpetology, volume 12. 457 pp.

Tinkle, D.W. 1969. The concept of reproductive effort and its relation to the evolution of life histories of lizards. The American Naturalist 103: 501-516.

Townsend, D.S., M.M. Stewart and F.H. Pough. 1984. Male parental care and its adaptive significance in a neotropical frog. Animal Behaviour 32: 421-431.

Townsend, D.S. 1986. The costs of male parental care and its evolution in a neotropical frog. Behavioral Ecology and Sociobiology 19: 187-195.

Townsend, D.S. 1996. Patterns of parental care in frogs of the genus *Eleutherodactylus*. Pp. 229-239 In: Contributions to West Indian herpetology: a tribute to Albert Schwartz. R. Powell and R.W. Henderson (eds), Society for the Study of Amphibians and Reptiles, Ithaca, New York. Contributions to Herpetology, volume 12. 457 pp.

Trivers, R.L. 1972. Parental investment and sexual selection. Pp. 136-179. In: *Sexual selection and the descent of man*, B. Campbell (ed.), Aldine Press, Chicago. 378 pp.

Trueb, L. and M.J. Tyler. 1974. Systematics and evolution of the Greater Antillean hylid frogs. Occasional Papers of the Museum of Natural History, University of Kansas 24: 1-60.

Trueb, L. and D.C. Cannatella. 1986. Systematics, morphology and phylogeny of the genus *Pipa* (Anura: Pipidae). Herpetologica 42: 412-449.

Trueb, L. and D. Massemin. 2001. The osteology and relationships of *Pipa aspera* (Amphibia: Anura: Pipidae) with notes on its natural history in French Guiana. Amphibia-Reptilia 22: 33-54.

Tullberg, B.S., M. Ah-King and H. Temrin. 2002. Phylogenetic reconstruction of parental-care systems in the ancestors of birds. Philosophical Transactions of the Royal Society of London, Series B 357: 251-257.

Tyler, M.J. and M. Davies. 1979. Foam nest construction by Australian leptodactylid frogs (Amphibia, Anura, Leptodactylidae). Journal of Herpetology 13: 509-510.

Tyler, M.J. and D.B. Carter. 1981. Oral birth of the young of the gastric brooding frog *Rheobatrachus silus*. Animal Behaviour 29: 280-282.

Ueda, H. 1986. Reproduction of *Chirixalus eiffingeri* (Boettger). Scientific Reports of the Laboratory in Amphibian Biology, Hiroshima University 8: 109-116.

van Dijk, D.E. 1997. Parental care in *Hemisus* (Anura: Hemisotidae). South African Journal of Zoology 32: 56-57.

van Wijngaarden, R. and F. Bolaños. 1992. Parental care in *Dendrobates granuliferus* (Anura: Dendrobatidae) with a description of the tadpole. Journal of Herpetology 26: 102-105.

Vaz-Ferreira, R. and A. Gehrau. 1974. Proteccion de la prole en leptodactylidos. Revista de Biologicá del Uruguay 2: 59-62.

Wager, V.A. 1986. *Frogs of South Africa*. Delta Books Ltd. Craighall. 183 pp.

Wake, M.H. 1978. The reproductive biology of *Eleutherodactylus jasperi* (Amphibia, Anura, Leptodactylidae), with comments on the evolution of live-bearing systems. Journal of Herpetology 12: 121-133.

Wake, M.H. 1980. The reproductive biology of *Nectophrynoides malcomi* (Amphibia: Bufonidae), with comments on the evolution of reproductive modes in the genus *Nectophrynoides*. Copeia 1980: 193-209.

Wassersug, R., 1971. On the comparative palatability of some dry-season tadpoles from Costa Rica. American Midland Naturalist 86: 101-109.

Wassersug, R.J., K.J. Frogner and R.F. Inger. 1981. Adaptations for life in tree holes by rhacophorid tadpoles from Thailand. Journal of Herpetology 15: 41-52.

Wells, K.D. 1977. The social behavior of anuran amphibians. Animal Behavior 25: 666-693.

Wells, K.D. 1981. Parental behavior in male and female frogs. Pp. 184-197 In R.D. Alexander and D.W. Tinkle (eds), *Natural Selection and Social Behavior,* Chiron Press, New York. 532 pp.

Wells, K.D. and K.M. Bard. 1988. Parental behavior of an aquatic-breeding tropical frog, *Leptodactylus bolivianus.* Journal of Herpetology 22: 361-364.

Werren, J.H., Gross, M.R. and Shine, R. 1980. Paternity and the evolution of male parental care. Journal of Theoretical Biology 82: 619-632.

Weygoldt, P. 1987. Evolution of parental care in dart poison frogs (Amphibia: Anura: Dendrobatidae). Z. zool. Syst. Evolut.-forsch. 25: 51-67.

Weygoldt, P. and S. Plotsch. 1991. Observations on mating, oviposition, egg sac formation, and development in the egg-brooding frog, *Fritziana goeldi.* Amphibia-Reptilia 12: 67-80.

Wieczorek, A.M., R.C. Drewes, and A. Channing. 2000. Biogeography and evolutionary history of *Hyperolius* species: application of molecular phylogeny. Journal of Biogeography 27: 1231-1243.

Wiley, E.O. 1981. Phylogenetics: the theory and practice of phylogenetic systematics. Wiley Interscience, New York. 439 pp.

Williams, G.C. 1966a. *Adaptation and Natural Selection: a Critique of Some Current Evolutionary Thought.* Princeton University Press, Princeton, 307 pp.

Williams, G.C. 1966b. Natural selection, the costs of reproduction, and a refinement of Lack's principle. The American Naturalist 100: 687-690.

Williams, G.C. 1975. *Sex and Evolution.* Princeton University Press, Princeton, i-xii + 200 pp.

Zimmermann, E. and H. Zimmermann. 1984. Durch nachtzucht erhalten: Baumsteigerfrosche *Dendrobates quinquevittatus* und *D. reticulatus.* Aquarien Magazin 18: 35-41.

Zimmermann, E. and H. Zimmermann. 1988. Ethotaxonomie und zoographische Artenggruppenbildung bei Pfeilgiftfroschen (Anura: Dendrobatidae). Salamandra 24: 125-146.

CHAPTER 9

Development

Ronald Altig

9.1 INTRODUCTION

The Class Amphibia includes three living Orders: Gymnophiona (the caecilians) Caudata (the salamanders) and Anura (the frogs). Members of this monophyletic (Trueb and Cloutier 1991) group lay large, yolky eggs without shells or extraembryonic membranes. Ambystomatid salamanders and several genera of frogs from several families have been favored subjects from the beginnings of developmental studies. External fertilization in these anuran genera allows the parentage of laboratory crosses to be controlled, a number of informative mutants have been described, and selected taxa are easy to culture in the laboratory.

The concept of development covers an enormous number of research fields that advances with the aid of a diverse array of techniques. This work was descriptive before moving on to experimental embryology and continues presently in many studies of molecular development and genetics. Embryological observations started, of course, with Aristotle, and such persons as Spallanzani, Roux, and Spemann continued the embryological tradition. The earliest descriptive treatises on tadpoles from Europe (e.g., Gesner 1586; Swammerdam 1737-1738; Boulenger 1892) were followed by workers from North America (e.g., Hinckley 1881; Wright 1929) and elsewhere in the world. The Introduction in McDiarmid and Altig (1999a) gives a larger historical account, and the subjects in this discussion and others are covered in more detail in other chapters of that book.

One should not assume that the following discussion covers anuran development uniformly. The research emphasis always has been on the *Rana-Xenopus* axis because of the ease of using these taxa in laboratories in Europe and North America; many unique morphological and developmental situations have been neglected. The development of very few endotrophs (= developmental

Department of Biological Sciences, Mississippi State University, Mississippi State, Mississippi 39762-5759, USA energy derived entirely from vitellogenic yolk or other parentally produced materials) is known in any detail. Of the approximately 4800 species of frogs, perhaps 75% of them have a free-living, eating tadpole (= exotrophs - free-living tadpole consumes various materials not derived from a parent) while the remainder are endotrophs. With some descriptive information on perhaps 75% of those with a tadpole, we have an overall database that is considerably wider than deep but inadequate in specific ways in either dimension. We know how a tadpole develops and how it metamorphoses into a frog, but we lack a lot of information on details that are unique to the tadpole. Many studies remain to be done on suites of characters involving various parts of the buccopharyngeal structures, mouthparts, nares, operculum, spiracle, vent tube, and viscera.

What is a tadpole? Larval frogs occur in almost every conceivable freshwater habitat from a few milliliters of water in bromeliad cisterns to large rivers and lakes and from sea level to over 3500 m; ephemeral pools are probably the most common sites. Developmental rates of species vary (e.g., Buchholz and Hayes 2000) regardless of temperature, and larval duration ranges from a few days to several years depending on species, latitude and elevation (i.e., duration of acceptable temperatures), and ecological situations (review by Alford 1999). At metamorphosis a froglet (= juvenile or metamorph, not adult as that implies sexual maturity) is produced. A general discussion of a typical tadpole will serve as an introduction.

A tadpole has composite head-body (Fig. 9.1) and a tail with a central axis of segmented muscles and unsupported (= no fin rays as in fishes) fins but no vertebrae. Neuromasts (= pressure sensing lateral line structures) occur in the skin, eyes are always present, and external nares usually are present. Water pumped in through the mouth passes over internal gills and food trapping structures before it exits through a spiracle(s). Contents of the lengthy, spiraled intestine exit through a vent tube of various configurations in close association with the ventral fin.

The jaw cartilages are very short so that the jaw gape lies well anterior to the eyes, the upper (suprarostral) and lower jaw (Meckel's cartilage and infrarostral) elements do not articulate, there are two joints in the lower jaw, and the upper jaw is operated by movements of the lower jaw. Jaw cartilages have serrated, keratinized sheaths for cutting, gouging and biting. An oral disc roughly comprised of upper and lower labia surrounds the mouth with most margins free and edged in fleshy papillae in various configurations. Many, small, keratinized labial teeth occur in transverse rows on the upper and lower labia. The development and variations of many of these features are discussed below.

The assignment of a sequential designation based on developmental attainment of specific morphological landmarks, known as staging, allows comparisons among specimens without the necessity for considering time or absolute size. The progressive development of the hind limb from a simple bud to a paddle to a fully developed limb throughout most of the tadpole period supplies the primary cues for staging tadpoles. At least 45 staging tables (Duellman and Trueb 1986: 128) have been prepared, but the tables of Gosner

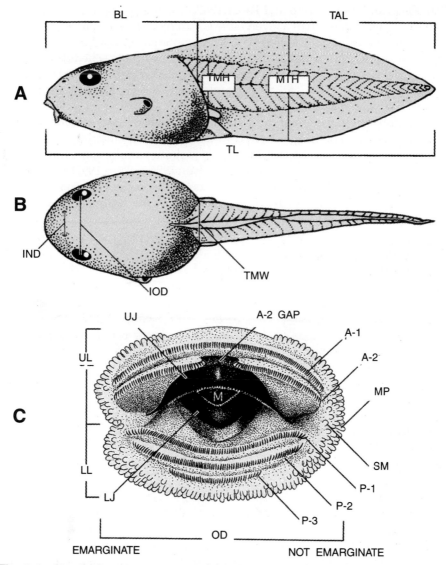

Fig. 9.1 Standard measurements and terminology of a typical tadpole in A. lateral and B. dorsal views and the C. oral apparatus. Abbreviations: A-1 and A-2 = first and second anterior labial tooth rows; A-2 gap = medial gap in labial tooth row A-2, AL = anterior labium, BL = body length, IND = internarial distance, IOD = interorbital distance, LJ = lower jaw sheath, LP = lateral process of upper jaw sheath, M = mouth, MP = marginal papillae, MTH = maximum tail height, OD = oral disc diameter, P-1 - P-3 = first, second, and third lower labial tooth rows, PL = posterior labium, SM = submarginal papillae, STN = snout to nares, STO = snout to orbit, STS = snout to spiracle, TAL = tail length, TH = tail height, TL = total length, TMH = tail muscle height, TMW = tail muscle width, and UJ = upper jaw sheath. Modified from Altig, R. and McDiarmid, R. W. 1999. Body plan: Development and Morphology. Pp. 24-51. In R. W. McDiarmid and R. Altig (eds), *Tadpoles: The Biology of Anuran Larvae*. University of Chicago Press, Chicago, Figs. 3.1 and 3.6.

(1960: *Rana*) and Nieuwkoop and Faber 1967: *Xenopus*) perhaps are used most commonly. The resolution of a given table is based on the number of stages per developmental time, but considerable developmental discordance with size, stage, and other features developing on their own variable schedules make it unlikely that one could prepare a universal staging table of equal accuracy for all species and situations. All staging references in this chapter are by the Gosner (1960; Fig. 9.2) table.

In the following commentary, I tried to emphasize an ecoevolutionary perspective of larval development in contrast to a treatise covering descriptive, experimental, and (see chapter **10** of this volume) molecular studies that appear in the appropriate textbooks (e.g., Carlson 1996; Gilbert 1997). Comments intended to stimulate future research are inserted. Because of the common lack of information, part of the discourse remains in the realm of descriptive

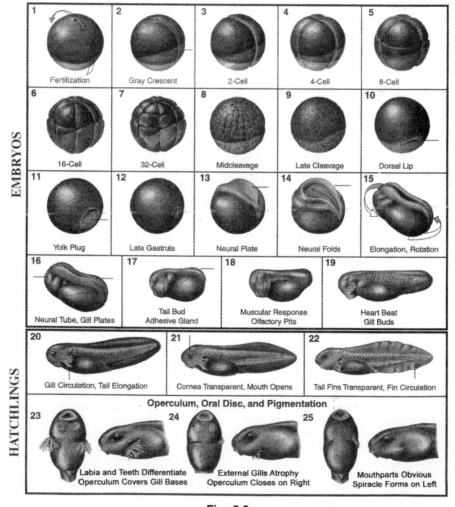

Fig. 9.2

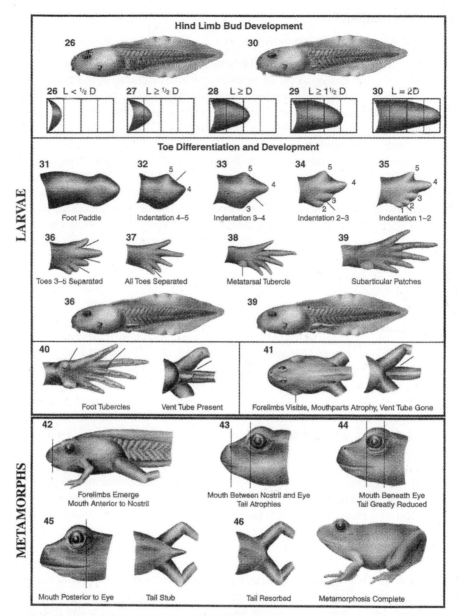

Fig. 9.2 Staging table of anuran embryos and tadpoles. Gosner's (1960) table as modified from McDiarmid, R. W., and Altig, R. 1999b. Research: Materials and Techniques. Pp. 7-23. In R. W. McDiarmid and R. Altig (eds), *Tadpoles: The Biology of Anuran Larvae*. University of Chicago Press, Chicago, Fig. 2.1.

morphology. Major sections include embryology, postembryology and metamorphosis, and development of taxa that lack a free-living tadpole stage. A general discussion and a literature cited complete the presentation.

9.2 COMMENTARY

9.2.1 Embryological Development

The literally hundreds of papers published on many aspects of observational and experimental embryology of frogs cover such diverse subjects as normal development, chimera formation, regeneration, teratology, and tissue transplants. Modern molecular studies reveal developmental and genetic mechanisms for these data from a preceding era. Only a brief summary of major events is presented.

Frog development can be divided into zygote (= fertilized egg), embryo (stages 1-24), tadpole (stages 25-41), metamorph (stages 42-46), and adult. "Ovum" refers specifically to the gamete of ovarian origin, and "egg" refers to the composite unit of ovum plus mucopolysaccharide jelly layers of oviducal origin. Except in a few cases, fertilization is external while the male and female are closely associated in amplexus. Parental care of eggs common in forms that lay terrestrial or aerial eggs is uncommon in those that lay aquatic eggs in either lentic or lotic sites.

Several holoblastic (= cleavage furrows pass entirely through the ovum) unequal (= all cleavage furrows do not pass through the equator) cleavage divisions (= cell division without interdivisional growth of daughter cells) cellularize the originally unicellular zygote and result in incrementally more cells of progressively smaller size. During blastulation, the solid zygote reforms into a hollow sphere, the blastula, and zygotic genes are activated about halfway through this process. By a complicated series of cellular migrations, gastrulation reorganizes the embryonic tissues and results in delimiting the three major germ layers in their proper positions - outer ectoderm (skin and nervous system), inner endoderm (gut and its derivatives), and intermediate mesoderm (notochord and skeletomuscular systems).

Neurulation then occurs along the dorsal surface of the embryo and sets aside the neuroectoderm, as the precursor for the central nervous system, chromatophores, and numerous other structures, from the ectoderm. The formation of a plateau (= neural plate) along the dorsal surface is followed by neural folds forming at the edge of the plateau that then bend medially and fuse in a cephalocaudal wave to form the hollow nerve cord at about stage 16. Next, somitogenesis (the formation of serially arranged blocks of mesoderm along the nervous system and notochord) occurs in a cephalocaudal wave, the basic bauplan is in place, and organogenesis continues.

During organogenesis, the eyes and nasal structures form, the gut and its derivatives differentiate, and the general body and early skeletal elements arrive at their definitive condition that is maintained until metamorphosis. Kemp (1951) followed gut development and showed how the shape and size of the abdominal cavity modifies the coiling pattern. There are numerous modifications of this general developmental scheme. Some ova are reported to be divide meroblastically and influences of ovum size (Williamson and Bull 1989), yolk density (Chipman *et al.* 1999) and genome size (Chipman *et al.* 2001; Goin *et al.* 1968) have been noted.

Starting at about stage 18 and usually disappearing by stage 25, an adhesive gland of various shapes occurs slightly posterior to the developing mouthparts. This transient, sticky gland stabilizes the embryo before it can swim, and epidermal cilia scattered throughout the surface of the epidermis allow the hatchling to glide over surfaces while lying on its side. The distribution of hatching glands on the top of the head delimits the margins of the head and allows the embryo to escape from the egg jelly.

At early stages the general body shape and shape of the yolk mass of frog and salamander embryos are similar; the body is somewhat elongate with the presumptive vent at the distal end of the yolk mass. In frogs the yolk mass subsequently shortens considerably at about stage 21, seemingly as a presage to the greatly shortened trunk of adult frogs.

The opening of the vent lies parallel with the axis of the ventral fin (i.e., medial) or opens to the right (dextral) or uncommonly to the left (sinistral) of the fin axis. There are many detailed differences in the vent tube and its association with the ventral fin and limb buds (Johnston and Altig 1986; Altig and McDiarmid 1999) that need further studies.

External gills usually occur at least on arches III-IV in stages 19-24, and these eventually atrophy and are overgrown by the operculum. The operculum is a fleshy sheet originating from the hyoid arch, and the pattern of its growth and fusion with surrounding ectoderm forms the various spiracle configurations; single and on the left side (= sinistral) is the most common position, but single and midventral (on the chest, *Ascaphus*; on the abdomen or near the vent, microhylids) and dual and lateral (pipids and rhinophrynids) are other examples. The shape of the spiracular opening, the length and projecting direction of the spiracular tube, and the presence or extent of the medial wall are common variations. As is so often the case with studies of tadpole developmental morphology, detailed studies of operculum closure would probably reveal informative systematic differences. Various environmental (e.g., oxygen tension) and ecological factors influence the rate and timing of operculum closure.

External nares occur from early embryology throughout tadpole stages in most taxa, although most microhylids do not develop nares until metamorphosis. The opening is commonly circular and positioned at various sites between the eyes and snout tip. The margins are often smooth but sometimes are ornamented with a tube or one to several papillae.

Eye positions are classified as dorsal (= no part of the eye involved in a dorsal silhouette) or lateral (= some part of the eye included in the dorsal silhouette), and there are ecological correlates; benthic forms have dorsal eyes and nektonic forms have lateral eyes. The curvature of the cornea (= greater in lateral eyes) and lenticular protrusion (= greater in lateral eyes) varies along with position.

A tadpole is not much more than an ephemeral eating machine that swims (see Wassersug 1989); complicated mouthparts harvest various materials from the environment, elaborate buccopharyngeal structures selectively capture material from that which enters the mouth, and an intestine-dominated viscera rapidly processes food. The mouthparts, buccal papillae, buccopharyngeal feeding structures, and viscera are discussed in sequence.

The complicated mouthparts of tadpoles are mostly external to the buccal cavity, and the development (Fig. 9.3) of only a few cases involving rather simple mouthparts is known (e.g., Marinelli and Vagnetti 1988; Thibaudeau and Altig 1988; Tubbs *et al.* 1993). The dimple-like stomodeum appears at the anteroventral surface of the head at about stage 19-20 as the first indication of mouth formation. The oropharyngeal membrane across this dimple will eventually rupture to open the mouth to the buccopharyngeal cavity. A labial pad elevated around the stomodeum and with slight indications of an eventual upper and lower labium are the first indications of mouthpart formation (see Catania *et al.* 1999 for a strange parallel formation pattern in an insectivore mammal). The upper and lower jaw sheaths then form, and the edges of the pad become free from the local surface and papillae form along the edge of the disc in various patterns. Tooth ridges appear at about this time and then labial teeth erupt on the ridges in the approximate order of A-1 - P-2- A-2 - P-1 - P-3. The development of tooth rows of species with less than 2 (upper labium)/ 3 (lower labium) rows usually appears on a shortened version of this scheme, and those with more than 2/3 rows continue the alternate addition of rows.

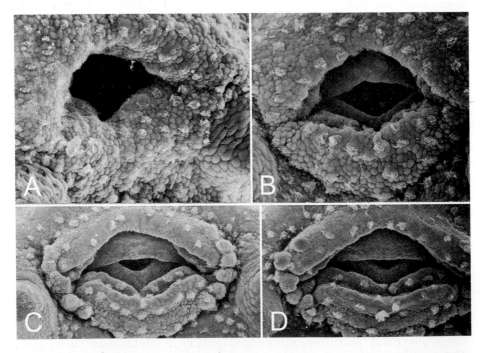

Fig. 9.3 Sequential stages in the development of the oral apparatus of the tadpole of *Hyla chrysoscelis*. A. stage 21, B. stage 23, C. stage 24, and D. stage 25, early. Modified from Thibaudeau, D. G. and Altig, R. 1988. Sequence of ontogenetic development and atrophy of the oral apparatus of six anuran tadpoles. Journal of Morphology 197: 63-69, Fig. 1.

The maximum known is 17/21, and a number of forms, often suspension feeders, lack labial teeth and jaw sheaths. Metamorphic atrophy of the oral apparatus occurs roughly in reverse of ontogenetic formation but with less precision.

The ontogeny and function of the several, highly variable series of variously-shaped papillae on the roof and floor of the buccal cavity (Fig. 9.4) have not been well examined. They are assumed to be sensory in some regard. Wassersug (1976a, b) formulated a terminology for these structures, and many subsequent papers (e.g., Chou and Lin 1997; Echeverría and Montanelli 1992a, b; Wassersug and Heyer 1988) reveal numerous variations that seem correlated systematically or ecologically.

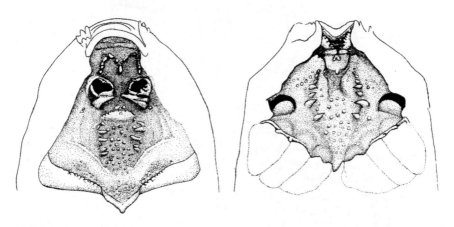

Fig. 9.4 An example of buccopharyngeal papillae on the (left) roof and (right) floor of a ranid tadpole, *Rana taipehensis*. Modified from Chou, W.-H. and Lin, J.-Y. 1997. Tadpoles of Taiwan. Special Publication of the National Museum of Natural Science (7): 1-98, Fig. 41.

Food particles that enter with the water stream entering the mouth are captured by a complicated series of food gathering structures; these structures are quite stable throughout tadpole ontogeny, and their early development has not been studied in detail. Food particles are captured in mucus and these strands are swallowed. Variations in these structures at least partially dictate what the animal is capable of capturing (Wassersug 1972; Seale and Wassersug 1979; Seale and Beckvar 1980; Seale 1982), and the removal efficiency and particles sizes captured are quite phenomenal.

The viscera of a tadpole (Fig. 9.5) is typical of vertebrates except that the gut is extremely long and usually packaged in a vertical spiral-within-a-spiral arrangement. Peristalsis is absent to weak, and the few data available suggest that studies of comparative morphology would reveal systematic differences (Nodzenski *et al.* 1989).

Embryos derived from pigmented ova have some sort of coloration from the beginning, and whether the coloration is uniform, as common in bufonids,

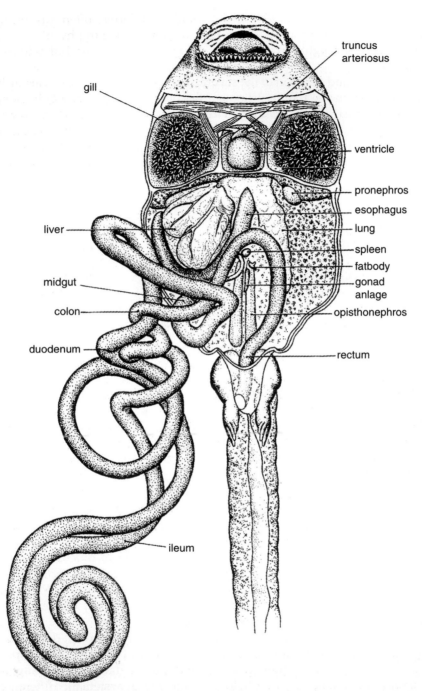

Fig. 9.5 An example of typical tadpole viscera based on *Rana temporaria*. Modified from Viertel, B. and Richter, S. 1999. Anatomy: Viscera and Endocrines. Pp. 92-148. In R. W. McDiarmid and R. Altig (eds), *Tadpoles: The Biology of Anuran Larvae*. University of Chicago Press, Chicago, Fig. 5.12.

microhylids, pelobatids, and ranids, or in some more complex pattern as in hylids is somewhat family dependent. Embryos derived from nonpigmented ova, as in many stream breeders and many aerial ova from various families, may remain so until well after hatching. Altig and Channing (1993) attempted to standardize terminology as applicable to tadpole colorations. Most colorations are muted colors that range from uniform to mottled; ocelli or simply black tail tips occur in a diversity of stream and pond tadpoles, and stripes and bands occur in some cases. Caldwell (1982) and McCollom and van Buskirk (1996) are among the few papers that directly address the functions of tadpole colorations. There is much to be learned about tadpole coloration, and the more prolific information on fishes and invertebrates will serve as an informational base.

9.2.2 Postembryological Development

The attainment of the tadpole stage at stage 25 (= operculum closure) persists until stage 41 (= immediately prior to front leg eruption) and essentially marks the end of embryology. From fertilization to stage 25 involves mostly developmental changes with some growth, while stages 25-41 involve primarily growth with development primarily limited to the limbs. Four stages occur during metamorphosis, the second major period of developmental changes, and when it is complete, a froglet (= metamorph or juvenile; not an adult) is produced.

The mouthparts typically are fully formed by stage 25, although taxa with many tooth rows may continue to add rows for several additional stages. All four limb buds soon develop and continue to differentiate throughout the tadpole stages. Forelimb buds develop at various sites (Starrett 1973) beneath the operculum at about the same time that hind limb buds appear in the crevice between the posteroventral body wall and tail musculature. If it is proper to extrapolate Gosner's (1960) table for use on the front limbs, it appears (unpublished data) that the development of the arms lag that of the legs by about 1 stage. They eventually erupt through the operculum in various patterns; in species with sinistral spiracles, the left arm exits through the spiracle.

Hind limbs with the sole usually facing medially or sometimes dorsally start out as a small limb bud that first elongates before attaining a paddle-shape. From the paddle, digits and upper and lower limb segments differentiate, and along with continued growth, the limb eventually flexes into the normal frog position. If toe pads and intercalary discs are present, they can be discerned by about stage 35.

The general growth pattern of the body and tail (Fig. 9.6) based on size vs. stage is similar in all cases. Of course there are variations in beginning points because of ovum and therefore embryo size, slopes of various parts of the lines because of the interactions of species-specific body proportions and growth rates with temperature and various ecological factors (e.g., density and site duration), and end points because of species-specific size differences and larval duration. Much of the growth during the tadpole stages is isometric (Strauss and Altig 1992; see discussion of plasticity below), and in rare tadpoles

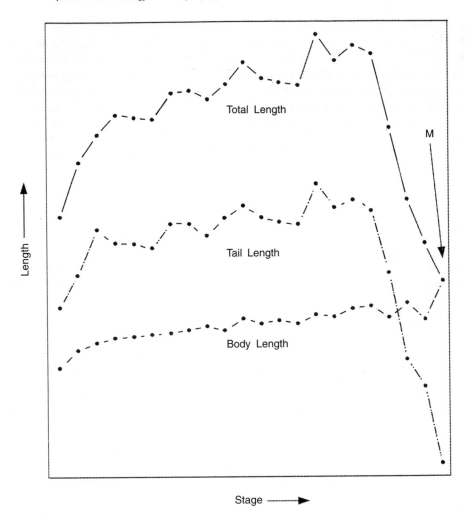

Fig. 9.6 A general scheme of growth in body, tail and total length in an exotrophic anuran from embryonic stages through metamorphosis. M = metamorphosis, where total length and body length are equal. Modified from Duellman, W. E. 1970. *The Hylid Frogs of Middle America*. Museum of Natural History, University of Kansas Monograph (1): 1-753, 72 plates, Fig. 204.

that become notably bigger than normal for a given species, it appears that the isometric trajectory is maintained (unpublished data).

There are many variations in the bodies and fins that have both systematic and ecological significance. Benthic and most stream forms have depressed (= cross-section wider than high) bodies with low fins that terminate near the tail-body junction and dorsal eyes, nektonic forms have circular or compressed (= higher than wide) bodies and often tall fins that may extend well onto the body and sometimes the abdomen and lateral eyes, and some suspension feeders have depressed bodies with low fins and lateral eyes. Fin tips range from broadly rounded to pointed and may have a flagellum (= an elongate posterior portion of the tail that is often pigmented differently from the rest of the tail and is sometimes independently mobile).

All features of the intricate mouthparts of tadpoles seemingly have an infinite number of variations, but the development of some of these structures (see above) has not been studied. Data show that one can expect closely-related members of a genus to have similar mouthparts, and in this light, the labial teeth of *Hoplobatrachus rugulosus* are an enigma. They are shaped differently and form and grow differently from those of close relatives and essentially are unique within Anura. Different but equally odd labial teeth-like structures occur in an undescribed *Phyllodytes* from Brazil and *Mantidactylus lugubris* (Glaw and Vences 1992) from Madagascar. Other species in each of these genera appear to have typical teeth. Very detailed histological examinations including cellular marking or molecular studies may provide a way to hypothesize if these odd structures evolved from labial teeth (or vice versa). Or are they de novo developmental novelties or atavisms (e.g., Hall 1984; Peters 1991; Stiassny 1992); truth of either of the latter two possibilities would suggest mutations in regulatory genes. In fact, alterations that are truly atavistic have to have a genetic basis, and one has to wonder how often these arrangements allow formation of new mouthpart configurations. The laborious hybridization among close relatives and carried to the more informative F_2 generation would supply at least rudimentary knowledge of the genetics involved. This discussion assumes that the systematics of the taxa involved are correct, but having such extreme novelties sitting within a large realm of morphological normality certainly is an intriguing situation that indicates interesting changes in genetic programming. The mere fact that novelties are rare suggests that the mechanisms that cause the novelties are uncommonly attained or that the novelties perform properly only in specific but uncommon circumstances. At the same time, we must proceed carefully in our genetic naiveté because there is the possibility that all or parts of the development of the mouthparts are somehow pleiotropic (see also Rice 1998) or similarly associated developmentally with another set of features. Grillitsch and Grillitsch (1989) found that tooth rows differed in the number of defects, and Rowe *et al.* (1996, 1998) found that long-lived *Rana catesbeiana* tadpoles had more defects than those of the short-lived tadpole of *Hyla cinerea* when both were collected from a coal-ash pond. More such data probably would provide informative data on developmental constraints of various species.

On the scale of general developmental events, anuran development is quite stable, and with sufficient data, one can often identify tadpoles by species-specific features. All features (e.g., size, coloration, or structure) related to species or sex recognition are lacking. Our understanding of geographic variations of tadpole features is minimal (e.g., Savage 1960; Gollman and Gollman 1995, 1996; several studies on New World pelobatids, see below), but most often one can assume a different species is involved if major differences are seen (e.g., an extra tooth row, but differences in coloration must be treated cautiously).

On a different scale, developmental variability or plasticity is a pervasive theme in most facets of larval anuran development. This phenomenon can be examined from three perspective: a complex array of ecological influences; cannibal morphotypes in the tadpoles of the pelobatid genus *Spea*, and predator-induced changes. A number of studies address the ecological variations in growth and metamorphosis (e.g., Wilbur and Collins 1973; Wassersug 1975, 1986; Werner 1986).

Many of these cases involve changes in the onset and offset timing of development and duration of these events. Fig. 9.7A illustrates possible patterns. The mean times of occurrence of characters A-E are shown by dark lines extending beyond the margins of the box, and this sequence and timing data would be used in a staging table. The onset and offset variation of each one, except character 5 which always occurs very punctually, is shown by various patterns of gradients indicating changes in the probability of starting or stopping. Because there is a sequence and because the correct formation of a prior feature might be required for the formation of the next feature, changes in developmentally distant features can affect an entire cascade of events. These kinds of variations cause considerable developmental discordance with size and stage. For example, it is common to find an embryo at one stage with mouthparts more developed than one at a more advanced stage; the development of mouthparts and the various characters used for staging are commonly discordant. This general concept always begs a question: is the variation outlined above related in any way to the subject being a larval form that goes through a profound metamorphosis with most larval features not contributing to the adult morphology?

The occurrence of cannibal morphotypes (i.e., hypertrophied jaw musculature and modified mouthparts) in the pelobatid genus *Spea* is the most profound postembryonic change in morphology, and uniquely, in this case the prey affects the predator. The occurrence of typical detritivore and cannibal morphs of *Spea* tadpoles from the same clutch in the same pond has been known for a long time, but Pfennig (1990) was the first to derive a comprehensive hypothesis for its cause; he proposed that a tadpole that ate fairy shrimp gained sufficient thyroxine to alter its growth trajectory. Those that did not remained typical detritivores. The developmentally exciting part of this is obvious - the tadpole cannot feed until stage 25 and by that time, the mouthparts are formed. The gain of the extraneous stimulus then altered the mouthparts, jaws, and jaw musculature tremendously well after those structures have developed embryologically. In this light and with metamorphosis yet to follow, one has to question when frog "embryology" is actually completed. Several studies by Smirnov show that bony elements can be added quite late in adult ontogeny (1994, 1997) and that larval duration can affect adult osteology (1992).

Present genetic knowledge takes us well beyond the older one-genotype-one-phenotype perspective, and the diversity in degree and kind of developmental plasticity caused by predator-induced changes that are being

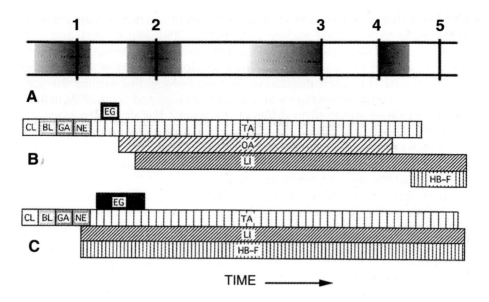

Fig. 9.7 A. A schematic illustration of developmental plasticity of a typical tadpole as influenced by genetic controls and environmental (e.g., temperature and oxygen), ecological (e.g., population density and presence of specific predators) or phylogenetic influences and their interactions. Developmental events 1-5 occur in a sequence (dark vertical lines extending outside the bar), but variations in the onset and offset of these events (shaded gradients in various patterns) can vary around the imposed convenience of a descriptive mean used in a staging table. B-C. Hypothetical illustrations of changes in developmental sequence that might occur in the development of a typical tadpole vs. a direct developer. Abbreviations: BL = blastulation, CL = cleavage, EG = external gills, GA = gastrulation, HB-F = formation of head and body of a frog, LI = limbs, NE = neurulation, OA = oral apparatus, and TA = tail. A is original; B-C from Thibaudeau, G. and Altig, R. 1999. Endotrophic anurans: Development and Evolution. Pp. 170-188. In R. W. McDiarmid and R. Altig (eds), *Tadpoles: The Biology of Anuran Larvae*. University of Chicago Press, Chicago, Fig. 7.3.

revealed simply is amazing. By stage 25 the tadpole morphotype is present; embryology is finished, and thus changes in normal developmental trajectories might seem unlikely. Until recently, one would assume that the remainder of ontogeny was governed more or less by some specific growth schedule. Cues, assumed to be primarily olfactory, obtained by prey species during predator-prey associations promote changes in nonmorphological and morphological features relative to control individuals. The recognition of predator-induced changes in tadpoles seemingly started with McCollum (1993) and since then many papers have shown similar responses. The species, their local ecology, and the type of predator (Relyea 2001) are modifiers. Some behavioral changes increase survivorship but decrease growth (e.g., remaining quiescent or hidden and thus reducing feeding time). Morphological changes can include growth rate, body shape, and coloration. The cost: benefit interaction and effects on metamorphs (Alvarez and Nicieza 2002; Relyea 2001; van Buskirk and Saxer 2001) have been addressed, but overall there are many more questions to be asked in this field of research. Do direct-developing embryos also respond? Are presumed ecological differences in morphology, like within the same species growing up in flowing vs. still water (Jennings and Scott 1993), actually a result of being subjected to different predators? For various reasons, one might not suspect that predator-induced changes affect the mouthparts, but in light of the discussion of cannibal morphs above, I would like to see verification of that just to make sure that the body features that are plastic and mouthpart development are not coupled in any way.

There are other ecological events that alter the genotype x environment interaction during development. Hatching does not occur at a set stage or time, and variations in hatching time are stimulated by ecological factors like oxygen tension (Petranka *et al.* 1983) and predator presence (e.g., Warkentin 1995). Developmental arrest can occur at various times during ontogeny if conditions are unfavorable (Bradford and Seymour 1985; Downie 1994a, b), and stunting occurs at least in laboratory populations (Petranka 1989, 1995).

Metamorphosis is a second period of profound developmental change that occurs at the end of the tadpole stages. The skull, vertebral column and limbs, major sensory organs and the viscera are the major continuations of structures from the tadpole into the frog. Metamorphosis can recover a frog morphotype from a variety of tadpole morphotypes, and the duration of the many and profound changes can occur quite rapidly. Three major events with many individual and interactive parts are occurring at about the same time - mouthparts atrophy in an approximate reverse order as ontogenetic formation and a frog mouth is formed along with eye elevation and head restructuring; the tail atrophies and the posterior end of the body restructures into that of a frog; and various parts of the viscera, especially the gut, atrophy and restructure profoundly. Through it all, the gonads are inactive.

The endocrinology of metamorphosis is fairly well understood, and in a loose sense, a tadpole metamorphoses when it gets to the proper size. Defining what is proper relative to many modifying factors and understanding the interplay of these factors is difficult, and a number of papers listed above address these interactions. The skull is restructured and a number of bones appear during this period (Hall and Hanken 1993, Trueb and Hanken 1992).

In summary, the morphological diversity of exotrophic anuran tadpoles is very large and some place one can find a tadpole in every kind of freshwater habitat. As an attempt to synthesize the vast amount of information and allow hypotheses to be made more easily, Altig and Johnston (1989; also McDiarmid and Altig 1999b) generated ecomorphological guilds that integrated morphology and ecology to the exclusion of taxon. Similar morphologies among various taxa were assumed to indicate tadpoles of similar ecology. Other categories could now be added, but the general picture is the same as noted several times above - there is a large realm of similarity accompanied by a small number of novelties. With the present database, one would think that we can safely talk about certain generalities and assume that drastic exceptions and morphological surprises are unlikely. Although this seems a safe assumption, there are surprises, and assuming the morphology of a tadpoles based on that of a presumed relative can be a logical trap.

9.2.3   Development without a Tadpole

Some form of endotrophy occurs in about 600 species in 90 genera and 11 families of anurans, and even though this phenomenon has been known for over 225 years, it remains relatively poorly known throughout the taxonomic diversity. Most descriptive studies (summary in Altig and Thibaudeau 1999) are incomplete and sometimes do little more than verify some form of endotrophy, and the only comprehensive staging table, implying sufficient developmental data are available for an endotroph is based on *Eleutherodactylus coqui* (Townsend and Stewart 1985). The ease of maintaining and breeding this frog in captivity has surely contributed to this research asymmetry.

Eggs of endotrophs are always large relative to the size of the adult and compared with egg sizes laid by similar sized adults that have a free-living tadpole. The eggs are particular yolky and normally nonpigmented, the jelly layers usually are tightly adpressed to the gamete, and these eggs occur in many sorts of moist sites but never in bodies of water; parental care is common. Slow development can demand several months to hatching, and of the many variations in this developmental scheme, there are two primary routes. In one case, nidicolous, a nonfeeding tadpole morphotype of various types is produced that eventually metamorphoses without feeding. In the other a froglet hatches from the egg jellies without going through tadpole-like stages in the egg jellies. A quite different developmental sequence (Fig. 9.7B) compared with typical frogs includes heterochronic shifts in timing of various events.

Eleutherodactylus coqui is the best studied endotroph, and Elinson (1987a) and Elinson *et al.* (1990) are pertinent general references. Fertilization and development can occur in an aqueous environment in some cases (Elinson 1987b), and thyroxin is an important metamorphic agent as in exotrophs (Callery and Elinson 2000). Early body wall development (Elinson 1994, Elinson and Fang 1998) and leg development varies from that of typical frogs. Some molecular genetics (Fang and Elinson 1996) show patterns of gene expression (e.g., distalless) and alterations in developmental patterning (Fang and Elinson 1999).

Five genera and 64 species, 44 in the genus *Gastrotheca*, in the hylid subfamily Hemiphractinae in Central and South America exhibit some form of endotrophic development. It is rare to have exotrophic and endotrophic development in one genus, but *Gastrotheca* is one exception. In either case, eggs are fertilized and placed in the female's dorsal pouch by the male during amplexus. After sufficient development in the pouch of some forms, tadpoles are released by the female into lentic sites where they continue typical development. In other cases, and there is no systematic pattern within the genus of the two forms of development, the embryos are retained until small froglets are released from the pouch. Most research on development in this group involves *Gastrotheca riobambae*, a species that produces free-living tadpoles. Many interesting studies remain to be done, but several developmental oddities in this group include differences in cleavage, somitogenesis, and gill structure. Ribosomal gene amplification is unusual in the multinucleate oocytes (Macgregor and Del Pino 1982), and a modified cleavage pattern results in the formation of an embryonic disk (Del Pino and Elinson 1984; Elinson and Del Pino 1985) similar to that of birds. The embryos are ureotelic in contrast to ammoniotely of typical amphibians (Alcocer *et al.* 1992), and limb development is modified (Hanken *et al.* 2001). Other pertinent citations include Duellman and Maness (1980), De Albuja *et al.* (1983), Wassersug and Duellman (1984), Hanken *et al.* (1997), Del Pino and Medina (1998) and Mitchell (2001).

9.3 DISCUSSION

Two summary comments on the progression of developmental studies end this chapter. There is much we do not know about the anuran tadpole, and I urge enterprising young developmental biologists to team up with similarly-minded systematists. These teams have to be strong enough to break free of the intoxication of adult frogs and focus on the more fascinating and perhaps more informative tadpoles (e.g., might microhylid tadpoles be more productive research subjects than the adults?) in order to make many exciting discoveries. When we understand the evolution of tadpole characters, we will have a better chance to understand the evolution of the tadpole stage and surely of frogs in general.

The common advent of new techniques can make a discussion of future research projects quite exciting, and I wish to emphasize an old idea that would make future endeavors most profitable. Even if such studies are illogically considered passé, much more descriptive embryology is needed so that we understand the developmental diversity that surely has systematic significance. Large strides in actually understanding developmental mechanisms probably will be made only with the inclusion of molecular techniques. In this light, it would help if interdisciplinary boundaries were more diffuse or if at least there was a more active collaboration and communication among research fields. The fullest scientific benefit will be realized only when the marriage of the capabilities of developmental biologists and evolutionists are accepted more openly (see Wagner 2000, 2001, Svoboda and Reenstra 2002). The use of "model organisms" is the largest deterrent to this productive pairing. I do understand what "model systems" attempt to do and why they are adopted by some, but even a partial broadening of this specific form of tunnel vision would greatly enhance scientific progress. I would gladly trade some of the supposed depth that models seek for a little breadth with a phylogenetic foundation. The Fly, The Frog, The Mouse, and The Worm are spontaneously generated entities of convenience, SGECs I presume, that exist in people's minds but should not occur in the science they pursue, they violate the tenets of proper scientific study, and they regress the perspectives of their proponents.

9.4   LITERATURE CITED

Alcocer, I., Santacruz, X., Steinbeisser, H., Thierauch, K. H. and Del Pino, E. M. 1992. Ureotelism as the prevailing mode of nitrogen excretion in larvae of the marsupial frog *Gastrotheca riobambae* (Fowler)(Anura, Hylidae). Comparative Biochemistry and Physiology 101A: 229-231.

Alford, R. A. 1999. Ecology: resource use, competition, and predation. Pp. 240-278. In R. W. McDiarmid and R. Altig (eds), *Tadpoles: The Biology of Anuran Larvae*. University of Chicago Press, Chicago.

Altig, R. and Channing, A. 1993. Hypothesis: functional significance of colour and pattern of anuran tadpoles. Herpetological Journal 3: 73-75.

Altig, R. and Johnston, G. F. 1989. Guilds of anuran larvae: relationships among developmental modes, morphologies, and habitats. Herpetological Monographs (3): 81-109.

Altig, R. and McDiarmid, R. W. 1999. Body plan: development and morphology. Pp. 24-51. In R. W. McDiarmid and R. Altig (eds), *Tadpoles: The Biology of Anuran Larvae*. University of Chicago Press, Chicago.

Alvarez, D. and Nicieza, A. G. 2002. Effects of induced variation in anuran larval development on postmetamorphic energy reserves and locomotion. Oecologia 131: 186-195.

Boulenger, G. A. 1892 (1891). A synopsis of the tadpoles of the European batrachians. Proceedings of the Zoological Society of London 1891: 593-627, plates 45-47.

Bradford, D. F. and Seymour, R. S. 1985. Energy conservation during the delayed-hatching period in the frog *Pseudophryne bibroni*. Physiological Zoology 58: 491-496.

Buchholz, D. R. and Hayes, T. B. 2000. Larval period comparison for the spadefoot toads *Scaphiopus couchii* and *Spea multiplicata* (Pelobatidae: Anura). Herpetologica 56: 455-468.

Caldwell, J. P. 1982. Disruptive selection: a tail color polymorphism in *Acris* tadpoles in response to differential predation. Canadian Journal of Zoology 60: 2818-2827.

Callery, E. M. and Elinson, R. P. 2000. Thyroid hormone-dependent metamorphosis in a direct developing frog. Proceedings of the National Academy of Science 97: 2615-2620.

Carlson, B. M. 1996. *Patten's Foundations of Embryology*. 6th ed. McGraw-Hill, Inc., New York.

Catania, K. C. Northcutt, R. G. and Kaas, J. H. 1999. The development of a biological novelty: a different way to make appendages as revealed in the snout of the star-nosed mole, *Condylura cristata*. Journal of Experimental Biology 202: 2719-2726.

Chipman, A. D., Haas, A. and Khaner, O. 1999. Variations in anuran embryogenesis: yolk-rich embryos of *Hyperolius puncticulatus* (Hyperoliidae). Evolution and Development 1: 49-61.

Chipman, A. D., Khaner, O., Haas, A. and Tchernov, E. 2001. The evolution of genome size: what can be learned from anuran development? Journal of Experimental Zoology 291: 365-374.

Chou, W.-H. and Lin, J.-Y. 1997. Tadpoles of Taiwan. Special Publication of the National Museum of Natural Science (7): 1-98.

De Albuja, C. M., Campos, M. and Del Pino, E. M. 1983. Role of progesterone on oocyte maturation in the egg-brooding hylid frog *Gastrotheca riobambae* (Fowler). Journal of Morphology 227: 271-276.

Del Pino, E. M. and Elinson, R. P. 1984. A novel development pattern for frogs: gastrulation produces an embryonic disk. Nature 306: 589-591.

Del Pino, E. M. and Medina, A. 1998. Neural development in the marsupial frog *Gastrotheca riobambae*. International Journal of Developmental Biology 42: 723-731.

Downie, J. R. 1994a. Developmental arrest in *Leptodactylus fuscus* tadpoles (Anura: Leptodactylidae). 1: Descriptive analysis. Herpetological Journal 4: 29-38.

Downie, J. R. 1994b. Developmental arrest in *Leptodactylus fuscus* tadpoles (Anura: Leptodactylidae). 2: Does a foam-borne factor block development? Herpetological Journal 4: 39-45.

Duellman, W. E. 1970. *The Hylid Frogs of Middle America*. Museum of Natural History, University of Kansas Monograph (1): 1-753, 72 plates.

Duellman, W. E. and Maness, S. J. 1980. The reproductive behavior of some hylid marsupial frogs. Journal of Herpetology 14: 213-222.

Duellman, W. E. and Trueb, L. 1986. *Biology of Amphibians*. McGraw-Hill, New York.

Echeverría, D. D. and Montanelli, S. B. 1992a. Estereomorfología del aparato bucal y cavidad oral de las larvas de *Ololygon fuscovaria* (Lutz, 1925)(Anura, Hylidae). Revista del Museo Argentino de Ciencias Naturales "Bernardino Rivadavia" 16: 3-13.

Echeverría, D. D. and Montanelli, S. B. 1992b (1995). Acerca del aparato bucal y de las fórmulas dentarias en *Odontophrynus americanus* (Duméril y Bibron, 1841) (Anura, Leptodactylidae). Physis 50B: 37-43.

Elinson, R. P. 1987a. Change in development patterns: embryos of amphibians with large eggs. Pp. 1-21. In R. A. Raff and E. C. Raff [eds.], *Development as an Evolutionary Process*. MBL Lectures in Biology 8. Alan R. Liss, New York.

Elinson, R. P. 1987b. Fertilization and aqueous development of the Puerto Rican terrestrial-breeding frog, *Eleutherodactylus coqui*. Journal of Morphology 193: 217-224.

Elinson, R. P. 1990. Direct development in frogs: wiping the recapitulationist slate clean. Seminars in Developmental Biology 1: 263-270.

Elinson, R. P. 1994. Leg development in a frog without a tadpole (*Eleutherodactylus coqui*). Journal of Experimental Zoology 270: 202-210.

Elinson, R. P. and Del Pino, E. M. 1985. Cleavage and gastrulation in the egg-brooding, marsupial frog, *Gastrotheca riobambae*. Journal of Embryology and Experimental Morphology 90: 223-232.

Elinson, R. P. and Fang, H. 1998. Secondary coverage of the yolk by the body wall in the direct developing frog, *Eleutherodactylus* coqui: an unusual process for amphibian embryos. Development, Genes and Evolution 208: 457-466.

Elinson, R. P., Townsend, D. S., Cuesta, F. C. and Eichhorn, P. 1990. A practical guide to the developmental biology of terrestrial-breeding frogs. Biological Bulletin 179: 163-177.

Fang, H. and Elinson, R. P. 1996. Patterns of distalless gene expression and inductive interaction in the head of the direct developing frog *Eleutherodactylus coqui*. Developmental Biology 179: 160-172.

Fang, H. and Elinson, R. P. 1999. Evolutionary alteration in anterior patterning: otx2 expression in the direct developing frog *Eleutherodactylus coqui*. Developmental Biology 205: 233-239.

Gesner, C. 1551-1604. *Historia Animalium*. Frankfurt: I. Wecheli.

Gilbert, S. F. 1997. *Developmental Biology*. 5th ed. Sinauer Associates, Inc., Sunderland, Massachusetts.

Glaw, R. and Vences, M. 1992. *A Fieldguide to the Amphibians and Reptiles of Madagascar*. Moos-Druck, Leverkusen.

Goin, O. B., Goin, C. J. and Bachmann, K. 1968. DNA and amphibian life history. Copeia 1968: 532-540.

Gollman, B. and Gollman, G. 1995. Morphological variation in tadpoles of the *Geocrinia laevis* complex: regional divergence and hybridization (Amphibia, Anura, Myobatrachidae). Journal of Zoological Systematics and Evolutionary Research 33: 32-41.

Gollman, B. and Gollman, G. 1996. Geographic variation of larval traits in the Australian frog *Geocrinia victoriana*. Herpetologica 52: 181-187.

Gosner, K. L. 1960. A simplified table for staging anuran embryos and larvae with notes on identification. Herpetologica 16: 183-190.

Grillitsch, B. and Grillitsch, H. 1989. Teratological and ontogenetic alterations to external oral structure in some anuran larvae (Amphibia: Anura: Bufonidae, Ranidae). Fortschritte Zoology 35: 276-282.

Hall, B. K. 1984. Developmental mechanisms underlying the formation of atavisms. Biological Review 59: 89-124.

Hall, B. K. and Hanken, J. 1993. Bibliography of skull development: 1937-1989. Pp. 378-577. In J. Hanken and B. K. Hall (eds), *The Skull: Vol. 1, Development*. University of Chicago Press, Chicago.

Hanken, J., Carl, T. F., Richardson, M. K., Olsson, L., Schlosser, G., Osabutey, C. K. and Klymkowsky, M. W. 2001. Limb development in a "nonmodel" vertebrate, the direct-developing frog *Eleutherodactylus coqui*. Journal of Experimental Zoology 291: 375-388.

Hanken, J., Jennings, D. H. and Olsson, L. 1997. Mechanistic basis of life-history evolution in anuran amphibians: direct development. American Zoologist 37: 160-171.

Hinckley, M. H. 1881. On some differences in the mouth structure of the anourous batrachians found in Milton, Mass. Proceedings of the Boston Society of Natural History 21: 307-315, plate 5.

Jennings, R. D. and Scott, Jr., N. J. 1993. Ecologically correlated morphological variation in tadpoles of the leopard frog, *Rana chiricahuensis*. Journal of Herpetology 27: 285-293.

Johnston, G. F. and R. Altig. 1986. Identification characteristics of anuran tadpoles. Herpetological Review 17: 36-37.

Kemp, N. E. 1951. Development of intestinal coiling in anuran larvae. Journal of Experimental Zoology 116: 259-287.

Marinelli, M. and Vagnetti, D. 1988. Morphology of the oral disc of *Bufo bufo* (Salientia: Bufonidae) tadpoles. Journal of Morphology 195: 71-81.

McCollum, S. A. 1993. Ecological consequences of predator-induced polyphenism in larval hylid frogs. Ph.D. Dissertation, Duke University, Durham, North Carolina.

McCollum, S. A. and Leimberger, J. D. 1997. Predator-induced morphological changes in an amphibian: predation by dragonflies affects tadpole color, shape, and growth rate. Oecologia 109: 615-621.

McCollum, S. A. and Van Buskirk, J. 1996. Costs and benefits of a predator-induced polyphenism in the gray treefrog *Hyla chrysoscelis*. Evolution 50: 538-593.

McDiarmid, R. W., and Altig, R. 1999a. (eds), *Tadpoles: The Biology of Anuran Larvae*, University of Chicago Press, Chicago.

McDiarmid, R. W., and Altig, N.R. 1999b. Research: materials and techniques. Pp. 7-23. In R. W. McDiarmid and R. Altig (eds), *Tadpoles: The Biology of Anuran Larvae*. University of Chicago Press, Chicago.

Macgregor, H. C. and Del Pino, E. M. 1982. Ribosomal gene amplification in multinucleate oocytes of the egg brooding hylid frog *Flectonotus pygmaeus*. Chromosoma 85: 475-488.

Mitchell, N. J. 2001. The energetics of endotrophic development in the frog *Geocrinia vitellina* (Anura: Myobatrachinae). Physiological and Biochemical Zoology 74: 832-842.

Nieuwkoop, P. D. and Faber, J. 1967. *Normal Table of Xenopus laevis (Daudin): A Systematical and Chronological Survey of the Development from the Fertilized Egg till the end of Metamorphosis.* 2nd ed. North-Holland Publishing Company, Amsterdam.

Nodzenski, E., Wassersug, R. J. and Inger, R. F. 1989. Developmental differences in visceral morphology of megophryine pelobatid tadpoles in relation to their body form and mode of life. Biological Journal of the Linnean Society 38: 369-388.

Peters, D. S. 1991. Behavior plus "pathology" - the origin of adaptations. Pp. 141-150. In N. Schmidt-Kittler and K. Vogel (eds), *Constructional Morphology and Evolution.* Springer-Verlag, Berlin.

Petranka, J. W. 1989. Chemical interference competition in tadpoles: does it occur outside laboratory aquaria? Copeia 1989: 921-930.

Petranka, J. W. 1995. Interference competition in tadpoles: are multiple agents involved? Herpetological Journal 5: 206-207.

Petranka, J. W., Just, J. J., and Crawford, E. C. 1983. Hatching of amphibian embryos; the physiological trigger. Science 217: 257-259.

Pfennig, D. 1990. The adaptive significance of an environmentally-cued developmental switch in an anuran tadpole. Oecologia 85: 101-107.

Relyea, R. A. 2001. The lasting effects of adaptive plasticity: predator-induced tadpoles become long-legged frogs. Ecology 82: 1947-1955.

Rice, S. H. 1998. The evolution of canalization and the breaking of von Baer's laws: modeling the evolution of development with epistasis. Evolution 52: 647-656.

Rowe, C. L., Kinney, O. M., Fiori, A. P. and Congdon, J. D. 1996. Oral deformities in tadpoles (*Rana catesbeiana*) associated with coal ash deposition: effects of grazing ability and growth. Freshwater Biology 36: 723-730.

Rowe, C. L., Kinney, O. M., Fiori, A. P. and Congdon, J. D. 1998. Oral deformities in tadpoles of the bullfrog (*Rana catesbeiana*) caused by conditions in a polluted habitat. Copeia 1998: 244-246.

Savage, J. M. 1960. Geographic variation in the tadpole of the toad *Bufo marinus*. Copeia 1960: 233-235.

Seale, D. B. 1982. Obligate and facultative suspension feeding in anuran larvae: feeding regulation in *Xenopus* and *Rana*. Biological Bulletin 162: 214-231.

Seale, D. B. and Beckvar, N. 1980. The comparative ability of anuran larvae (genera: *Hyla, Bufo,* and *Rana*) to ingest suspended blue-green algae. Copeia 1980: 495-503.

Seale, D. B. and R. J. Wassersug. 1979. Suspension feeding dynamics of anuran larvae related to their functional morphology. Oecologia 39: 259-272.

Smirnov, S. V. 1992. The influence of variation in larval period on adult cranial diversity in *Pelobates fuscus* (Anura: Pelobatidae). Journal of Zoology 226: 601-612.

Smirnov, S. V. 1994. Postmaturation skull development in *Xenopus laevis* (Anura, Pipidae): late-appearing bones and their bearing on the pipid ancestral morphology. Russian Journal of Herpetology 1: 21-29.

Smirnov, S. V. 1997. Additional dermal ossifications in the anuran skull: morphological novelties or archaic elements. Russian Journal of Herpetology 4: 17-27.

Starrett, P. H. 1973. Evolutionary patterns in larval morphology. p. 251-271. In J. L. Vial [Ed.], *Evolutionary Biology of the Anurans. Contemporary Research on Major Problems.* University of Missouri Press, Columbia.

Stiassny, M. L. J. 1992. Atavisms, phylogenetic character reversals, and the origin of evolutionary novelties. Netherlands Journal of Zoology 42: 260-276.

Strauss, R. E. and Altig, R. 1992. Ontogenetic body form changes in three ecological morphotypes of anuran tadpoles. Growth, Development and Aging 56: 3-16.

Svoboda, K. K. H. and Reenstra, W. R. 2002. Approaches to studying cellular signaling: a primer for morphologists. Anatomical Record 269: 123-139.

Swammerdam, J. 1737-1738. *Biblia Naturae.* II.

Thibaudeau, D. G. and Altig, R. 1988. Sequence of ontogenetic development and atrophy of the oral apparatus of six anuran tadpoles. Journal of Morphology 197: 63-69.

Thibaudeau, G. and Altig, R. D. Endotrophic anurans: Development and Evolution. Pp. 170-188. In R. W. McDiarmid and R. Altig (eds), *Tadpoles: The Biology of Anuran Larvae.* University of Chicago Press, Chicago.

Townsend, D. S. and Stewart, M. M. 1985. Direct development in *Eleutherodactylus coqui* (Anura: Leptodactylidae): a staging table. Copeia 1985: 423-436.

Trueb, L. and Cloutier, R. 1991. A phylogenetic investigation of the inter- and intrarelationships of the Lissamphibia (Amphibia: Temnospondyli). Pp. 223-313. In H. P. Schultze and L. Trueb [eds.], *Origins of the Higher Groups of Tetrapods: Controversy and Consensus.* Cornell University Press, Ithaca, New York.

Trueb, L. and Hanken, J. 1992. Skeletal development in *Xenopus laevis* (Anura, Pipidae). Journal of Morphology 214: 1-41.

Tubbs, L. O. E., Stevens, R., Wells, M. and Altig, R. 1993. Ontogeny of the oral apparatus of the tadpole of *Bufo americanus.* Amphibia-Reptilia 14: 333-340.

Van Buskirk, J. and Saxer, G. 2001. Delayed costs of an induced defense in tadpoles? Morphology, hopping, and development rate at metamorphosis. Evolution 55: 821-829.

Viertel, B. and Richter, S. 1999. Anatomy: Viscera and Endocrines. Pp. 92-148. In R. W. McDiarmid and R. Altig (eds), *Tadpoles: The Biology of Anuran Larvae.* University of Chicago Press, Chicago.

Wagner, G. P. 2000. What is the promise of developmental evolution? Part I: why is developmental biology necessary to explain evolutionary innovations? Journal of Experimental Zoology 288: 95-98.

Wagner, G. P. 2001. What is the promise of developmental evolution? Part II: a causal explanation of evolutionary innovations may be impossible. Journal of Experimental Zoology 291: 305-309.

Warkentin, K. M. 1995. Adaptive plasticity in hatching age: a response to predation risk trade-offs. Proceedings of the National Academy of Science 92: 3507-3510.

Wassersug, R. 1972. The mechanism of ultraplanktonic entrapment in anuran larvae. Journal of Morphology 137: 279-288.

Wassersug, R. 1975. The adaptive significance of the tadpole stage with comments on the maintenance of complex life cycles in anurans. American Zoologist 15: 405-417.

Wassersug, R. J. 1976a. Oral morphology of anuran larvae: terminology and general description. Occasional Papers of the Museum of Natural History, University of Kansas (48): 1-23.

Wassersug, R. J. 1976b. Internal oral features in *Hyla regilla* (Anura: Hylidae) larvae: an ontogenetic study. Occasional Papers of the Museum of Natural History, University of Kansas (49): 1-24.

Wassersug, R. 1986. How does a tadpole know when to metamorphose? A theory linking environmental and hormonal cues. Journal of Theoretical Biology 118: 171-181.

Wassersug, R. 1989. Locomotion in amphibian larvae (or "Why aren't tadpoles built like fishes?"). American Zoologist 29: 65-84.

Wassersug, R. J. and Duellman, W. E. 1984. Oral structures and their development in egg-brooding hylid frog embryos and larvae: evolutionary and ecological implications. Journal of Morphology 182: 1-37.

Wassersug, R. J. and Heyer, W. R. 1988. A survey of internal oral features of leptodactyloid larvae (Amphibia: Anura). Smithsonian Contributions in Zoology (457): 1-99.

Werner, E. E. 1986. Amphibian metamorphosis: growth rate, predation risk, and the optimal size at transformation. American Naturalist 128: 319-341.

Wilbur, H. M. and Collins, J. P. 1973. Ecological aspects of amphibian metamorphosis. Science 182: 1305-1314.

Williamson, I. and Bull, C. M. 1989. Life history variation in a population of the Australian frog *Ranidella signifera*: egg size and early development. Copeia 1989: 349-356.

Wright, A. H. 1929. Synopsis and description of North American tadpoles. Proceedings of the United States National Museum 74, 11 (2756): 1-70, plates 1-9.

CHAPTER 10

Molecular Development

Brian Key

10.1 INTRODUCTION

This review concentrates on the principal determinative events in early development that are responsible for polarizing the embryo and establishing the dorsal-ventral axis. The frog has a long history as an experimental model for these studies and it has led to elucidation of important principles underlying embryonic vertebrate development. This discussion will be restricted to creation of the two early germ layers, the endoderm and mesoderm. It will not delve into the complexities of gastrulation and convergent extension but instead will reveal the mechanisms underlying the polarization of the endoderm and its subsequent role in mesoderm induction and dorsal-ventral axis formation. While the classical studies of embryonic development in Anura used a variety of species, most modern molecular studies have taken advantage of *Xenopus*. Since this review covers the principal molecular events responsible for germ layer development, it will mostly refer to data derived from *Xenopus*.

10.2 TWO SEMINAL EXPERIMENTS HIGHLIGHT THE IMPORTANCE OF INDUCTIVE EVENTS

Fate mapping studies have revealed that the pigmented animal hemisphere of the embryo gives rise to the ectoderm, the yolky vegetal hemisphere forms the endoderm and an equatorial region separating the two develops into mesoderm (Dale and Slack 1987) (Fig. 10.1A). There are two seminal experiments which have laid the foundation for our understanding of early embryonic inductive events underlying the formation of these germ layers. First, when the vegetal pole is isolated from early embryos and placed in culture it spontaneously forms endoderm, indicating that development of this germ layer is cell autonomous (Nieuwkoop 1969a). Second, dorsal endodermal cells induce ectodermal cells to form dorsal mesoderm when these tissues are co-cultured

Department of Anatomy and Developmental Biology, School of Biomedical Sciences, University of Queensland, Brisbane 4067, Australia

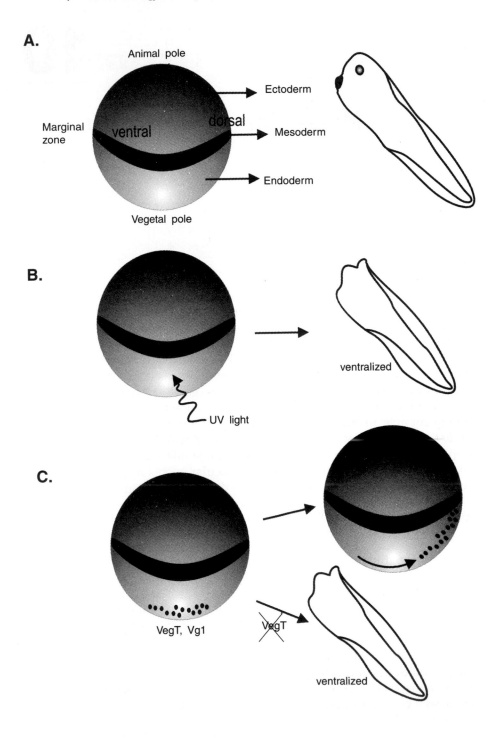

(Nieuwkoop 1969b). Similarly, ventral endodermal cells induce ectodermal cells to form ventral mesoderm, indicating that the early endoderm is polarized along its dorsoventral axis. This polarization is cell non-autonomously transferred between the two early germ layers.

10.3 EXPERIMENTAL APPROACHES TO QUESTIONS OF MOLECULAR DEVELOPMENT

What is the molecular basis of the initial formation and then subsequent polarization of the endoderm? The answer to this question is clearly dependent on the approach taken to assay the appearance of endoderm. The role of various molecules or sub-cellular events in endoderm formation can be assessed by examining the ability of endoderm to either induce mesoderm (which normally begins at approximately the 16 to 32 cell stage in *Xenopus*; Gimlich and Gerhart 1984) either in vitro or in vivo. The advantage of this approach is that it enables one to interrogate the role of early events in specifying later endoderm inductive ability. In addition it does not require specific molecular markers of endoderm. The difficulty, however, is being able to determine whether a phenotypic effect is directly, and not indirectly, involved in mesoderm induction.

There is the alternative strategy of examining either the morphological or molecular differentiation of endoderm derivatives in vitro. Probably the earliest that this tissue can be defined in this case is by the end of blastulation, when endodermal cells are histologically recognizable and selectively express endodermin and not other germ layer markers (Sasai *et al.* 1996). This endodermin expressing domain appears as a ring of cells incorporating the dorsal lip and blastopore (Sasai *et al.* 1996). This approach suffers from the availability of specific markers. Moreover, vegetal hemisphere derivatives may be well on the way to differentiating into endoderm before specific markers are expressed. Again interpretations based on results obtained from this approach are affected by an inability to distinguish between whether an event is involved in induction of the endoderm or is instead specifying its dorsoventral axis. This has been compounded by the realization that the events of induction and polarization are often intimately intertwined. For instance, induction may produce an initially ventralized germ layer which is then subsequently further polarized by inhibitory dorsalizing agents.

Fig. 10.1 Development of the dorsoventral axis in *Xenopus* is dependent on the cortical rotation of cytoplasmic RNA towards the dorsovegetal region. **A.** The fertilized egg consists of an animal and vegetal pole separated by a marginal ring of cells destined to form the mesoderm. The animal hemisphere gives rise to the ectoderm while the vegetal hemisphere forms the endoderm. **B.** Ultraviolet (UV) light irradiation of the vegetal hemisphere leads to a ventralized embryo lacking dorsoanterior structures. **C.** During normal development the animal pole contains a cortical condensation of VegT and Vg1 transcripts which are translocated towards the dorsal side of the embryo soon after fertilization. When the egg is depleted of VegT the animal becomes ventralized as following UV exposure. Original.

10.4 ENDODERM IS FORMED BY ACTION OF CELL AUTONOMOUS MATERNALLY-DERIVED RNA

The differentiation of endodermal cells from the vegetal hemisphere is a progressive event. Transplantation experiments revealed that vegetal cells do not become restricted to the fate of endodermal cells until late blastula stages (Heasman *et al.* 1984; Wylie *et al.* 1987). That is, these cells are capable of being diverted into other lineages up until just before gastrulation. Interestingly, masses of endodermal cells cultured in vitro develop into endoderm indicating that their determination is a cell autonomous event, independent of interaction from the animal pole (Wylie *et al.* 1987). Determination of cell fate is clearly dependent on cell-cell interactions since dispersed single vegetal cells either fail to express or express reduced levels of general endodermal markers such as the zygotically expressed Sox17α, Mix.1, GATA-4 and Mixer (Rosa 1989; Yasuo and Lemaire 1999; Chang and Hemmati-Brivanlou 2000). Only when sufficient tissue mass is cultured do endodermal cells develop (Wylie *et al.* 1987).

Mix.1 (homeodomain containing transcription factor) and Sox17a (transcription factor) are expressed just after the mid-blastula transition while Mixer (homeodomain containing transcription factor) and GATA-4 (transcription factor) turn on about 2-3 hours later (Yasuo and Lemaire 1999). (It should be noted here that throughout this chapter reference is made to numerous genes. In all cases, unless otherwise noted, expression means the presence of mRNA transcripts determined by either reverse transcription polymerase chain reaction, or Northern blot analysis or in situ hybridization. When appropriate it is indicated in parentheses whether the gene encodes a transcription factor, a growth factor or some other protein.) During the post-mid-blastula period (see below) both Sox17α and Mix.1 increase considerably (Yasuo and Lemaire 1999). Although isolated blastomeres failed to express Mix.1 and GATA-4, they do express Mix.1 and Sox17α at mid-blastula stage levels suggesting that the expression of these latter two markers was independent of cell-cell interactions. This was confirmed when isolated vegetal cells were reaggregated after the midblastula stage and levels of Mix.1 and Sox17α were rescued (Yasuo and Lemaire 1999). These data indicate that endoderm formation is dependent on cell autonomous events before the mid-blastula transition and then on cell non-autonomous interactions after this period.

Blocking the type-II activin (growth factor) receptor in embryos revealed that TGFβ (Transforming growth factors) ligands were responsible for the post-mid-blastula increase in Mix.1 and Sox17α and for the non-autonomous activation of Mixer and GATA-4. When protein synthesis was inhibited after the mid-blastula stage, Mixer expression was lost while Sox17α and Mix.1 remained low (Yasuo and Lemaire 1999). These results indicate that at least Mixer required the presence of TGFβ-like factors expressed in the late blastula period. Mixer expression in animal caps induces Sox17 and Xnr3 (Xenopus nodal-related growth factor) suggesting that Sox17 expression may be positively enhanced not only directly via TGFβ ligands but also indirectly through activation of Mixer (Henry and Melton 1998). Mixer expression is necessary for endoderm formation since its ectopic expression causes ectopic formation of endoderm and its loss-of-function inhibits endoderm development (Henry and Melton 1998). GATA-4 expression, however, appears to also depend on maternal factors other than the TGFβ ligands (Yasuo and Lemaire 1999). Sox17α also seems to be activated by maternal factors other than VegT (transcription factor of T-box family) since low levels of this transcript are found in embryos depleted of VegT (Xanthos et al. 2001). One possibility is that the TGFβ ligand Vg1, which is present in the early vegetal hemisphere, may also be activating endodermal programs. Vg1 stimulates endoderm marker expression in animal caps (Yasuo and Lemaire 1999) and in particular it turns on Sox17 even when Mixer is inhibited. A role for Vg1 is also supported by the fact that the RNA binding protein *Xenopus* Bicaudal-C is turned on by Vg1 and this protein upregulates Mix.1 and Sox17 (Wessely and De Robertis 2000). Furthermore, when Vg1 activity is blocked dorsal markers of endoderm are lost from vegetal pole explants (Joseph and Melton 1998).

Endoderm formation is dependent on the presence of maternal T-box transcription factor VegT in the egg cytoplasm. Depletion of maternal stores of VegT transcripts using anti-sense oligonucleotides prevents endoderm formation (Zhang et al. 1998) and expression of VegT in animal caps generates endoderm (Horb and Thomsen 1997). The homeobox genes Bix1 and Bix4 are expressed in the vegetal hemisphere and are induced in animal caps by VegT, as are Sox17α, Mix.1 and Mixer (Tada and Smith 1998; Tada et al. 1998; Casey et al. 1999; Chang et al. 2000; Xanthos et al. 2001). Exogenous Bix4 is also able to rescue the expression of endodermal markers in VegT depleted embryos (Casey et al. 1999). Together these results suggest that the Bix genes are downstream of VegT and upstream of Sox17α and Mix.1 and that they are responsible for the VegT-mediated cell autonomous expression of Sox17α and Mix.1 in vegetal hemisphere cells (Yasuo and Lemaire 1999; Chang and Hemmati-Brivanlou 2000). Perturbations to Bix1, Bix4, and Mix.1 all cause defects to endoderm formation, supporting their important role in the VegT triggered endodermal program (Lemaire et al. 1998; Tada et al. 1998; Casey et al. 1999).

The zygotically expressed TGFβ family members Xnr1 (*Xenopus* Nodal related 1), Xnr2, Xnr4 and Derriere are prime candidates for the post-midblastula transition mediated upregulation of Sox17α and Mix.1 and for the activation of GATA-4 and Mixer since each of these factors can rescue VegT depleted embryos (Kofron et al. 1999). These TGFβ family members are also responsible for the mesoderm-inducing activity of endoderm (Clements et al. 1999). VegT activates the expression of Xnr1, Xnr2, Xnr4 and Derriere in animal caps (Clements et al. 1999; Xanthos et al. 2001) and blocking TGFβ signaling in vivo perturbs expression of endoderm markers (Chang and Hemmati-Brivanlou 2000). Xnr1, Xnr2 and Derriere strongly induce Mixer and Mix.1 and differentially induce Sox17α and GATA-4 when expressed in animal caps (Yasuo and Lemaire 1999). The expression of these TGFβ ligands are controlled by both maternal and zygotic factors and they appear to act in a positive feedback loop to control their own expression (Clements *et al.* 1999; Yasuo and Lemaire 1999). Together, these data support the role of Xnr1, Xnr2, Xnr4 and Derriere as key regulators of endoderm formation.

Interactions between the growth factor BMP-4 and its antagonist Chordin are believed to be involved in both induction and patterning of the gastrulating endoderm. BMP-4 is expressed in the ventral marginal zone (Fainsod *et al.* 1994; Holley *et al.* 1995) and diffuses into the animal and vegetal hemispheres where it inhibits endoderm formation, as defined by expression of endodermin (Sasai *et al.* 1996). Chordin, expressed in the organizer, is believed to neutralize BMP-4 action in the vegetal hemisphere leading to endoderm, whereas its action in the animal hemisphere is complemented by FGF (fibrobalst growth factor) signaling to produce neural tissue (Sasai *et al.* 1996; Piccolo *et al.* 1996). Despite this evidence for BMP-2 and Chordin in endoderm formation it may be that these molecules are associated with later specification events rather than early induction, as dominant-negative BMP receptors do not block endoderm development from vegetal caps in vitro (Chang and Hemmati-Brivanlou 2000). When antagonists of BMP-4 are injected into ventral marginal zone, only ventral-dorsal patterning is affected; the initial formation of mesoderm proceeds normally (Eimon and Harland 1999). Thus, the initial induction of mesoderm appears to be separable from later patterning events which polarize the mesoderm.

10.4.1 Summary of Endoderm Formation

Maternal VegT and Vg1 are probably at the top of a hierarchy of signals that initially act cell autonomously prior to the mid-blastula period to increase two key factors: Mix.1 and Sox17α. Levels of these proteins are then further increased in the immediate post-midblastula period due to zygotic TGFβ ligands. Positive feedback loops further activate these TFGβ ligands leading to endoderm formation by the beginning of gastrulation (Fig. 10.2).

10.5 POLARIZATION OF THE ENDODERM ESTABLISHES THE DORSAL-VENTRAL AXIS

When dorsal portions of the vegetal mass of blastula were cultured together with animal poles, dorsal mesodermal structures such as notochord and muscle were induced (Boterenbrood and Nieuwkoop 1973). Because this region of the vegetal pole of the embryo led to the differentiation of dorsal tissues, similar to that produced by the Spemann-Mangold organizer it was referred to as the "organizer of the organizer" or "Nieuwkoop's center" (Gerhart 2001). The Spemann-Mangold organizer refers to a small patch of tissue from the upper blastopore lip that has the ability to generate a secondary embryo when transplanted to a different region of another embryo (Spemann and Mangold 2001; Snader and Faessler 2001). In contrast, ventral vegetal mass induces the formation of ventral mesodermal structures such as blood and mesenchyme in these Nieuwkoop recombinant explants. These results indicate that the vegetal mass is not homogenous and contains an asymmetric localization of factors responsible for establishing the dorsal-ventral axis. The molecular pathways underlying the establishment of the dorsal-ventral axis are discussed below and Figure 10.3 provides an overview of these interactions.

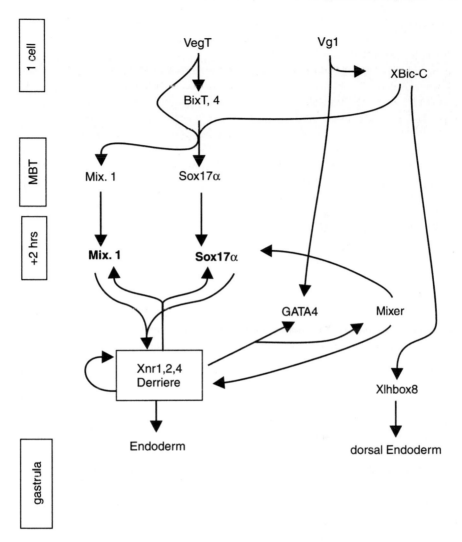

Fig 10.2 Molecular pathway leading to endoderm formation. VegT and most likely Vg1 lie at the top of a hierarchy of genes which predominantly act as activators leading to the expression of the TGFβ ligands Xnr1, Xnr2, Xnr4 and Derriere. Low levels of Mix.1 and Sox17a are present at the mid-blastula transition (MBT). Within two hours the levels of these two factors have been considerably increased through the action of the TFGβs and GATA4 and Mixer. Original.

10.5.1 Cortical Rotation is Essential for Normal Development

Polarization of the vegetal mass is dependent on the cortical rotation of vegetal cytoplasm that occurs within an hour of fertilization. This rotation moves the yolky cortical cytoplasm to underneath the pigmented equatorial region and produces a grey crescent of cortical cytoplasm that is opposite the site of sperm entry in the animal pole. This grey crescent defines the site of the future

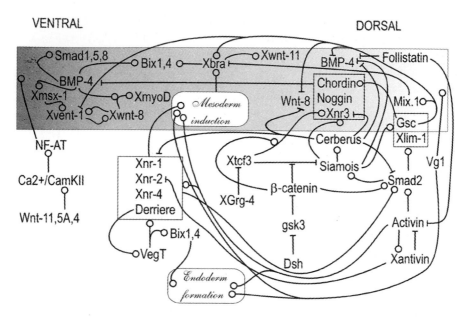

Fig. 10.3 The network of genes and factors responsible for the formation and patterning of the mesoderm. The mesoderm is represented as a rectangular box displaying ventral–dorsal spatial patterns of gene expression and interactions. Positive interactions are represented by a line ending in a circle whereas inhibitory or negative interactions are represented by a line ending in a bar. Original.

blastopore and together with the sperm entry site demarcates the dorsal-ventral axis of the embryo. If the animal pole region is removed within 30 minutes of fertilization then embryos form without a dorsal axis (Sakai 1996). However, if this region is removed after cortical rotation has displaced the cytoplasm to the dorsal side then normal development occurs. If this dorsal cytoplasm is then injected ectopically into another embryo it induces a secondary dorsal axis indicating that dorsal determinants are located in the vegetal pole and then translocated into the dorsal vegetal region during cortical rotation (Yuge *et al.* 1990). These dorsal determinants are localized to the egg cortex and require contact with the cytoplasm of the equatorial region in order to be active (Kageura 1997).

When the vegetal pole of the fertilized egg is irradiated with ultraviolet (UV) light the cortical cytoskeleton is disrupted, cortical rotation is perturbed and animals become ventralized (Malacinski *et al.* 1975; Gerhart *et al.* 1989) (Fig. 10.1B). These embryos can be rescued either by transplantation of dorsal vegetal cells from untreated animals (Gimlich 1986) or the vegetal injection of mRNA for several different signaling molecules indicating that regional localization of mRNAs in the embryo are important for axis formation (see below). Similar treatments of oocytes revealed that germ cell determinants are localized within unfertilized eggs (Elinson and Pasceri 1989). During oogenesis the egg is polarized and in many frogs a definitive axis is apparent by the localization of pigment to the animal pole. In addition to this visual pigment, the vegetal pole of the egg contains a diffuse cortical and subcortical localization of VegT mRNA as well as cortical Vg1 mRNA (Fig. 10.1C). Their restricted distribution in the germ plasm, the region of the egg cytoplasm that contains determinants important for the formation of the germ layers of the future embryo, is maintained by localization elements or protein binding motifs in the 3′UTRs of the RNA (Bubunenko *et al.* 2002; Kwon *et al.* 2002). This cortical localization of RNAs is dependent on the integrity of the microfilaments (Kloc and Etkin 1995) and is essential for germ layer formation.

Vg1 appears to be involved not only in endoderm formation but also in dorsal mesoderm patterning (Joseph and Melton 1998). It is not however involved in development of the organizer (Fagotto *et al.* 1997). The use of mutant Vg1 ligands that disrupt endogenous Vg1 activity causes ventralization of the embryo without loss of general mesoderm markers such as Xbra (*Xenopus* brachyury; member of the T-box family of transcription factors) (Joseph and Melton 1998). Loss of Vg1 also directly affected expression of endodermal markers in the vegetal hemisphere. The signaling pathway mediating Vg1 activity is unclear but evidence does indicate some interaction with the β-catenin pathway (see below) since exogenous Vg1 can rescue embryos depleted of β-catenin (member of a transcriptional activator complex which is translocated to the nucleus and regulates gene expression) (Wylie *et al.* 1996). It is unlikely that Vg1 interacts with the activin signaling pathway since embryos expressing a truncated type II activin receptor produce embryos with very different phenotypes to the ventralized embryos affected by mutant Vg1 (Dyson and Gurdon 1997). The dorsal signaling mediated by Vg1 is hard to reconcile with the widespread distribution of Vg1 in vegetal hemisphere unless it is proposed that cortical rotation activates a small amount of Vg1 in the dorsovegetal region (Thomsen and Melton 1993).

10.5.2 Maternal RNA, Zygotic Factors and Endoderm Polarization

The fate of germ layers is radically perturbed following depletion of maternal RNA stores. When VegT is reduced in oocytes and then subsequently fertilized in vitro, embryos develop severe dorso-anterior body axis defects with notable loss of head morphology (Zhang *et al.* 1998), a phenotype reminiscent of the ventralized animals produced by UV treatment (Fig. 10.1C). This result highlights the principle that the *earliest polarity is imposed on the embryo along the dorsoanterior axis by the asymmetric disposition of maternal mRNA and factors in the unfertilized egg.* VegT depleted embryos undergo normal cleavage and early blastulation and only exhibit defects during gastrulation (Zhang *et al.* 1998). This is consistent with the role of VegT as a transcription factor acting on the zygotic genome.

A curious phenomenon occurs during the mid-blastula stages of *Xenopus* development that is critical for subsequent patterning of the embryo. Blastulation begins at stage 7 (~4 hours) in *Xenopus* and lasts until the end of stage 9 (~9 hours). The mid-blastula transition occurs between the 12th (~4,224 cell stage) and 13th round of cell division when blastomeres no longer divide synchronously (Masui and Wang 1998). It is during this transition that expression of the zygotic genome begins (Newport and Kirschner 1982). Thus, the mid-blastula period represents a transition from regulation of early development by maternal RNA and protein to regulation via the zygotic genome.

VegT appears to be central to activating the zygotic genome and ensuring normal development of the anterior-posterior axis of the embryo. Although VegT depleted embryos do not form a blastopore, vegetal cells do migrate into the embryo (Zhang *et al.* 1998). The primary defect appears to be lack of formation of endoderm by vegetal cells and subsequent abnormal mesoderm differentiation. Consequently, equatorial cells which are normally destined to form mesoderm and undergo convergent extension movements instead take on the capacity to form ectoderm. Dorsal vegetal cells only induce dorsal organizer genes in Nieuwkoop recombinant experiments after the midblastula transition has occurred (Nagano *et al.* 2000). The above observations are significant because they reveal that mesoderm induction involves zygotic inducing signals synthesized under the influence of maternal transcription factors. Thus, mesoderm-inducing signals may not be present in the maternal vegetal cytoplasm but instead this plasm contains RNA/factors that drive the subsequent transcriptional machinery necessary for zygotic expression.

VegT may be acting to regulate the TGFβ family members Xnr1, Xnr2, Xnr4 and Derriere since each of these factors can rescue VegT depleted embryos (Kofron *et al.* 1999). Some of these factors also appear to be regulated by β-catenin. In β-catenin-depleted animals the expression of Xnr1, 2 and 4 was markedly reduced and their temporal wave of expression along the dorsal-ventral axis of the vegetal hemisphere was lost (Xanthos *et al.* 2002).

There is considerable correlative evidence supporting a central role of Xnr1, 2 and 4 as mesoderm-inducing factors. VegT activates the expression of these ligands as well as Derriere in animal caps (Clements *et al.* 1999). Blocking the function of TGFβ family signals (including BMP and Vg1) with a generic dominant-negative activin receptor prevented the expression of global and dorsal markers of mesoderm. Mesoderm induction is also blocked in the presence of a mutant form of Cerberus that specifically antagonizes only Nodal factors indicating that these TGFβ ligands are essential for general mesoderm induction (Agius *et al.* 2000).

The function of VegT RNA goes beyond just the restricted expression of VegT protein in the vegetal mass after the mid-blastula transition. The regional localization of VegT RNA is essential for the restricted distribution of other RNAs such as Vg1, Bicaudal-C (Wessely and DeRobertis 2000) and the signaling factor Wnt11 (Ku and Melton 1993). Depletion of VegT RNA not only caused a dispersal of these other RNAs, but also led to loss of Vg1 protein. Thus, VegT could have pleiotropic effects by affecting the localization of other cortical RNAs. Vg1 is involved in endoderm and mesoderm formation (Joseph and Melton 1998), Wnt11 affects convergent extension during gastrulation (Tada *et al.* 2002) and Bicaudal-C is an inducer of endoderm (Wessely and De Robertis 2000). However, it appears that VegT is probably more directly involved in developmental events since the expression of mesodermal markers is rescued by the injection of exogenous VegT mRNA into vegetal blastomeres at the 8-cell stage. If maternal VegT mRNA was merely regulating development by indirectly maintaining the vegetal localization of other RNAs in the oocyte then the phenotype would not have been rescued by its late injection. When VegT protein was down-regulated by antisense morpholinos, which both leaves VegT mRNA intact and does not affect the cortical localization of Bicaudal-C and Wnt11, development of the anterior-posterior axis continued to be disrupted as when the VegT mRNA was degraded (Heasman *et al.* 2001). These results highlight the principle that *development of the endoderm is dependent on zygotic transcription controlled by maternal transcription factors.*

10.5.3 β-catenin is in a Critical Signaling Pathway for Dorsalization

The cortical rotation of cytoplasm is also believed to move β-catenin stabilizing agents at the pole region into the dorsal vegetal hemisphere, the so-called Nieuwkoop's center, where it leads to a local increase in β-catenin levels (Rowning *et al.* 1997). While the nature of the catenin stabilizing agent is uncertain it most likely involves the Dishevelled mediated inhibition of gsk-3 (glycogen synthase kinase-3), a negative regulator of β-catenin stability (Miller *et al.* 1999). β-catenin increases on the dorsal side of the embryo by the two-cell stage and these elevated levels lead to increased nuclear localization of β-catenin in dorsal blastomeres between the 16-32 cell stages (Larabell *et al.* 1997). These increases are paralleled by concomitant dorsal increases in Dishevelled achieved through microtubule-assisted translocation of Dishevelled enriched vesicles from the vegetal pole during cortical rotation (Miller *et al.* 1999). Dishevelled inhibits gsk-3, which normally phosphorylates β-catenin and targets it for degradation (Yost *et al.* 1996). Thus, the cortical rotation of Dishevelled inhibits gsk-3 and leads to the stabilization of β-catenin.

There is strong supporting evidence indicating that β-catenin is a key mediator of dorsalization. β-catenin is maternally expressed and exogenously injected β-catenin induces secondary axes in *Xenopus*. More importantly, reduction of β-catenin mRNA in oocytes prevents normal axis formation (Heasman *et al.* 1994). The function of β-catenin is non-cell-autonomous indicating that cells expressing β-catenin induce other cells to become dorsalized (Wylie *et al.* 1996). The function of β-catenin as a dorsalizing signal is only active prior to the mid-blastula transition since zygotically transcribed β-catenin fails to rescue embryos depleted of maternal β-catenin (Wylie *et al.* 1996). Interestingly, vegetal cells do not release a dorsalizing signal until after the midblastula stage indicating that this factor is zygotically transcribed under the influence of maternal β-catenin (Wylie *et al.* 1996). β-catenin seems to be upstream of Vg1, siamosis (transcription factor), and Noggin (secreted factor) since each of these factors are able to rescue embryos depleted of maternal β-catenin (Wylie *et al.* 1996). Injection of a dominant negative BMP2/4 receptor also rescues the mutant phenotype indicating that BMP2/4 signaling acts as a negative regulator downstream of β-catenin (Wylie *et al.* 1996).

Depletion of maternal XTcf3 RNA produced a dorsalized embryo which overexpressed organizer/dorsal mesoderm-specific genes inappropriately throughout the dorsal-ventral axis (Houston *et al.* 2002). This suggested that XTcf3 was normally repressing organizer gene expression and it was only when this molecule was deleted that organizer genes were selectively expressed. One possibility is that β-catenin binds XTcf3 in the dorsal vegetal mass to de-repress organizer gene transcription. This is consistent with the observation that when the equatorial regions of β-catenin depleted embryos are cultured in isolation they express global and ventral markers of mesoderm but lack dorsal markers (Zanthos *et al.* 2002). Although de-repression is necessary, it is not sufficient for dorsalization (Houston *et al.* 2002). Deletion of XTcf3 in oocytes prior to fertilization allows expression of dorsal genes. However, when VegT is also deleted these genes are not expressed indicating that VegT is also required for their activation.

10.5.4 Nieuwkoop Experiments Reveal Overlapping Molecular Interactions

An insight into the overlap of the molecular pathways underlying mesoderm induction has come from a series of elegant Nieuwkoop recombinant experiments using vegetal masses and equatorial cells from embryos depleted of VegT and/or β-catenin (Fig. 10.4). When equatorial cells from VegT-depleted embryos are isolated in culture they fail to express either global or dorsal-specific markers of mesoderm (Kofron *et al.* 1999). In Nieuwkoop recombinants, wildtype vegetal masses rescue VegT-depleted equatorial cells through a VegT dependent pathway (Kofron *et al.* 1999). Similarly, VegT-depleted vegetal masses fail to induce animals caps to form mesoderm (Zhang *et al.* 1998). Together these experiments indicate that VegT regulates the expression of mesoderm-inducing signals in the vegetal hemisphere. When equatorial cells were isolated from β-catenin-depleted embryos and cultured in isolation they expressed global but not dorsal markers of mesoderm (Xanthos *et al.* 2002). This revealed that VegT was sufficient to induce mesoderm and that β-catenin acted subsequently to dorsalize the mesoderm. That VegT-depleted vegetal masses (with normal β-catenin) could not rescue β-catenin-depleted equatorial cells confirmed that both VegT and β-catenin pathways converge to dorsalize mesoderm (Xanthos *et al.* 2002).

Fig. 10.4 Summary of Nieuwkoop recombinant experiments examining the roles of VegT and β-catenin in induction and patterning of the mesoderm (Zhang *et al.* 1998; Koffron *et al.* 1999; Xanthos *et al.* 2002). Marginal zone cells from wild-type embryos or embryos depleted of either VegT or β-catenin were co-cultured in various combinations with vegetal masses from these same embryos. Cultures from panel A and B indicate that the vegetal mass is essential for induction and patterning. Panel C indicates that that VegT is responsible for this behavior. Panels D and E indicate that β-catenin in the marginal zone is not necessary for induction or patterning as long as wild-type vegetal mass is present (not shown is that wild-type marginal cells autonomously express global and dorsal mesoderm markers). Panel F reveals that VegT present in the vegetal mass is sufficient for inductive events even in the absence of β-catenin in the marginal zone.

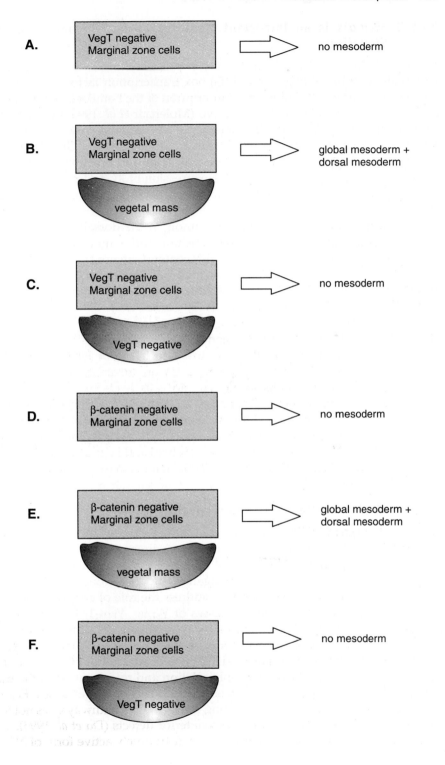

10.5.5 Siamois is an Important Mediator of Activin and β-catenin Signaling

β-catenin binds to XTcf-3, a maternally expressed *Xenopus* homolog of the vertebrate high mobility group (HMG) box transcription factor genes, resulting in its nuclear translocation and transcription of the homeobox gene siamois in the dorsovegetal region of the embryo (Molenaar *et al.* 1996; Brannon *et al.* 1997). In the absence of β-catenin, XTcf-3 represses siamois expression (Brannon *et al.* 1997). The maternally expressed transcriptional repressor Groucho (XGrg-4) inhibits transcription of XTcf-3 (Roose *et al.* 1998) and probably acts together with β-catenin to regulate siamois. Overexpression of XGrg-4 prevents normal axis formation (Roose *et al.* 1998), as does a dominant negative form of XTcf-3 (Molenaar *et al.* 1996).

Several lines of evidence implicate siamois in the dorsalization pathway: siamois appears after the midblastula transition and is more abundant in the dorsal vegetal hemisphere of Nieuwkoop's center (Lemaire *et al.* 1995; Brannon *et al.* 1997); normal activity of XTcf-3 is essential for formation of the dorsoanterior axis (Molenaar *et al.* 1996); injection of exogenous siamois transcripts induces body axis duplication (Lemaire *et al.* 1995; Carnac *et al.* 1996; Brannon *et al.* 1997); and dominant negative siamois blocks dorsal development (Fan and Sokol 1997; Kessler 1997). A related homeobox gene Twin appears not only to be co-regulated with siamois but also to share many of its same activities (Laurent *et al.* 1997). Siamois is upstream of the organizer-specific genes since ectopic expression of siamois in animal cap cells leads to upregulation of goosecoid, Xnr3 and Chordin (Carnac *et al.* 1996). The activin/Smad2 (activated Smad proteins in the cytoplasm enter the nucleus and together with cofactors form complexes which regulate transcription) signaling pathway acts co-operatively with the β-catenin pathway (Schohl and Fagotto 2002) to significantly upregulate siamois (Crease *et al.* 1998). Siamois in turn enhances expression of goosecoid and Chordin.

10.5.6 Wnt Signaling and Polarization

Although the Wnt signaling pathways involving β-catenin are clearly involved in axis formation there is little evidence for a direct role of Wnt ligands in activating these pathways. Most of the evidence for Wnt activity comes from ectopic injection of exogenous Wnt ligands and not from loss-of-function approaches which would more directly address the role of endogenous Wnts. There are two functionally distinct classes of Wnts: Wnt-1, -3A, -8 and −8b stimulate the "canonical" signaling pathway involving β-catenin (Smith *et al.* 2000) while Wnt-11, -5A and −4 stimulate the "Wnt/Ca²⁺" pathway which involves an inositol triphosphate mediated rise in intracellular Ca^{2+} (Tada *et al.* 2002). This Ca^{2+} rise leads to dephosphorylation and translocation of the transcription factor NF-AT into the nucleus where it acts on target genes. Ectopic expression of exogenous Wnts mediating the Wnt/Ca²⁺ pathway does not lead to axis duplication but does cause dorsoanterior defects (Du *et al.* 1995). This is supported by recent observations that a constitutively active form of NF-AT ventralizes embryos while a dominant negative NF-AT induced a secondary dorsal axis (Saneyoshi *et al.* 2002). These results suggest that activated NF-AT has a dual role in both stimulating ventral cell fates and inhibiting dorsalization. This was further confirmed by the fact that both constitutively active NF-AT (Saneyoshi *et al.* 2002) and exogenous Wnt 5A (Torres *et al.* 1996) can rescue the duplication of the dorsal axis produced by ectopic expression of Wnt-8.

Xwnt-11 is maternally expressed and has a cytoplasmic distribution similar to that of Vg1 (Ku and Melton 1993). It is not a direct mesoderm inducer since it is unable to stimulate animal caps to form mesoderm (Ku and Melton 1993). While the canonical signaling pathway mediates dorsalization, the canonical Wnt ligands may not directly participate in activation of dorsal β-catenin since overexpression of a dominant negative Wnt-8 (Kuhl *et al.* 2000) had no effect on dorsal axis formation (Hoppler *et al.* 1996). There is however evidence that endogenous Xwnt-5A and −11 are active in ventralization since a dominant negative form of Xwnt-11 blocks ventral gene expression and causes dorsalization of ventral cells (Kuhl *et al.* 2000). Moreover, Xwnt-5 and −11 are present in the egg and seem to be responsible for the ventral accumulation of Ca^{2+}/calmodulin-dependent protein kinase II (CamKII), an enzyme which lies upstream of NF-AT and whose activation is necessary for specifying ventral cell fate (Ku and Melton 1993; Kuhl *et al.* 2000). Taken together, the Wnt/Ca^{2+} pathway ventralizes the embryo as well as antagonizes the dorsal pathway while the canonical Wnt pathway dorsalizes the embryo.

10.5.7 TGFβ Ligands Induce Mesoderm and Dorsalize Embryos

Xnr3 inhibits mesoderm induction by binding to BMP-4 (Hansen *et al.* 1997). Moreover, siamois also causes animal cap cells to repress the ventralizing gene BMP-4 and to secrete a mesodermal dorsalizing activity that is distinct from Noggin or Chordin (Carnac *et al.* 1996). While siamois upregulates the anterior inducing genes Cerberus, Frizzled-b1 (Wnt membrane receptor) and Xlim1 (homeodomain transcription factor) in vegetal blastomeres this activity is dependent on the presence of endogenous TGFβ signals (Engleka and Kessler 2001). It appears that siamois upregulates the homeobox genes goosecoid and Xlim1 which then upregulate expression of Chordin (Sasai *et al.* 1994; Kodjabachian *et al.* 2001). Cerberus is a secreted protein that is highly expressed in the organizer and anterior endoderm where it is believed to pattern the anterior axis (Bouwmeester *et al.* 1996) by antagonizing Nodal, BMP and Wnt signals (Glinka *et al.* 1997; Piccolo *et al.* 1999). Cerberus appears to downregulate the activity of signals involved in trunk formation so as to facilitate head induction (Glinka *et al.* 1997; Piccolo *et al.* 1999).

The role of TGFβ members is consistent with observations that activin (Asashima *et al.* 1991; Nishimatsu *et al.* 1992; Fukui *et al.* 1994; Dyson and Gurdon 1997), Vg1 (Dale *et al.* 1989; Kessler and Melton 1995), Derriere (Sun *et al.* 1999; White *et al.* 2002), and Xnr1 and Xnr2 (Aguis *et al.* 2000; Onuma *et al.* 2002) are all located in the vegetal blastomeres and are capable of inducing mesoderm. The use of mutant activin receptors indicates that activin-like ligands are probably the strongest candidate as a mesoderm inducing factor secreted by the Nieuwkoop center (Gurdon *et al.* 1994; Dyson and Gurdon 1997; Chang *et al.* 1997; Asashima *et al.* 1999). Activin has a tendency to induce dorsoanterior structures in vivo (Thomsen *et al.* 1990) and this appears to be dose dependent (Green and Smith 1990; Green *et al.* 1992) and partly mediated via Xnrs (Osada and Wright 1999). The TGFβ ligands bind to receptors which phosphorylate and activate Smad proteins which then in turn bind DNA binding proteins and modulate gene expression. Smads 1, 5 and 8 are activated by BMP signaling and induce ventral mesoderm whereas Smad 2 transduces activin signaling (Faure *et al.* 2000).

While activin-like ligands are present in *Xenopus* unfertilized eggs and blastula (Asashima *et al.* 1991), the phosphorylated and hence activated forms of Smad2 are not detected until after zygotic transcription begins at the mid-blastula transition, at the same time as when activin mRNA is first detected in *Xenopus* embryos (Dohrmann et a. 1993). This Smad2 phosphorylation is dependent on zygotic transcription and relies on the activity of VegT (Lee *et al.* 2001). Smad2 phosphorylation is under tight regulatory control with molecules such as antivin and Cerberus acting as antagonists, while β-catenin accelerates the VegT mediated activation of Smad2 (Lee *et al.* 2001). That mesoderm induction begins after the midblastula stage when Smad2 is phosphorylated is consistent with a role of activin in the early phase of this event.

Although early studies (Slack 1991) indicated that mesoderm induction is not blocked by exogenous Follistatin, an activin inhibitor, more recent analyses indicate that inhibition is concentration dependent (Fukui *et al.* 1993; Fukui *et al.* 1994; Hemmati-Brivanlou *et al.* 1994; Marchant *et al.* 1998). Activin induces Xantivin in the marginal zone during gastrulation which also antagonizes the mesoderm inducing activity of both Xnr2 and activin (Cheng *et al.* 2000; Tanegashima *et al.* 2000). It is unlikely that activin is acting alone as a mesoderm inducer. Activin probably acts in conjunction with nodal factors since blocking their function with a mutant form of Cerberus inhibits mesoderm formation (Agius *et al.* 2000).

10.5.8 Xwnt and the Ventralization Pathway

Activin induces Xbra, Mix.1, goosecoid and Xlim-1 and represses expression of Xwnt-8 in the organizer, most likely via goosecoid (Taira *et al.* 1992; Christian and Moon 1993; Steinbeisser *et al.* 1993; Kinoshita and Asashima 1995; Latinkic and Smith 1999; Yao and Kessley 2001). This is important since overexpression of Xwnt-8 in the organizer after the midblastula transition ventralizes these cells (Christian and Moon 1993). Dominant negative forms of Xwnt-8 further reveal that Xwnt-8 function is primarily concerned with late mesoderm events such as muscle differentiation (Hoppler *et al.* 1996). The restricted expression of Xwnt-8 in the ventral mesoderm during gastrula, achieved by repression from Cerberus and goosecoid (Yao and Kessler 2001), is essential for the patterning of the ventrolateral mesoderm, such as expression of muscle XmyoD (muscle specific transcription factor) and somite differentiation (Hoppler and Moon 1998).

This patterning is also dependent on the co-operative activity of BMP-4 (Hoppler and Moon 1998). The ventral expression of Xwnt-8 appears to be positively regulated by Xvent-1 (homeobox transcription factor) in a reciprocal feedback loop (Hoppler and Moon 1998). Xvent-1 itself is upregulated by BMP-4 signaling and it is capable of ventralizing dorsal mesoderm (Gawantka et al. 1995). BMP-4 induces the homeobox gene Xmsx-1 which then activates Xvent-1 (Yamamoto et al. 2000). Xmsx-1 expression overlaps that of BMP-4 and inhibition of Xmsx-1 causes the formation of an ectopic head (Yamamoto et al. 2001). Thus, dorsal inhibition of BMP-4/Xmsx-1 signaling is required for head formation. Goosecoid and Xvent-1 act in a negative regulatory loop, either directly or indirectly, to suppress each others activity in the ventral and dorsal mesoderm respectively (Gawantka et al. 1995).

At high concentrations of activin, goosecoid represses transcription of Xbra directly whereas Mix.1 represses Xbra indirectly, most probably by activating goosecoid (Latinkic and Smith 1999). Xbra, which is an early response gene to mesoderm induction (Smith et al. 1991), is essential for regulation of gastrulation movements and this effect appears to be mediated by its activation of Xwnt-11 (Tada and Smith 2000). Although Xwnt-11 function appears to be primarily involved in gastrulation movements rather than with mesoderm specification, ectopic expression of Xbra in animal caps causes formation of ventrolateral mesoderm derivatives (Saka et al. 2000). The expression of Xwnt-11 mirrors that of Xbra during gastrula stages and over expression of a dominant negative Xwnt-11, like dominant negative Xbra (Conlon and Smith 2000), inhibits convergent extension movements (Tada and Smith 2000). This late role of Xwnt-11 in mesoderm development is consistent with its ability, like Xwnt-8, to rescue UV ventralized embryos albeit at considerably higher doses than Xwnt-8 (Ku and Melton 1993). Overexpression of Bix1 in whole embryos resulted in defects in gastrulation movements and loss of anterior structures, consistent with this transcription factor being downstream of Xbra (Tada et al. 1998).

10.5.9 Mesoderm Induction and Ventralization are Separate Events

BMP-4 is expressed in the ventral equatorial zone and has strong ventral mesoderm inducing activity (Dale et al. 1992; Fainsod et al. 1994), but only during gastrulation, after mesoderm has already been induced (Jones et al. 1996). Members of the TGF-β family other than BMP-4 are most likely the early mesoderm inducing factors since injection of a dominant negative activin receptor in *Xenopus* embryos, which inhibits signaling via a number of TGF-β ligands, blocks mesoderm formation (Hemmati-Brivanlou and Melton 1992). Several BMP antagonists including Chordin, Noggin and Follistatin are expressed by the organizer and are able to dorsalize ventral mesoderm (Sasai et al. 1994; Smith et al. 1993; Fainsod et al. 1997). Goosecoid also reduces dorsal BMP-4 activity but whether this is a direct or an indirect effect through Chordin remains to be determined (Fainsod et al. 1994). These results have led to a model by where the mesoderm is induced by a TGFβ ligand and then ventralized by BMP-4. BMP antagonists secreted by the organizer (which has been induced by Nieuwkoop's center) neutralize the BMP-4 allowing dorsalization of the mesoderm. Interestingly, injection of dominant negative BMP receptors leads to dorsalization of ventral mesoderm indicating that either the organizer expresses an active dorsalizing factor in addition to BMP antagonists or that the default state of the mesoderm is perhaps dorsal (Maeno *et al.* 1994; Graff *et al.* 1994; Suzuki *et al.* 1994).

10.6 SUMMARY AND CONCLUSIONS

The classical recombinant experiments involving pieces of vegetal hemisphere mass being cultured together with animal pole revealed the significance of regional differences in the vegetal tissue for polarizing the mesoderm. These results were later complemented by approaches using cell transplants, cell cytoplasm injections and ultimately RNA injections of specific factors. Such strategies led to our current and reasonably detailed understanding of the signaling pathways underlying inductive events between the vegetal and animal hemispheres, at least in *Xenopus*.

There has been a tendency in the literature to reduce these complex interactions into a series of discrete and sometimes seemingly mutually exclusive events. Two-step and three step models of induction are often presented as simplified versions of these development programs. However it is clear from the plethora of data in this field that the actual interactions are highly complex and yet the approaches adopted to tease out their intricacies are often surprisingly crude and designed to understand a signal molecule's role during developmental time-frames which include multiple milestone events. Nonetheless, having said this it is still possible to identify key events and present them as major contributors to the formation of the endoderm and mesoderm.

Endoderm induction is driven by the maternal transcription factor VegT which activates the expression of the endoderm specific genes Mix.1 and Sox17 which are essential for the subsequent development of endoderm derivatives as well as for mesoderm inductive behavior. The expression of these phenotypic characteristics of endoderm are enhanced by VegT's activation of Nodal related factors. These same factors induce what could be considered a nonpolarized mesoderm which is then subsequently polarized by the action of β-catenin and activin in the dorsal endoderm and by BMP-4 and the Wnt/Ca^{2+} pathway in the ventral endoderm. It is very difficult to separate these events from a general mesoderm induction since they appear to be occurring concomitantly. In the ventral marginal zone BMP-4 activates a cascade that culminates in the activation of Wnt8 while in the dorsal marginal zone antagonists are expressed which act directly and indirectly to reduce BMP-4 and Xwnt-8. In the dorsoventral region TGFβs act in conjunction with activin and β-catenin pathways to dorsalize the mesoderm. Thus, the TGFβ ligands have an essential role that includes both mesoderm induction as well as patterning.

What has finally emerged from all the experimental data is a dynamic picture of the cooperative activity of biochemical networks involving both activators and repressors of gene expression as well as of protein activity. These interacting components are intricately intertwined, acting in separate and overlapping pathways to simultaneously sculpture two primary germ layers and to set the stage for the subsequent differentiation of specific tissue types.

10.7 ACKNOWLEDGEMENTS

The author would like to thank the all members of the Neurodevelopment Laboratory for their assistance and perseverance as I wrote this Chapter. Kendra Coufal and Christina Claxton provided valuable administrative support during the formulation of this manuscript. The laboratory is funded by both the Australian Research Council and the National Health and Medical Research Council of Australia.

10.8 LITERATURE CITED

Agius, E., Oelgeschlager, M., Wessely, O., Kemp, C., De Robertis, E.M. 2000. Endodermal Nodal-related signals and mesoderm induction in *Xenopus*. Development 127: 1173-1183.

Asashima, M., Kinoshita, K., Ariizumi, T., Malacinski, G.M. 1999. Role of activin and other peptide growth factors in body patterning in the early amphibian embryo. International Review of Cytology 191: 1-52.

Asashima, M., Nakano, H., Uchiyama, H., Sugino, H., Nakamura, T., Eto, Y., Ejima, D., Nishimatsu, S., Ueno, N., Kinoshita, K. 1991. Presence of activin (erythroid differentiation factor) in unfertilized eggs and blastulae of *Xenopus laevis*. Proceeding of the National Academy of Sciences of the United States of America 88: 6511-6514.

Boterenbrood, E.C. and Nieuwkoop, P.D. 1973. The formation of meso-endoderm in the urodelean amphibians. V. Its regional induction by the endoderm. Roux' Archives 173: 319-332.

Bouwmeester, T., Kim, S., Sasai, Y., Lu, B., De Robertis, E.M. 1996. Cerberus is a head-inducing secreted factor expressed in the anterior endoderm of Spemann's organizer. Nature 382: 595-601.

Brannon, M., Gomperts, M., Sumoy, L., Moon, R.T. and Kimelman, D. 1997. A β-catenin/XTcf-3 complex binds to the *siamois* promoter to regulate dorsal axis specification in *Xenopus*. Genes and Development 11: 2359-2370.

Brannon, M., Brown, J.D., Bates, R., Kimelman, D., Moon, R.T. 1999. XCtBP is a XTcf-3 co-repressor with roles throughout *Xenopus* development. Development 26: 3159-3170.

Bubunenko, M., Kress, T.L., Vempati, U.D., Mowry, K.L. and King, M.L. 2002. A consensus RNA signal the directs germ layer determinants to the vegetal cortex of *Xenopus* oocytes. Developmental Biology 248: 82-92.

Carnac, G., Kodjabachian, L., Gurdon, J.B., Lemaire, P. 1996. The homeobox gene Siamois is a target of the Wnt dorsalisation pathway and triggers organiser activity in the absence of mesoderm. Development 122: 3055-3065.

Casey, E.S., Tada, M., Fairclough, L., Wylie, C.C., Heasman, J., Smith, J.C. 1999. Bix4 is activated directly by VegT and mediates endoderm formation in *Xenopus* development. Development 126: 4193-4200.

Chang, C. and Hemmati-Brivanlou, A. 2000. A post-mid-blastula transition requirement for TGFbeta signaling in early endodermal specification. Mechanisms of Development 90: 227-235.

Chang, C., Wilson, P.A., Mathews, L.S., Hemmati-Brivanlou, A. 1997. A *Xenopus* type I activin receptor mediates mesodermal but not neural specification during embryogenesis. Development 124: 827-837.

Cheng, A.M., Thisse, B., Thisse, C., Wright, C.V. 2000. The lefty-related factor Xatv acts as a feedback inhibitor of nodal signaling in mesoderm induction and L-R axis development in *Xenopus*. Development 127: 1049-1061.

Christian, J.L. and Moon, R.T. 1993. Interactions between Xwnt-8 and Spemann organizer signaling pathways generate dorsoventral pattern in the embryonic mesoderm of *Xenopus*. Genes and Development 7: 13-28.

Clements, D., Friday, R.V. and Woodland, H.R. 1999. Mode of action of VegT in mesoderm and endoderm formation. Development 126: 4903-4911.

Conlon, F.L. and Smith, J.C. 1999. Interference with brachyury function inhibits convergent extension, causes apoptosis, and reveals separate requirements in the FGF and activin signalling pathways. Developmental Biology 213: 85-100.

Crease, D.J., Dyson, S., Gurdon, J.B. 1998. Cooperation between the activin and Wnt pathways in the spatial control of organizer gene expression. Proceeding of the National Academy of Sciences of the United States of America 95: 4398-4403.

Dale, L., Howes, G., Price, B.M., Smith, J.C. 1992. Bone morphogenetic protein 4: a ventralizing factor in early *Xenopus* development. Development 115: 573-585.

Dale, L., Matthews, G., Tabe, L., Colman, A. 1989. Developmental expression of the protein product of Vg1, a localized maternal mRNA in the frog *Xenopus laevis*. EMBO Journal 8: 1057-1065.

Dale, L. and Slack, J.M. 1987. Regional specification within the mesoderm of early embryos of *Xenopus* laevis. Development 100: 279-295.

Dohrmann, C.E., Hemmati-Brivanlou, A., Thomsen, G.H., Fields, A., Woolf, T.M., Melton, D.A. 1993. Expression of activin mRNA during early development in *Xenopus laevis*. Developmental Biology 157: 474-483.

Du, S.J., Purcell, S.M., Christian, J.L., McGrew, L.L., Moon, R.T. 1995. Identification of distinct classes and functional domains of Wnts through expression of wild-type and chimeric proteins in *Xenopus* embryos. Molecular and Cellular Biology 15: 2625-2634.

Dyson, S. and Gurdon, J.B. 1997. Activin signalling has a necessary function in *Xenopus* early development. Current Biology 7: 81-84.

Eimon, P.M. and Harland, R.M. 1999. In *Xenopus* embryos, BMP heterodimers are not required for mesoderm induction, but BMP activity is necessary for dorsal/ventral patterning. Developmental Biology 216: 29-40.

Elinson, R.P. and Pasceri, P. 1989. Two UV-sensitive targets in dorsoanterior specification of frog embryos. Development 106: 511-518.

Engleka, M.J. and Kessler, D.S. 2001. Siamois cooperates with TGFbeta signals to induce the complete function of the Spemann-Mangold organizer. International Journal of Developmental Biology 45: 241-250.

Fagotto, F., Guger, K., Gumbiner, B.M. 1997. Induction of the primary dorsalizing center in *Xenopus* by the Wnt/GSK/beta-catenin signaling pathway, but not by Vg1, Activin or Noggin. Development 124: 453-460.

Fainsod, A., Deissler, K., Yelin, R., Marom, K., Epstein, M., Pillemer, G., Steinbeisser, H., Blum, M. 1997. The dorsalizing and neural inducing gene follistatin is an antagonist of BMP-4. Mechanisms of Development 63: 39-50.

Fainsod, A., Steinbeisser, H., De Robertis, E.M. 1994. On the function of BMP-4 in patterning the marginal zone of the *Xenopus* embryo. EMBO Journal 13: 5015-5025.

Fan, M.J. and Sokol, S.Y. 1997. A role for Siamois in Spemann organizer formation. Development 124: 2581-2589.

Faure, S., Lee, M.A., Keller, T., ten Dijke, P., Whitman, M. 2000. Endogenous patterns of TGFbeta superfamily signaling during early *Xenopus* development. Development 127: 2917-2931.

Fukui, A., Nakamura, T., Sugino, K., Takio, K., Uchiyama, H., Asashima, M., Sugino, H. 1993. Isolation and characterization of *Xenopus* follistatin and activins. Developmental Biology 159: 131-139.

Fukui, A., Nakamura, T., Uchiyama, H., Sugino, K., Sugino, H., Asashima, M. 1994. Identification of activins A, AB, and B and follistatin proteins in *Xenopus* embryos. Developmental Biology 163: 279-281.

Gawantka, V., Delius, H., Hirschfeld, K., Blumenstock, C., Niehrs, C. 1995. Antagonizing the Spemann organizer: role of the homeobox gene Xvent-1. EMBO Journal 14: 6268-6279.

Gerhart, J. 2001. Evolution of the organizer and the chordate body plan. International Journal of Developmental Biology 45: 133-153.

Gerhart, J., Danilchik, M., Doniach, T., Roberts, S., Rowning, B., Stewart, R. 1989. Cortical rotation of the *Xenopus* egg: consequences for the anteroposterior pattern of embryonic dorsal development. Development 107 Suppl: 37-51.

Gimlich, R.L 1986. Acquisition of developmental autonomy in the equatorial region of the *Xenopus* embryo. Developmental Biology 115: 340-352.

Gimlich R.L and Gerhart, J.C. 1984. Early cellular interactions promote embryonic axis formation in *Xenopus laevis*. Developmental Biology 104: 117-130.

Glinka, A., Wu, W., Onichtchouk, D., Blumenstock, C., Niehrs, C. 1997. Head induction by simultaneous repression of Bmp and Wnt signalling in *Xenopus*. Nature 389: 517-519.

Graff, J.M., Thies, R.S., Song, J.J., Celeste, A.J., Melton, D.A. 1994. Studies with a *Xenopus* BMP receptor suggest that ventral mesoderm-inducing signals override dorsal signals in vivo. Cell 79: 169-179.

Green, J.B., New, H.V., Smith, J.C. 1992. Responses of embryonic *Xenopus* cells to activin and FGF are separated by multiple dose thresholds and correspond to distinct axes of the mesoderm. Cell 71: 731-739.

Green, J.B. and Smith, J.C. 1990. Graded changes in dose of a *Xenopus* activin A homologue elicit stepwise transitions in embryonic cell fate. Nature 347: 391-394.

Gurdon, J.B., Harger, P., Mitchell, A., Lemaire, P. 1994. Activin signalling and response to a morphogen gradient. Nature 371: 487-492.

Hansen, C.S., Marion, C.D., Steele, K., George, S., Smith, W.C. 1997. Direct neural induction and selective inhibition of mesoderm and epidermis inducers by Xnr3. Development 124: 483-492.

Heasman, J., Crawford, A., Goldstone, K., Garner-Hamrick, P., Gumbiner, B., McCrea, P., Kintner, C., Noro, C.Y., Wylie, C. 1994. Overexpression of cadherins and underexpression of beta-catenin inhibit dorsal mesoderm induction in early *Xenopus* embryos. Cell 79: 791-803.

Heasman, J., Wessely, O., Langland, R., Craig, E.J. and Kessler, D.S. 2001. Vegetal localization of maternal mRNAs is disrupted by VegT depletion. Developmental Biology 240: 377-386.

Heasman, J., Wylie, C.C., Hausen, P., Smith, J.C. 1984. Fates and states of determination of single vegetal pole blastomeres of *X. laevis*. Cell 37: 185-194.

Hemmati-Brivanlou, A., Kelly, O.G., Melton, D.A. 1994. Follistatin, an antagonist of activin, is expressed in the Spemann organizer and displays direct neuralizing activity. Cell 77: 283-295.

Hemmati-Brivanlou, A. and Melton, D.A. 1992. A truncated activin receptor inhibits mesoderm induction and formation of axial structures in *Xenopus* embryos. Nature 359: 609-6014.

Henry, G.L. and Melton, D.A. 1998. Mixer, a homeobox gene required for endoderm development. Science 281: 91-96.

Holley, S.A., Jackson, P.D., Sasai, Y., Lu, B., De Robertis, E.M., Hoffmann, F.M., and Ferguson, E.L. 1995. A conserved system for dorsal-ventral patterning in insects and vertebrates involving sog and chordin. Nature 376: 249-253.

Horb, M.E. and Thomsen, G.H. A vegetally localized T-box transcription factor in *Xenopus* eggs specifies mesoderm and endoderm and is essential for embryonic mesoderm formation. Development 124: 1689-1698.

Hoppler, S., Brown, J.D., Moon, R.T. 1996. Expression of a dominant-negative Wnt blocks induction of MyoD in *Xenopus* embryos. Genes and Development. 10: 2805-2817.

Houston, D.W., Kofron, M., Resnik, E., Langland, R., Destree, O., Wylie, C. and Heasmen, J. 2002. Repression of organizer genes in dorsal and ventral *Xenopus* cells mediated by maternal XTcf3. Development 129: 4015-4025.

Hudson, C., Clements, D., Friday, R.V., Stott, D., Woodland, H.R. 1997. Xsox17alpha and -beta mediate endoderm formation in *Xenopus*. Cell 91: 397-405.

Iemura, S., Yamamoto, T.S., Takagi, C., Uchiyama, H., Natsume, T., Shimasaki, S., Sugino, H., Ueno, N. 1998. Direct binding of follistatin to a complex of bone-morphogenetic protein and its receptor inhibits ventral and epidermal cell fates in early *Xenopus* embryo. Proceedings of the National Academy of Sciences of the United States of America 95: 9337-9342.

Jones, C.M., Dale, L., Hogan, B.L., Wright, C.V., Smith, J.C. 1996. Bone morphogenetic protein-4 (BMP-4) acts during gastrula stages to cause ventralization of *Xenopus* embryos. Development 122: 1545-1554.

Joseph, E.M. and Melton, D.A. 1998. Mutant Vg1 ligands disrupt endoderm and mesoderm formation in *Xenopus* embryos. Development 125: 2677-2685.

Kageura, H. 1997. Activation of dorsal development by contact between the cortical dorsal determinant and the equatorial core cytoplasm in eggs of *Xenopus laevis*. Development 124: 1543-1551.

Kessler, D.S. 1997. Siamois is required for formation of Spemann's organizer. Proceedings of the National Academy of Sciences of the United States of America 94: 13017-13022.

Kessler, D.S. and Melton, D.A. 1995. Induction of dorsal mesoderm by soluble, mature Vg1 protein. Development 121: 2155-2164.

Kinoshita, K. and Asashima, M. 1995. Effect of activin and lithium on isolated *Xenopus* animal blastomeres and response alteration at the midblastula transition. Development 121: 1581-1589.

Kloc, M. and Etkin, M. 1995. Two distinct pathways for the localization of RNAs at the vegetal cortex in *Xenopus* oocytes. Development 121: 287-297.

Kodjabachian, L., Karavanov, A.A., Hikasa, H., Hukriede N.A., Aoki, T., Taira, M., Dawid, I.B. 2001. A study of Xlim1 function in the Spemann-Mangold organizer. International Journal of Developmental Biology 45: 209-218.

Kofron, M., Demel, T., Xanthos, J., Lohr, J., Sun, B., Sive, H., Osada, S.I., Wright, C., Wylie, C. and Heasman, J. 1999. Mesoderm induction in *Xenopus* is a zygotic event regulated by maternal VegT via TGFβ growth factors. Development 126: 5759-5770.

Ku, M. and Melton, D.A. 1993. *Xwnt-11*: a maternally expressed *Xenopus wnt* gene. Development 119: 1161-1173.

Kuhl, M., Sheldahl, L.C., Malbon, C.C., Moon, R.T. 2000. Ca(2+)/calmodulin-dependent protein kinase II is stimulated by Wnt and Frizzled homologs and promotes ventral cell fates in *Xenopus*. Journal of Biological Chemistry 275: 12701-12711.

Kwon, S., Abramson, T., Munro, T.P., John, C.M., Kohrmann, M. and Schnapp, B.J. 2002. UUCAC- and Vera-dependent localization of VegT RNA in *Xenopus* oocytes. Current Biology 12: 558-564.

Larabell, C.A., Torres, M., Rowning, B.A., Yost, C., Miller, J.R., Wu, M., Kimelman, D. and Moon, R.T. 1997. Establishment of teh dorso-ventral axis in *Xenopus* embryos is presaged by early asymmetries in β-catenin that are modulated by the Wnt signaling pathway. The Journal of Cell Biology 136: 1123-1136.

Latinkic, B.V. and Smith, J.C. 1999. Goosecoid and mix.1 repress Brachyury expression and are required for head formation in *Xenopus*. Development 126: 1769-1779.

Laurent, M.N., Blitz, I.L., Hashimoto, C., Rothbacher, U., Cho, K.W. 1997.The *Xenopus* homeobox gene twin mediates Wnt induction of goosecoid in establishment of Spemann's organizer. Development 124: 4905-4916.

Lee, M.A., Heasman, J., Whitman, M. 2001. Timing of endogenous activin-like signals and regional specification of the *Xenopus* embryo. Development 128: 2939-2952.

Lemaire, P., Darras, S., Caillol, D., Kodjabachian, L. 1998. A role for the vegetally expressed *Xenopus* gene Mix.1 in endoderm formation and in the restriction of mesoderm to the marginal zone. Development 125: 2371-2380.

Lemaire, P., Garrett, N., Gurdon, J.B. 1995. Expression cloning of Siamois, a *Xenopus* homeobox gene expressed in dorsal-vegetal cells of blastulae and able to induce a complete secondary axis. Cell 81: 85-94.

Maeno, M., Ong, R.C., Suzuki, A., Ueno, N., Kung, H.F. 1994. A truncated bone morphogenetic protein 4 receptor alters the fate of ventral mesoderm to dorsal mesoderm: roles of animal pole tissue in the development of ventral mesoderm. Proceedings of the National Academy of Sciences of the United States of America 91: 10260-10264.

McDowell, N. and Gurdon, J.B. 1999. Activin as a morphogen in *Xenopus* mesoderm induction. Seminars in Cell and Developmental Biology 10: 311-317.

McKendry, R., Hsu, S.C., Harland, R.M., Grosschedl, R. 1997. LEF-1/TCF proteins mediate wnt-inducible transcription from the *Xenopus* nodal-related 3 promoter. Developmental Biology 192: 420-431.

Malacinski, G.M., Benford, H. and Chung, H.M. 1975. Association of an ultraviolet irradiation sensitive cytoplasmic localization with the future dorsal side of the amphibian egg. Journal of Experimental Zoology 191: 97-110.

Malacinski, G.M., Brothers, A.J., Chung, H.M. 1977. Destruction of components of the neural induction system of the amphibian egg with ultraviolet irradiation. Developmental Biology 56: 24-39.

Marchant, L., Linker, C., Mayor, R. 1998. Inhibition of mesoderm formation by follistatin. Development Genes and Evolution 208: 157-160.

Masui, Y. and Wang, P. 1998. Cell cycle transition in early embryonic development of *Xenopus laevis*. Biology of the Cell 90: 537-548.

Miller, J.R., Rowning, B.A., Larabell, C.A., Yang-Snyder, J.A., Bates, R.L. and Moon, R.T. 1999. Establishment of the dorsal-ventral axis in *Xenopus* embryos coincides with the dorsal enrichment of dishevelled that is dependent on cortical rotation. The Journal of Cell Biology 146(2): 427-437.

Molenaar, M., Brian, E., Roose, J., Clevers, H., Destree, O. 2000. Differential expression of the Groucho-related genes 4 and 5 during early development of *Xenopus* laevis. Mechanisms of Development 91: 311-315.

Molenaar, M., van de Wetering, M., Oosterwegel, M., Peterson-Maduro, J., Godsave, S., Korinek, V., Roose, J., Destree, O., Clevers, H. 1996. XTcf-3 transcription factor mediates beta-catenin-induced axis formation in *Xenopus* embryos. Cell 86: 391-399.

Moon, R.T., Brown, J.D., Torres, M. 1997. WNTs modulate cell fate and behavior during vertebrate development. Trends in Genetics 13: 157-162.

Nagano, T., Ito, Y., Tashiro, K., Kobayakawa, Y., Sakai, M. 2000. Dorsal induction from dorsal vegetal cells in *Xenopus* occurs after mid-blastula transition. Mechanisms of Development 93: 3-14.

Newport, J. and Kirschner, M. 1982. A major developmental transition in early *Xenopus* embryos: I. Characterization and timing of cellular changes at the midblastula stage. Cell 30: 675-686.

Nieuwkoop, P.D. 1969b. The formation of the mesoderm in the urodelean amphibians. I. The Induction by the endoderm. Roux' Archives 163: 341-373.

Nieuwkoop, P.D. 1969b. The formation of the mesoderm in the urodelean amphibians. II. The origin of dorso-ventral polarity of the mesoderm. Roux' Archives 163: 298-315.

Nishimatsu, S., Iwao, M., Nagai, T., Oda, S., Suzuki, A., Asashima, M., Murakami, K., Ueno, N. 1992. A carboxyl-terminal truncated version of the activin receptor mediates activin signals in early *Xenopus* embryos. FEBS Letters 312: 169-173.

Nishimatsu, S., Suzuki, A., Shoda, A., Murakami, K., Ueno, N. 1992. Genes for bone morphogenetic proteins are differentially transcribed in early amphibian embryos. Biochemical and Biophysical Research Communications 186: 1487-1495.

Onuma, Y., Takahashi, S., Yokota, C., Asashima, M. 2002. .Multiple nodal-related genes act coordinately in *Xenopus* embryogenesis. Developmental Biology 241: 94-105.

Osada, S.I. and Wright, C.V. 1999. *Xenopus* nodal-related signaling is essential for mesendodermal patterning during early embryogenesis. Development 126: 3229-3240.

Piccolo, S., Agius, E., Leyns, L., Bhattacharyya, S., Grunz, H., Bouwmeester, T., De Robertis, E.M. 1999. The head inducer Cerberus is a multifunctional antagonist of Nodal, BMP and Wnt signals. Nature 397: 707-710.

Piccolo, S., Sasai, Y., Lu, B. and De Robertis, E.M. 1996. Dorsoventral patterning in *Xenopus*: inhibition of ventral signals by direct binding of chordin to BMP-4. Cell 86: 589-598.

Roose, J., Molenaar, M., Peterson, J., Hurenkamp, J., Brantjes, H., Moerer, P., van de Wetering, M., Destree, O., Clevers, H. 1998. The *Xenopus* Wnt effector XTcf-3 interacts with Groucho-related transcriptional repressors. Nature 395: 608-6012.

Rosa, F.M. 1989. Mix.1, a homeobox mRNA inducible by mesoderm inducers, is expressed mostly in the presumptive endodermal cells of *Xenopus* embryos. Cell 57: 965-974.

Rowning, B.A., Wells, J., Wu, M., Gerhard, J.C., Moon, R.T. and Larabell, C.A. 1997. Microtubule-mediated transport of organelles and localization of β-catenin to the future dorsal side of *Xenopus* eggs. Proceedings of the National Academy of Sciences of the USA 94: 1224-1229.

Saka, Y., Tada, M., Smith, J.C. 2000. A screen for targets of the *Xenopus* T-box gene Xbra. Mechanisms of Development 93: 27-39.

Sakai, M. 1996. The vegetal determinants required for the Spemann organizer move equatorially during the first cell cycle. Development 122: 2207-2214.

Sander, K. and Faessler, P.E. 2001. Introducing the Spemann-Mangold organizer: e xperiments and insights that generated a key concept in developmental biology. International Journal of Developmental Biology 45: 1-11.

Saneyoshi, T., Kume, S., Amasaki, Y., Mikoshiba, K. 2002. The Wnt/calcium pathway activates NF-AT and promotes ventral cell fate in *Xenopus* embryos. Nature 417: 295-299.

Sasai, Y., Lu, B., Piccolo, S., De Robertis, E.M. 1996. Endoderm induction by the organizer-secreted factors chordin and noggin in *Xenopus* animal caps. EMBO Journal 15: 4547-4555.

Sasai, Y., Lu, B., Steinbeisser, H., Geissert, D., Gont, L.K., De Robertis, E.M. 1994. *Xenopus* chordin: a novel dorsalizing factor activated by organizer-specific homeobox genes. Cell 79: 779-790.

Schmidt, J.E., Suzuki, A., Ueno, N., Kimelman, D. 1995. Localized BMP-4 mediates dorsal/ventral patterning in the early *Xenopus* embryo. Developmental Biology 169: 37-50.

Schohl, A. and Fagotto, F. 2002. Beta-catenin, MAPK and Smad signaling during early *Xenopus* development. Development 129: 37-52.

Slack, J.M. 1991. The nature of the mesoderm-inducing signal in *Xenopus*: a transfilter induction study. Development 113: 661-669.

Smith, J.C., Conlon, F.L., Saka, Y., Tada, M. 2000. Xwnt11 and the regulation of gastrulation in *Xenopus*. Philosophical Transactions of the Royal Society of London Series B: Biological Sciences 355: 923-930.

Smith, W.C., Knecht, A.K., Wu, M., Harland, R.M. 1993. Secreted noggin protein mimics the Spemann organizer in dorsalizing *Xenopus* mesoderm. Nature 361: 547-549.

Smith, J.C., Price, B.M., Green, J.B., Weigel, D., Herrmann, B.G. 1991. Expression of a *Xenopus* homolog of Brachyury (T) is an immediate-early response to mesoderm induction. Cell 67: 79-87

Spemann, H. and Mangold, H. 2001. Induction of embryonic primordial by implantation of organizers from a different species. International Journal of Developmental Biology 45: 13-38.

Steinbeisser, H., De Robertis, E.M., Ku, M., Kessler, D.S., Melton, D.A. 1993. *Xenopus* axis formation: induction of goosecoid by injected Xwnt-8 and activin mRNAs. Development 118: 499-507.

Sun, B.I., Bush, S.M., Collins-Racie, L.A., LaVallie, E.R., DiBlasio-Smith, E.A., Wolfman, N.M., McCoy, J.M., Sive, H.L. 1999. derriere: a TGF-beta family member required for posterior development in *Xenopus*. Development 126: 1467-1482.

Suzuki, A., Thies, R.S., Yamaji, N., Song, J.J., Wozney, J.M., Murakami, K., Ueno, N. 1994. A truncated bone morphogenetic protein receptor affects dorsal-ventral patterning in the early *Xenopus* embryo. Proceedings of the National Academy of Sciences of the United States of America 91: 10255-10259.

Tada, M., Casey, E.S., Fairclough, L., Smith, J.C. 1998. Bix1, a direct target of *Xenopus* T-box genes, causes formation of ventral mesoderm and endoderm. Development 125: 3997-4006.

Tada, M., Concha, M. and Heisenberg, C. 2002. Non-canonical Wnt signalling and regulation of gastrulation movements. Cell and Developmental Biology 13: 251-260.

Tada, M., Smith, J.C. 2000. Xwnt11 is a target of *Xenopus* Brachyury: regulation of gastrulation movements via Dishevelled, but not through the canonical Wnt pathway. Development 127: 2227-2238.

Taira, M., Jamrich, M., Good, P.J., Dawid, I.B. 1992. The LIM domain-containing homeo box gene Xlim-1 is expressed specifically in the organizer region of *Xenopus* gastrula embryos. Genes and Development 6: 356-366.

Tanegashima, K., Yokota, C., Takahashi, S., Asashima, M. 2000. Expression cloning of

Xantivin, a *Xenopus* lefty/antivin-related gene, involved in the regulation of activin signaling during mesoderm induction. Mechanisms of Development 99: 3-14.

Thomsen, G.H. and Melton, D.A. 1993. Processed Vg1 protein is an axial mesoderm inducer in *Xenopus*. Cell 74: 433-441.

Thomsen, G., Woolf, T., Whitman, M., Sokol, S., Vaughan, J., Vale, W., Melton, D.A. 1990. Activins are expressed early in *Xenopus* embryogenesis and can induce axial mesoderm and anterior structures. Cell 63: 485-493.

Torres, M.A., Yang-Snyder, J.A., Purcell, S.M., DeMarais, A.A., McGrew, L.L., Moon, R.T. 1996. Activities of the Wnt-1 class of secreted signaling factors are antagonized by the Wnt-5A class and by a dominant negative cadherin in early *Xenopus* development. Journal of Cell Biology 133: 1123-1137.

Wessely, O. and De Roberts, E.M. 2000. The *Xenopus* homologue of *Bicaudal-C* is a localized maternal mRNA that can induce endoderm formation. Development 127: 2053-2062.

White, R.J., Sun, B.I., Sive, H.L., Smith, J.C. 2002. Direct and indirect regulation of derriere, a *Xenopus* mesoderm-inducing factor, by VegT. Development 129: 4867-4876.

Wylie, C., Kofron, M., Payne, C., Anderson, R., Hosobuchi, M., Joseph, E., Heasman, J. 1996. Maternal beta-catenin establishes a 'dorsal signal' in early *Xenopus* embryos. Development 122: 2987-2996.

Wylie, C.C., Snape, A., Heasman, J., Smith, J.C. 1987. Vegetal pole cells and commitment to form endoderm in *Xenopus* laevis. Developmental Biology 119: 496-502.

Yamamoto, T.S., Takagi, C., Hyodo, A.C., Ueno, N. 2001. Suppression of head formation by Xmsx-1 through the inhibition of intracellular nodal signaling. Development 128: 2769-2779.

Yamamoto, T.S., Takagi, C., Ueno, N. 2000. Requirement of Xmsx-1 in the BMP-triggered ventralization of *Xenopus* embryos. Mechanisms of Development 91: 131-141.

Yao, J. and Kessler, D.S. 2001. Goosecoid promotes head organizer activity by direct repression of Xwnt8 in Spemann's organizer. Development 128: 2975-2987.

Yasuo, H. and Lemaire, P.A. 1999. two-step model for the fate determination of presumptive endodermal blastomeres in *Xenopus* embryos. Current Biology 9: 869-879.

Yost, C., Torres, M., Miller, J.R., Huang, E., Kimelman, D., Moon, R.T. 1996. The axis-inducing activity, stability, and subcellular distribution of beta-catenin is regulated in *Xenopus* embryos by glycogen synthase kinase 3. Genes and Development 10: 1443-1454.

Yuge, M., Kobayakawa, Y., Fujisue, M., Yamana, K. 1990. A cytoplasmic determinant for dorsal axis formation in an early embryo of *Xenopus laevis*. Development 110: 1051-1056.

Xanthos, J.B., Kofron, M., Tao, Q., Schaible, K., Wylie, C., Heasman, J. 2002. The roles of three signaling pathways in the formation and function of the Spemann Organizer. Development 129: 4027-4043.

Xanthos, J.B., Kofron, M., Wylie, C., Heasman, J. 2001. Maternal VegT is the initiator of a molecular network specifying endoderm in *Xenopus laevis*. Development 128: 167-180.

Zhang, J., Houston, D.W., King, M.L., Payne, C., Wylie, C. and Heasman, J. 1998. The role of maternal VegT in establishing the primary germ layers in *Xenopus* embryos. Cell 94: 515-524.

Index